Semiconducting Metal Oxide Thin-Film Transistors

Semiconducting Metal Oxide Thin-Film Transistors

Edited by
Ye Zhou
Institute for Advanced Study, Shenzhen University, Nanshan, Shenzhen, Guangdong, People's Republic of China

IOP Publishing, Bristol, UK

ISBN 978-0-7503-2556-1 (ebook)
ISBN 978-0-7503-2554-7 (print)
ISBN 978-0-7503-2557-8 (myPrint)
ISBN 978-0-7503-2555-4 (mobi)

DOI 10.1088/978-0-7503-2556-1

Version: 20210101

IOP ebooks

British Library Cataloguing-in-Publication Data: A catalogue record for this book is available from the British Library.

Published by IOP Publishing, wholly owned by The Institute of Physics, London

IOP Publishing, Temple Circus, Temple Way, Bristol, BS1 6HG, UK

US Office: IOP Publishing, Inc., 190 North Independence Mall West, Suite 601, Philadelphia, PA 19106, USA

Contents

Preface

Semiconducting metal oxide thin-film transistors are promising candidates for functional electronic devices. They have attracted considerable attention owing to their superior electrical performance, high transparency, excellent stability and uniformity. Semiconducting metal oxides usually exhibit higher carrier mobilities than those of amorphous Si and most organic transistors, and better device uniformity than that of polycrystalline Si transistors.

In this book, we will first introduce the concept and basic working mechanism of semiconducting metal oxide thin-film transistors. We then focus on a series of metal oxide thin films with desirable electrical and optical properties for various device applications. The typical devices such as logic circuits, LED drivers, photodetectors, tactile imaging, transistor memories, artificial synaptic devices and transparent transistors will be discussed.

I would like to express my gratitude to all the fellow authors who have contributed in this book. I also want to acknowledge Robert Trevelyan and Caroline Mitchell at IOP Publishing, for all the help during the book editorial process, and for the excellent experience of working with them. I want to thank all the readers for their interest in our book. I hope that our book can be useful as a reference guide for researchers and students who work in the field of metal oxides and metal oxide-based transistors.

Ye Zhou

Editor biography

Ye Zhou

Professor Ye Zhou is an IAS Fellow and group leader in the Institute for Advanced Study, Shenzhen University. He received his B.S. (2008) in Electronic Science and Engineering from Nanjing University, M.S. (2009) in Electronic Engineering from Hong Kong University of Science and Technology and PhD (2013) in Physics and Materials Science from City University of Hong Kong. His research interests include flexible electronics, nano-composite materials and nano-scale devices for technological applications such as logic circuits, memories, photonics and sensors. He has 16 granted USA/China patents and published more than 120 papers in journals such as Science, Chemical Reviews, Nature Electronics, Nature Communications, Advanced Materials, etc. More than 40 of his works have been highlighted as cover pages or frontispieces. He is the Associate Editor of IEEE Access, Applied Nanoscience and Science and Technology of Advanced Materials, Editorial/Community Board Member of Materials Horizons, Multifunctional Materials, Chemistry, PLOS ONE and Chemistry Proceedings, and served as the Guest Editor of Nanoscale Horizons, Frontiers in Chemistry, Frontiers in Physics and Polymer International.

List of contributors

Thomas D Anthopoulos
King Abdullah University of Science and Technology (KAUST), Thuwal 23955-6900, Kingdom of Saudi Arabia

Te-Wei Chiu
Department of Materials and Mineral Resources Engineering, National Taipei University of Technology, Taipei 106, Taiwan

Sunbin Deng
State Key Laboratory of Advanced Displays and Optoelectronics Technologies, The Hong Kong University of Science and Technology, Hong Kong, People's Republic of China

Hendrik Faber
King Abdullah University of Science and Technology (KAUST), Thuwal 23955-6900, Kingdom of Saudi Arabia

Can Fu
School of Electronic Science and Applied Physics and Anhui Provincial Key Laboratory of Advanced Materials and Devices, Hefei University of Technology, Hefei, Anhui 230009, People's Republic of China

Jie Jiang
School of Physics and Electronics, Central South University, Changsha, Hunan 410083, People's Republic of China

Hoi-Sing Kwok
State Key Laboratory of Advanced Displays and Optoelectronics Technologies, The Hong Kong University of Science and Technology, Hong Kong, People's Republic of China

Yichun Liu
Key Laboratory for UV Light-Emitting Materials and Technology, Ministry of Education, Northeast Normal University, People's Republic of China

Shuhai Liu
Lanzhou University, Lanzhou, Gansu 730000, People's Republic of China

Weizhen Liu
Key Laboratory for UV Light-Emitting Materials and Technology, Ministry of Education, Northeast Normal University, People's Republic of China

Lin-Bao Luo
School of Electronic Science and Applied Physics and Anhui Provincial Key Laboratory of Advanced Materials and Devices, Hefei University of Technology, Hefei, Anhui 230009, People's Republic of China

Yanran Li
School of Physics and Electronics, Central South University, Changsha, Hunan 410083, People's Republic of China

Xiaoliang Liu
School of Physics and Electronics, Central South University, Changsha, Hunan 410083, People's Republic of China

Yong Qin
Lanzhou University, Lanzhou, Gansu 730000, People's Republic of China

Subramanian Sakthinathan
Department of Materials and Mineral Resources Engineering, National Taipei University of Technology, Taipei 106, Taiwan

Fengyun Wang
College of Physics and Cultivation Base for State Key Laboratory, Qingdao University, Qingdao 266071, People's Republic of China

Xiaokun Wen
Huazhong University of Science and Technology, Wuhan, People's Republic of China

Man Wong
State Key Laboratory of Advanced Displays and Optoelectronics Technologies, The Hong Kong University of Science and Technology, Hong Kong, People's Republic of China

Zhihe Xia
State Key Laboratory of Advanced Displays and Optoelectronics Technologies, Hong Kong University of Science and Technology, Hong Kong, People's Republic of China

Haiyang Xu
Key Laboratory for UV Light-Emitting Materials and Technology, Ministry of Education, Northeast Normal University, People's Republic of China

Yan Yan
College of Electronics and Information Engineering, Shenzhen University, Nanshan, Shenzhen, Guangdong, People's Republic of China

Emre Yarali
King Abdullah University of Science and Technology (KAUST), Thuwal 23955-6900, Kingdom of Saudi Arabia

Emre Yengel
King Abdullah University of Science and Technology (KAUST), Thuwal 23955-6900, Kingdom of Saudi Arabia

Zhi Ye
College of Information Science and Electronic Engineering, Zhejiang University, Hangzhou 310027, People's Republic of China

Shi-Rui Zhang
Institute for Advanced Study, Shenzhen University, Nanshan, Shenzhen, Guangdong, People's Republic of China

Meng Zhang
Institute of Microscale Optoelectronics (IMO), Shenzhen University, Nanshan, Shenzhen, Guangdong, People's Republic of China

Cen Zhang
Key Laboratory for UV Light-Emitting Materials and Technology, Ministry of Education, Northeast Normal University, People's Republic of China

Ye Zhou
Institute for Advanced Study, Shenzhen University, Nanshan, Shenzhen, Guangdong, People's Republic of China

Hongliang Zu
College of Physics and Cultivation Base for State Key Laboratory, Qingdao University, Qingdao 266071, People's Republic of China

Ning Zhang
College of Information Science and Electronic Engineering, Zhejiang University, Hangzhou 310027, People's Republic of China

Wenfeng Zhang
Huazhong University of Science and Technology, Wuhan, People's Republic of China

Chapter 1

Introduction to semiconducting metal oxides

Shi-Rui Zhang and Ye Zhou

Semiconducting metal oxide (SMO) materials are promising candidates for various electronic devices. They have attracted considerable attention owing to their superior electrical performance, high transparency, excellent stability and uniformity. In this chapter, we mainly introduce some basic ideas of SMOs.

1.1 Background

Oxide, containing at least one oxygen and one other element, is one of the most abundant compounds on Earth, existing in gas, liquid and solid all over the place. A very common example of liquid oxides in room temperature is hydrogen oxide (H_2O) which we call water, used as a universal solvent and necessary for life on our blue planet. Carbon dioxide (CO_2) is a kind of gas oxide familiar to people, known as an ingredient of photosynthesis and one of the products of respiration. Apart from water and gas oxide, solid oxides also play an important role in our daily life. For instance, silicon oxide (SiO_2), also called silica, has widespread applications in daily necessities, architectures and technologies. The most traditional type of colorless glass is quartz, in which silicon oxide is the single constituent in amorphous state. With excellent optical properties and chemical stability, quartz is not only utilized to manufacture optical fibers used in cable television and communication systems but also applied to the processing of diverse semiconductor devices [1]. Since the advent of the first MOSFET in the late 1950s, solid oxides have been proved to be indispensable in the modern semiconductor industry [2, 3]. The aforementioned silicon oxide acts as an insulating layer in the metal-oxide-semiconductor (MOS) structure, the core of MOSFET. MOSFET is a kind of device with three terminals: source, drain and gate. The MOS structure inside the device can be regarded as a planar capacitor where the semiconductor is basically p-type or n-type silicon and the metal layer could be conductive polycrystalline silicon besides metals. At the beginning, MOS structure can be fabricated by growing a layer of silicon oxide through thermal deposition on the silicon wafer and then depositing the metal layer.

doi:10.1088/978-0-7503-2556-1ch1

When talking about the function of oxide layer in MOS structure and even in the whole device, we could simply say that it performs a dielectric property, preventing electrons or holes from tunneling from semiconductor to metal gate, contributing to the formation of channel near the interface between oxide and semiconductor.

Alhough oxides were usually thought to be insulators, some metal oxides such as simple binary oxide including ZnO, In_2O_3, SnO_2, etc [4], and complicated oxide Zn–Sn–O, In–Sn–O [5, 6] actually have properties of semiconductors. Before going deep into talking about SMOs, a widely used device should be introduced. As the derivative of MOSFET, thin-film transistors (TFTs) have wide application in electronics. TFT works based on the same field effect mechanism as MOSFET. In general, TFT is fabricated by deposited semiconductor materials, dielectric materials and conducting materials in the form of thin film on a certain substrate [7–9]. The primary application of TFT is in display products such as liquid crystal display (LCD) and organic light-emitting diode (OLED) which are based on active matrix (AM) [10, 11]. TFT is widely used in the manufacture of active matrix liquid crystal display (AMLCD), occupying the largest segment of output of the panel display industry. In the meantime, active matrix organic light-emitting diode (AMOLED) is in rapid development. Generally speaking, TFT used in AMLCD and AMOLED can be made by using a very wide variety of semiconducting materials, but commonly used material is silicon including single crystalline silicon, microcrystalline silicon, low temperature polysilicon (LTPS) and amorphous silicon. Facing the advent of next generation display technologies, there are still some disadvantages existing in current silicon based TFT for display. To begin with, amorphous silicon based TFTs have relatively inadequate mobility, lower than 1 $cm^2\ V^{-1} \cdot s^{-1}$, limiting the operating speed of TFT devices as higher mobility could lead to high current which makes capacitive loads quickly charged or discharged [12, 13]. Besides, though LTPS based TFTs possess mobility about 100 $cm^2\ V^{-1} \cdot s^{-1}$, the existence of lots of grain boundaries leads to the difference of electrical characteristics between TFT units [14, 15]. What's more, silicon based TFTs are opaque to visible light and temperature of deposition is generally high, not applicable to the trend of transparent display and flexible panels. By using transparent electrodes such as indium tin oxide (ITO) and transparent semiconductors we can make TFTs totally transparent. In this condition, semiconducting metal oxide becomes an ideal candidate due to its sufficient mobility and characteristics of transparency under visible light, attracting great attention from researchers. In 2003, several groups adopted ZnO as n-channel to fabricate highly transparent TFTs [16–18]. It's worth noting that these ZnO-based TFTs could reach mobility larger than 1 $cm^2\ V^{-1} \cdot s^{-1}$, symbolizing the beginning of application of SMOs as the semiconducting channel layer in TFTs.

1.2 Fundamentals of semiconducting metal oxide

As mentioned above, oxide contains at least one oxygen and one other element, existing in solid, liquid and gas phase. Unlike molecular oxides, SMO generally has the form of solid and belongs to the family of non-stoichiometric compounds, which means proportions of elemental composition cannot be exactly represented by a ratio of small natural numbers. As a kind of ionic solid, SMO has strong ionic bonds connecting positive metallic ions and negative oxygen ions. As for SMO, inside the materials, the 's' electronic shell is filled inherently while the 'd' shell is incompletely filled, owing to which, SMO possesses superior thermal/chemical stability and unique properties such as tunable energy bands, high dielectric constants and so on [19]. And these unique properties endow SMO capabilities in different applications. SMO is a very large group of materials, so they differ from each other in crystalline structure, inherent defects, energy band level and other physical aspects. For example, in terms of crystalline structure, ZnO could exist in hexagonal wurtzite or cubic zincblende. However, under ambient conditions, hexagonal wurtzite is more stable and common with the space group of $P6_3mc$ [20]. The lattice constants of ZnO in wurtzite form are $a = 3.25$ Å and $c = 5.2$ Å, where the ratio of c/a is 1.633, closed to ideal value. In contrast, In_2O_3 crystallizes in the form of cubic with lattice constant of 10.08 Å while SnO_2 belongs to rutile tetragonal and lattice constants are $a = 4.737$ Å, $c = 3.185$ Å [4]. These various crystal structures lead to different unique properties of SMO.

For oxides, several kinds of defects exist widely inside materials, including point defects, line defects, plane defects and volume defects. In fact, defects exist in all crystalline solids which are not perfect though they exhibit periodic crystal structures [21, 22]. Point defects are also called zero-dimensional defects and can be mainly divided into three kinds, including vacancy defects, interstitial defects and Frenkel pairs when there is no heteroatom [23, 24]. Vacancy defects refer to vacant sites that should be occupied by certain atoms in a totally perfect crystal. And because of the stability of crystal structure around the vacancy, the neighboring atoms of the vacancies will not collapse easily, guaranteeing the existence of these vacancies. What's more, once a certain atom moves into the vacant site, the vacancy would move in the opposite direction and occupy the site where the atom used to be located. In contrast, when atoms appear in a region where there are commonly no atoms existing, interstitial defects show up with generally high energy configuration. The last kind of defect seems to be a combination of vacancy defects and interstitial defects, which means an atom occupies an interstitial site and leaves a vacancy, called Frenkel defects. As mentioned above, SMO belongs to ionic solids, owing to which, ionic bonds are more easily broken than covalent bonds so SMO usually has more intrinsic defects compared to elemental semiconductor materials [25]. For SMO, there are six classic point defects such a site of metallic atom (M) replaced with oxygen atom (O), a site of O replaced with M, no atom on a site of M, no atom on a site of O, M or O existing in an interstitial region [19]. On account of the existence of intrinsic defects like oxygen vacancies and metallic interstitial, undoped SMO generally exhibits n-type semiconducting properties. But for certain SMOs, the origin of n-type semiconducting is still controversial. For instance, there is another explanation for n-type conductivity of ZnO instead of inherent defects in the absence of intentional doping [26]. By using first-principle calculation and experiment, hydrogen is identified and characterized in ZnO bulk, being thought to be responsible plausibly as its presence is inevitable in most growth and annealing processes [27–30].

For solid-state physics, band theory is a significant concept, describing the range of energy levels that electrons could stay at (bands) and could not stay at (band gap). Pauli exclusion principle finally leads to the formation of valence band and conduction band [31, 32]. Many physical properties of solids including resistance, absorption and so on can be well explained by band theory. In short, band gap means a range of energy where no electronic states exist and band gap energy of a semiconductor is the minimum energy that the electrons required to transfer from valence band into conduction band. According to band theory, electrons cannot move freely inside the solid if the valence band is totally full and the conduction band is completely empty. Once electrons transfer from valence band to conduction band, they will be able to move freely, which means the formation of current. External heat or a photon with energy higher than band gap energy could excite electrons so that they can leave the valence band and transit into the conduction band, leaving orbital holes in the valence band. For most SMOs, the band gap is generally wide (>3 eV). For example, the band gap of ZnO, In_2O_3, SnO_2 is about 3.4 eV, 3.5 eV, 3.6 eV [4]. Owing to the wide-band gap, SMO materials are able to endure large electric field and have high breakdown voltages, which means that SMO-based electronic devices could be operated under high temperature and power. What's more, the wide band gap also endows SMOs with high optical transparency in the visible region of the electromagnetic spectrum, which is a desired characteristic in transparent electronics.

In addition to band theory, impurity level is another vital parameter of semiconductors. Impurity level has significant impact on carrier concentration, microstructure and related aspects of a semiconductor. Generally, through doping, the intentional introduction of extra compositions into an intrinsic semiconductor alters the impurity level, resulting in the variation of electrical and optical properties. From the perspective of band theory, doping introduces extra energy levels within the band gap and the introduced energy levels will be closed to valence band or conduction band when species of dopant are different. Specifically, electron donor or n-type doping brings in energy levels near the conduction band while electron acceptor or p-type doping brings in ones near the valence band. As for ZnO, n-type doping can be achieved by replacing with Al, Ga, In and other group-III elements or by substituting group-VII elements for O while p-type doping is commonly thought to be difficult [4, 33, 34]. From point of view of carrier concentration, regardless of n-type or p-type doping, doped semiconductors would have higher carrier concentration compared to their intrinsic state because both electrons in n-type semiconductors and holes in p-type semiconductors contribute to conductivity significantly. Some highly doped semiconductors even possess conductivity comparable to metals. The reason impurity level of semiconductor is importants is that it has a direct impact on the carrier mobility, which is another crucial parameter of semiconductors when they are applied to electronic devices. Carrier mobility is related to two kinds of scattering inside material: lattice scattering (phonon scattering) and ionized impurity scattering [35]. In terms of lattice scattering, when temperature is higher than absolute zero, atoms inside semiconductor will vibrate and create acoustic waves (phonons), which break the ideal periodic potential. As a consequence, vibrating atoms interact with moving carriers by scattering them, causing the reduction of mobility. In other words, higher temperature results in smaller mobility and vice versa. Regarding ionized impurity scattering, electron donors or acceptors are commonly ionized when being doped into semiconductor. Coulombic interaction will be produced when moving electrons or holes approach these doped impurities, leading to scattering of carriers. Under constant temperature, increasing impurity level normally lowers the carrier mobility.

1.3 Application of semiconducting metal oxide

SMOs are widely used in many fields, especially in electronics. The electrode may be the simplest application of SMO. For example, indium tin oxide (ITO) is a highly doped n-type semiconductor with low resistivity and high carrier concentration. Besides, ITO also possesses typical wide band gap and high transmittance in the range of visible light. These unique electrical and optical properties make ITO an ideal candidate for a transparent conducting electrode [36]. For example, ITO can not only be drain, source and gate electrode in TFT, but can also be used as anode in OLED [6, 37, 38]. Furthermore, due to the transparent conductive property as well as ease of being deposited as thin film, ITO is used to make transparent conductive coating for display technologies and photovoltaics [39–41].

As mentioned above, TFT is one of the important applications of SMO. The introduction of SMO could break the limitation of traditional silicon based TFT. In 2003, Hoffman *et al* [17] investigated transparent thin-film transistor (TTFT) devices by using ZnO as n-channel. In this research, highly transparent n-type ITO served as drain, source and gate electrode at the same time with aluminum–titanium oxide (ATO) deposited as gate insulator. These TTFTs showed optical transmission of 75% in the visible light region of the electromagnetic spectrum and drain current ON/OFF of about 10^7. Most importantly, this research also demonstrated that these ZnO based devices' channel mobilities could be more than 1 $cm^2\ V^{-1}\cdot s^{-1}$ and reached up to 2.5 $cm^2\ V^{-1}\cdot s^{-1}$. Nomura *et al* [18] fabricated TFTs in single-crystalline highly doped oxide semiconductor, $InGaO_3(ZnO)_5$. The devices exhibited field-effect mobility of about 80 $cm^2\ V^{-1}\cdot s^{-1}$ due to less grain boundary and scattering in a single crystal as well as high doping density. In fact, in the modern display industry, amorphous In–Ga–Zn–O (a-IGZO)-based TFTs perform excellently due to high mobility and high uniformity, so a-IGZO is a kind of material being widely used in actual mass production [42–47].

The use of SMO-based TFT was naturally extended to the area of flash memory devices as they adopted TFT as fundamental structure. In 2017, Hota *et al* [48] constructed a transparent flash memory by combining amorphous oxide with high-κ oxide materials by the sol–gel method. As a kind of n-type semiconductor, ZnO is applied to form the channel. Ta_2O_5 functions as the charge trapping layer and tunneling dielectric. For low voltage operation, Al_2O_3 plays the role of blocking dielectric layer. ITO is the electrode. With program and erase, the maximum memory window of ~10.7 V is demonstrated. Flash memory performed well in durability and reliability tests, holding data for more than 10^4 s and having approximately 100 program/erase cycles. With the development of non-volatile memory devices, resistive random access memory (ReRAM) is thought to be one of the potential candidates for next-generation devices. So many materials exhibit resistive switching phenomenon which is required in ReRAM. In fact, resistive switching phenomenon was first found on Al_2O_3 by Hickmott in 1962 [49]. Hickmott studied five metal–oxide–metal sandwiches such as Al–SiO_2–Au, Al–Al_2O_3–Au, Al–Ta_2O_3–Au, Al–ZrO_2–Au and Al–TiO_2–Au under direct-current-voltage, in which Al_2O_3 was investigated extensively. The switching times of these oxides are below 1 μs, indicating the excellent performance of oxide in resistive switching. Up to now, numerous kinds of SMO have been investigated. For example, the highest ON/OFF ratio ($>10^9$) was realized in the research of TaO_x and GeO_x devices [50, 51]. HfO_x and TaO_x based devices both reached nanosecond level in program speed [52–54]. ReRAMs based on other metal oxides such as CuO_x [55–57], CoO_x [58, 59], NiO_x [60], etc, have also been studied systematically. The underlying mechanism of resistive switching behavior of metal oxides is mainly put down to the formation and rupture of conductive filaments (CFs) or the changes of valence states caused by the migration of oxygen vacancy [61, 62]. For SMO-based ReRAM, owing to the production and the migration of oxygen vacancies inside metal oxide film, conductive filaments can evolve and resistive switching or memory function can be realized without active metal electrode and metal conductive filaments. Commonly speaking, 'electroforming' process is required when the first resistive switching happens in SMO-based ReRAM. In this process, some oxygen atoms inside oxides will lose electrons and oxygen vacancies will be produced under high applied voltage accompanied with generating O_2. And then the redistribution of these oxygen vacancies causes the changes of resistance states of the metal oxide film under different applied voltages, resulting in the memory function of ReRAM.

In conclusion, we have listed some major applications of SMO and discussed them in limited space, but we don't enumerate all the applications as they are extremely wide. In this book, more detailed discussion in SMOs and their application in different fields will be presented.

References

[1] Fanderlik I 2013 *Silica Glass and Its Application* (Amsterdam: Elsevier)

[2] Nicollian E H and Brews J R 1982 *MOS/Metal Oxide Semiconductor/Physics and Technology* (New York: Wiley) p 920

[3] Lee M L, Fitzgerald E A, Bulsara M T, Currie M T and Lochtefeld A 2005 Strained Si, SiGe, and Ge channels for high-mobility metal-oxide-semiconductor field-effect transistors *J. Appl. Phys.* **97** 011101

[4] Batzill M and Diebold U 2005 The surface and materials science of tin oxide *Prog. Surf. Sci.* **79** 47–154

[5] Rajachidambaram J S, Sanghavi S, Nachimuthu P, Shutthanandan V, Varga T, Flynn B, Thevuthasan S and Herman G S 2012 Characterization of amorphous zinc tin oxide semiconductors *J. Mater. Res.* **27** 2309–17

[6] Kim H, Gilmore C M, Piqué A, Horwitz J S, Mattoussi H, Murata H, Kafafi Z H and Chrisey D B 1999 Electrical, optical, and structural properties of indium–tin–oxide thin films for organic light-emitting devices *J. Appl. Phys.* **86** 6451–61

[7] Brody T P 1984 The thin film transistor—a late flowering bloom *IEEE Trans. Electron Devices* **31** 1614–28

[8] Weimer P 1962 The TFT a new thin-film transistor *Proc. IRE* **50** 1462–9

[9] Brody T P and Kunig H E 1966 A high-gain InAs thin-film transistor *Appl. Phys. Lett.* **9** 259–60

[10] Kuo Y 2013 Thin film transistor technology—past, present, and future *Interface Mag.* **22** 55–61

[11] Fortunato E, Barquinha P and Martins R 2012 Oxide semiconductor thin-film transistors: a review of recent advances *Adv. Mater.* **24** 2945–86

[12] Sirringhaus H 2014 25th anniversary article: organic field-effect transistors: the path beyond amorphous silicon *Adv. Mater.* **26** 1319–35

[13] Wager J F 2003 Applied physics. Transparent electronics *Science* **300** 1245–6

[14] Wager J F 2014 Flat-panel-display backplanes: LTPS or IGZO for AMLCDs or AMOLED displays? *Inf. Disp.* **30** 26–9

[15] Nathan A, Chaji G R and Ashtiani S J 2005 Driving schemes for a-Si and LTPS AMOLED displays *J. Disp. Technol.* **1** 267–77

[16] Carcia P F, McLean R S, Reilly M H and Nunes G 2003 Transparent ZnO thin-film transistor fabricated by rf magnetron sputtering *Appl. Phys. Lett.* **82** 1117–9

[17] Hoffman R L, Norris B J and Wager J F 2003 ZnO-based transparent thin-film transistors *Appl. Phys. Lett.* **82** 733–5

[18] Nomura K, Ohta H, Ueda K, Kamiya T, Hirano M and Hosono H 2003 Thin-film transistor fabricated in single-crystalline transparent oxide semiconductor *Science* **300** 1269–72

[19] Deng Y 2019 Understanding semiconducting metal oxide gas sensors *Semiconducting Metal Oxides for Gas Sensing* ed Y Deng (Singapore: Springer) pp 1–22

[20] Özgür Ü, Alivov Y I, Liu C, Teke A, Reshchikov M A, Doğan S, Avrutin V, Cho S J and Morkoç H 2005 A comprehensive review of ZnO materials and devices *J. Appl. Phys.* **98** 041301

[21] Watkins G 1997 Native defects and their interactions with impurities in silicon *MRS Online Proc. Libr. Arch.* **469** 139

[22] Smyth D M 2000 *Defect Chemistry of Metal Oxides* (Oxford: Oxford University Press) p 304

[23] Crawford J H and Slifkin L M 2013 *Point Defects in Solids: General and Ionic Crystals* (Berlin: Springer)

[24] Spaeth J-M, Niklas J R and Bartram R H 2012 *Structural Analysis of Point Defects in Solids: An Introduction to Multiple Magnetic Resonance Spectroscopy* (Berlin: Springer)

[25] Fisher E R, Elkind J L, Clemmer D E, Georgiadis R, Loh S K, Aristov N, Sunderlin L S and Armentrout P B 1990 Reactions of fourth-period metal ions (Ca^+–Zn^+) with O_2: metal-oxide ion bond energies *J. Chem. Phys.* **93** 2676–91

[26] Look D C, Hemsky J W and Sizelove J R 1999 Residual native shallow donor in ZnO *Phys. Rev. Lett.* **82** 2552–5

[27] Van De Walle C G 2000 Hydrogen as a cause of doping in zinc oxide *Phys. Rev. Lett.* **85** 1012–5

[28] Janotti A and Van de Walle C G 2007 Hydrogen multicentre bonds *Nat. Mater.* **6** 44–7

[29] Janotti A and Van de Walle C G 2009 Fundamentals of zinc oxide as a semiconductor *Rep. Prog. Phys.* **72** 126501

[30] Cox S F *et al* 2001 Experimental confirmation of the predicted shallow donor hydrogen state in zinc oxide *Phys. Rev. Lett.* **86** 2601–4

[31] Kittel C and McEuen P 1976 *Introduction to Solid State Physics* (New York: Wiley)

[32] Holgate S A 2009 *Understanding Solid State Physics* (Boca Raton, FL: CRC Press)

[33] Kato H, Sano M, Miyamoto K and Yao T 2002 Growth and characterization of Ga-doped ZnO layers on a-plane sapphire substrates grown by molecular beam epitaxy *J. Cryst. Growth* **237-239** 538–43

[34] Norton D P, Heo Y W, Ivill M P, Ip K, Pearton S J, Chisholm M F and Steiner T 2004 ZnO: growth, doping and processing *Mater. Today* **7** 34–40

[35] Neamen D A 2012 *Semiconductor Physics and Devices: Basic Principles* (New York: McGraw-Hill)

[36] Farhan M S, Zalnezhad E, Bushroa A R and Sarhan A A D 2013 Electrical and optical properties of indium-tin oxide (ITO) films by ion-assisted deposition (IAD) at room temperature *Int. J. Precis. Eng. Manuf.* **14** 1465–9

[37] Kim H, Piqué A, Horwitz J S, Mattoussi H, Murata H, Kafafi Z H and Chrisey D B 1999 Indium tin oxide thin films for organic light-emitting devices *Appl. Phys. Lett.* **74** 3444–6

[38] Kawachi G, Kimura E, Wakui Y, Konishi N, Yamamoto H, Matsukawa Y and Sasano A 1994 A novel technology for a-Si TFT-LCD's with buried ITO electrode structure *IEEE Trans. Electron Devices* **41** 1120–4

[39] Du J, Chen X-L, Liu C-C, Ni J, Hou G-F, Zhao Y and Zhang X-D 2014 Highly transparent and conductive indium tin oxide thin films for solar cells grown by reactive thermal evaporation at low temperature *Appl. Phys.* A **117** 815–22

[40] Frach P, Glöß D, Goedicke K, Fahland M and Gnehr W M 2003 High rate deposition of insulating TiO_2 and conducting ITO films for optical and display applications *Thin Solid Films* **445** 251–8

[41] Kim S I, Cho S H, Choi S R, Oh M C, Jang J H and Song P K 2009 Crystallization and electrical properties of ITO:Ce thin films for flat panel display applications *Thin Solid Films* **517** 4061–4

[42] Nomura K, Ohta H, Takagi A, Kamiya T, Hirano M and Hosono H 2004 Room-temperature fabrication of transparent flexible thin-film transistors using amorphous oxide semiconductors *Nature* **432** 488–92

[43] Yabuta H, Sano M, Abe K, Aiba T, Den T, Kumomi H, Nomura K, Kamiya T and Hosono H 2006 High-mobility thin-film transistor with amorphous $InGaZnO_4$ channel fabricated by room temperature RF-magnetron sputtering *Appl. Phys. Lett.* **89** 112123

[44] Jeong J K, Jeong J H, Yang H W, Park J-S, Mo Y-G and Kim H D 2007 High performance thin film transistors with cosputtered amorphous indium gallium zinc oxide channel *Appl. Phys. Lett.* **91** 113505

[45] Jeong J H, Yang H W, Park J-S, Jeong J K, Mo Y-G, Kim H D, Song J and Hwang C S 2008 Origin of subthreshold swing improvement in amorphous indium gallium zinc oxide transistors *Electrochem. Solid-State Lett.* **11** H157

[46] Kim G H, Kim H S, Shin H S, Ahn B D, Kim K H and Kim H J 2009 Inkjet-printed InGaZnO thin film transistor *Thin Solid Films* **517** 4007–10

[47] Kikuchi Y, Nomura K, Yanagi H, Kamiya T, Hirano M and Hosono H 2010 Device characteristics improvement of a-In–Ga–Zn–O TFTs by low-temperature annealing *Thin Solid Films* **518** 3017–21
[48] Hota M K, Alshammari F H, Salama K N and Alshareef H N 2017 Transparent flash memory using single Ta_2O_5 layer for both charge-trapping and tunneling dielectrics *ACS Appl. Mater. Interfaces* **9** 21856–63
[49] Hickmott T W 1962 Low-frequency negative resistance in thin anodic oxide films *J. Appl. Phys.* **33** 2669–82
[50] Tsuruoka T, Terabe K, Hasegawa T and Aono M 2010 Forming and switching mechanisms of a cation-migration-based oxide resistive memory *Nanotechnology* **21** 425205
[51] Kurnia F, Liu C, Jung C U and Lee B W 2013 The evolution of conducting filaments in forming-free resistive switching Pt/TaO_x/Pt structures *Appl. Phys. Lett.* **102** 152902
[52] Torrezan A C, Strachan J P, Medeiros-Ribeiro G and Williams R S 2011 Sub-nanosecond switching of a tantalum oxide memristor *Nanotechnology* **22** 485203
[53] Miranda E A, Walczyk C, Wenger C and Schroeder T 2010 Model for the resistive switching effect in HfO_2 MIM structures based on the transmission properties of narrow constrictions *IEEE Electron Device Lett.* **31** 609–11
[54] Chen Y Y, Goux L, Clima S, Govoreanu B, Degraeve R, Kar G S, Fantini A, Groeseneken G, Wouters D J and Jurczak M 2013 Endurance/retention trade-off on HfO_2/metal cap 1T1R bipolar RRAM *IEEE Trans. Electron Devices* **60** 1114–21
[55] Chen A, Haddad S, Wu Y C, Fang T N, Kaza S and Lan Z 2008 Erasing characteristics of Cu_2O metal-insulator-metal resistive switching memory *Appl. Phys. Lett.* **92** 013503
[56] Hu P, Li X Y, Lu J Q, Yang M, Lv Q B and Li S W 2011 Oxygen deficiency effect on resistive switching characteristics of copper oxide thin films *Phys. Lett.* A **375** 1898–902
[57] Yin M, Zhou P, Lv H B, Tang T A, Chen B A, Lin Y Y, Bao A and Chi M H 2008 Enhancement of endurance for CuxO based RRAM cell *2008 9th Int. Conf. Solid-State and Integrated-Circuit Technology (Beijing)* pp 917–20
[58] Kwak J S, Do Y H, Bae Y C, Im H and Hong J P 2010 Reproducible unipolar resistive switching behaviors in the metal-deficient CoO_x thin film *Thin Solid Films* **518** 6437–40
[59] Gao X, Guo H, Xia Y, Yin J and Liu Z 2010 Unipolar resistive switching characteristics in Co_3O_4 films *Thin Solid Films* **519** 450–2
[60] Yoshida C, Kinoshita K, Yamasaki T and Sugiyama Y 2008 Direct observation of oxygen movement during resistance switching in NiO/Pt film *Appl. Phys. Lett.* **93** 042106
[61] Sawa A 2008 Resistive switching in transition metal oxides *Mater. Today* **11** 28–36
[62] Li C *et al* 2017 Direct observations of nanofilament evolution in switching processes in HfO2-based resistive random access memory by *in situ* TEM studies *Adv. Mater.* **29** 1602976

IOP Publishing

Chapter 2

Fundamentals of metal-oxide thin-film transistors

Meng Zhang, Sunbin Deng, Yan Yan, Man Wong and Hoi-Sing Kwok

Nowadays, thin-film transistors (TFTs) are employed in a wide range of applications including but not limited to displays, sensors, memories, flexible electronics, and artificial intelligence. For example, the introduction of TFTs into flat panel display (FPD) backplanes has enabled an active-matrix (AM) addressing scheme for higher image quality and lower power consumption, and the blooming of the FPD industry is promoting the rapid development and commercialization of TFT technologies. Among the various technologies, TFTs with metal oxide (MO) semiconductor channels have drawn great attention and successfully entered the ranks of the mainstream in recent years. In this chapter, we will introduce the relevant fundamental background, the content of which covers device architecture classification, MO semiconductor design, reliability analysis, and possible solutions for performance improvement.

2.1 Overview of thin-film transistor technology

2.1.1 History and mainstream technologies

Since the field effect phenomenon was discovered in the 1930s and the first TFT was demonstrated in the 1960s, TFTs have been developing and evolving for more than half a century. As summarized in figure 2.1, there have emerged four mainstream TFT technologies over the years: hydrogenated amorphous silicon (a-Si:H) TFTs, low-temperature polysilicon (LTPS) TFTs, MO TFTs and organic TFTs, classified according to their material attributes and the processing techniques of the active channels. Each technology has its own pros and cons, and their overall comparison is presented in table 2.1 [1, 2].

Among the technologies, a-Si:H TFT technology has been widely used in the AM-FPD industry owing to its low manufacturing cost and high yield after long-term development, but the insufficient mobility, large leakage current and poor

doi:10.1088/978-0-7503-2556-1ch2

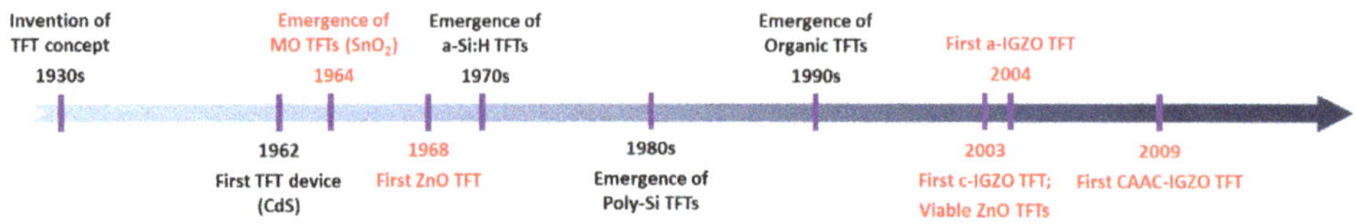

Figure 2.1. Milestones of TFT technologies with MO TFTs highlighted.

Table 2.1. Overall comparison of the a-Si:H, LTPS, MO and organic TFT technology.

TFT type	a-Si:H	LTPS	MO	Organics
Bandgap (eV)	~1.8	~1.1	>3	1–3
Density of deep states (cm^{-3} eV^{-1})	$\sim 10^{18}$	$\sim 10^{18}$	$\sim 10^{17}$	$>10^{14}$
Mobility (cm^2 V^{-1} s^{-1}) (polarity)	0.5–1 (N)	~150 (N), ~80 (P)	5–50 (N)	0.1–5 (P)
Subthreshold swing (V dec^{-1}) (variation)	0.4–0.5 (~0.2)	0.2–0.4 (~0.1)	0.1–0.3 (~0.02)	0.1–1 (0.01–0.2)
Leakage current (A)	$\sim 10^{-12}$	$\sim 10^{-11}$	$<10^{-13}$	$\sim 10^{-12}$
On/off ratio	$\sim 10^{7}$	$\sim 10^{7}$	$>10^{10}$	$\sim 10^{6}$
Long-term reliability	Poor	Good	Fair	Poor
Spatial uniformity (threshold voltage variation)	Fair (~0.5 V @ Gen-4.5*)	Poor (~1 V @ Gen-4.5*)	Good (~0.2 V @ Gen-4.5*)	Good (no exact data)
Transparency	Low	Low	High	Low
Process temperature (°C)	150–350	400–600	RT~350	RT~200
Flexibility	Poor	Poor	Fair	Good
Mask counts	4–5	5–11	4–7	4–5
Manufacturing cost	Low	High	Low	Low
Yield	High	Medium	Medium	High

* Gen-4.5 refers to the generation number of the FPD panel production line, and the size of a Gen-4.5-line glass substrate is 73 cm by 92 cm.

stability mean it struggles as the demanded driving capability increases. The LTPS TFT technology is equipped with enhanced device characteristics and reliability, so it has become prevalent in recent years. However, the electrical non-uniformity issue because of grain boundaries limits its application in large-area electronics, which is one of the most promising directions for TFTs. Different from its inorganic competitors, organic TFT technology is more popular in flexible and printable electronics. Unfortunately, its ambient reliability needs to be further improved before mass manufacturing is possible. As another emerging choice, MO TFT technology has many merits, such as reasonable electrical properties, high transparency in the visible light band, low-temperature and low-cost manufacturing, etc. For MO TFTs, amorphous indium–gallium–zinc oxide (a-IGZO) is one of the most prevalent channel materials [1–5]. Compared with conventional a-Si:H TFTs, the superiority of a-IGZO TFTs mainly lies in their relatively high mobility (5–10 $cm^2 V^{-1} s^{-1}$) and obviously low off-state current ($<10^{-13}$ A). These properties are of great value in high-performance and low-power-consumption applications such as mobile and wearable apparatus. Although the LTPS TFT technology can also provide high mobility, its spatial non-uniformity of electrical performance across a large area is a serious issue [6, 7]. In contrast, MO TFTs are compatible with a wider range of substrate sizes. Thanks to its advantages, the MO TFT technology has successfully become a mainstream choice and has even been implemented in practical products such as AM-FPD panels [8]. However, more efforts are required to further improve its electrical performance and reliability before this technology is able to occupy a dominant position in the market.

2.1.2 Operation mechanism

A TFT is basically a three-terminal field-effect transistor (FET). It always consists of a substrate, gate electrode (GE), gate insulator (GI), active channel (AC) and source/drain (S/D) electrode, and also usually contains an interlayer dielectric (ILD), passivation (PV) layer, etc. The substrate provides support and isolation for the other parts of the TFT. The source/drain electrodes make contact with the semiconducting active channel, while the gate electrode is separated from the active channel via the gate insulator layer. As a field-effect device, a TFT's gate electrode can manipulate the capacitive injection of carriers near the gate insulator/active channel interface. Then, its channel resistivity, and hence current flow, between the source/drain electrodes can be modulated indirectly.

As a microelectronics technology, TFTs share many similarities in design, fabrication, and operation with metal-oxide-semiconductor field-effect transistors (MOSFETs), which are the first choice in high-performance applications like microprocessors. However, there are several intrinsic differences between these two technologies [1]. Firstly, the formation of the interfacial conducting channels is realized by carrier accumulation in TFTs, whereas it relies on carrier inversion in MOSFETs. TFTs have no p–n junctions in their source/drain regions like MOSFETs do. For MO TFTs in particular, most MO semiconductor channels are intrinsically n-type as a result of the low formation energy of native donors, and there is a high density of subgap states near the valence band maximum (VBM). Even when a very negative gate bias is applied, these near-VBM states can firmly pin the Fermi level of channels, resulting in no carrier inversion. Therefore, the MO TFT technology is typically non-complementary due to a lack of high-performance p-type devices [9]. Secondly, TFTs are always fabricated on large-scale glass or plastic substrates and require extra semiconducting thin films as their channel layers, while MOSFETs are directly implemented on silicon wafers by employing their own single-crystal silicon as channels. Though the overall electrical characteristics of TFTs are inferior, they outperform MOSFETs in terms of substrate scalability and manufacturing cost and are capable of supporting applications such as flat panel displays, sensors, and memories. Thirdly, several processes in the fabrication of MOSFETs need to be conducted at a temperature of over 1000 °C, which exceeds the melting point of glass

(600–650 °C) and plastic (<400 °C) substrates. Such high temperatures are not applicable for TFT fabrication, which requires low-temperature processes.

2.1.3 Device structure

In general, a TFT structure can be defined as staggered or coplanar, referring to Weimer's definition [10]. Specifically, if its source/drain electrodes and gate electrode are on opposite sides of the active channel, the device is staggered. Otherwise, it is coplanar. Moreover, according to the stacking sequence, TFTs can be further divided into a normal (top-gate) architecture or inverted (bottom-gate) architecture, as shown in figure 2.2. For MO TFTs, the inverted staggered architecture and the normal coplanar architecture are the two major structural configurations.

The inverted staggered structure, which is also described as a bottom-gate top-contact (BG-TC) structure, is frequently seen in hydrogenated amorphous silicon (a-Si:H) TFTs due to their easier processing and hence less degraded electrical performance. Since the a-Si:H channel is light sensitive, its underlying metallic gate can act as a light-shielding layer and protect the channel layer from backlight exposure [1]. Considering processing compatibility, the inverted staggered architecture was also adopted in early-stage AM-FPD products employing MO TFTs. As shown in figures 2.3(a) and (b), such a structural configuration can be further divided into etch-stopper (ES) type and back-channel-etched (BCE) type [11]. In ES-type MO TFTs, the channel layer is capped with an extra etch-stop layer before S/D deposition so that the back-channel surface can be isolated from the following S/D electrode patterning processes to guarantee good device reliability. In contrast, BCE-type TFT technology is free of an etch-stop layer, intrinsically reducing the photolithography mask count to enable more cost-effective manufacturing. Meanwhile, it also miniaturizes the parasitic capacitance and device footprint to realize applications with higher transistor density [11, 12]. Figures 2.4(a) and (b),

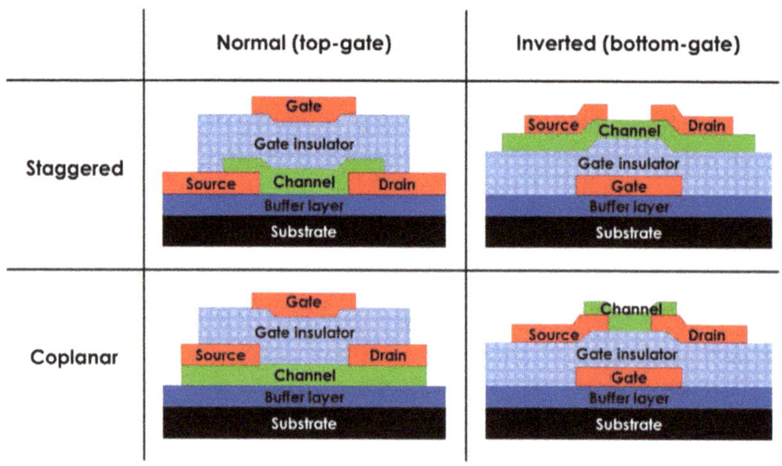

Figure 2.2. Schematic diagrams of TFT structures classified according to Weimer's definition and the deposition sequence.

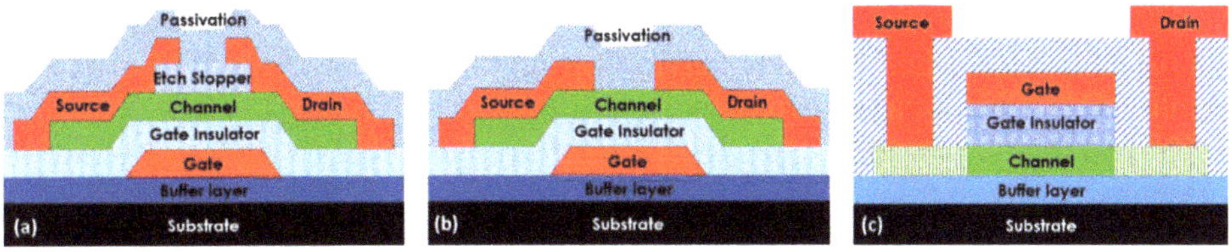

Figure 2.3. Cross-sectional diagrams of three major structural configurations in MO TFTs. (a) ES-type structure. (b) BCE-type structure. (c) TG-SA structure.

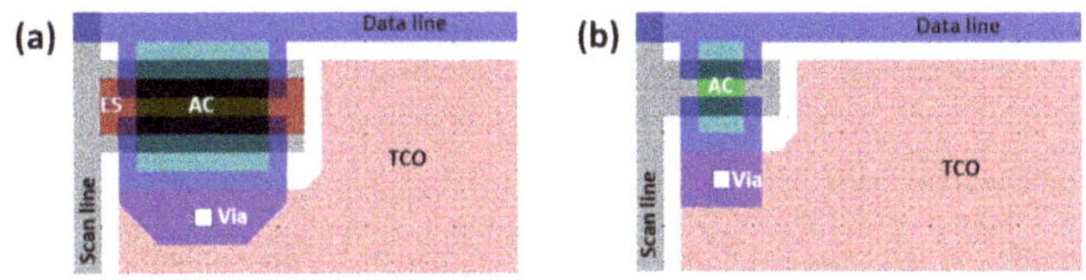

Figure 2.4. Layout illustration of (a) an ES-type MO TFT and (b) a BCE-type or TG-SA MO TFT in the 1T1C pixel circuit of AMLCD panels.

respectively, show the layout of an ES-type TFT and a BCE-type TFT in a one-transistor-one-capacitor (1T1C) pixel circuit of active-matrix liquid crystal display (AMLCD) panels. When the design rule is defined as 2 μm, the minimum allowed channel length for the ES-type TFT should be 6 μm if a safe alignment margin is taken into consideration, whereas the result is only 2 μm for the BCE-type TFT. To maintain an equivalent driving current, the ratio of the channel width over the channel length should remain the same. Therefore, the BCE-type TFT will occupy a smaller area, and then the AMLCD panel can have a higher aperture ratio or higher resolution. However, the BCE-type TFT also has some limitations. For example, its S/D electrode patterning is not perfect at present because most etch processes inevitably bring residues or cause damage to the channels, giving rise to urgent performance and reliability issues [13–15].

Regardless of whether the device is ES-type or BCE-type, it is impossible to avoid the intrinsic overlaps between the gate electrodes and the S/D electrodes. These overlaps introduce undesirable parasitic capacitance, limiting the operation speed of the circuits [16]. To reduce the parasitic capacitance in MO TFTs, a suggested solution is to adopt a special type of normal coplanar structure, the top-gate self-aligned (TG-SA) structure, as shown in figure 2.3(c). Originally, this architecture was often found in low-temperature polysilicon (LTPS) TFTs, where the gates are employed as hard masks and the exposed polysilicon is heavily doped as self-aligned conductive S/D regions. In principle, TG-SA devices have four main advantages. Firstly, an extra photolithography step for defining the S/D electrodes is no longer required, enabling more cost-effective manufacturing. Secondly, the underlying substrates and the upper gate insulators can act as *in situ* protection layers for the sandwiched channels. Thirdly, the undesirable parasitic capacitance, as previously mentioned, can be eliminated. This is indeed helpful to reduce the RC delay of signals and hence increase the operation speed of circuits. Also, for self-emitting (such as active-matrix organic light-emitting diode (AMOLED)) displays, the luminescence non-uniformity issue can be well addressed by suppressing the pixel voltage shift (also called the kickback/feedthrough voltage) as a result of capacitive coupling [11]. Finally, as shown in figure 2.4(b), TG-SA TFTs also provide stronger device scalability, like BCE-type devices, and are favorable to realize applications with high density of transistors. Despite their advantages, however, there exist many technical issues in the fabrication of TG-SA MO TFTs. For example, how to form self-aligned source/drain regions with durably high conductivity has been a challenge. Current strategies such as hydrogen doping and plasma treatment cannot guarantee devices with satisfactory electrical performance, thermal/electrical reliability and spatial uniformity across a large area at the same time [17–20].

2.1.4 Key static parameters

The electrical performance of TFTs can be evaluated by several static parameters, field-effect mobility (μ), threshold voltage (V_{th}), subthreshold swing (SS) and on-off ratio. They can be extracted from the transfer ($I_{ds}-V_{gs}$) and output ($I_{ds}-V_{ds}$) curves of TFTs, as plotted in figure 2.5.

Mobility (μ) is one of the key parameters to describe carrier transport in TFTs. It can directly determine the drain current (I_{ds}) and transit time (t_{sd}) (or operation frequency) of devices according to the following theoretical equations:

$$I_{ds} = \frac{W}{L}\mu C_{ox}\left[(V_{gs} - V_{th})V_{ds} - \frac{1}{2}V_{ds}^2\right], \tag{2.1}$$

and

$$t_{sd} = \frac{L^2}{\mu V_{ds}}, \tag{2.2}$$

where W, L, and C_{ox} are the channel width, channel length, and gate insulator capacitance per unit area, respectively.

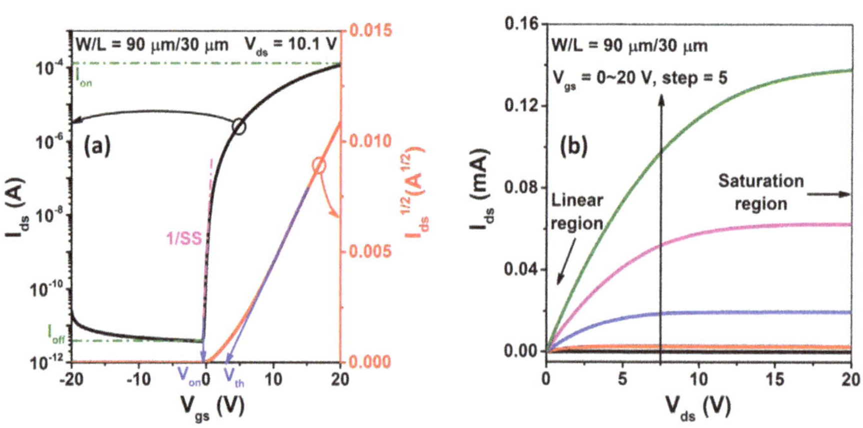

Figure 2.5. (a) Transfer ($I_{ds}-V_{gs}$) and (b) output ($I_{ds}-V_{ds}$) curves of a typical n-type MO TFT.

In the linear region of MO TFTs ($V_{ds} \ll V_{gs}-V_{th}$), μ can be specifically defined as the linear field-effect mobility (μ_{fe}), which is obtained from the TFT transconductance (g_m) and expressed as

$$\mu_{fe} = \frac{\frac{dI_{ds}}{dV_{gs}}}{C_{ox}\frac{W}{L}V_{ds}} = \frac{g_m}{C_{ox}\frac{W}{L}V_{ds}}. \tag{2.3}$$

In the saturation region of MO TFTs ($V_{ds} > V_{gs} - V_{th} > 0$), equation (2.1) can be further rewritten as

$$I_{ds} = \frac{1}{2}\frac{W}{L}\mu_{sat}C_{ox}(V_{gs} - V_{th})^2, \tag{2.4}$$

where μ is donated by the saturation mobility (μ_{sat}), which is derived from the following equation:

$$\mu_{sat} = \frac{\left(\frac{d\sqrt{I_{ds}}}{dV_{gs}}\right)^2}{\frac{1}{2}\frac{W}{L}C_{ox}}. \tag{2.5}$$

Threshold voltage (V_{th}) theoretically corresponds to the gate bias (V_{gs}) for which a conductive source-to-drain channel is formed at the gate insulator/active channel interface by carrier accumulation. In practice, there are many methods to extract V_{th}. For example, V_{th} can be determined by extrapolating a straight line in the I_{ds} – V_{gs} plot at a low V_{ds} (the linear region) or in the $(I_{ds})^{1/2}$–V_{gs} plot at a high V_{ds} (the saturation region). For simplicity, it is sometimes defined as the V_{gs} corresponding to a specific I_{ds} (e.g. $\frac{W}{L} \times 10^{-8}$ A). V_{th} is an essential indicator of the device operation mode. For n-type TFTs, enhancement- and depletion-mode devices have a positive V_{th} and negative V_{th}, respectively. Along with V_{th}, the turn-on voltage (V_{on}) is another common parameter, and it is equal to the minimum V_{gs} where I_{ds} is larger than the off-state current [21]. In a log I_{ds}–V_{gs} plot, V_{on} is the value of V_{gs} corresponding to the onset of a continuous increase of I_{ds}. When V_{gs} is lower than V_{on}, TFT channels should be completely turned off. Although V_{th} is not the same as V_{on}, their difference is small in MO TFTs considering their steep subthreshold swing.

Subthreshold swing (SS) is related to the amount of V_{gs} change required to increase I_{ds} by a decade. Its definition is expressed as

$$SS = \left(\left.\frac{d\log(I_{ds})}{dV_{gs}}\right|_{max}\right)^{-1}. \tag{2.6}$$

Since the relationship between I_{ds} and V_{gs} in the subthreshold region is

$$I_{ds} = I_{ref} \exp\left[\frac{q(V_{gs} - V_{th})}{nk_BT}\right], \tag{2.7}$$

$$n = 1 + \frac{C_d + C_{it}}{C_{ox}}, \tag{2.8}$$

where I_{ref}, n, q, k_B, T, C_d, C_{it} and C_{ox} are the effective subthreshold reference current at V_{th}, ideal factor, electron charge, Boltzmann's constant, absolute temperature, depletion capacitance, interfacial trap capacitance, and gate insulator capacitance, respectively. SS can be further expressed as

$$SS = \ln(10)\frac{k_BT}{q}\left(1 + \frac{C_d + C_{it}}{C_{ox}}\right) = 2.3\frac{k_BT}{q}\left(1 + \frac{C_d + C_{it}}{C_{ox}}\right). \tag{2.9}$$

This means that SS is an indicator of the total trap density in the channel bulks and at the gate insulator/active channel interfaces. For classical TFTs, the value of SS has a fundamental lower limit, which is $(2.3k_BT)$ mV/decade or ~60 mV/decade at room temperature (300 K). Such a theoretical limit is famous for the Boltzmann tyranny [22].

On–off ratio refers to the ratio of the maximum I_{ds} over the minimum I_{ds}. The maximum I_{ds} is determined by many aspects such as semiconductor mobility, channel dimensions, gate insulator capacitance density, etc. The minimum I_{ds} is primarily limited by the dark current of channels, gate leakage current, or measurement limit. As an eligible switch, like in AMLCD panel pixels, TFTs should have an on–off ratio of over 5×10^5. Since the off-state current of MO TFTs is generally below the measurement limit (~10^{-14} A), the resultant on–off ratio ought to be much higher than the requirement, implying that MO TFTs should be very promising as low-power-consumption electronic switches.

2.2 Metal oxide semiconductors for thin-film transistors

Owing to merits such as reasonable electrical properties, excellent optical transparency, low manufacturing cost, and low temperature, MO semiconductors have become attractive TFT channel candidates in recent years. According to figure 2.1, the history of studying MO semiconductors in TFTs is not short, with over half a century passing since the first report of MO TFTs in 1964 [23] and the first report of single-crystalline lithium-doped ZnO TFTs in 1968 [24]. In fact, MO TFTs emerged earlier than their competitors; however, most efforts at the early stage were devoted to binary MO semiconductors, such as ZnO, SnO_2, and In_2O_3. Since the fabricated MO devices were always below expectation, interest in the alternative a-Si:H TFT technology bloomed along with the growth of personal computers in the early 1990s, and researchers gradually transferred their enthusiasm from MO TFTs to silicon-based TFTs.

Fortunately, in the early 2000s, Hosono *et al* opened a new window on the MO TFT field by shifting from binary to multicomponent MO semiconductors, proposing single-crystalline indium–gallium–zinc oxide (c-IGZO) in 2003 [25] and then a-IGZO in 2004 [4] and making remarkable progress in the development of MO TFT technology. Since then, a great number of amorphous oxide semiconductors (AOSs) have been developed as high-performance MO TFT channels via compositional approaches [26–30]. In 2009, Yamazaki *et al* proposed a special type of crystalline multicomponent MO semiconductor, *c*-axis-aligned crystal indium–gallium–zinc oxide (CAAC-IGZO), which not only retains the advantages of a-IGZO, but also reaches an extremely low off-state current of $\sim 10^{-24}$ A μm^{-1} in the corresponding TFTs [31]. In 2012, together with Sharp, Yamazaki *et al* commercialized this outstanding material by releasing television and smartphone panels. These milestones confirm the enormous potential of MO semiconductors as high-performance TFT channels. In this section, we will discuss both the binary and multicomponent MO semiconductor groups in sequence to help readers understand the underlying reasons for the rise of MO TFTs.

2.2.1 Binary metal oxide semiconductors

Among binary MO semiconductors, ZnO has been intensively investigated for many years. Theoretically, wurtzite ZnO is an n-type semiconductor with a wide and direct band gap ($E_g = 3.4$ eV at 300 K), so it is highly transparent in the visible light band. Meanwhile, it has been found that the intrinsic electrical properties of ZnO are at a medium level. For example, the ideal electron Hall mobility (μ_{Hall}) in single crystalline ZnO can reach as high as 200 $cm^2\ V^{-1}\ s^{-1}$, while its intrinsic carrier concentration is less than $10^6\ cm^{-3}$ [32]. Additionally, zinc is quite abundant in the Earth's upper continental crust, leading to an extremely low raw material price. This is beneficial for low-cost electronics. These merits of ZnO mean that its implementation in TFTs was always eagerly expected. After a long incubation period, remarkable progress was indeed made in 2003, when Hoffman *et al* [33], Masuda *et al* [34], and Carcia *et al* [35] individually reported their own well-performing ZnO TFTs. Notably, the ZnO channel layer in Carcia's work was deposited using a radio-frequency magnetron sputtering technique at room temperature, and the following processes were also free of high temperature. This was valuable for the development and promotion of ZnO TFTs. Afterwards, research in the field of MO TFTs became attractive once again. Only one year later, fully transparent ZnO TFTs with room-temperature processes were demonstrated by Fortunato *et al* [36], and their mobility could reach nearly 20 $cm^2\ V^{-1}\ s^{-1}$.

Despite their many advantages, ZnO TFTs also have many inevitable drawbacks that hinder their further development. Firstly, many native point defects such as oxygen vacancies (V_o) and zinc interstitials (I_{Zn}) exist in the ZnO channels, as shown in figure 2.6(a) [37]. They release free electrons and make the ZnO channels unintentionally n-type doped. Without effective control of native point defects, it is hard to switch off the devices with a low off-state current. Meanwhile, these defects, especially V_o, are usually associated with device instability. Therefore, few ZnO TFTs with excellent reliability have been produced so far. Secondly, the sputtered ZnO films are typically polycrystalline (pc) with a preferential grain growth along the *c*-axis, even on amorphous substrates (figure 2.6(b)). Like the

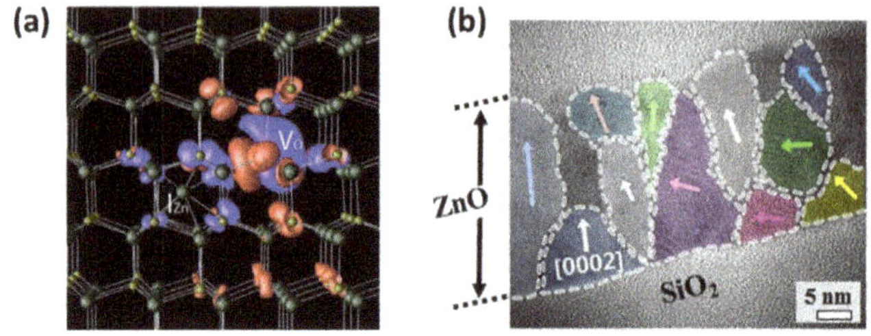

Figure 2.6. Illustration of (a) atomic arrangements and native defects. (b) Nanocrystals and grain boundaries in pc-ZnO films. The arrows indicate the [002] direction of each nanocrystals. (a) [38] (2009) Copyright. With permission of Springer. (b) [39] John Wiley © Sons. © (2016) WILEY-VCH Verlag GmbH & Co., KGaA, Weinheim.

LTPS TFT technology, the high density of grain boundaries deteriorates the electrical uniformity of the devices and hence limits their large-size applications. Meanwhile, due to atomic disorder, many native defects are generated at the grain boundaries, degrading the pc-ZnO TFTs in both electrical performance and reliability.

In_2O_3 is another common n-type binary MO semiconductor, which crystallizes with a bixbyite cubic structure. Like ZnO, In_2O_3 is natively doped by oxygen vacancies; however, their formed donor levels are closer to the conduction band minumim (CBM) (0.005–0.03 eV) in comparison to those in ZnO (0.01–0.05 eV) [40], so more free electrons can be easily released into In_2O_3 TFT channels with the Fermi level approaching the conduction band. In other words, the corresponding TFTs normally have to be operated in depletion mode or even perform no switching behavior at all. In principle, the CBM of both In_2O_3 and ZnO is primarily composed of sphere ns orbitals of metal cations, but the difference is that the 5s orbital radius of indium cations (180 pm) is larger than the 4s orbital radius of zinc cations (~137 pm) [41]. This means that the neighboring indium cations can overlap with each other more easily, providing better-defined paths for electron transport, so In_2O_3 should thus possess higher carrier mobility. However, the reality is that normally deposited In_2O_3 thin films at room temperature suffer from grain boundaries, which considerably limit electron transport, reduing mobility to a similar level to that in pc-ZnO thin films. As such, it is necessary to suppress the grain boundaries before mobility can be boosted. However, increasing thin film crystallinity is not the only solution for In_2O_3.

In 2014, Buchholz *et al* reported their study on Hall mobility variation of In_2O_3 thin films deposited at different temperatures [42]. It was found that the crystallinity could keep increasing with the rise in deposition temperature. However, an unexpected peak emerged in the range between −50 °C and 100 °C, and the local extremum value at 0 °C was surprisingly comparable to the results at >400 °C, where highly crystalline In_2O_3 thin films were deposited. According to the grazing incidence x-ray diffraction (GIXRD) results, the In_2O_3 thin film deposited at 0 °C was in a hybrid phase (hp), where crystal grains were embedded in an amorphous matrix. But fortunately, the corner-shared InO polyhedra in the hp-In_2O_3 were found to reach a maximum fraction of >80% via extended x-ray absorption fine structure (EXAFS) measurements and *ab initio* molecular dynamics (MD) liquid-quench simulations [43]. Basically, an indium–indium skeleton is formed in In_2O_3 by a network of InO polyhedra, which can exist in non-, corner-, edge- or face-shared form. Among these interconnected InO polyhedra, the corner-shared type shows the best spatial extension ability due to its longest indium–indium distance (~3.5 Å) with a small angle (~115°). Therefore, the long-distance chaining of the corner-shared InO polyhedra can offer the most efficient percolation conduction path. This is beneficial to forming long-distance directional chains for electron transport in the hp-In_2O_3 thin films, leading to a high Hall mobility. Moreover, the hybrid-phase thin films also possess a low density of grain boundaries since their crystallinity is relatively low, so there is no worry about electrical non-uniformity of the corresponding TFTs. Above all, it is expected that the hybrid phase in In_2O_3 can promote the further development of MO semiconductors as high-performance TFT channels. However, binary In_2O_3 is relatively expensive. It would be much better if such a crystal morphology (or microstructure) could be applied to other low-cost material systems such as multicomponent MO semiconductors.

2.2.2 Multicomponent metal oxide semiconductors

The drawbacks of binary MO semiconductors can be partially addressed by their multicomponent variants. In 2003, Hosono and coworkers reported a transparent MO TFT based on a single-crystalline $InGaO_3(ZnO_5)$ (or c-IGZO) channel [25]. The enhancement-mode device exhibited impressive electrical performance with a field-effect mobility of ~80 $cm^2\ V^{-1}\ s^{-1}$ and an on–off current ratio of 10^6. However, the growth of the single-crystalline phase required a thermal annealing treatment at 1400 °C. This was not favorable for the subsequent promotion of c-IGZO as TFT channels, which are usually deposited on glass or plastic substrates, but this work is still very meaningful, revealing the attractive potential of multicomponent MO semiconductors as high-performance TFT channels. In the following year, the authors further announced a transparent a-IGZO TFT. Although the previous crystalline channel was replaced by its amorphous counterpart, the devices fabricated using near room-temperature processes could perform well, with a saturation mobility of ~9 $cm^2\ V^{-1}\ s^{-1}$ [4].

In addition to the a-IGZO channel, the authors of [4] also proposed a theory about AOS materials to guide the design of multicomponent MO semiconductor channels. This AOS theory includes two core ideas, which are described as follows:

(1) The AOS materials should contain at least one type of post-transition-metal cation with an $(n-1)d^{10}ns^0$ ($n \geqslant 4$) electronic configuration [44]. For a-IGZO with ionic bonds, this refers to indium cation (In^{3+}), the sphere 5s orbital radius of which is as large as 180 pm [41]. In contrast, the 4s orbital radius of zinc cations is only about 137 pm. Thus, the ns orbitals of the neighboring post-transition-metal cations are more likely to overlap with each other, forming the CBM. Since the orbital overlaps are relatively insensitive to atomic randomness, they enable the formation of well-defined

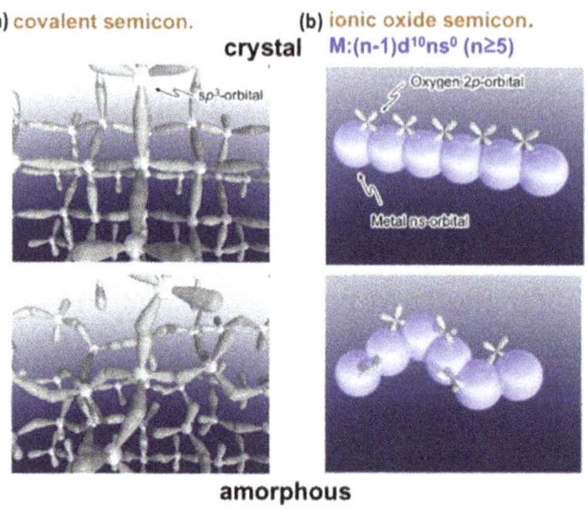

Figure 2.7. Schematic orbital drawing for carrier transport paths composed of (a) strongly directive sp^3 orbitals in covalent semiconductors and (b) overlapped sphere ns orbitals in AOS materials with post-transistion-metal cations. [4] (2004) Copyright. With permission of Springer.

paths for electron percolation conduction, so the electron mobility in the AOS materials can be maintained at a reasonably high level. However, for a-Si:H with covalent bonds, the CBM is mainly composed of strongly directive sp^3 orbitals. When the silicon atoms are highly disordered, the magnitude of orbital overlap will be severely degraded, with a large bond length and angle fluctuation, resulting in the deterioration of carrier transport paths. Therefore, the carrier mobility in a-Si:H is inevitably suppressed in comparison with that in single-crystalline silicon, as shown in figure 2.7.

(2) Apart from the post-transition-metal cations, other types of cations with different ionic radii and oxygen coordinates need to be incorperated as well. Then, the mixtures can exhibit a reduced degree of crystallization, and most of them deposited at room temperature can present an amorphous phase. Meanwhile, the incoporation of multiple cations also provides more freedom to researchers in tailoring material properties. Regarding a-IGZO, it has been reported that the addition of gallium cations can not only help to amorphize it, but also enlarge the material bandgap, acting as carrier stabilizers. This is becasue Ga_2O_3 has a wider band gap (~4 eV) [1] and a stronger metal–oxygen bond (~2 eV) [45] than In_2O_3 and ZnO.

Initiated by the AOS theory, many other advanced multicomponent MO semicondutors with the amorphous phase have been developed. In terms of ternary compounds, one of representative material systems is amorphous indium–zinc oxide (a-IZO) [26, 46, 47]. Compared with a-IGZO, there is a lack of gallium cations. Therefore, the carrier concentration in a-IZO thin films is usually higher owing to the easier formation of donor-like oxygen vacancies. This enables a carrier mobility boost in the correponding TFTs according to the percolation condution theory [48]. However, it may also increase the off-state current and cause the photobias reliability of the devices to deteriorate. Moreover, indium is relatively expensive, and its price has maintained a high level over the past 20 years [49], so its excessive inclusion will raise the raw material cost. For the sake of low-cost electronics, amorphous tin–zinc oxide (a-TZO) is another promsing choice among ternary compounds [50–52]. This is because the tin element is abundant on the Earth, and meanwhile tin cations can also follow indium cations, providing their large 5s orbitals for electron transport. However, prior to their inclusion in practical applications, a-TZO TFTs need to be further improved in terms of their reliability [11].

The major difference of quaternary and pentagonal compounds from ternary compounds is the involvement of more metal cations, most of which act as carrier stabilizers. Apart from gallium, such metal cations also include hafnium [53–55], zirconium [56], aluminum [57–59], lanthanum [60], etc. With carrier stabilizers, the TFTs can be easier to switch off with limited leakage current and improved reliability, but their carrier mobility may be reduced at the same time. Therefore, there is a tradeoff between mobility and reliability in light of the percolation conduction theory. However, when it comes to amorphous indium–tin–zinc oxide (a-ITZO), such a tradeoff seems to be better balanced. This is presumably because tin cations are multivalent in the thin films, where the 5s orbitals of Sn^{4+} cations are responsible for electron transport, while the carrier stabilizers are served by Sn^{2+} cations [11, 61]. So far, many a-ITZO TFTs with both high mobility (>30 $cm^2\ V^{-1}\ s^{-1}$) and robust stability have been reported [28, 62, 63]. These results imply that a-ITZO is another highly promising quaternary MO semiconductor as high-performance TFT channels.

With the prevalence of AOS materials over a long period, increasingly more researchers have failed to differentiate the AOS materials from multicomponent MO semiconcutors. Herein, we would like to clarify that they are not theoretically identical concepts. Strictly speaking, AOS materials should be a part of the multicomponent MO semiconductor family with the amorphous phase as a prerequisite. A typical example is CAAC-IGZO, which is a multicomponent MO semiconcutor with a particular crystal morphology, but it does not belong the group of AOS materials. More importantly, Yamazaki and coworkers reveal that the extremely low leakage current ($\sim10^{-24}$ A μm^{-1}) in CAAC-IGZO TFTs is actually attributed to the CAAC phase [64–66]. Unlike room-temperature processed a-IGZO TFTs, CAAC-IGZO TFTs need to be fabricated with a greater thermal budget, but still within the reasonable range that common TFT substrates can accept. Other examples of multicomponent MO semiconductors that are beyond AOS materials include but are not limited to nanocrystalline IGZO (nc-IGZO) [67, 68], cloud-aligned composite IGZO (CAC-IGZO) [69, 70], and hybrid-phase ITZO (hp-ITZO) [71, 72].

2.3 Reliability of metal oxide thin-film transistors

The current prevalence of MO TFT technology is due to is many advantages. However, its reliability issues have not yet been well addressed. When encountering stress from electrical bias, light exposure, temperature, and/or the ambient environment, MO TFTs always suffer from undesirable degradation, where a shift of threshold voltage, a deterioration of SS, a variation of on-/off-state current, and hump and hysteresis phenomena may be observed after stress. This may considerably hinder MO TFTs in practical applications. For example, a 0.1 V shift in threshold voltage in a driving TFT is reported to bring ~20% brightness variation in its corrsponding AMOLED pixel [73]. Before the reliability of devides are enhanced, it is necessary to understand the underlying degradation mechanisms. Therefore, this section will analyze the impacts of each kind of stress in detail.

2.3.1 Ambient stress

In the early stages of MO TFT development, most such devices were out of passivation, so their electrical characteristics could be significantly affected by atmospheric ambient interactions, which are mainly related to the oxygen and moisture chemisorption on the back-channel surface of MO semiconductors. To be more specific, the adsorbed oxygen molecules are acceptors to trap electrons in the channels, and the adsorption process can be described by the following chemical reaction formula [74]:

$$O_2(\text{gas}) + 2e^- \rightarrow 2O^-(\text{solid}). \tag{2.10}$$

Then, the phenomonon that a-IGZO TFTs present a more positive ΔV_{th} with the increase of oxygen partial pressure in ambient can be faithfully explained according to Le Chatelier's principle [75]. In addition, a more positive ΔV_{th} is also available when a positive gate bias is applied. This is because the oxygen adsorption process will be accelerated due to the accumulation of electrons in the channels, as illustrated in figure 2.8(a) [74]. On the other hand, the role of water molecules in the chemisorption process is found to be double-faced depending on the thickness of the channels. For thicker a-IGZO channels (⩾100 nm), the adsorbed water molecules are electron donors, leading to a negative ΔV_{th}. For thinner a-IGZO channels (⩽70 nm), more acceptor-like traps are generated after the moisture adsorption. Indeed, the negative shift of ΔV_{th} is well suppressed, while the value of SS is severely deteriorated at the same time. Such a difference in the trap property

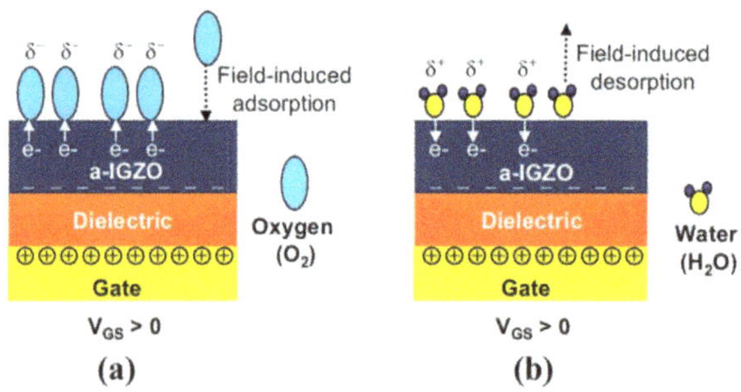

Figure 2.8. Schematic drawing of (a) the adsorption process of oxygen molecules and (b) the desorption process of moisture on the back-channel surface of a-IGZO TFTs under a positive gate bias. Reprinted from [74], with the permission of AIP Publishing.

is mainly attributed to the screening of intrinsic traps in the a-IGZO channels [74, 76]. The moisture chemisorption process can also be affected by the electric field. For instance, a positve gate bias promotes the desorption process of donor-like water molecules on the back-channel surface of a-IGZO TFTs, and then their V_{th} is positively shifted (figure 2.8(b)). To screen the ambient stress, in recent years, both inorganic and organic materials have been successfully developed to passivate MO TFTs. Although the reliability issues of MO TFTs cannot be completely addressed by introducing passivation layers, their ambient robustness is considerably enhanced.

2.3.2 Gate bias stress

As electronic switches, TFT devices may have to experience various electrical bias stresses. Among them, a negative/positive gate bias stress (N/PBS) in the dark at room temperature is commonly used to initially examine device robustness against bias stresses. The N/PBS for testing is implemented by applying a negative/positive direct-current (dc) bias on the gate electrodes of TFTs and meanwhile connecting their source/drain electrodes to the ground for a period. Since only a parallel shift with no SS degradation is usually observed in the tranfer curves of MO TFTs after N/PBS tests, the degradation mechanism is mainly associated with charge trapping. Empirically, the value of ΔV_{th} can be described as a function of stress time (t) referring to the following stretched-exponential equation:

$$\Delta V_{\text{th}} = \Delta V_{\text{th},\infty}\left\{1 - \exp\left[-\left(\frac{t}{\tau_{\text{trap}}}\right)^{\beta}\right]\right\}, \tag{2.11}$$

where $\Delta V_{\text{th},\infty}$, τ_{trap} and β denote the ΔV_{th} at infinite time, initial V_{th} before stress, characteristic trapping time and stretched-exponential exponent, respectively. Moreover, τ_{trap} can be determined by

$$\tau_{\text{trap}} = \tau_0 \exp\left(\frac{E_{\text{trap}}}{k_{\text{B}}T}\right), \tag{2.12}$$

where τ_0, E_{trap}, k_{B} and T represent the thermal pre-factor for emission over the barrier, average effective energy barrier to surpass for charge trapping, Boltzmann's constant and absolute temperature, respectively. In principle, the value of ΔV_{th} in the NBS tests should be negative, whereas it comes to be positive in the PBS tests. This is attributed to positive/negative charge (hole/electron) trapping at the gate insulator/active channel interfaces or in the gate insulator bulks during the N/PBS tests. Considering that most MO TFTs have n-type channels because of electron accumulation, they are usually faced with a more severe PBS-induced degradation.

Meanwhile, there also exists a charge de-trapping process in devices, where the captured holes/electrons are released from the sites at the gate insulator/active channel interfaces or in the gate insulator bulks. The ΔV_{th} related to charge de-trapping follows the following equation:

$$\Delta V_{\text{th}} = \Delta V_{\text{th,max}} \exp\left[-\left(\frac{t}{\tau_{\text{detrap}}}\right)^{\beta}\right], \tag{2.13}$$

where $\Delta V_{\text{th,max}}$ and τ_{detrap} are the maximum ΔV_{th} before recovery and characteristic de-trapping time, respectively. Furthermore, τ_{detrap} is equal to

$$\tau_{\text{detrap}} = \tau_0 \exp\left(\frac{E_{\text{detrap}}}{k_{\text{B}}T}\right), \tag{2.14}$$

where E_{detrap} represents the average effective energy barrier to surpass for charge de-trapping. In fact, the de-trapping process and the trapping process simultaneously exist in TFTs, and they are always competing with each other. Therefore, when the gate bias stress is removed, the de-trapping process will occupy a dominant position between them, and then the devices enter a recovery phase. As a result, their behaviors can gradually return towards the initial status before the stress. Certainly, if there is charge injection into the gate insulator bulks, the room-temperature recovery may fail to provide sufficiently high activation energy to overcome the charge de-trapping barrier, so there will be a V_{th} gap remaining after long-term recovery. Under this circumstance, extra stimuli are required to further activate the de-trapping process.

2.3.3 Illumination stress

Despite a wide bandgap exceeding 3 eV in plenty of MO semiconductor channels, they are still sensitive to incident light with a lower photon energy due to the existence of subgap states [77–79]. In many practical applications, MO TFTs suffer from not only electrical bias stress but also illumination stress at the same time. Particularly for switching TFTs in AMLCD pixels, the n-type devices must keep an 'off' state within a large portion of a frame period, so they suffer from a critical negative bias illumination stress (NBIS), which lasts nearly 1000 times longer than their experienced positive bias illumination stress (PBIS). However, even within the same stress time, the degradation under the NBIS operation is actually worse than that under the PBIS operation. Thus, the NBIS-induced instability has been a challenging issue for MO TFTs.

Since the NBIS-induced ΔV_{th} is negative, it is believed that electrons are donated to the conduction band of channels and/or positive charges are trapped at the gate insulator/active channel interfaces or in the gate insulator bulks. Considering that the room-temperature mobility ($\sim 10^{-27}$ cm^2 V^{-1} s^{-1}) of ionized oxygen vacancies ($V_o^{+/2+}$) [80] is much lower than that of holes ($\sim 10^{-5}$ cm^2 V^{-1} s^{-1}) [81], holes ought to be more possible positive charges. The origin of NBIS-induced degradation it is still debatable. Currently, at least three types of models, the oxygen photo-desorption model [82], the photo-created hole trapping model [83–85], and the oxygen vacancy photo-ionization model [80, 86–89], have been proposed.

The first model associates the oxygen chemsorption effect with the NBIS instability. It is said that photo irradiation can promote the desorption process of adsorbed oxygen on the back-channel surface, leading to the release of extra delocalized electrons into the channels and hence the negative shift of the threshold voltage. However, though most of today's devices have been well passivated to block ambient interactions, NBIS-induced degradation still emerges in these devices, implying the existence of other mechanisms.

The second model involves hole trapping theory. As shown in figure 2.9(a), holes are photo-excited in the valence band via the direct or indirect transition of electrons, and their generation process can be intuitively expressed as

$$h\nu \rightarrow e^- + h^+, \tag{2.15}$$

where h and ν are the Planck constant and incident light frequency, respectively, e^- denotes an electron, and h^+ is a hole. Under the NBIS-initiated electric field, the holes will drift towards the gate insulator/active channel interfaces. Some of them will be trapped at the interfaces or injected into the gate insulator bulks, repelling the depletion of electrons in the channels.

In the photo-created hole trapping model, researchers are apt to regard photo-excited electron–hole pairs as the source of holes. However, in the third model, holes are believed to originate from the photo-ionization of oxygen vacancies. This model includes a two-step process (figure 2.9(b)). The first step is the transition of oxygen vacancies from neutron states (V_O) to positively charged metastable states ($V_O{}^{2+}$), and each oxygen vacancy can release two delocalized electrons to the conduction band of channels. The relevant reaction formula is

$$V_O + h\nu \rightarrow V_O^+ + e^- \rightarrow V_O^{2+} + 2e^-, \tag{2.16}$$

where V_O^+ and V_O^{2+} represent a single-charged and a double-charged oxygen vacancy (V_o), respectively. This step requires an incident photon energy of ~2.3 eV for a-IGZO channels. If the photon energy exceeds ~2.8 eV, the second step can take place. In this step, holes are generated by transferring electrons from the valance band to the $V_O{}^{2+}$ states, namely,

$$V_O^{2+} + h\nu \rightarrow V_O + 2h^+. \tag{2.17}$$

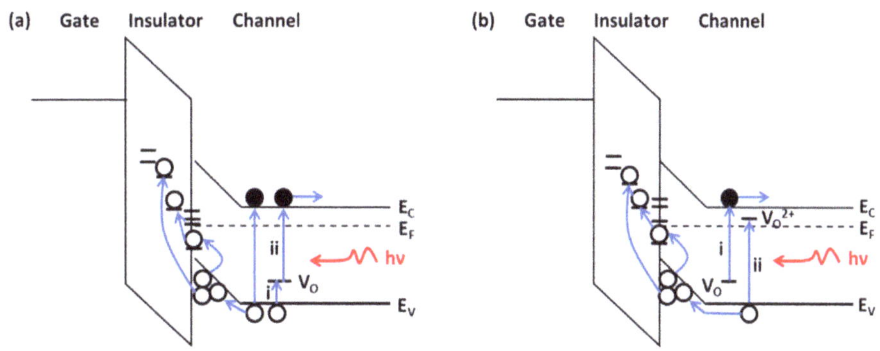

Figure 2.9. Schematic band diagram to explain (a) the photo-created hole trapping model and (b) the oxygen vacancy photo-ionization model.

Then, these holes are attracted by the NBIS-initiated electric field and get trapped at the gate insulator/active channel interfaces or in the gate insulator bulks, as mentioned previously.

Although the above three models were proposed based on experimental observations, none of them are now able to explain or predict every NBIS-induced degradation phenomenon. Therefore, development of a unified model to simulate the degradation/reliability of MO TFTs remains an important topic.

In the PBIS tests, the photo-excitation phenomenon also exists, but it is electrons rather than holes that will move towards the gate insulator/active channel interfaces and then be trapped under a positive gate bias. More importantly, the light-induced electron de-trapping process is found to overwhelm the simultaneous electron trapping process [90]. This means that the PBIS-induced degradation should be slighter with a smaller ΔV_{th} in comparison with the PBS case.

2.3.4 Thermal stress

Despite the NBIS instability being a challenging issue, it can be well addressed by introducing an additional light-shelding layer to avoid light illumination on the MO semiconductor channel regions. However, considering the self-heating phenomenon and the poor thermal dissipation in many electrical applications, it is inevitable for MO TFTs to suffer from thermal stress, often together with electrical bias stress. Generally speaking, the degradation mechanism under negative/positive bias temperature stress (N/PBTS) is similar to the analysis in the N/PBS case, where the hole/electron trapping process dominates. According to equation (2.12), the increase in temperature reduces the characteristic trapping time τ_{trap}. In other words, the holes/electrons are endowed with more energy to surpass the barrier for charge trapping through thermionic emission, so the probability of trapping increases. Since the majority of carriers in most MO TFTs are electrons, the PBTS-induced degradation due to electron trapping usually presents a positive ΔV_{th} [91–93], and the shift is larger than that in the PBS case. Sometimes a negative or bidirectional ΔV_{th} is also reported [94–96], and this is inconsistent with the electron trapping mechanism. Such an abnormal shift always emerges together with a degradation of the subthreshold swing or a hump phenomenon in the transfer curves, so there are likely to be other mechanisms, for example, the PBTS-induced generation of donor-like defects around the Fermi level or the existence of parasitic transistors. In the NBTS case, the population of holes is very limited in the channels, so the NBTS-induced degradation is found to be slight.

2.4 Approaches towards high performance

Frankly speaking, high performance is an ambiguous description that can be endowed with different meanings for different objects. In terms of MO TFTs, the content of high performance may include relatively high mobility (>25 $\mathrm{cm^2\ V^{-1}\ s^{-1}}$), nearly-zero threshold voltage, substantially steep subthreshold swing (<0.15 V/decade), extremely high on–off ratio (>10^{10}), and excellent reliability. Although the representive a-IGZO TFTs are prevalent in both academia and industry at the current stage, there is a large space for investigators to further improve their performance. In the following, we would like to share some of our efforts to enable high-performance MO TFTs through active channels, gate electrodes, gate insulators, and source/drain regions.

2.4.1 Active channels

No matter whether for AM-FPD backplanes or other electrical applications, increasingly higher carrier mobility in MO TFTs is demanded to enable larger driving current, higher frequency response and lower power consumption. For instance, ultra-high-definition (UHD) displays with an 8k × 4k resolution and a frame rate of 120 Hz require a TFT mobility exceeding 16 $cm^2\ V^{-1}\ s^{-1}$ if no advanced signal addressing scheme is employed in the panels [5]. However, the prevalent a-IGZO channels with a common atomic ratio (In:Ga:Zn = 1:1:1 at%) can only offer a typical mobility of 5–10 $cm^2\ V^{-1}\ s^{-1}$. Therefore, the design of advanced MO semiconductor (beyond a-IGZO) channels with boosted mobility and good spatial uniformity has been a popular topic in the field of MO TFTs.

In AOS materials, improvement in carrier mobility is typically realized via a compositional approach. Figure 2.10 systematically presents the influence of element composition on electron transport properties in the amorphous In_2O_3–Ga_2O_3–ZnO system [97, 98]. It can be seen that the electron mobility is closely associated with the content of indium cations. For example, the thin films can achieve a maximum Hall mobility of 39 $cm^2\ V^{-1}\ s^{-1}$ when their indium content exceeds 70%, and many groups have even directly employed a-IZO [26, 46, 47] or its analogues [99–102] as TFT channels or one stack among multi-layered channels [103–106]. However, the excess incorporation of indium cations may raise the free carrier population in the channels, leading to depletion-mode devices with high off-state current. Meanwhile, indium is so expensive that it will raise material cost and hence lose one of the key advantages of MO TFTs. More importantly, Steward *et al* predicted a theoretical upper carrier mobility limit in amorphous semicondutors no

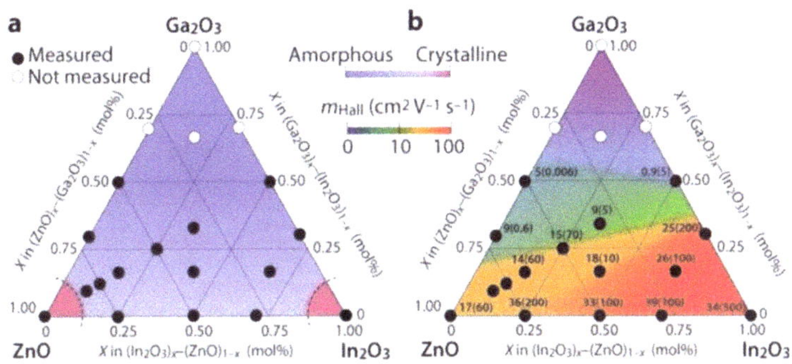

Figure 2.10. (a) Crystal morphology distribution and (b) electron transport properties in an In_2O_3–Ga_2O_3–ZnO system. In (b), the values outside parentheses denote the electron Hall mobility ($cm^2\ V^{-1}\ s^{-1}$) density, and the values inside parentheses denote the electron concentration ($\times10^{18}\ cm^{-3}$). [98] (2010) Copyright. With permission of Springer.

matter how the material composition is adjusted [81, 107, 108]. Therefore, we move beyond only considering the AOS materials and propose a microstructural approach to acquire higher mobility with good spatial uniformity. Such an approach can cooperate with the composition approach in general multicomponent MO semiconductors, which are not required to be in their amorphous phase.

As mentioned in the last section, the binary In_2O_3 thin films can achieve a Hall mobility peak in their hybrid phase, which is also hoped to be applicable to multicomponent MO semiconductors. By co-sputtering a polycrystalline indium–tin oxide (pc-ITO) target and a polycrystalline zinc oxide (pc-ZnO) target at room temperature, we could successfully deposit a hybrid-phase indium-tin-zinc oxide (hp-ITZO) thin film [71, 72, 109], which contained a number of columnar nanocrystals embedded in an amorphous matrix with blurry grain boundaries, as shown in figures 2.11(a) and (b). Similar to In_2O_3, the ITZO thin film achieved a Hall mobility peak in its hybrid phase (sample B in figure 2.11(c) and (d)) when compared with its counterparts with different crystallinities but the same composition. Moreover, this experimentally measured result (31.2 $cm^2\ V^{-1}\ s^{-1}$) in hp-ITZO indeed surpassed the

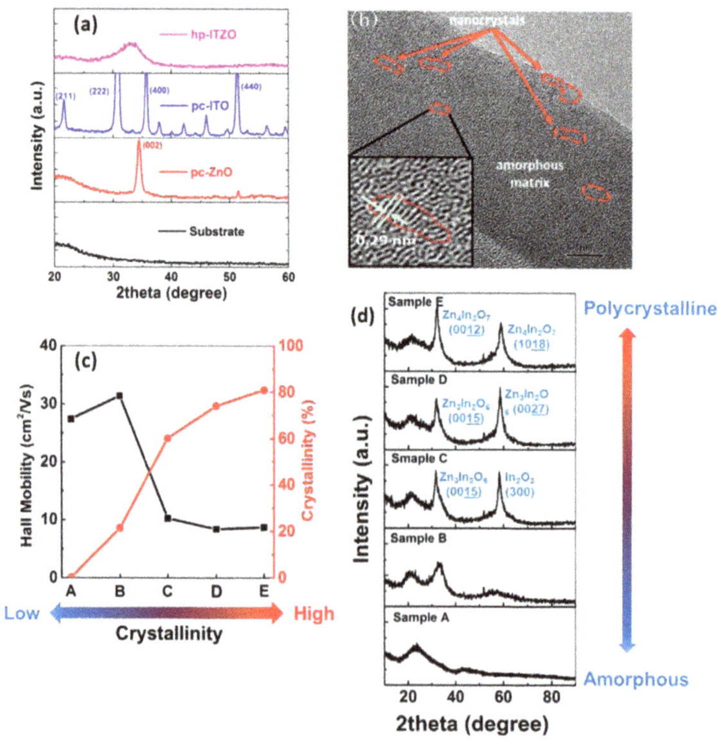

Figure 2.11. (a) XRD spectra of pc-ITO thin film, pc-ZnO thin film, hp-ITZO thin film and silicon wafer coated with thermally grown SiO_2. The thickness of thin films is 50 nm. (b) Cross-sectional high-resolution TEM image of hp-ITZO thin film. Inset: magnified image with nanocrystalline regions, which are surrounded by red dashed lines. (c) Plot of Hall mobility versus the crystallinity of ITZO thin films with the same composition. (d) XRD spectra of ITZO thin films with different crystallinities. (b) [109] John Wiley & Sons. © (2018) WILEY-VCH Verlag GmbH & Co., KGaA, Weinheim.

theoretically calculated upper mobility limit (26.3 $cm^2\ V^{-1}\ s^{-1}$) in its amorphous counterpart. This was attributed to the effective long-distance chaining of the InO/SnO polyhedra in hp-ITZO, and then the material potential for electron transport was completely dug out. In other words, the hp-ITZO thin film had a lower amount of shallow tail states, so its mobility deviated less from the intrinsic mobility without traps.

Afterwards, the hp-ITZO thin films were applied to inverted staggered TFTs with lift-off S/D electrodes [72]. Figure 2.12 exhibits the excellent electrical characteristics of the devices. Their typical μ_{fe}, V_{th} and on/off ratio could reach 27.3 $cm^2\ V^{-1}\ s^{-1}$, 0.5 V and over 10^9, respectively. In addition, the value of SS was as low as 89 mV/decade, indicating that the total trap density in the channel bulks and at the gate insulator/active channel interfaces was maintained at a relatively low level. Although the mobility difference between our hp-ITZO TFTs and other a-ITZO TFTs was not large [110–112], the hp-ITZO channels contained lower indium content (<40%). This could lead to normally-on devices with a measurement-noise-level off-state current and probably reduce manufacturing cost by maximizing mateiral potential. Apart from enhanced device behaviors, the hp-ITZO channels also well maintained device uniformity across a large area. This was because their nanocrystal size was almost ignorable in comparison with the channel length. Therefore, there was no need to be concerned when the amorphous channels were replaced by the hybrid-phase channels. In addition, the ambient robustness of the hp-ITZO channels was very impressive as well. The devices could be normally operated after storage in an atmospheric environment with a relative humidity of 50% and a temperature of 25 °C for over 26 months. The ΔV_{th} was as small as −0.6 V without SS deterioration. Overall, the outstanding performance exactly reflects the validity of the microstructural approach to realize higher mobility in MO TFTs.

The microstructural approach is also applicable to other multicomponent MO semiconductors. For example, hybrid-phase tin–zinc oxide (hp-TZO) thin films are available by simultaneously reactively sputtering a Sn target and a ZnO target in an oxygen-rich atmosphere. The corresponding inverted staggered TFTs, which are operated in enhancement mode, can easily present a μ_{fe} of >25 $cm^2\ V^{-1}\ s^{-1}$.

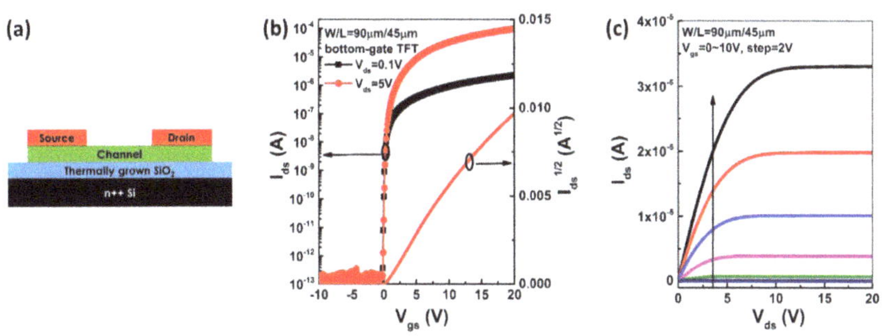

Figure 2.12. (a) Schematic diagram, (b) transfer curves and (c) output curves of the inverted staggered TFTs with the hp-ITZO channels and the lift-off S/D electrodes. © (2017) IEEE. Reprinted, with permission, from [113].

Since the hp-TZO channels are free of indium cations, their material cost can be considerably reduced.

2.4.2 Gate insulators

For TFTs, especially those with process-sensitive MO semiconductor channels, gate insulators are one of the core device components. Although the microstructural approach provides a solution to effectively boost carrier mobility while maintaining a low population of free carriers in the channels, the choice of gate insulator materials [114, 115] and their deposition processes [116, 117] will significantly affect device performance and reliability [115, 118, 119]. For example, SiO_2 layers created using a plasma-enhanced chemical vapor deposition (PECVD) technique with a gaseous silane source (SiH_4–SiO_2) are often selected as the gate insulators in MO TFTs. However, their deposition process is always accompanied by unexpected hydrogen (H) doping, especially in the fabrication of top-gate MO TFTs. This causes electrons to be released into the channels, causing a rise in off-state current and a negative shift of the threshold voltage in the devices. PECVD SiO_2 using a liquid tetraethyl-orthosilicate source (TEOS-SiO_2) is another common gate insulator, but hydroxyl (OH) groups and water molecules are rich inside the TEOS-SiO_2 bulks and at the gate insulator/active channel interfaces [120]. These by-products may bring severe hysteresis [121] as well as a large number of trap sites in the devices [122].

To improve the quality of single-layer PECVD SiO_2 gate insulators, we tried to deposit the SiH_4–SiO_2 layer on the top of the TEOS-SiO_2 layer, as shown in figure 2.13(a), and the stacked gate insulators were unexpectedly optimized [123]. In the stacks, both the H and OH content was found to be reduced due to their mutual interactions during the following thermal processes. The Al/SiO_2/n++ Si capacitors revealed that the electrical quality of the PECVD SiO_2 stacks was strengthened compared with that of the single-layer TEOS-SiO_2 or SiH_4–SiO_2 (figure 2.13(b)). The high-quality stacks were then applied to normal staggered hp-ITZO TFTs (figure 2.2). Their nearly-zero V_{th} reflects that the undesirable doping that accompanies the PECVD process was well suppressed. Then, the limited free carrier population in the channels together with the low gate leakage current maintained the

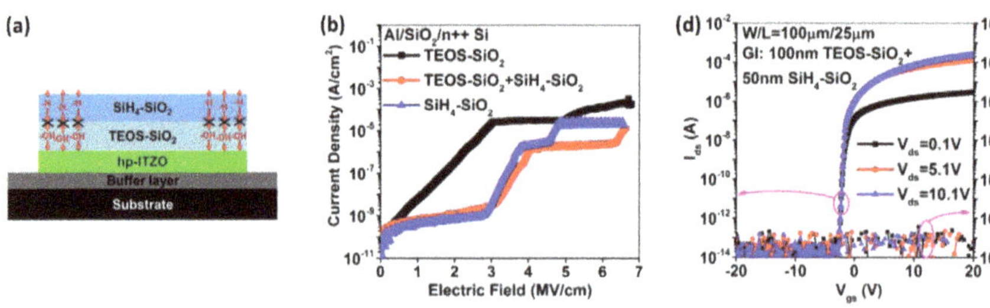

Figure 2.13. (a) Illustration of mutual interactions between H groups and OH groups in the gate insulator stacks. (b) Leakage current density of different gate insulators. (c) Typical transfer curves of the normal staggered hp-ITZO TFTs with the stacked PECVD SiO_2 gate insulators. © (2019) IEEE. Reprinted, with permission, from [123].

TFT off-state current at the noise level of the measurement system (~10^{-14} A). In addition, the extracted subthreshold swing, as shown in figure 2.13(c), was as low as 96.9 mV/decade, indicating the extremely high quality of the interfaces between the stacked PECVD SiO_2 gate insulators and the hp-ITZO active channels. Furthermore, the obtained TFTs also exhibited very robust ambient and electrical reliability.

Apart from stacked PECVD SiO_2, high-dielectric-constant (high-k) insulators such as Al_2O_3 [124–126], HfO_2 [127–129], ZrO_2 [130–132], Ta_2O_5 [133–135], and AlO_x:Nd [136–138] have been widely adopted for high-performance MO TFTs. Theoretically, high-k materials provide larger gate insulator capacitance density or smaller effective oxide thickness (EOT) than the common PECVD SiO_2 with the same thickness, so more carriers can accumulate at the gate insulator/active channel interfaces for higher mobility and steeper subthreshold swing. This means that the same drain current can be achieved with a lower gate bias. Moreover, under the same capacitance density, thicker high-k gate insulators are helpful to suppress the gate leakage current. To sum up, we know that high-k gate insulators in MO TFTs are promising in low-power-consumption electronics.

2.4.3 Gate electrodes

For a long period, studies on gate electrodes have examined techniques for low-resistance interconnections such as copper processes. From the perspective of circuits and systems, Cu gate processes are indeed promising in the reduction of signal RC delay and the increase of circuit speed [11]. Beyond this, people pay no more attention on the gate electrodes.

In fact, it is found that the gate electrodes are also very useful in the V_{th} modification of MO TFTs, and there are at least two underlying mechanisms. The first is related to the band bending of the active channels due to the Fermi level misalignment between the gate electrodes and the active channels. This is well known by the following equation [140]:

$$V_{gs} = \Phi_{ms} - \frac{Q_i}{C_{ox}} + \Psi_s + \frac{\varepsilon_s}{C_{ox}} E(\Psi_s), \tag{2.18}$$

where $\Phi_{ms} = \Phi_m - \Phi_s$, Q_i, C_{ox}, ε_s, Ψ_s, and $E(\Psi_s)$ are the work function difference between the gate electrodes and the channels, effective positive charges in the gate insulator bulks and at the gate insulator/active channel interfaces, gate insulator capacitance per unit area, dielectric constant of the active channels, surface potential, and its corresponding electric field, respectively. However, according to our study [139], when Al gates were replaced by ITO gates in top-gate hp-ITZO TFTs (figure 2.2), the actual ΔV_{th} (~1.7 V) was found to be much larger than the theoretical result (~0.4 V) induced by the work function difference. Further study revealed that not only the work function of the gate electrodes but also their permeability for hydrogen diffusion out of and oxygen diffusion into the MO semiconductor channels during post-annealing could efficiently modulate V_{th} (figure 2.14). This is of great value for high-mobility MO TFTs, which are always

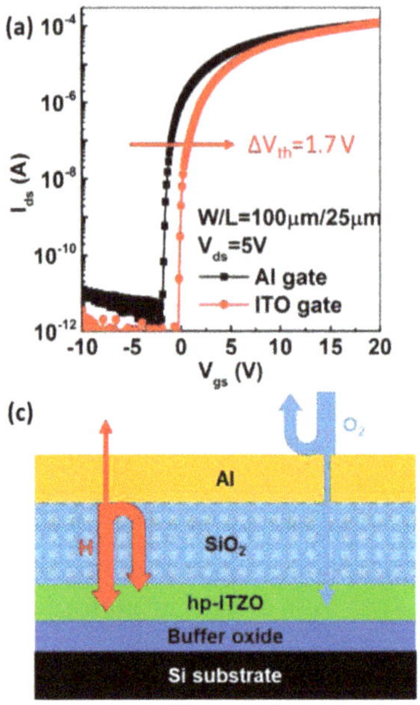

Figure 2.14. (a) Typical transfer curves of the top-gate hp-ITZO TFTs with Al and ITO gate electrodes. Schematic diagrams of thin-film-stacked samples (b) without gate electrodes, (c) with Al gate electrodes, and (d) with ITO gate electrodes. The arrow width indicates the degree of hydrogen diffusion out of or oxygen diffusion into the hp-ITZO channels. © (2018) IEEE. Reprinted, with permission, from [139].

operated in their depletion mode because of the tradeoff between carrier mobility and concentration. Such a new perspective provides a powerful tool to obtain the normally-off devices with little performance degradation and even monolithically integrate them with their normally-on counterparts for low-power-consumption electronics.

2.4.4 Source/drain regions

The source/drain regions are another key aspect to considerably impact the electrical performance of MO TFTs. For example, the carrier mobility is closely related to the contact resistance or the Schottky barrier in the source/drain regions [141–143]. Nowadays, self-aligned MO TFTs are prevalent in industry due to the intrinsically eliminated parasitic capacitance between the gate electrodes and the source/drain regions. However, the formation of self-aligned source/drain regions with durable low resistance has been a challenging issue, and needs to be addressed as soon as possible. So far, there have emerged at least four possible solutions: metal reaction, plasma treatment, cover-dependent annealing, and selective ion doping.

The metal reaction method was initially reported by Morosawa *et al* in 2011 [144]. They sputtered 5 nm thick Al or Ti as the reactive metal after gate electrode patterning. Subsequently, a thermal annealing treatment was carried out in an oxygen-containing atmosphere at 300 °C. Then, the thin Al or Ti layer could react with the exposed a-IGZO regions and form Al_2O_3 or TiO_2 by consuming oxygen atoms in a-IGZO. As a result, oxygen vacancy defects were formed and electrons could be released, reducing the width-normalized parasitic source-to-drain resistance to only 2 Ω-cm. Later on, other types of reactive metals such as Ca [145] and Mg [146] were developed. However, since the reactive metal layer in such devices is too thin to precisely control its deposition process and film quality, they often encounter non-reproducibility and non-uniformity issues especially across a large area. Thus, more attention needs to be paid before the metal reaction method can be applied to mass production in the future.

The plasma treatment method generally employs non-hydrogen or hydrogen-rich plasma to gain the conductivity of source/drain regions. The principle of non-hydrogen (such as Ar) plasma treatment [147] is based on ion bombardment, which can preferentially break relatively weak bonds like In–O in a-IGZO. Then, due to the positive mixing enthalpy and high surface energy, indium atoms will segregate at the surface to form conductive source/drain regions. For the hydrogen-rich (such as NH_3 [148]) plasma treatment, together with ion bombardment, hydrogen species as shallow donors can contribute to the resistivity reduction in MO semiconductors. However, there are also several drawbacks that come with the plasma treatments. For example, ion bombardment may cause damage at the sidewall of gate insulators to increase the gate leakage current, and the segregation of indium or other post-transition-metal atoms may result in large back channel leakage. Although the subsequent thermal processes can repair the gate insulator and redistribute the segregated atoms, the exposed source/drain regions will likely return to a high-resistivity state at the same time. Particularly for the devices treated by hydrogen-rich plasma, the fast hydrogen diffusion during post-annealing is also challenging, as mentioned. Above all, a lack of strong thermal immunity has been a serious issue so far for self-aligned MO TFTs that are fabricated via the plasma treatment method.

The cover-dependent annealing method [149, 150] is based on the phenomenon that MO semiconductors with gas-impermeable coverings are shown to be highly conductive after thermal oxidation owing to the generation of donor-like defects. For example, the resistivity of a-IGZO capped by a SiO_2/SiN_x stack can reach the vicinity of 10^{-2} Ω-cm after annealing [151]. To form the self-aligned source/drain regions, a gas-impermeable SiN_x cover layer was firstly deposited on the top of passivation layer, and then defined through back-side photolithography using metallic bottom-gate electrodes as masks. After the exposed SiN_x removal and thermal annealing in an oxygen atmosphere, the MO regions corresponding to the projection of the cover layer became conductive, while other regions with no cover layer returned to an intrinsically high resistivity state. These two MO states could then be employed in the high-performance self-aligned MO TFTs as their source/drain and channel regions, respectively. Since the cover-dependent annealing method involves 'vertical' and 'lateral' oxidation processes, there was a misalignment between the edge of the cover layer and the source/drain regions. However, this misaligned length (ΔL_A) was adjustable, and it could be minimized to about 0.25 μm after 4 h oxygen annealing at 400 °C. The bottom-gate self-aligned TFTs with either a-IGZO or a-ITZO channels exhibited relatively high electrical performance and robust device reliability, but their manufacturing efficiency and cost need to be

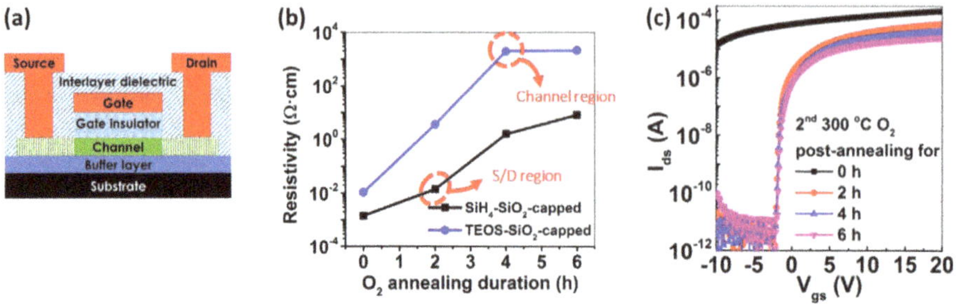

Figure 2.15. (a) Schematic diagram of the self-aligned hp-ITZO TFTs with the SiH_4–SiO_2-capped source/drain regions. (b) Resistivity of the SiH_4–SiO_2-capped or the TEOS-SiO_2-capped hp-ITZO thin films as a function of oxygen annealing duration at 300 °C. (c) Transfer curves of the self-aligned hp-ITZO TFTs after different post-annealing durations. © (2017) IEEE. Reprinted, with permission, from [157].

further optimized considering that long-term high-temperature thermal annealing treatments in a pure oxygen atmosphere are included at present.

The selective ion doping method is well established in the fabrication of LTPS TFTs, but the conventional dopants, such as boron, phosphorus and arsenic in silicon, and the following activation processes do not well match the MO semiconductor channels. Currently, hydrogen doping along with the deposition of SiO_2 or SiN_x interlay dielectrics is a common solution in the fabrication of self-aligned MO TFTs [16, 152]. However, the following thermal processes will promote the diffusion of hydrogen species towards the intrinsic channel regions, resulting in the devices having a negatively shifted threshold voltage or even no switching behavior. For us, this issue can be easily addressed by delicately designing the recipe for SiH_4–SiO_2 interlay dielectric deposition, where not only hydrogen but also silicon dopants are doped into the MO semiconductors. Since silicon ions are effective donors in ZnO [153–156], the hp-ITZO thin films capped by a SiH_4–SiO_2 layer were found to keep a low-resistivity state even after long-term post-annealing, as shown in figure 2.15. Considering the low diffusion coefficient of silicon dopants, the SiH_4–SiO_2-capped hp-ITZO thin films could be employed as highly conductive and thermally stable source/drain regions in self-aligned MO TFTs. The fabricated devices exhibited an average μ_{fe} of 19.56 $cm^2\ V^{-1}\ s^{-1}$, V_{th} of −1.65 V, and on–off ratio of over 10^7. Particularly, thanks to their strong immunity to the thermal budget, these devices were sufficiently annealed in the end, leading to a SS of 105 mV/decade. This value is likely the steepest result among the self-aligned MO TFTs ever reported. In addition, the devices were also very robust against ambient and electrical stresses [157].

2.5 Summary

In this chapter, the fundamentals of MO TFTs, including technology advantages, operation mechanism, device structures, key parameters, binary and multicomponent MO semiconductor materials, and reliability issues under various stresses have been introduced comprehensively. Effective modifications from the aspects of active channels, gate insulators, gate electrodes, and source/drain regions have also been proposed to improve the performance of MO TFTs. It is believed that high-performance MO TFTs with a relatively high mobility, nearly-zero threshold voltage, substantially steep subthreshold swing, extremely high on–off ratio, and excellent ambient/electrical reliability are of high value in a wide range of electronic applications.

References

[1] Fortunato E, Barquinha P and Martins R 2012 Oxide semiconductor thin-film transistors: a review of recent advances *Adv. Mater.* **24** 2945–86

[2] Nathan A, Lee S, Jeon S and Robertson J 2014 Amorphous oxide semiconductor TFTs for displays and imaging *J. Display Technol.* **10** 917–27

[3] Park J S, Maeng W-J, Kim H-S and Park J-S 2012 Review of recent developments in amorphous oxide semiconductor thin-film transistor devices *Thin Solid Films* **520** 1679–93

[4] Nomura K, Ohta H, Takagi A, Kamiya T, Hirano M and Hosono H 2004 Room-temperature fabrication of transparent flexible thin-film transistors using amorphous oxide semiconductors *Nature* **432** 488

[5] Kamiya T, Nomura K and Hosono H 2010 Present status of amorphous In–Ga–Zn–O thin-film transistors *Sci. Technol. Adv. Mater.* **11** 044305

[6] Zhou W *et al* 2012 Bridged-grain solid-phase-crystallized polycrystalline-silicon thin-film transistors *IEEE Electron Device Lett.* **33** 1414–6

[7] Deng S, Chen R, Zhou W, Ho J Y L, Wong M and Kwok H S 2016 Fabrication of high-performance bridged-grain polycrystalline silicon TFTs by laser interference lithography *IEEE Trans. Electron Devices* **63** 1085–90

[8] Wager J F 2016 Oxide TFTs: a progress report *Inf. Display* **31** 16–21

[9] Wang Z, Nayak P K, Caraveo-Frescas J A and Alshareef H N 2016 Recent developments in p-Type oxide semiconductor materials and devices *Adv. Mater.* **28** 3831–92

[10] Wallmark J T and Johnson H 1966 *Field-Effect Transistors: Physics, Technology and Aplications* (Upper Saddle River, NJ: Prentice Hall)

[11] Kwon J Y and Jeong J K 2015 Recent progress in high performance and reliable n-type transition metal oxide-based thin film transistors *Semicond. Sci. Technol.* **30** 024002

[12] Yeom H-I *et al* 2016 60-3: distinguished paper: oxide vertical TFTs for the application to the ultra high resolution display *SID Symp. Dig. Tech. Pap.* **47** 820–22

[13] Xu H *et al* 2011 High performance indium-zinc-oxide thin-film transistors fabricated with a back-channel-etch-technique *Appl. Phys. Lett.* **99** 253501

[14] Luo D *et al* 2013 Effects of etching residue on positive shift of threshold voltage in amorphous indium–zinc-oxide thin-film transistors based on back-channel-etch structure *IEEE Trans. Electron Devices* **61** 92–7

[15] Park Y, Um J, Mativenga M and Jang J 2015 Modification of electrode-etchant for sidewall profile control and reduced back-channel corrosion of inverted-staggered metal-oxide TFTs *ECS J. Solid State Sci. Technol.* **4** Q124–9

[16] Kang D H, Kang I, Ryu S H and Jang J 2011 Self-aligned coplanar a-IGZO TFTs and application to high-speed circuits *IEEE Electron Device Lett.* **32** 1385–7

[17] Chen R, Zhou W, Zhang M, Wong M and Kwok H S 2012 Self-aligned indium–gallium–zinc oxide thin-film transistor with source/drain regions doped by implanted arsenic *IEEE Electron Device Lett.* **34** 60–2

[18] Chen R, Zhou W, Zhang M, Wong M and Kwok H S 2012 Self-aligned indium–gallium–zinc oxide thin-film transistor with phosphorus-doped source/drain regions *IEEE Electron Device Lett.* **33** 1150–2
[19] Wu C-H, Hsieh H-H, Chien C-W and Wu C-C 2009 Self-aligned top-gate coplanar In-Ga-Zn-O thin-film transistors *J. Disp. Technol.* **5** 515–9
[20] Kim S *et al* 2009 Source/drain formation of self-aligned top-gate amorphous GaInZnO thin-film transistors by NH_3 plasma treatment *IEEE Electron Device Lett.* **30** 374–6
[21] Allemang C R *et al* 2020 High-performance zinc tin oxide TFTs with active layers deposited by atomic layer deposition *Adv. Electron. Mater.* **6** 2000195
[22] Zhirnov V V and Cavin R K 2008 Negative capacitance to the rescue? *Nat. Nanotechnol.* **3** 77–8
[23] Klasens H and Koelmans H J S 1964 A tin oxide field-effect transistor *Solid-State Electron.* **7** 701–2
[24] Boesen G and Jacobs J E 1968 ZnO field-effect transistor *Proc. IEEE* **56** 2094–5
[25] Nomura K, Ohta H, Ueda K, Kamiya T, Hirano M and Hosono H 2003 Thin-film transistor fabricated in single-crystalline transparent oxide semiconductor *Science* **300** 1269–72
[26] Yaglioglu B, Yeom H Y, Beresford R and Paine D C 2006 High-mobility amorphous In_2O_3–10wt%ZnO thin film transistors *Appl. Phys. Lett.* **89** 062103
[27] Chiang H, Wager J, Hoffman R, Jeong J and Keszler D A 2005 High mobility transparent thin-film transistors with amorphous zinc tin oxide channel layer *Appl. Phys. Lett.* **86** 013503
[28] Ryu M K, Yang S, Park S-H K, Hwang C-S and Jeong J K 2009 High performance thin film transistor with cosputtered amorphous Zn–In–Sn–O channel: combinatorial approach *Appl. Phys. Lett.* **95** 072104
[29] Kim C-J *et al* 2009 Amorphous hafnium-indium-zinc oxide semiconductor thin film transistors *Appl. Phys. Lett.* **95** 252103
[30] Cho D H *et al* 2009 Transparent oxide thin-film transistors composed of Al and Sn-doped zinc indium oxide *IEEE Electron Device Lett.* **30** 48–50
[31] Yamazaki S *et al* 2019 Crystalline IGZO ceramics (crystalline oxide semiconductor)-based devices for artificial intelligence *Int. J. Ceram. Eng. Sci.* **1** 6–20
[32] Özgür Ü *et al* 2005 A comprehensive review of ZnO materials and devices *J. Appl. Phys.* **98** 11
[33] Hoffman R, Norris B J and Wager J 2003 ZnO-based transparent thin-film transistors *Appl. Phys. Lett.* **82** 733–5
[34] Masuda S, Kitamura K, Okumura Y, Miyatake S, Tabata H and Kawai T 2003 Transparent thin film transistors using ZnO as an active channel layer and their electrical properties *J. Appl. Phys.* **93** 1624–30
[35] Carcia P, McLean R, Reilly M and Nunes G Jr 2003 Transparent ZnO thin-film transistor fabricated by rf magnetron sputtering *Appl. Phys. Lett.* **82** 1117–9
[36] Fortunato E M *et al* 2005 Fully transparent ZnO thin-film transistor produced at room temperature *Adv. Mater.* **17** 590–4
[37] Kim Y S and Park C H 2009 Rich variety of defects in ZnO via an attractive interaction between O vacancies and Zn interstitials: origin of n-type doping *Phys. Rev. Lett.* **102** 2009
[38] 2009 Zinc oxide: perfect pair of defects *NPG Asia Mater.* (https://doi.org/10.1038/asiamat.2009.216)

[39] Ahn C H, Kim S H, Kim Y H and Cho H K 2016 Extremely thin Al_2O_3 surface-passivated nanocrystalline ZnO thin-film transistors designed for low process temperature *J. Am. Ceram. Soc.* **99** 1305–10
[40] Barquinha P, Martins R, Pereira L and Fortunato E 2012 *Transparent Oxide Electronics: From Materials to Devices* (New York: Wiley)
[41] Jun T, Kim J, Sasase M and Hosono H 2018 Material design of p-type transparent amorphous semiconductor, Cu-Sn-I *Adv. Mater.* **30** 1706573
[42] Buchholz D B *et al* 2014 The structure and properties of amorphous indium oxide *Chem. Mater.* **26** 5401–11
[43] Medvedeva J E, Buchholz D B and Chang R P H 2017 Recent advances in understanding the structure and properties of amorphous oxide semiconductors *Adv. Electron. Mater.* **3** 1700082
[44] Hosono H, Kikuchi N, Ueda N and Kawazoe H 1996 Working hypothesis to explore novel wide band gap electrically conducting amorphous oxides and examples *J. Non-Cryst. Solids* **198** 165–9
[45] Kamiya T, Nomura K and Hosono H 2010 Subgap states, doping and defect formation energies in amorphous oxide semiconductor a-InGaZnO4 studied by density functional theory *Phys. Status Solidi* a **207** 1698–703
[46] Itagaki N *et al* 2008 Zn-In-O based thin-film transistors: compositional dependence *Phys. Status Solidi* a **205** 1915–9
[47] Spiehl D, Haming M, Sauer H M, Bonrad K and Dorsam E 2015 Engineering of flexo- and gravure-printed indium– zinc-oxide semiconductor layers for high-performance thin-film transistors *IEEE Trans. Electron Devices* **62** 2871–7
[48] Lee S *et al* 2011 Trap-limited and percolation conduction mechanisms in amorphous oxide semiconductor thin film transistors *Appl. Phys. Lett.* **98** 203508
[49] Lokanc M, Eggert R and Redlinger M 2015 The availability of indium: the present, medium term, and long term *Report* National Renewable Energy Lab. (NREL), Golden, CO (United States)
[50] Son Y, Frost B, Zhao Y and Peterson R L J N E 2019 Monolithic integration of high-voltage thin-film electronics on low-voltage integrated circuits using a solution process *Nat. Electron.* **2** 540–8
[51] Wang G *et al* 2019 Implementation of self-aligned top-gate amorphous zinc tin oxide thin-film transistors *IEEE Electron Device Lett.* **40** 901–4
[52] Kim Y-H, Han J-I and Park S K 2011 Effect of zinc/tin composition ratio on the operational stability of solution-processed zinc–tin–oxide thin-film transistors *IEEE Electron Device Lett.* **33** 50–2
[53] Ghaffarzadeh K *et al* 2010 Persistent photoconductivity in Hf–In–Zn–O thin film transistors *Appl. Phys. Lett.* **97** 143510
[54] Jeong W H *et al* 2010 Investigating addition effect of hafnium in InZnO thin film transistors using a solution process *Appl. Phys. Lett.* **96** 093503
[55] Maeng W *et al* 2011 The effect of active-layer thickness and back-channel conductivity on the subthreshold transfer characteristics of Hf–In–Zn–O TFTs *IEEE Electron Device Lett.* **32** 1077–9
[56] Park J S, Kim K, Park Y G, Mo Y G, Kim H D and Jeong J K 2009 Novel ZrInZnO thin-film transistor with excellent stability *Adv. Mater.* **21** 329–33

[57] Ye Z, Yue S, Zhang J, Li X, Chen L and Lu J 2016 Annealing treatment on amorphous InAlZnO films for thin-film transistors *IEEE Trans. Electron Devices* **63** 3547–51
[58] Cho D-H *et al* 2008 Transparent Al–Zn–Sn–O thin film transistors prepared at low temperature *Appl. Phys. Lett.* **93** 142111
[59] Cho D-H *et al* 2008 Transparent oxide thin-film transistors composed of Al and Sn-doped zinc indium oxide *IEEE Electron Device Lett.* **30** 48–50
[60] Park J C, Kim S W, Kim C J and Lee H-N 2012 Low-temperature fabrication and characteristics of lanthanum indium zinc oxide thin-film transistors *IEEE Electron Device Lett.* **33** 685–7
[61] Kim S H, Ahn C H, Yun M G, Cho S W and Cho H K 2014 Anomalous tin chemical bonding in indium-zinc-tin oxide films and their thin film transistor performance *J. Phys. D: Appl. Phys.* **47** 485101
[62] Ji Hun S, Kwang Suk K, Yeon Gon M, Rino C and Kyeong Jeong J 2014 Achieving high field-effect mobility exceeding 50 cm/Vs in In-Zn-Sn-O thin-film transistors *IEEE Electron Device Lett.* **35** 853–5
[63] Kim M-G *et al* 2010 High-performance solution-processed amorphous zinc-indium-tin oxide thin-film transistors *JACS* **132** 10352–64
[64] Kimizuka N and Yamazaki S 2016 *Physics and Technology of Crystalline Oxide Semiconductor CAAC-IGZO: Fundamentals* (New York: Wiley)
[65] Yamazaki S and Tsutsui T 2017 *Physics and Technology of Crystalline Oxide Semiconductor CAAC-IGZO: Application to Displays* (New York: Wiley)
[66] Yamazaki S and Fujita M 2016 *Physics and Technology of Crystalline Oxide Semiconductor CAAC-IGZO: Application to LSI* (New York: Wiley)
[67] Yamazaki S *et al* 2014 Properties of crystalline In–Ga–Zn-oxide semiconductor and its transistor characteristics *Japan J. Appl. Phys.* **53** 04ED18
[68] Kim G H, Du Ahn B, Shin H S, Jeong W H, Kim H J and Kim H J 2009 Effect of indium composition ratio on solution-processed nanocrystalline InGaZnO thin film transistors *Appl. Phys. Lett.* **94** 233501
[69] Yamazaki S, Shima Y, Hosaka Y, Okazaki K and Koezuka J 2016 Achievement of a high-mobility FET with a cloud-aligned composite oxide semiconductor *Japan J. Appl. Phys.* **55** 115504
[70] Koezuka J *et al* 2017 24-1: invited paper: flexible OLED display using C-axis-aligned-crystal/cloud-aligned composite oxide semiconductor technology and laser separation technology *SID Symp. Dig. Tech. Pap.* **48** 329–32
[71] Deng S *et al* 2016 High-performance staggered top-gate thin-film transistors with hybrid-phase microstructural ITO-stabilized ZnO channels *Appl. Phys. Lett.* **109** 182105
[72] Deng S *et al* 2017 Investigation of high-performance ITO-stabilized ZnO TFTs with hybrid-phase microstructural channels *IEEE Trans. Electron Devices* **64** 3174–82
[73] In H-J and Kwon O-K 2009 External compensation of nonuniform electrical characteristics of thin-film transistors and degradation of OLED devices in AMOLED displays *IEEE Electron Device Lett.* **30** 377–9
[74] Jeong J K, Won Yang H, Jeong J H, Mo Y-G and Kim H D 2008 Origin of threshold voltage instability in indium-gallium-zinc oxide thin film transistors *Appl. Phys. Lett.* **93** 123508
[75] Kang D *et al* 2007 Amorphous gallium indium zinc oxide thin film transistors: sensitive to oxygen molecules *Appl. Phys. Lett.* **90** 192101

[76] Park J-S, Jeong J K, Chung H-J, Mo Y-G and Kim H D J A P L 2008 Electronic transport properties of amorphous indium-gallium-zinc oxide semiconductor upon exposure to water *Appl. Phys. Lett.* **92** 072104
[77] Chen T-C, Kuo Y, Chang T-C, Chen M-C and Chen H-M 2017 Mechanism of a-IGZO TFT device deterioration—illumination light wavelength and substrate temperature effects *J. Phys. D: Appl. Phys.* **50** 42LT02
[78] Chuang C S *et al* 2008 P-13: photosensitivity of amorphous IGZO TFTs for active-matrix flat-panel displays *SID Symp. Dig. Tech. Pap.* **39** 1215–8
[79] Jeon S *et al* 2012 Gated three-terminal device architecture to eliminate persistent photoconductivity in oxide semiconductor photosensor arrays *Nat. Mater.* **11** 301
[80] Li J, Lu L, Chen R, Kwok H-S and Wong M J 2017 A physical model for metal–oxide thin-film transistor under gate-bias and illumination stress *IEEE Trans. Electron Devices* **65** 142–9
[81] Stewart K A, Yeh B-S and Wager J F 2016 Amorphous semiconductor mobility limits *J. Non-Cryst. Solids* **432** 196–9
[82] Yang S *et al* 2010 Improvement in the photon-induced bias stability of Al–Sn–Zn–In–O thin film transistors by adopting AlO_x passivation layer *Appl. Phys. Lett.* **96** 213511
[83] Lee K-H *et al* 2009 The effect of moisture on the photon-enhanced negative bias thermal instability in Ga–In–Zn–O thin film transistors *Appl. Phys. Lett.* **95** 232106
[84] Ji K H *et al* 2010 Comparative study on light-induced bias stress instability of IGZO transistors with SiN_x SiO_2 gate dielectrics *IEEE Electron Device Lett.* **31** 1404–6
[85] Shin J H *et al* 2009 Light effects on the bias stability of transparent ZnO thin film transistors *ETRI J.* **31** 62–4
[86] Kwon J-Y *et al* 2010 The impact of gate dielectric materials on the light-induced bias instability in Hf–In–Zn–O thin film transistor *Appl. Phys. Lett.* **97** 183503
[87] Huang X *et al* 2012 Electrical instability of amorphous indium-gallium-zinc oxide thin film transistors under monochromatic light illumination *Appl. Phys. Lett.* **100** 243505
[88] Migliorato P, Delwar Hossain Chowdhury M, Gwang Um J, Seok M and Jang J 2012 Light/negative bias stress instabilities in indium gallium zinc oxide thin film transistors explained by creation of a double donor *Appl. Phys. Lett.* **101** 123502
[89] Lee S, Nathan A, Jeon S and Robertson J 2015 Oxygen defect-induced metastability in oxide semiconductors probed by gate pulse spectroscopy *Sci. Rep.* **5** 1–10
[90] Hsieh T-Y, Chang T-C, Chen T-C and Tsai M-Y 2014 Review of present reliability challenges in amorphous In-Ga-Zn-O thin film transistors *ECS J. Solid State Sci. Technol.* **3** Q3058–70
[91] Oh S, Baeck J H, Bae J U, Park K-S and Kang I B 2016 Effect of interfacial excess oxygen on positive-bias temperature stress instability of self-aligned coplanar InGaZnO thin-film transistors *Appl. Phys. Lett.* **108** 141604
[92] Kim D H *et al* 2017 21-4: distinguished paper: experimental decomposition of the positive bias temperature stress-induced instability in self-aligned coplanar InGaZnO thin-film transistors and its modeling based on the multiple stretched-exponential functions *SID Symp. Dig. Tech. Pap.* **48** 298–301
[93] Ji K H *et al* 2010 The effect of density-of-state on the temperature and gate bias-induced instability of InGaZnO thin film transistors *J. Electrochem. Soc.* **157** H983
[94] Jin S, Kim T-W, Seol Y-G, Mativenga M and Jang J 2014 Reduction of positive-bias-stress effects in bulk-accumulation amorphous-InGaZnO TFTs *IEEE Electron Device Lett.* **35** 560–2

[95] Yang J *et al* 2017 Investigation of an anomalous hump phenomenon in via-type amorphous In-Ga-Zn-O thin-film transistors under positive bias temperature stress *Appl. Phys. Lett.* **110** 143508
[96] Im H, Song H, Jeong J, Hong Y and Hong Y 2015 Effects of defect creation on bidirectional behavior with hump characteristics of InGaZnO TFTs under bias and thermal stress *Japan. J. Appl. Phys.* **54** 03CB03
[97] Hosono H 2006 Ionic amorphous oxide semiconductors: material design, carrier transport, and device application *J. Non-Cryst. Solids* **352** 851–8
[98] Kamiya T and Hosono H 2010 Material characteristics and applications of transparent amorphous oxide semiconductors *NPG Asia Mater.* **2** 15
[99] Kyungsoo J, Raja J, Youn-Jung L, Doyoung K and Junsin Y 2013 Effects of carrier concentration, indium content, and crystallinity on the electrical properties of indium-tin-zinc-oxide thin-film transistors *IEEE Electron Device Letters* **34** 1151–3
[100] Saji K J, Jayaraj M K, Nomura K, Kamiya T and Hosono H 2008 Optical and carrier transport properties of cosputtered Zn–In–Sn–O films and their applications to TFTs *J. Electrochem. Soc.* **155** H390–5
[101] Shao Y, Xiao X, Wang L, Liu Y and Zhang S 2014 Anodized ITO thin-film transistors *Adv. Funct. Mater.* **24** 4170–5
[102] Le Y, Shao Y, Xiao X, Xu X and Zhang S 2016 Indium–tin–oxide thin-film transistors with *in situ* anodized Ta_2O_5 passivation layer *IEEE Electron Device Lett.* **37** 603–6
[103] Jang Yeon K and Kyeong J J 2015 Recent progress in high performance and reliable n-type transition metal oxide-based thin film transistors *Semicond. Sci. Technol.* **30** 024002
[104] Jeon S *et al* 2010 Low-frequency noise performance of a bilayer InZnO–InGaZnO thin-film transistor for analog device applications *IEEE Electron Device Letters* **31** 1128–30
[105] Chung Y J, Kim U K, Hwang E S and Hwang C S 2014 Indium tin oxide/InGaZnO bilayer stacks for enhanced mobility and optical stability in amorphous oxide thin film transistors *Appl. Phys. Lett.* **105** 013508
[106] Wang S-L *et al* 2012 High mobility thin film transistors with indium oxide/gallium oxide bi-layer structures *Appl. Phys. Lett.* **100** 063506
[107] Stewart K A and Wager J F 2016 69-2: oxide-TFT mobility limits and CMOS feasibility *SID Symp. Dig. Tech. Pap.* **47** 944–6
[108] Stewart K A and Wager J F 2016 Thin-film transistor mobility limits considerations *J. Soc. Inf. Disp.* **24** 386–93
[109] Deng S *et al* 2018 P-20: towards high-performance and cost-effective top-gated oxide TFTs with hybrid-phase microstructural channels *SID Symp. Dig. Tech. Pap.* **49** 1252–5
[110] Pan T-M, Peng B-J, Her J-L and Lou B-S 2017 Effect of In and Zn content on structural and electrical properties of InZnSnO thin-film transistors using an Yb_2TiO_5 gate dielectric *IEEE Trans. Electron Devices* **64** 2233–8
[111] Cheng M-H, Zhao C, Huang C-L, Kim H, Nakata M and Kanicki J 2016 Amorphous InSnZnO thin-film transistor voltage-mode active pixel sensor circuits for indirect x-ray imagers *IEEE Trans. Electron Devices* **63** 4802–10
[112] Raja J *et al* 2016 Improved data retention of InSnZnO nonvolatile memory by H_2O_2 treated Al_2O_3 tunneling layer: a cost-effective method *IEEE Electron Device Lett.* **37** 1272–5
[113] Deng S *et al* 2017 Investigation of high-performance ITO-stabilized ZnO TFTs with hybrid-phase microstructural channels *IEEE Trans. Electron Devices* **64** 3174–82

[114] Chiu C, Chang S and Chang S-J 2010 High-performance a-IGZO thin-film transistor using Ta_2O_5 gate dielectric *IEEE Electron Device Lett.* **31** 1245–7

[115] Ma P *et al* 2018 Low voltage operation of IGZO thin film transistors enabled by ultrathin Al_2O_3 gate dielectric *Appl. Phys. Lett.* **112** 023501

[116] Kim J, Fuentes-Hernandez C, Potscavage W Jr, Zhang X-H and Kippelen B 2009 Low-voltage InGaZnO thin-film transistors with Al_2O_3 gate insulator grown by atomic layer deposition *Appl. Phys. Lett.* **94** 142107

[117] Rembert T, Battaglia C, Anders A and Javey A 2015 Room temperature oxide deposition approach to fully transparent, all-oxide thin-film transistors *Adv. Mater.* **27** 6090–5

[118] Ji K H *et al* 2010 Comparative study on light-induced bias stress instability of IGZO transistors with SiN_x and SiO_2 gate dielectrics *IEEE Electron Device Lett.* **31** 1404–6

[119] Lee J-M, Cho I-T, Lee J-H, Cheong W-S, Hwang C-S and Kwon H-I 2009 Comparative study of electrical instabilities in top-gate InGaZnO thin film transistors with Al_2O_3 and Al_2O_3/SiN_x gate dielectrics *Appl. Phys. Lett.* **94** 222112

[120] Seshan K and Schepis D 2018 *Handbook of Thin Film Deposition* (Burlington, MA: William Andrew)

[121] Lu L, Li J and Wong M 2014 A comparative study on the effects of annealing on the characteristics of zinc oxide thin-film transistors with gate-stacks of different gas-permeability *IEEE Electron Device Lett.* **35** 841–3

[122] Park J-S, Jeong J K, Chung H-J, Mo Y-G and Kim H D 2008 Electronic transport properties of amorphous indium-gallium-zinc oxide semiconductor upon exposure to water *Appl. Phys. Lett.* **92** 072104

[123] Deng S *et al* 2019 Gate insulator engineering in top-gated indium-tin-oxide-stabilized ZnO thin-film transistors**40** 1104–7

[124] Kim J, Fuentes-Hernandez C, Potscavage W Jr, Zhang X-H and Kippelen B 2009 Low-voltage InGaZnO thin-film transistors with Al_2O_3 gate insulator grown by atomic layer deposition *Appl. Phys. Lett.* **94** 142107

[125] Ma P *et al* 2018 Low voltage operation of IGZO thin film transistors enabled by ultrathin Al_2O_3 gate dielectric *Appl. Phys. Lett.* **112** 023501

[126] Chen R, Zhou W, Zhang M, Wong M and Kwok H S 2013 Self-aligned top-gate InGaZnO thin film transistors using SiO_2/Al_2O_3 stack gate dielectric *Thin Solid Films* **548** 572–5

[127] Shao Y, Xiao X, He X, Deng W and Zhang S 2015 Low-voltage a-InGaZnO thin-film transistors with anodized thin HfO_2 gate dielectric *IEEE Electron Device Lett.* **36** 573–5

[128] Ma P *et al* 2018 Half-volt operation of IGZO thin-film transistors enabled by ultrathin HfO_2 gate dielectric *Appl. Phys. Lett.* **113** 063501

[129] Yuan L, Zou X, Fang G, Wan J, Zhou H and Zhao X 2011 High-performance amorphous indium gallium zinc oxide thin-film transistors with $HfO_xN_y/HfO_2/HfO_xN_y$ tristack gate dielectrics *IEEE Electron Device Lett.* **32** 42

[130] Shan F, Liu A, Liu G, Meng Y, Fortunato E and Martins R 2014 Low-voltage high-stability InZnO thin-film transistor using ultra-thin solution-processed ZrO_x dielectric *J. Disp. Technol.* **11** 541–6

[131] Lee J S, Chang S, Koo S-M and Lee S Y 2010 High-performance a-IGZO TFT with ZrO_2 gate dielectric fabricated at room temperature *IEEE Electron Device Lett.* **31** 225–7

[132] Park J C *et al* 2014 Comparative study of ZrO_2 and HfO_2 as a high-k dielectric for amorphous InGaZnO thin film transistors *J. Nanoelectron. Optoelectron.* **9** 67–70

[133] Chang T, Chiu C, Weng W, Chang S-J, Tsai T and Huang Z J A P L 2012 High responsivity of amorphous indium gallium zinc oxide phototransistor with Ta_2O_5 gate dielectric *Appl. Phys. Lett.* **101** 261112

[134] Chiu C, Chang S and Chang S-J J I E D L 2010 High-performance a-IGZO thin-film transistor using Ta_2O_5 gate dielectric *IEEE Electron Device Lett.* **31** 1245–7

[135] Shao Y, Xiao X, Wang L, Liu Y and Zhang S 2014 Anodized ITO thin-film transistors *Adv. Funct. Mater.* **24** 4170–5

[136] Ning H *et al* 2018 Facile room temperature routes to improve performance of IGZO thin-film transistors by an ultrathin Al_2O_3 passivation layer *IEEE Trans. Electron Devices* **65** 537–41

[137] Weng S *et al* 2019 High-performance amorphous zinc–tin–oxide thin-film transistors with low tin concentration *IEEE J. Electron Devices Soc.* **7** 632–7

[138] Zhong W, Yao R, Liu Y, Lan L and Chen R 2020 Effect of self-assembled monolayers (SAMs) as surface passivation on the flexible a-InSnZnO thin-film transistors *IEEE Trans. Electron Devices* **67** 3157–62

[139] Deng S *et al* 2018 Threshold voltage adjustment in hybrid-microstructural ITO-stabilized ZnO TFTs via gate electrode engineering *IEEE Electron Device Lett.* **39** 975–8

[140] Chen C-L *et al* 2016 A physics-based model of threshold voltage for amorphous oxide semiconductor thin-film transistors *AIP Adv.* **6** 035025

[141] Wang W *et al* 2015 Analysis of the contact resistance in amorphous InGaZnO thin film transistors *Appl. Phys. Lett.* **107** 063504

[142] Kim W-S, Moon Y-K, Kim K-T, Lee J-H and Park J-W 2010 An investigation of contact resistance between metal electrodes and amorphous gallium–indium–zinc oxide (a-GIZO) thin-film transistors *Thin Solid Films* **518** 6357–60

[143] Lee S, Park H and Paine D C 2012 The effect of metallization contact resistance on the measurement of the field effect mobility of long-channel unannealed amorphous In–Zn–O thin film transistors *Thin Solid Films* **520** 3769–73

[144] Morosawa N, Ohshima Y, Morooka M, Arai T and Sasaoka T 2012 Novel self-aligned top-gate oxide TFT for AMOLED displays *J. Soc. Inform. Display* **20** 47–52

[145] Nag M *et al* 2015 Low-temperature formation of source–drain contacts in self-aligned amorphous oxide thin-film transistors *J. Inform. Display* **16** 111–7

[146] Peng H *et al* 2020 Top-gate amorphous indium-gallium-zinc-oxide thin-film transistors with magnesium metallized source/drain regions *IEEE Trans. Electron Devices* **67** 1619–24

[147] Park J *et al* 2008 Self-aligned top-gate amorphous gallium indium zinc oxide thin film transistors *Appl. Phys. Lett.* **93** 053501

[148] Kim S *et al* 2009 Source/drain formation of self-aligned top-gate amorphous-GaInZnO thin film transistors by NH_3 plasma treatment *IEEE Electron Device Lett.* **30** 374

[149] Xia Z, Lu L, Li J, Kwok H-S and Wong M 2018 A bottom-gate metal-oxide thin-film transistor with self-aligned source/drain regions**65** 2820–6

[150] Xia Z, Lu L, Li J, Kwok H-S and Wong M 2019 Self-aligned elevated-metal metal-oxide thin-film transistors for displays and flexible electronics *2019 IEEE Int. Electron Devices Meeting (IEDM)* (Piscataway, NJ: IEEE) pp 8.4.1–4

[151] Lu L, Li J, Feng Z, Kwok H S and Wong M 2016 Elevated-metal–metal-oxide thin-film transistor: technology and characteristics *IEEE Electron Device Lett.* **37** 728–30

[152] Nag M *et al* 2015 P-6: impact of buffer layers on the self-aligned top-gate a-IGZO TFT characteristics *SID Symp. Dig. Tech. Pap.* **46** 1139–42

[153] Minami T, Sato H, Nanto H and Takata S 1986 Highly conductive and transparent silicon doped zinc oxide thin films prepared by RF magnetron sputtering *Japan J. Appl. Phys.* **25** L776

[154] Das A, Misra P and Kukreja L 2009 Effect of Si doping on electrical and optical properties of ZnO thin films grown by sequential pulsed laser deposition *J. Phys. D: Appl. Phys.* **42** 165405

[155] Clatot J, Campet G, Zeinert A, Labrugère C, Nistor M and Rougier A 2011 Low temperature Si doped ZnO thin films for transparent conducting oxides *Sol. Energy Mater. Sol. Cells* **95** 2357–62

[156] Howard D, Marchand P, Carmalt C, Parkin I and Darr J 2017 Si-doped zinc oxide transparent conducting oxides; nanoparticle optimisation, scale-up and thin film deposition *J. Mater. Chem.* C **5** 8796–801

[157] Deng S *et al* 2017 Hybrid-phase microstructural ITO-stabilized ZnO TFTs with self-aligned coplanar architecture *IEEE Electron Device Lett.* **38** 1676–9

Chapter 3

Preparation of metal oxide thin film by pulsed laser deposition

Te-Wei Chiu and Subramanian Sakthinathan

Thin-film materials have been used for various applications in electronic products, solar energy, military weapon systems, computer components, sensors, nano-electronics and robotics technology. In this chapter, we focus on thin-film materials fabricated by laser-based technologies for the preparation of multifunctional thin films with a large range of applications. The materials CeO_2/MgO, ZnO/STO, ZnO/LAO, ZnO/sapphire, VO_2/sapphire, PZT/YSZ/Si, STO/TiO_2/CeO_2/YSZ/Si, SBN/LNO/CeO_2/YSZ/Si, and VO_2/ZnO buffer/glass have been prepared by pulsed laser deposition (PLD). The thin film structural properties were examined with x-ray diffraction, atomic force microscopy, and cross-sectional transmission electron microscopy. Moreover, we discuss the PLD parameters, such as oxygen pressure, substrate temperature, laser wavelength, and target-to-substrate distance, which affect the crystallinity and functional properties of the thin film. Furthermore, highly crystalline and high quality epitaxial thin films have been obtained under optimized fabrication conditions. Thin films of various compositions were prepared by different plasma mixtures in PLD. In the PLD system, the ambient gas influences the formation of thin film. This chapter discusses the formation of SBN thin film on Pt/Ti/SiO_2/Si substrate in the presence of O_2 or N_2O gases. On the other hand, laser annealing is a method of heat tempering the surface that causes an oxidation process on the material surface. ZnO and $CuCrO_2$:Mg thin films were fabricated by PLD with *in situ* laser annealing. Aurora PLD for enhancing the activation and ionization of the ablated particles during the PLD process is also discussed.

3.1 Introduction

3.1.1 Preparation of metal oxides by pulsed laser deposition

Thin films have been used in several applications, such as in electronics, sensors, spintronic-based valves, insulator substrates and components for many electronic

doi:10.1088/978-0-7503-2556-1ch3

products [1, 2]. Currently, controlled preparation of thin films can be challenging due to the need for high deposition rates and well-defined interfaces between thin-film layers and efficient stoichiometric mass transfer. Techniques such as electron beam evaporation (EB), chemical vapor deposition (CVD), magnetron plasma sputtering, molecular beam epitaxy (MBE), radiofrequency sputtering and pulsed laser deposition (PLD) have been used to prepare thin films with fascinating properties. MBE allows the production of thin films of extremely high quality; however, due to the ultrahigh vacuum required, the deposition rate is very low and the apparatus is expensive. On the other hand, in the sputtering system, the working pressure is between 0.1 mTorr to 100 mTorr for plasma generation. The working pressure of the PLD system has a very wide range because the energy source for deposition is introduced from outside the vacuum chamber, so deposition can occur under a wide range of ambient working pressures, which is important in growing metal oxides and controlling the oxidation state of the thin film. Moreover, PLD is suitable for depositing multiple components. Its composition is relatively easier to reproduce than those of MBE, CVD, and sputtering. In CVD and MBE, every element must be controlled independently to control the composition. The sputtering rates are usually different for different elements [3–5].

However, unlike most other thin film deposition methods, PLD meets almost all the criteria for the best deposition characteristics, such as the energy of the atomic flux, substrate temperature, absolute arrival rates of atoms for compound films, multilayer structures, heteroepitaxial growth, and superior electrical and structural properties [6, 7]. Moreover, PLD features simplicity, flexibility, fast response, energetic evaporation, congruent evaporation, a significant range of parameters, a wide range of ambient gas pressures, a carbon-free source, high deposition rates, a degree of control over stoichiometry, and the capability to overcome high melting points [8, 9]. But PLD also has disadvantages, such as the need for high process temperatures, highly corrosive precursors, the limited area of deposition, and subsequent thermal treatments for crystallization [10–12]. The first attempt to prepare a thin film was made in 1965 by Turner and Smith, but true development was achieved by Dijjkamp *et al* in 1987. In the last two decades, PLD has successfully been employed for the deposition of large-scale layered materials, and metal oxide thin films [13, 14].

3.1.2 Apparatus

A PLD system consists of a laser, vacuum chamber, target holder, rotator, and substrate heater with pressure controllers, and other system instruments. Thin film deposition can be performed in reactive environments if the partial pressure of ozone and oxygen is carefully controlled. The substrate is equipped with mirrors, attenuators, and lenses to focus the laser beam on the target. The target absorbs the laser pulse and ejects a small amount of material. The plasma is deposited onto the substrate facing the target with a separation distance of 3–5 cm. The substrate temperature can be varied from room temperature, depending on the type of heater [15] (figure 3.1).

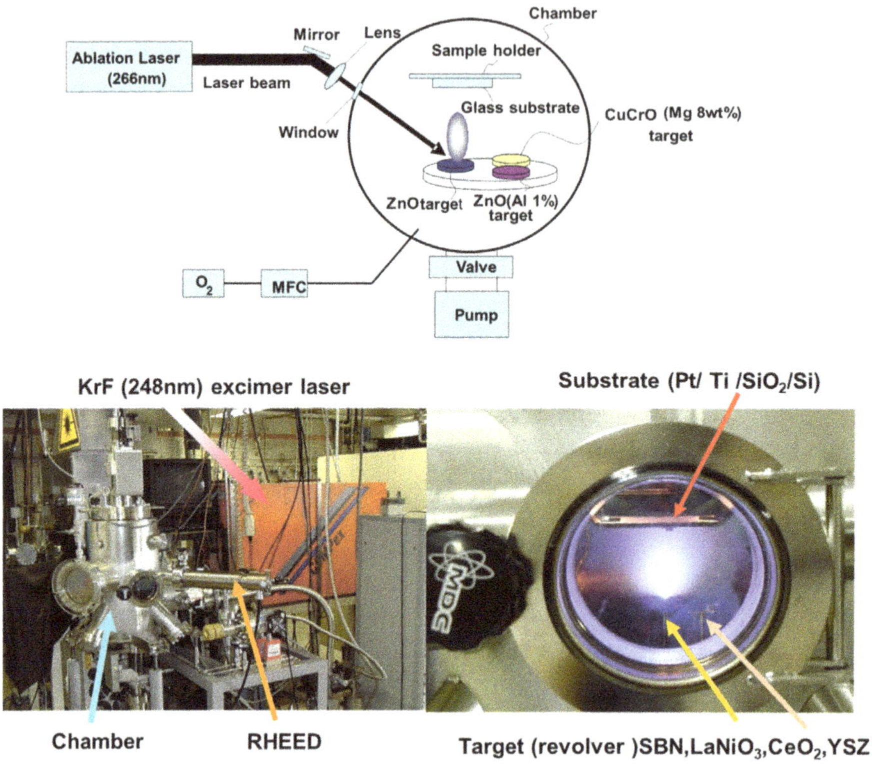

Figure 3.1. Experimental setup for pulsed laser deposition (PLD).

In PLD, a high-power pulsed laser beam is focused on a small area of the target material inside a vacuum chamber. The material to be deposited is vaporized from the target and deposited as a thin film on a substrate. Ultraviolet excimer lasers delivering 0.1–1 J/pulse of 15–45 ns duration are commonly used with 1–100 Hz repetition rates to produce deposition rates 0.1–100 nm/min. External high-energy laser pulses hitting the target create a plume of ceramic liberated from the target, which consists of a variety of molecular species. This whole deposition process is generally performed in a high vacuum chamber, and the laser pulse is provided by an external laser source. When the laser beam is absorbed by the target and the laser energy is converted to electronic excitation, the result is evaporation, plasma formation and ablation. The ejected materials expand into the surrounding vacuum in the form of a plume containing many energetic species, including atoms, molecules, electrons, ions, clusters, and molten globules. Generally, the process of PLD has four stages:

1. Laser absorption and laser ablation on the target material.
2. Dynamic plasma materials.
3. Laser ablation material deposition on the substrate.
4. Nucleation and film growth of the substrate.

In the PLD chamber, the target material evaporated by the laser source is a rotating disc attached to a support. However, some factors that influence the deposition rate are as follows:

1. Target material.
2. Repetition rate of the laser.
3. Pulse energy of the laser.
4. Temperature of the substrate.
5. Type of gas in the chamber (oxygen, argon).
6. Distance from the target to the laser substrate.

The system for the laser-ablation process has a combination oven/laser-vaporization apparatus [16–18].

3.1.2.1 Laser

Generally, in PLD, a high power pulsed laser such as a KrF or ArF excimer laser is used, as are frequency-quadrupled Nd:YAG lasers operated at $\lambda = 248$, 193, and 266 nm, respectively. The laser beam, repetition frequency, pulse duration, laser wavelength, and substrate distance affect the properties of the prepared thin films.

3.1.2.2 Reflection high energy electron diffraction (RHEED)

PLD could equipped with reflection high energy electron diffraction (RHEED). The use of RHEED has become common since the discovery of RHEED intensity variation by Joyce and co-workers [19, 20]. RHEED is limited to low background pressures, and the surface structure can be monitored during the deposition of oxides at higher pressures. In addition, oscillations in intensity allow accurate control of the growth rate and layer-by-layer formation. RHEED has a small angle of incidence that causes the geometry of the set-up to look like a reflection. The penetration depth of high energy electrons is several tens of nanometers due to the small angle of incidence. The diffraction pattern comes from a small number of lattice planes. This is advantageous for the investigation of the epitaxial growth of thin films.

The growth of a single-crystalline layer on a single-crystalline substrate is called epitaxy. In this process, the structural information is transferred from the growing layer to the substrate. This structural information comes from the upper lattice planes atoms of the substrate, which are investigated by RHEED. RHEED is also used to determine the crystallographic and morphological orientation of film growth [21–23].

Figure 3.2 illustrates the apparatus of the RHEED. The electron gun produces an electron beam, which passes through a small slot and is focused on a fluorescence screen by the lens. The basic principle of RHEED is that a high-energy electron beam enters a surface at a grazing incident angle. These high-energy electron beams can penetrate several hundreds of nanometers into any material. However, the electrons interact with the top layer of atoms, which makes the technique very surface sensitive. In RHEED, the incident electron beam is targeted at the surface of the sample and then reflected onto a phosphorescent screen. The reflected electrons form a diffraction pattern consisting of bright spots and streaks on the viewing screen. The scattered electrons collected by the screen show the diffraction pattern of the crystal structure of

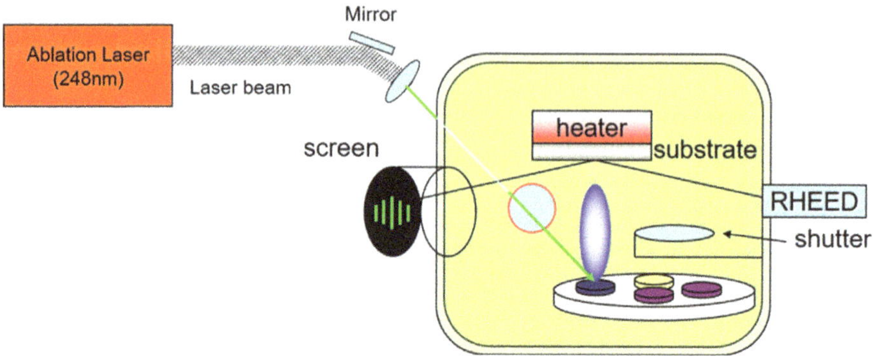

Figure 3.2. Schematic diagram of RHEED apparatus.

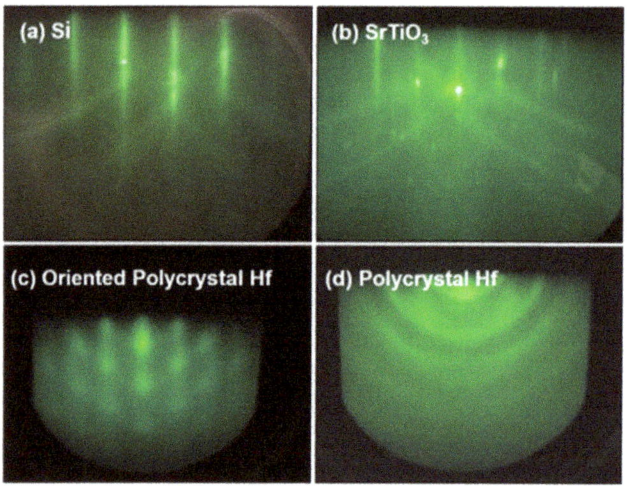

Figure 3.3. RHEED patterns of (a) Si, (b) $SrTiO_3$, (c) oriented polycrystal Hf and (d) polycrystal Hf.

the surface and also contain information about the surface morphology. RHEED is sensitive to variations in morphology and roughness during thin film growth, so it is used to monitor the thickness of thin film in real time [24, 25]. Figure 3.3 shows RHEED patterns of (a) Si(100) substrate, (b) $SrTiO_3$(100) substrate, (c) preferentially oriented polycrystal Hf thin film, and (d) polycrystal Hf thin film [54].

3.2 Orientation control

3.2.1 Epitaxy

Metal oxides have various properties, such as ferroelectricity, depending on the crystalline direction, so to obtain the best performance of metal oxide, the orientation must be controlled. When deposited crystalline thin films form with a well-defined orientation with respect to the crystalline substrate, these new layers are called epitaxial layers. The main application of epitaxial growth is in the semiconductor industry. Epitaxial thin films have a specific orientation with the lattice direction of substrate surface. For example, the lattice of Si(100) surface is square and Si(111) is trigonal. The term *homoepitaxy* refers to a material of the same type as the substrate growing on a crystalline substrate, and *heteroepitaxy* refers to a material of not the same type as the substrate growing on a substrate. Epitaxial thin films with high mobility values and high crystalline quality can be obtained under optimized preparation conditions. Moreover, several techniques can be used for the epitaxial growth of thin films, including CVD, MBE, and PLD. In this chapter, we discuss the different kinds of epitaxial growth of thin films that can be achieved with PLD.

3.2.1.1 Oxide single-crystal substrate

Single-crystal substrates are used for epitaxy growth of metal oxide thin films. It is relatively easy to obtain metal oxides with tetragonal or cubic structure with epitaxial growth on cubic single-crystal substrate such as MgO, $SrTiO_3$, $LaAlO_3$, etc. On the other hand, metal oxides with hexagonal structure can be obtained with epitaxial growth on a hexagonal plane such as c-cut sapphire substrate. For example, CeO_2 will be cubic on ZnO and hexagonal on hexagon epitaxial growth on c-cut sapphire. On the other hand, VO_2 grown *b*-axis oriented on c-cut sapphire epitaxial grown thin film could obtain the highest quality of uniformity crystallinity. However, single-crystal substrates are expensive and it is hard to get large size substrate. Thus, we try to use a buffer layer to obtain high quality metal oxide thin film without high cost single-crystal substrates.

CeO_2 thin films can be deposited on single-crystal MgO(001) substrates by PLD to obtain high-quality metal oxide thin films. CeO_2 is an excellent material for optical devices because of its good transmittance in the visible spectrum and high refractive index. CeO_2 thin films can also be grown on MgO substrate by PLD. Figure 3.4 shows the diffraction spectra of CeO_2/MgO films, which contain several narrow peaks, identified as the (00*l*) diffraction lines of the CeO_2 phase. These peaks indicate that CeO_2 films have a preferential orientation and are highly textured along the direction of the MgO substrate surface. Increasing the deposition temperature leads to better crystallinity in CeO_2/MgO films [26, 27, 31].

Vanadium oxide (VO_2) is used in optoelectronics materials due to its ultrafast optical and electrical properties [28–30]. VO_2 thin films having excellent photoluminescence properties have been successfully prepared on sapphire substrates by PLD under different oxygen partial pressures [31]. Figure 3.4 presents diffraction studies of VO_2 thin films fabricated on sapphire substrates under pressures of 0.80, 2.67, and 6.67 Pa, respectively. Substrate diffractions at 2θ of 41.67° and 90.71° were obtained for sapphire (006) and (0012), respectively. Furthermore, two diffraction peaks located at 39.9° and 86.1° correspond to the (020) and (040) planes of monoclinic VO_2. The x-ray diffraction (XRD) patterns for the thin films produced under the three different pressures are similar, but increasing the O_2 pressure during deposition decreases the intensity of VO_2(020) and (040). Moreover, no other peaks appear in the diffraction studies, indicating that the VO_2 films are highly oriented with (010) planes parallel to the sapphire substrate.

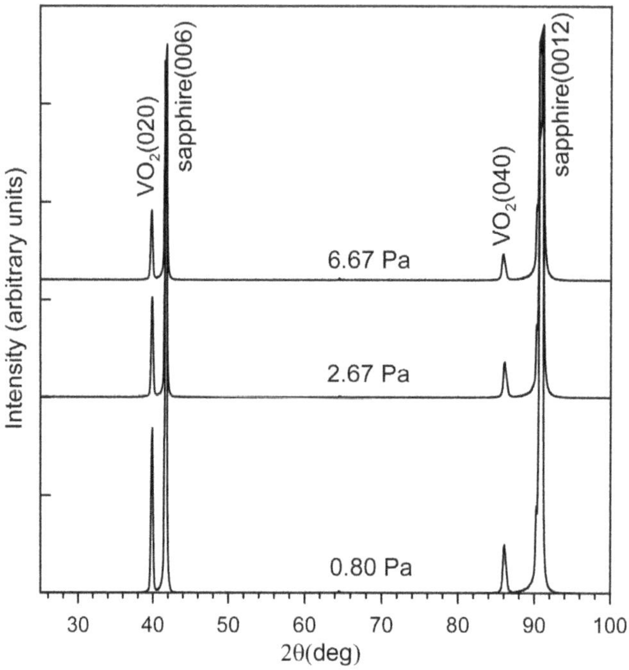

Figure 3.4. XRD studies of VO_2 thin films deposited on sapphire substrates under various O_2 partial pressures. Reprinted from [46], Copyright (2010), with permission from Elsevier.

3.2.2 Buffer layer

3.2.2.1 STO/CeO_2/YSZ

$SrTiO_3$ (STO) has a cubic perovskite structure, a high relative dielectric constant, and high thermal stability. Epitaxial STO films on Si can be used as ferroelectric thin films for electronic devices and as metal ferroelectric insulators in semiconductors. Based on the previous literature, STO can be epitaxially grown on Si using a single layer, such as MgO, Sr, Ce-doped ZrO_2 (CZO), yttria-stabilized zirconia (YSZ) and CeO_2/YSZ. However, the epitaxial relationship between STO and Si depends on the materials of the buffer layers. Hence, suitable buffer layers are needed to control the orientation of STO film on Si [32–34].

Yamada *et al* reported the fabrication of STO/(TiO_2)/CeO_2/YSZ/Si film by PLD with KrF excimer laser. The laser energy and repetition rate were varied between 0.7 to 1.5 J cm^{-2} and 1–7 Hz, respectively. The targets were fixed at a distance of 55 mm from the substrate and rotated continuously during deposition. The targets for YSZ, CeO_2, and STO were made using the standard ceramic synthesis technique. The diffraction spectrum of STO film directly deposited on CeO_2/YSZ/Si is presented in figure 3.5(a). The spectrum exhibits high intensity STO (110) and (220) peaks and a very low intensity STO(002) peak. This indicates that the STO/CeO_2/YSZ/Si film is preferentially (110) oriented. Figure 3.5(b) shows that each set can be distinguished into three peaks; the sharpest one, which is narrow and high in intensity, is located at the center of each set of peaks, and the low intensity peaks are located near the

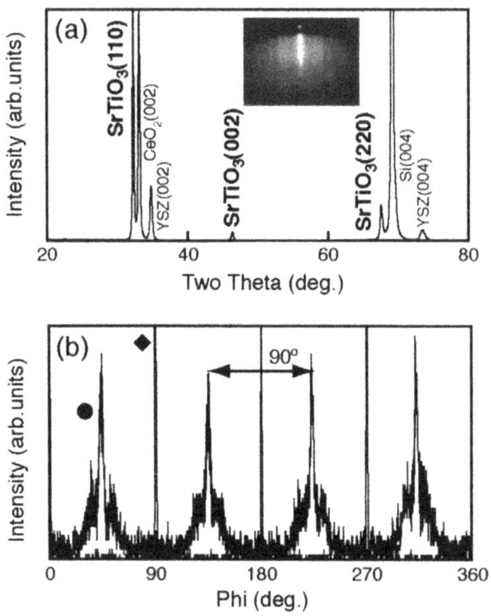

Figure 3.5. XRD patterns of STO film directly deposited on CeO_2/YSZ/Si. (a) $2\theta/\omega$ scan and RHEED pattern along CeO_2[100] direction (inset) and (b) STO (111) (●) and CeO_2(222) (♦) Φ scans. Reprinted from [36], with the permission of AIP Publishing.

central peak. These diffraction patterns indicate that the alignment of STO(110) is mainly STO[110]‖CeO_2[100] with a slight amount of STO[111]‖CeO_2[110]. From these diffraction studies, it is clear that the STO film grew epitaxially on the CeO_2(001) surface with the preferred (110) orientation under the deposition conditions of the study.

Figure 3.6(a) shows the diffraction spectrum of STO film deposited on 1 ML TiO_2-covered CeO_2/YSZ/Si substrate at $2\theta/\omega$. The diffraction spectrum shows high intensity (002) and (001) peaks, indicating (001) orientation. Furthermore, figure 3.6(b) presents STO(110)‖CeO_2(100) epitaxial cooperation, indicating the growth of STO on CeO_2. These results indicate that TiO_2 ML on the CeO_2(001) surface induced epitaxial growth of STO film with (001) orientation. Figure 3.7(a) shows STO film grown on 2 ML TiO_2-covered CeO_2/YSZ/Si film, and the rocking curve of the (111) peak is visible. Figure 3.7(b) exhibits the (111) oriented STO films epitaxially grown with an STO[011]‖CeO_2[100] relationship. Nevertheless, STO/CeO_2/YSZ/Si film has an epitaxial relationship in spite of the growth of TiO_2 on CeO_2(001) surfaces at TiO_2 thicknesses exceeding 1 ML [35, 36].

Yamashita *et al* reported the deposition of STO thin film on $(La_{0.5}Sr_{0.5})CoO_3$/CeO_2 buffered yttria-stabilized zirconia (YSZ) (001) substrates. The diffraction patterns of STO/($(La_{0.5}Sr_{0.5})CoO_3$)LSCO/CeO_2/YSZ (001) thin film with and without a magnetic field. These diffraction studies found that the diffraction angle of STO(002) shifts to a lower angle when a magnetic field is applied. The out-of-plane lattice parameter of STO thin film deposited with a magnetic field is larger than that of film deposited without a magnetic field [37].

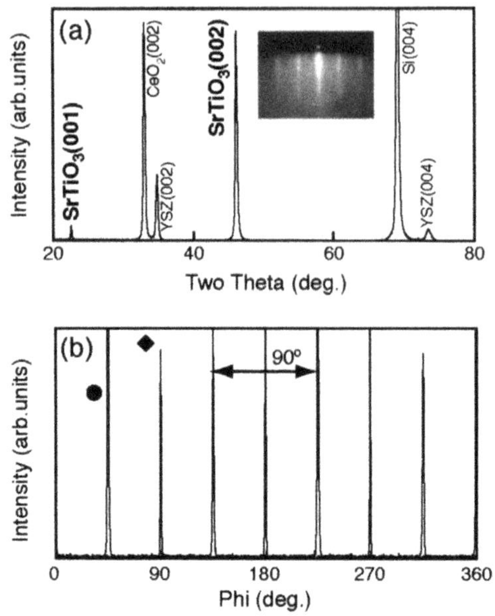

Figure 3.6. XRD and RHEED patterns of STO film deposited on 1 ML TiO_2-covered CeO_2/YSZ/Si. (a) $2\theta/\omega$ scan and RHEED pattern along CeO_2[100] direction (inset), and (b) STO (111) (●) and CeO_2 (222) (♦) Φ scans. Reprinted from [36], with the permission of AIP Publishing.

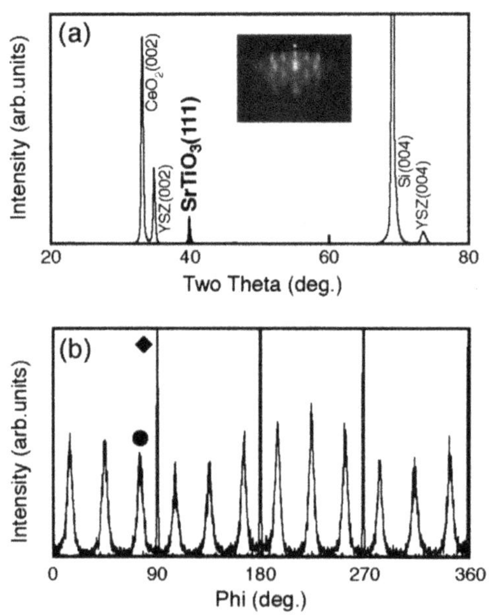

Figure 3.7. XRD and RHEED patterns of STO film deposited on 2 ML TiO_2-covered CeO_2/YSZ/Si: (a) $2\theta/\omega$ scan and RHEED pattern along CeO_2[100] direction (inset), and (b) STO (110) (●) and CeO_2(222) (♦). Reprinted from [36], with the permission of AIP Publishing.

3.2.2.2 LNO–CeO_2–YSZ

Chiu *et al* reported the deposition of strontium barium niobate (SBN) thin films on epitaxial LaNiO (LNO)/CeO/YSZ/Si(100) substrates by PLD. Epitaxial-like highly (001) textured SBN thin films were obtained on LNO/CeO/YSZ/Si(100) in plane orientations with respect to the unit cell of the Si substrate. SBN materials have a tetragonal tungsten bronze structure and are attractive materials for electrooptics, holographic storage, spatial light modulators, pyroelectric detectors, surface acoustic wave devices and beam steering due to their excellent photorefractive effect and optical properties. SBN thin films have been prepared by a variety of deposition methods, such as metal-organic chemical vapor deposition, sol–gel, PLD and RF magnetron sputtering. Among these methods, PLD offers some advantages, such as a high deposition rate, less stringent requirements, and the formation of thin films with suitable stoichiometry [38–44].

The interaction between the SBN crystallites and substrate can be enhanced by the use of epitaxial structure LNO electrodes on Si(100) substrate via a CeO and YSZ buffer layer. The CeO_2/YSZ buffer layer has been used to realize the heteroepitaxial growth of several oxides, such as LaSrCoO(LSCO), $YBa_2Cu_3O_7$(YBCO), and LNO on Si(100). LNO has a pseudocubic perovskite structure with resistivity of approximately 10^{-5} Ω m. Therefore, LNO is a favorable material for the electrodes used in thin film preparation. Chiu *et al* studied the crystal structure and surface morphology of highly *c*-axis-oriented SBN thin films with epitaxial LNO via a CeO_2/YSZ bi-buffer layer [38].

Figure 3.8 shows the XRD spectrum of SBN/LNO/CeO_2/YSZ/Si(001) thin film. The thicknesses of the SBN, LNO, CeO, and YSZ layers were 300, 100, 10, and 10 nm, respectively. The diffraction spectra showed only (001) diffraction peaks for all layers, and no other peaks were observed. The x-ray pole figure measurement was identified in the plane orientation of the film. The results showed that YSZ films can be epitaxially grown on Si(100) substrate with its native oxide. They are (100) oriented with a cube-on-cube epitaxial relationship with the substrate. The CeO films were epitaxially grown on YSZ/Si(100) with (100) orientation and exhibited an epitaxial relationship with YSZ. Figure 3.8 also shows RHEED patterns of (a) YSZ/Si, (b) CeO_2, (c) LNO and (d) SBN. This RHEED was identified for *in situ* observation of the heteroepitaxial growth of SBN, LNO, CeO_2 and YSZ thin films [38]. Figure 3.9 shows AFM images of 300 nm thick SBN thin films deposited on Si substrates via YSZ and a CeO bi-buffer layer and LNO bottom electrode. The surface was crack- and pinhole-free. The rms value of thin film surface roughness was 2 nm, indicating that CeO, YSZ, and LNO layers drastically improved the surface morphology of SBN thin film [38, 45].

3.2.2.3 VO_2/ZnO/glass

Vanadium oxide (VO_2) has a phase transition from a semiconducting state to a metallic state at low temperature to high temperature. The phase changes are important for modifications to optical transmittance, electrical resistivity, and reflectance in the infrared region. Thin films made of VO_2 are thus excellent materials for technological applications such as un-cooled IR microbolometers, optical

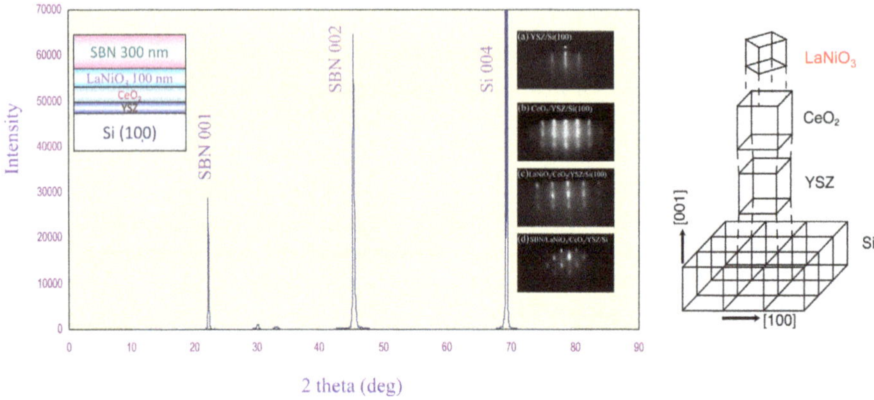

Figure 3.8. XRD pattern of the SBN/LNO/CeO_2/YSZ/Si (100) thin film. The thicknesses of SBN, LNO, CeO and YSZ are 300, 100, 10, and 10 nm, respectively. RHEED patterns of YSZ/Si, CeO_2, LNO, SBN.

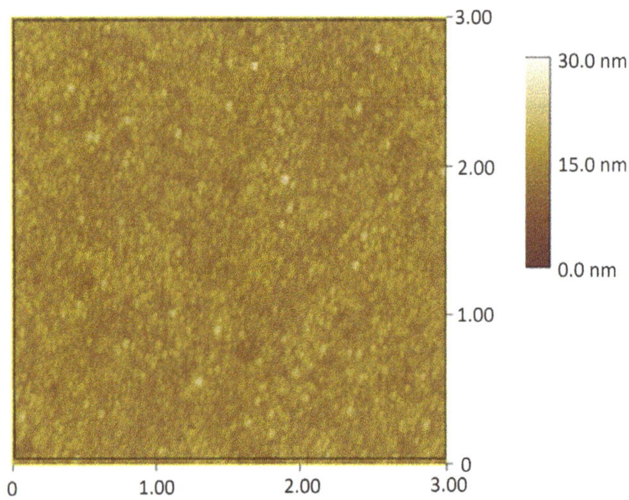

Figure 3.9. AFM image of the SBN/LNO/CeO_2/YSZ/Si (100) thin film. Reprinted from [38], Copyright (2003), with permission from Elsevier.

switching devices, optical recording devices, infrared sensors, switchable/tunable microwave devices, and energy efficient smart windows for buildings due to their high transparency under the transition temperature [46, 47]. Thin VO_2 thin films can be produced by various growth techniques, including sol–gel, sputtering, physical vapor deposition and PLD. Of the aforementioned methods, the PLD method has proven to be an excellent method for the deposition of oxide films [48–51]. Nagashima *et al* reported VO_2 thin films on a $TiO_2(001)$ single crystal substrate. Kim *et al* reported *b*-axis oriented VO_2 thin films on c-cut and r-cut sapphire. However, single-crystal substrates of TiO_2 and sapphire with large areas are difficult to prepare and expensive, so they are difficult to apply in smart windows. Therefore, a current topic of interest is the use of a buffer layer to modify the surface condition for thin-film growth. A high *c*-axis oriented ZnO buffer layer is used to control the thin- film growth orientation of VO_2 on glass substrates [52].

The ZnO buffers have been prepared on glass substrate under 1.33 Pa at 500 °C, and VO_2 thin films have been deposited on the ZnO buffers under various O_2 pressures at 500 °C for 60 min. Figure 3.10 presents the diffraction patterns of the VO_2 films deposited on a glass substrate by ZnO buffer with various O_2 pressures and the diffraction patterns of VO_2 film directly deposited on a glass substrate under 1.33 Pa O_2 ambient at 500 °C. As shown by the XRD spectra, when VO_2 thin film is directly deposited on glass substrates, it forms on the polycrystalline amorphous substrate surface. The spectrum shows an (011) peak located at 27.87° and a broad diffraction peak at 20°, representing the glass substrate. However, the ZnO (002) peak does not appear for all O_2 pressures due to its thin nature. The diffraction peak indicates a $VO_2(020)$ plane located at 39.87° and a $VO_2(040)$ plane located at 86.16° deposited under pressures of 1.33 and 2.67 Pa, respectively. However, an increase in O_2 pressure decreases the peak intensity of $VO_2(020)$, and the peak of $Zn_2V_2O_7(002)$ is visible at 19.18° for VO_2 thin films prepared under 2.67 Pa. The XRD spectra of $Zn_2V_2O_7(002)$, (004), and $VO_2(020)$ show 19.13°, 38.48°, and 39.90° for the film deposited under 4.00 Pa. When VO_2 thin films were deposited on 5 nm thick ZnO buffer under 5.33 and 6.67 Pa to form randomly-oriented $VO_2(011)$ thin film, the diffraction peak was at 27.90°. The *b*-axis oriented VO_2 film can be prepared under

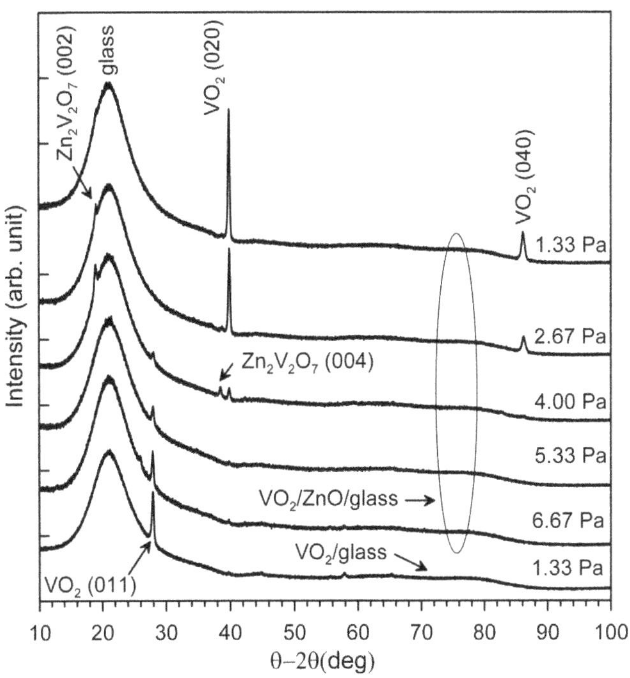

Figure 3.10. XRD patterns of the thin films formed by laser ablation of vanadium metal targets deposited on glass substrates with a 5 nm thick c-axis preferentially oriented ZnO layer under various O_2 pressures. Reprinted from [46], Copyright (2010), with permission from Elsevier.

1.33 Pa O_2 ambient using ZnO buffer. However, under 2.67 and 4.00 Pa, VO_2 film deposited on the *c*-axis oriented ZnO buffer forms $Zn_2V_2O_7$. Therefore, VO_2 thin films deposited under 5.33–6.67 Pa form an oriented polycrystalline VO_2 layer and inhibit $Zn_2V_2O_7$ [46].

Figure 3.11 shows thermal hysteresis loops of VO_2 thin film resistance on ZnO buffer under various O_2 atmospheres. VO_2 thin films deposited on ZnO buffer/glass substrates exhibit a significant change at transition temperatures. It was found that resistance increases by 5.2×10^2–1.9×10^2 at higher transition temperature with increasing O_2 pressure. The reflectance spectra of VO_2 film deposited on a glass substrate using ZnO buffers in the presence of different O_2 partial pressures are shown in figure 3.12. The reflectance spectrum at 80 °C shows high reflectance, but the reflectance decreases with increasing O_2 pressure during deposition. The VO_2 thin films deposited under 6.67 Pa on ZnO buffers exhibit degradation of both electrical resistivity and optical reflectance [46].

3.3 Ambient gas

The tetragonal tungsten bronze structure of $Sr_{1-x}Ba_xNb_2O_6$ (SBN) has excellent optical properties, such as a strong photorefractive effect and high electrooptic coefficient [53, 54]. Moreover, SBN thin films have potential applications in holographic storage, electrooptics, spatial light modulators, surface acoustic wave devices, pyroelectric detectors, and beam steering [55–57]. Therefore, SBN thin film

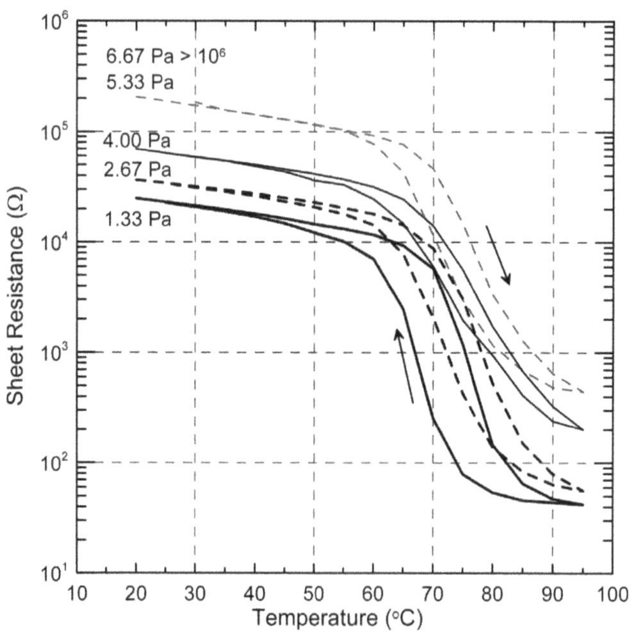

Figure 3.11. Thermal hysteresis loops of sheet resistance of VO_2 thin films deposited on glass substrates with a 5 nm thick ZnO buffer under various O_2 pressures. Reprinted from [46], Copyright (2010), with permission from Elsevier.

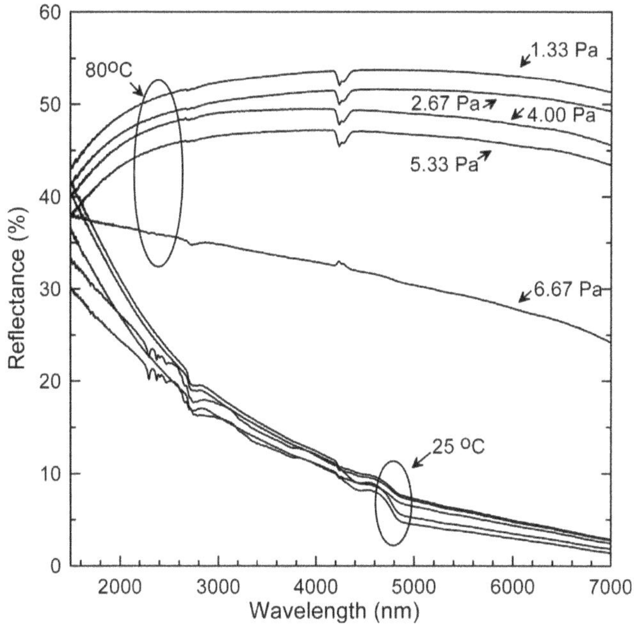

Figure 3.12. IR transmittance spectra of VO_2 thin films with a 5 nm thick c-axis preferentially oriented ZnO layer under various O_2 pressures. Reprinted from [46], Copyright (2010), with permission from Elsevier.

is important for fabricating sensor materials for close integration with Si technology. SBN thin films have been prepared by deposition methods such as metalorganic chemical deposition (MOCVD), sol–gel process, rf-magnetron sputtering, and PLD. Among these methods, PLD has advantages such as less stringent requirements, high deposition rates and proper stoichiometry of thin film formation. However, the crystallinity of SBN thin films is greatly influenced by the ambient gas pressure and laser energy density. Therefore, the optimization of thin film growth conditions is important. The ambient gas influences the phase structure, orientation and surface morphology of SBN thin film [58–60].

The SBN films were grown on Pt-coated Si ($Pt/Ti/SiO_2/Si(100)$) substrates by PLD using a KrF (248 nm wavelength) laser system with a pulse repetition rate of 7 Hz, at an angle of 45°. The heating temperature was varied from 700 °C to 800 °C using an infrared lamp heater. The SBN thin films were deposited under a controlled ambient atmosphere of oxygen or N_2O, and the pressure was varied from 0.27 to 26.67 Pa.

3.3.1 Effect of laser energy density under O_2 gas ambient

Figure 3.13 shows XRD spectra for SBN films on $Pt/Ti/SiO_2/Si(100)$ substrates in the presence of O_2 atmosphere as a function of laser energy. The deposition time, substrate temperature and O_2 gas pressure were 30 min, 6.67 Pa and 750 °C, respectively. The applied laser energy densities were 1.5, 1.32 and 1.97 J cm^{-2}. From the XRD studies, the tetragonal tungsten bronze phase structure of SBN thin films

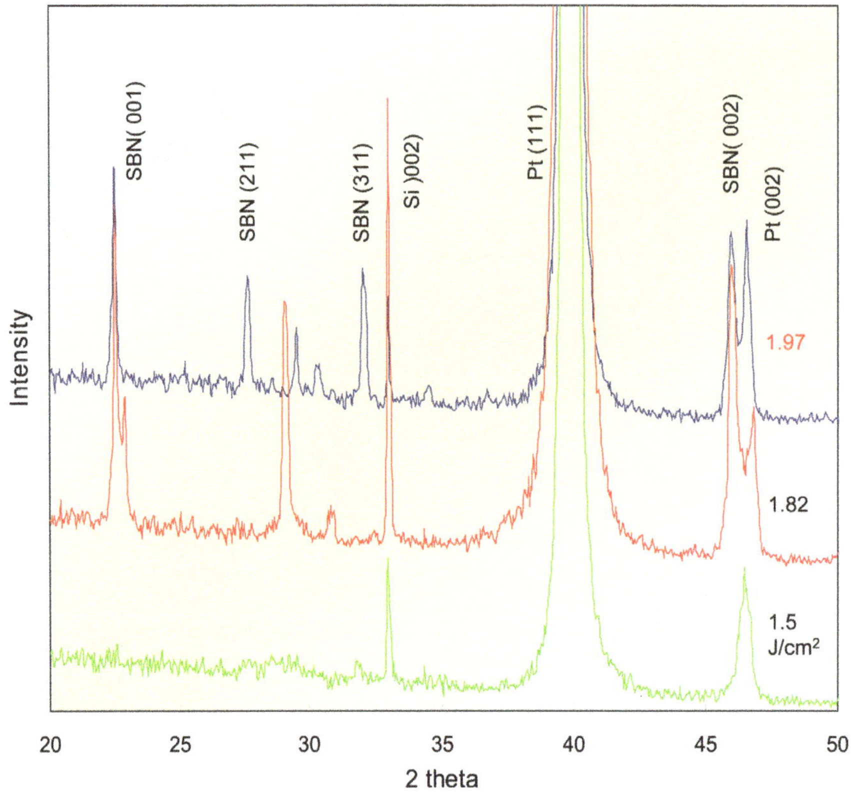

Figure 3.13. XRD patterns of SBN thin films deposited on Pt-coated Si substrates at various laser energy densities for 30 min at an O_2 pressure of 6.67 Pa and 750 °C.

was not obtained at a laser energy density of 1.5 J cm^{-2}. But the SBN peak was observed at a laser energy density above 1.32 J cm^{-2}. Moreover, impurity phases were present in the SBN thin films. The finding of these impurity phases is more difficult due to the presence of other intermediate compounds, such as the $SrNb_2O_6$ and $BaNb_2O_6$ phases. This poor crystallization of the thin film is the reason for the incomplete chemical reaction with oxygen radicals on the substrate. Furthermore, increasing the density of laser energy to 1.97 J cm^{-2} enhances the activity of ablated species with O_2 molecules to form pure tungsten bronze phase SBN thin films [61].

3.3.2 Effect of N_2O gas

SBN films have been deposited on $Pt/Ti/SiO_2/Si$ substrate in the presence of N_2O gas at a pressure of 2.67 Pa under laser beam with 1.5 J cm^{-2}. The diffraction spectra of SBN films deposited at 750 °C show (001), (311), (211), and (002) peaks. Figure 3.14 shows the XRD spectra of SBN films on $Pt/Ti/SiO_2/Si$ substrate as a function of N_2O gas pressure at 6.67 Pa for 30 min, with a substrate temperature of 750 °C. At the lower N_2O gas pressure, the formation of the tetragonal tungsten bronze SBN structure was inhibited. Moreover, the tetragonal tungsten bronze peaks were not

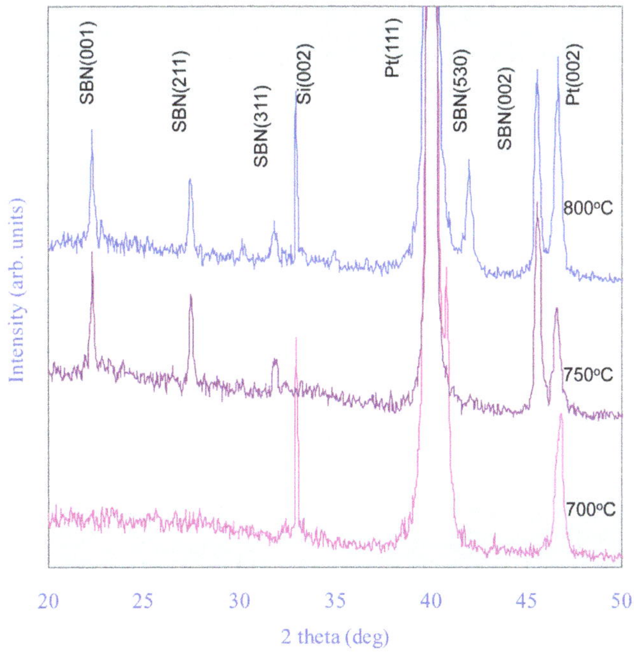

Figure 3.14. XRD patterns of SBN thin films deposited on Pt-coated Si substrates at various N_2O gas pressure for 30 min at 750 °C.

obtained in SBN deposited under higher N_2O gas pressure. When the SBN was deposited on Pt/Ti/SiO_2/Si substrate under N_2O gas pressure, the N_2O gas scattered from the target and lost kinetic energy, leading to less chemical reaction on the substrate. However, at N_2O gas pressure below 1.33–2.67 Pa, not much kinetic energy was lost to scattering of the ambient N_2O gas during the film formation on the substrate. Under this condition, the incident kinetic energy was sufficient to complete the chemical reactions to form SBN films with the tetragonal tungsten bronze structure. Figure 3.15 mentioned the formation ranges of tungsten bronze-based SBN film on Pt/Ti/SiO_2/Si substrate in the presence of O_2 or N_2O [61–63].

3.4 Laser annealing

Laser annealing is a method of heat tempering of the surface of a material with laser to cause oxidation. Laser-annealing techniques have been used in the semiconductor industry for the treatment of materials such as metal oxides, nanomaterials, and silicon. Moreover, laser annealing has numerous advantages over other thermal annealing treatments, including low-temperature processing, controllable crystallization and large-area fabrication [64–66].

Semiconductor transparent thin films are attracting interest in electronics due to their potential transparency. Moreover, the preparation of p–n junction devices is a fundamental step in electronic devices such as photovoltaic cells and diodes. Transparent oxides can be fabricated by PLD or sputtering. The substrate is heated

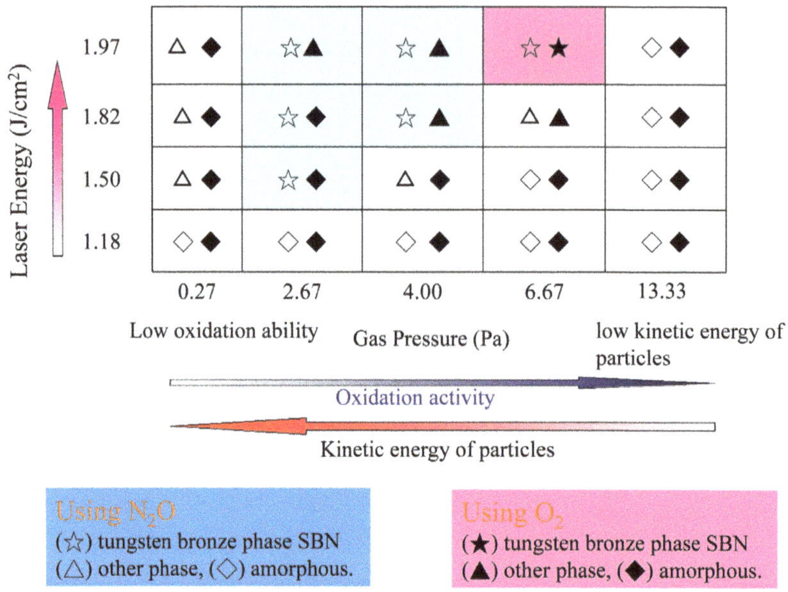

Figure 3.15. Formation range of tungsten bronze phase SBN.

to 400 °C–700 °C during the deposition of the films to obtain p–n junction devices with rectifying current voltages. This high-temperature thin film process is suitable for depositing film on heat-resistant substrates. However, plastic substrates cannot be used due to their poor thermal properties. Therefore, the process temperature must be reduced if materials with low melting temperatures are to be used as substrates [66–68].

To reduce the temperature, several processes have been used, including UV irradiation, seed layer, and catalyst methods, as well as laser annealing. Among these methods, laser annealing is considered the most suitable method for thin film deposition on plastic substrates. A short-duration pulse of laser irradiation can transmit energy to the target materials and avoid the effects of heating the substrate. Therefore, PLD irradiation of an oxide film on a substrate appears to be a suitable method for improving the crystallinity of the film on a substrate with minimal thermal damage. Chiu *et al* reported the electrical properties of microcrystalline n-ZnO and p-$CuCrO_2$:Mg film samples prepared by PLD with *in situ* laser annealing. A p–n heterojunction based n-ZnO and p-$CuCrO_2$:Mg film was fabricated on a glass substrate at room temperature. In this material, ZnO has the advantages of controllability of electrical conductivity, carrier mobility, and a wide bandgap (E_g = 3.4 eV). Furthermore, Mg-doped $CuCrO_2$ has the highest conductivity of 220 s cm^{-1} among ZnO transparent semiconductors [69, 70].

A PLD system with *in situ* laser annealing was developed to prepare ZnO and $CuCrO_2$ films (figure 3.16). The transparent materials were $CuCrO_2$ (E_g = 3.2 eV) and ZnO (E_g = 3.4 eV). Annealing and ablation were both performed with quadrupled Nd:YAG lasers operating at λ = 266 nm with a frequency of 20 Hz

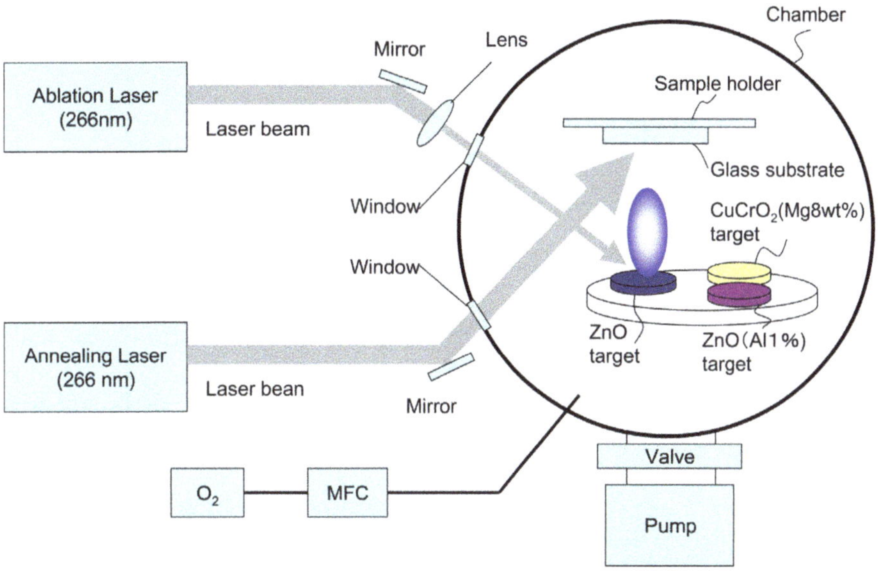

Figure 3.16. Schematic illustration of a PLD system employing *in situ* laser annealing. Reprinted from [66], Copyright (2008), with permission from Elsevier.

and laser pulse duration of 20 ns. The laser beam was focused through a quartz lens at the target at an incident angle of 40°. The laser energy density at the transparent film surface was controlled in the range of 0–30 mJ cm^{-2}. The deposition chamber was maintained at a pressure of 10^{-5} Pa for the deposition of each film. The $CuCrO_2$:Mg films were typically deposited at a pressure of 0.27 Pa O_2, and the ZnO films were typically deposited at a pressure of 1.3 Pa O_2. The thin film deposition rate was identified to be about 0.5 nm/pulse for both $CuCrO_2$:Mg and ZnO films. The glass substrate was placed above the target rotated at 20 rpm at a distance of 40 mm for uniform deposition. A series of $CuCrO_2$:Mg and ZnO films were deposited on glass substrates by varying the annealing laser power and substrate temperatures (T_s) [68].

3.4.1 ZnO thin films

Figure 3.17 shows the diffraction spectra of ZnO films deposited without laser annealing at various substrate temperatures. The diffraction patterns show only two peaks for the (002) plane and (004) plane. Furthermore, for samples deposited at a higher substrate temperature of 200 °C, the ZnO(002) plane peak increased rapidly [71, 72]. Figure 3.18 exhibits the XRD patterns of ZnO films prepared at various annealing laser power values. The diffraction spectra show only two diffraction peaks, the ZnO(002) peak and the ZnO(004). The ZnO(002) peak increased as annealing laser power increased. The ZnO films prepared with *in situ* laser annealing had microcrystalline properties similar to those of films prepared with substrate heating during the thin film deposition. Figure 3.18 shows the full width at half maximum in the ω scan, which was 13.2° for the samples obtained by *in situ* laser

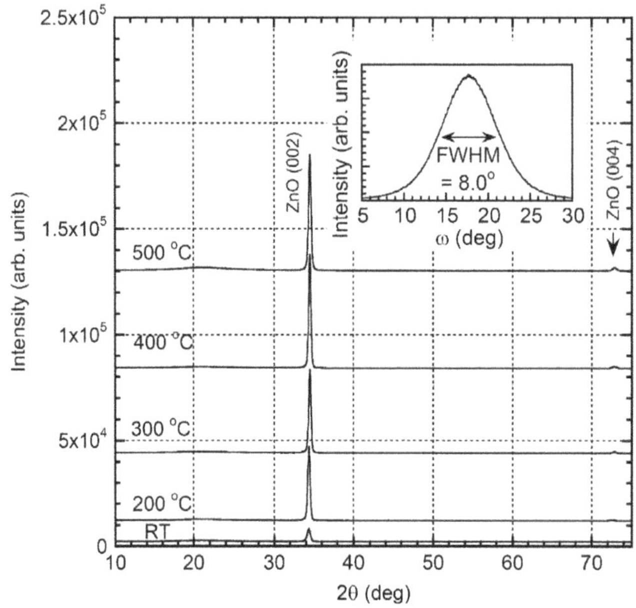

Figure 3.17. XRD patterns of ZnO films (thickness » 100 nm) deposited at various substrate temperatures without laser annealing. The inset shows the XRD rocking curves of the (002) peak of the film deposited at 400 °C. Reprinted from [66], Copyright (2008), with permission from Elsevier.

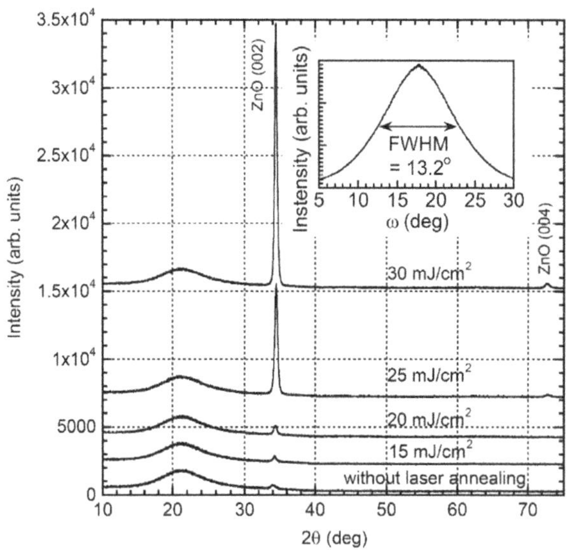

Figure 3.18. XRD patterns of the ZnO films prepared by *in situ* laser annealing at various annealing laser power values. The inset shows the XRD rocking curve of the (002) peak. Reprinted from [66], Copyright (2008), with permission from Elsevier.

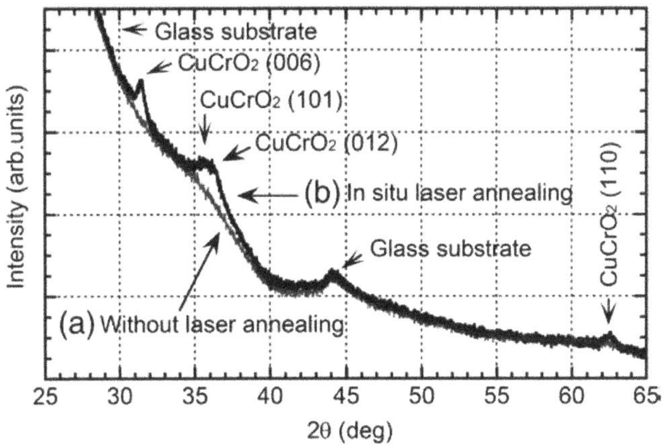

Figure 3.19. XRD patterns of $CuCrO_2$:Mg films deposited on glass substrates at room temperature (a) without laser annealing, and (b) with *in situ* laser annealing at 30 mJ/cm^2. Reprinted from [66], Copyright (2008), with permission from Elsevier.

annealing at 30 mJ cm^{-2}. On the other hand, FWHM value of 8.0° obtained for the ZnO film deposited at $T_s = 400$ °C. For that reason, the ZnO film obtained by *in situ* laser annealing exhibited a rather low *c*-axis orientation tendency as compared to those obtained by substrate heating. ZnO films are known to have an apparent tendency towards *c*-axis orientation when they are deposited on heated substrates. The crystallization generated by laser irradiation developed into a seed crystal that determined the crystalline structure of the film. On the other hand, laser irradiation from the front surface will generate crystallization cores mainly at the front surface, and they move outward during deposition. Therefore, the preference for *c*-axis orientation may be reduced in ZnO films obtained by laser irradiation from the front surface as compared to those obtained by substrate heating. Pulsed laser irradiation at 266 nm was consequently shown to effectively induce crystallization through optical excitation in oxide semiconductors such as ZnO. The crystal size in ZnO films deposited on glass substrates by PLD as a function of (a) laser power for *in situ* laser annealing and (b) substrate temperature. The crystal size increased with the substrate temperature and reached approximately 118 nm at 500 °C. The crystal size of the ZnO thin films increased with laser power. However, laser annealing was found to be less effective than substrate heating for producing thin films with larger crystal sizes [66, 73].

3.4.2 $CuCrO_2$:Mg thin films

$CuCrO_2$:Mg thin films have been deposited similarly to ZnO on glass by PLD with *in situ* laser annealing. Figure 3.19 shows the XRD patterns of the $CuCrO_2$:Mg films deposited on glass (a) with laser annealing and (b) without laser annealing. The $CuCrO_2$:Mg thin film deposited on glass substrate without laser annealing exhibited only broad peaks at 21° and 44°. Hence, the $CuCrO_2$:Mg film was considered to be amorphous. However, the sample deposited with *in situ* laser annealing showed

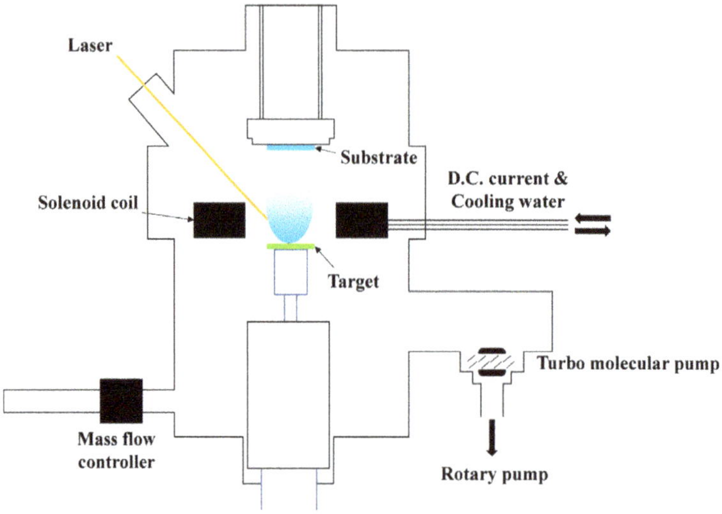

Figure 3.20. Schematic drawing of an air core coil installed in the PLD apparatus (Dynamic Aurora PLD).

diffraction peaks that were identified as $CuCrO_2$ delafossite and corresponded to the (006), (101), (012), and (110) planes. Therefore, laser irradiation was used during the delafossite-type of $CuCrO_2$ film deposition. It was concluded that the laser annealing PLD process can be used for the preparation of transparent semiconductor films having a $CuCrO_2$:Mg/ZnO/n^+-ZnO/glass structure without intentional heating of the substrate and with lower cost substrates of poor heat resistance [66, 74].

3.5 Aurora PLD

PLD is widely used to prepare thin-film materials. In Dynamic Aurora PLD, the ion-impingement of the film is enhanced through the recombination of electrons and ions in the plume by the magnetic field during thin film growth. Aurora PLD has been proposed for enhancing activation and ionization of the ablated particles during transport from the target to the substrate due to the magnetic fields [75, 76]. Dirnberger *et al* reported a Dynamic Aurora PLD system that generated a plume by laser ablation [77]. This Aurora PLD enhanced emission and ionization in the presence of a 43 mT magnetic field generated by 240 A DC current (figure 3.20). For nickel zinc ferrite (NZF-$Ni_{0.132}Zn_{0.139}Fe_{2.729}O_4$) film preparation, $SrTiO_3$(001) single crystals were used as substrates, and KrF excimer laser was used to ablate the targets. This substrate temperature was fixed at 600 °C, and oxygen pressure was 1.3 Pa. The film deposition was carried out for 30 min with a magnetic field of 43 mT [78, 79].

There were changes in thickness of NZF film with deposition temperature and without a magnetic field. Further, the deposition rate is significantly increased by applying the magnetic field due to the formation centripetal force. The concentration of ions is increased and the radius of the plume is decreased by the higher deposition rate. Moreover, the thickness ratio of NZF film increases with deposition temperature due to the higher efficiency of ionization and deposition [77, 80, 81].

There were changes in the composition of the film with and without a magnetic field as a function of substrate temperature. In the NZF film preparation, the Fe content was increased and that of Ni/Zn decreased for the application of a magnetic field on the materials. The results indicated that the composition of the NZF film approached that of the magnetic field target, based on the substrate temperature. The NZF films deposited above 200 °C were crystallized with (00l) orientation, and those crystallized below 100 °C were amorphous. Furthermore, x-ray pole figure measurements revealed that the NZF films underwent epitaxial growth on $SrTiO_3$(001). Hence, the thickness of the NZF film deposited in the magnetic field was twice that of the sample deposited without the magnetic field. In the Dynamic Aurora PLD preparation, the electromagnet is installed between the substrate and target to grow epitaxial NZF thin films [77].

3.6 Conclusion

In this chapter, we have discussed different kinds of thin films fabricated by pulsed laser deposition. The structural properties of the prepared thin films were analyzed with different physicochemical characterization techniques, and the effects of oxygen pressure, laser wavelength, substrate temperature, and target-to-substrate distance during the thin film preparation were discussed. Moreover, we noted high quality and high crystalline epitaxial thin films that could be obtained under optimized fabrication conditions. The production of thin films with high structural and morphological quality as well as high carrier mobility values obtained at suitable fabrication parameters was described. The importance of oxygen pressure in the growth and crystalline quality of the deposited films was explained, as were the effects of film thickness on the structural properties of films, the Aurora PLD method, and laser annealing methods for thin film preparation. The chapter ended with a discussion of the Dynamic Aurora PLD system and corresponding thin film thicknesses.

References

[1] Rossall A K, Berg J A V, Meehan D, Rajendiran S and Wagenaars E 2019 Analysis of plasma enhanced pulsed laser deposition of transition metal oxide thin films using medium energy ion scattering *Nucl. Instrum. Methods Phys. Res.* B **450** 274–8

[2] Ohnishi T, Koinuma H and Lippmaa M 2006 Pulsed laser deposition of oxide thin films *Appl. Surf. Sci.* **252** 2466–71

[3] Baig M K, Atiq S, Bashir S, Riaz S, Naseem S, Soleimani H and Yahya N 2016 Pulsed laser deposition of SMCO thin films for MEMS applications *J. Appl. Res. Technol.* **14** 287–92

[4] Xia H, Tang S B and Lu L 2007 Thin film microbatteries prepared by pulsed laser deposition *J. Korean Phys. Soc.* **51** 1055–62

[5] Kuwata N, Kumar, Toribami K, Suzuki T, Hattori T and Kawamura J 2006 Thin film lithium ion batteries prepared only by pulsed laser deposition *Solid State Ionics* **177** 2827–32

[6] Xia H, Lu L and Ceder G 2006 Substrate effect on the microstructure and electrochemical properties of $LiCoO_2$ thin films grown by PLD *J. Alloys Compd.* **417** 304–10
[7] Rajendiran S, Rossall A, Gibson A and Wagenaars E 2014 Modelling of laser ablation and reactive oxygen plasmas for pulsed laser deposition of zinc oxide *Surf. Coat. Technol.* **260** 417–23
[8] Henning R A, Uredat P, Simon C, Bloesser A, Cop P, Elm M T and Marschall R 2019 Characterization of MFe_2O_4 (M = Mg, Zn) thin films prepared by pulsed laser deposition for photo electrochemical applications *J. Phys. Chem.* C **123** 18240–7
[9] Lethy K J, Beena D, Pillai V P M and Ganesan V 2008 Bandgap renormalization in titania modified nanostructured tungsten oxide thin films prepared by pulsed laser deposition technique for solar cell applications *J. Appl. Phys.* **104** 033515
[10] Yan B, Liu J, Song B, Xiao P and Lu L 2013 Li-rich thin film cathode prepared by pulsed laser deposition *Sci. Rep.* **3** 3332
[11] Kitai S, Maida O, Kanashima T and Okuyama M 2003 Preparation and characterization of high-k praseodymium and lanthanoid oxide thin films prepared by pulsed laser deposition *Jpn. J. Appl. Phys.* **42** 247–53
[12] Guanjun Z, Rong C J, Rui C, Wen Y S and Yan M Z 2006 Preparation of $BiFeO_3$ thin films by pulsed laser deposition method *Trans. Nonferrous Met. Soc. China* **16** 123–5
[13] Ohshima T, Thareja R K, Ikegami T and Ebihara K 2003 Preparation of ZnO thin films on various substrates by pulsed laser deposition *Surf. Coat. Technol.* **169** 517–20
[14] Popescu C, Dorcioman G and Popescu A C 2016 *Laser Ablation Applied for Synthesis of Thin Films: Insights Into Laser Deposition Methods* (Rijeka: IntechOpen) (https://doi.org/10.5772/65124)
[15] Ma C and Chen C 2016 Pulsed laser deposition for complex oxide thin film and nano-structure *Advanced Nano Deposition Methods* ed Y Lin and X Chen 1st edn (New York: Wiley)
[16] Phuoc T X 2014 Complete green synthesis of gold nanoparticles using laser ablation in deionized water containing chitosan and starch *J. Mater. Sci. Nanotechnol.* **2** 1–7
[17] Mkrtychev O V, Privalov V E, Fotiadi A E and Shemanin V G 2015 Laser ablation studies of nanocomposites *St. Petersburg Polytech. Univ. J. Phys. Math.* **1** 82–6
[18] Yang G, Lin Q, Ding Y, Tian D and Duan Y 2015 Laser induced breakdown spectroscopy based on single beam splitting and geometric configuration for effective signal enhancement *Sci. Rep.* **5** 7625
[19] Harris J J, Joyce B A and Dobson P J 1981 Oscillations in the surface structure of Sn-doped GaAs during growth by MBE *Surf. Sci.* **103** L90–6
[20] Mahan J E, Geib K M, Robinson G Y and Long R G 1990 A review of the geometrical fundamentals of reflection high energy electron diffraction with application to silicon surfaces *J. Vac. Sci. Technol.* A **8** 3692–700
[21] Horio Y, Hashimoto Y and Ichimiya A 1996 A new type of RHEED apparatus equipped with an energy filter *Appl. Surf. Sci.* **100** 292–6
[22] Andrieu S and Avitaya F A 1991 Ga adsorption on Si (111) analyzed by RHEED and *in situ* ellipsometry *J. Cryst. Growth* **112** 146–52
[23] Ichimiya A and Cohen P I 2004 *Reflection High-Energy Electron Diffraction* (Cambridge: Cambridge University Press)
[24] Lu Z, Sun X, Xiang Y, Washington M, Wang G C and Lu T M 2017 Revealing the crystalline integrity of wafer-scale graphene on SiO_2/Si: an azimuthal RHEED approach *ACS Appl. Mater. Interfaces* **9** 23081–91

[25] Tang F, Parker T, Wang G C and Lu T M 2007 Surface texture evolution of polycrystalline and nanostructured films: RHEED surface pole figure analysis *J. Phys. D: Appl. Phys.* **40** R427–39
[26] Schroeder J L, Ingason A S, Rosen J and Birch J 2015 Beware of poor-quality Mgo substrates: a study of Mgo substrate quality and its effect on thin film quality *J. Cryst. Growth* **420** 22–31
[27] Casero R P, Roman R G S, Perriere J, Laurent A, Seiler W, Gergaud P and Keller D 1997 Epitaxial growth of CeO_2 on MgO by pulsed laser deposition *Appl. Surf. Sci.* **109** 341–4
[28] Morin F J 1959 Oxides which show a metal-to-insulator transition at the Neel temperature *Phys. Rev. Lett.* **3** 34
[29] Surnev S, Ramsey M G and Netzer F P 2003 Vanadium oxide surface studies *Prog. Surf. Sci.* **73** 117–65
[30] Qazilbash M M *et al* 2007 Mott transition in VO_2 revealed by infrared spectroscopy and nano-imaging *Science* **318** 1750–3
[31] Bian J, Miao L and Zhao S 2014 Structural and optical properties of VO_2 film grown on sapphire substrate by pulsed laser deposition *Adv. Mater. Res.* **912** 325–8
[32] Mori H and Ishiwara H 1991 Epitaxial growth of $SrTiO_3$ films on Si(100) substrates using a focused electron beam evaporation method *Japan J. Appl. Phys.* **30** L1415
[33] Moon B K and Ishiwara H 1994 Roles of buffer layers in epitaxial growth of $SrTiO_3$ films on silicon substrates *Japan J. Appl. Phys.* **33** 1472
[34] Sanchez F, Aguiar R, Trtik V, Guerrero C, Ferrater C and Varela M 1998 Epitaxial growth of $SrTiO_3$ (00h), (0hh) and (hhh) thin films on buffered Si (001) *J. Mater. Res.* **13** 1422–5
[35] Yamada T, Wakiya N, Shinozaki K and Mizutani N 2002 Growth mechanism of $SrTiO_3$ thin film on CeO_2(001) surface *Key Eng. Mater.* **137** 228–9
[36] Yamada T, Wakiya N, Shinozaki K and Mizutani N 2003 Epitaxial growth of $SrTiO_3$ films on CeO_2/yttria-stabilized zirconia/Si (001) with TiO_2 atomic layer by pulsed-laser deposition *Appl. Phys. Lett.* **83** 4815–7
[37] Yamashita K, Sakamaki Y, Sakamoto N, Shinozaki K, Suzuki H and Wakiya N 2011 Ferroelectricity of $SrTiO_3$ thin films prepared by dynamic-aurora pulsed laser deposition *Key Eng. Mater.* **485** 11–4
[38] Chiu T W, Wakiya N, Shinozaki K and Mizutani N 2003 Growth of highly (001)-textured strontium barium niobate thin films on epitaxial $LaNiO_3/CeO_2$/YSZ/Si (100) *Thin Solid Films* **426** 62–7
[39] Zhang Y L, Xie X S, Mo D, Pun E Y B and Shi L P 1996 Growth and photorefractive properties of Mn-doped $(Kna)_{0.1}(Sr_{0.6}Ba_{0.4})_{0.9}Nb_2O_6$ crystals *J. Appl. Phys.* **79** 8835–7
[40] Chen C J, Xu Y, Xu R and Mackenzie J D 1991 Ferroelectric and pyroelectric properties of strontium barium niobate films prepared by the sol-gel method *J. Appl. Phys.* **69** 1763–5
[41] Copetti C A, Soltner H, Schubert J, Zander W, Hollricher O, Buchal C, Schulz H, Tellmann N and Klein N 1993 High quality epitaxy of $YBa_2Cu_3O_{7-x}$ on silicon-on-sapphire with the multiple buffer layer YSZ/CeO_2 *Appl. Phys. Lett.* **63** 1429–31
[42] Haakenaasen R, Fork D K and Colovchenko J A 1994 High quality crystalline $YBa_2Cu_3O_{7-\delta}$ films on thin silicon substrates *Appl. Phys. Lett.* **64** 1573–5
[43] Li Y, Linzen S, Seidel P, Machalett F, Schmidl F, Schneidewind H, Schmauder T, Cihar R and Schaller S 1995 Epitaxial growth of $Yba_2Cu_3O_7$-delta thin films on $CoSi_2$/Si substrates with combined CeO_2/YSZ buffer layers *Inst. Phys. Conf. Ser.* **148** 911–4

[44] Sanchez F, Ferrater C, Alcobe X, Bassas J and Garcıa Cuenca M V 2001 Pulsed laser deposition of epitaxial $LaNiO_3$ thin films on buffered Si (100) *Thin Solid Films* **384** 200–5
[45] Chiu T W, Wakiya N, Shinozaki K and Mizutani N 2002 Effects of ambient gas and film thickness on orientation and surface morphology of $Sr_{0.5}Ba_{0.5}Nb_2O_6$ thin films prepared by pulsed laser deposition *J. Ceram. Soc. Jpn.* **110** 368–72
[46] Chiu T W, Tonooka K and Kikuchi N 2010 Influence of oxygen pressure on the structural, electrical and optical properties of VO_2 thin films deposited on ZnO/glass substrates by pulsed laser deposition *Thin Solid Films* **518** 7441–4
[47] Hood P J and DeNatale J F 1991 Millimeter-wave dielectric properties of epitaxial vanadium dioxide thin films *J. Appl. Phys.* **70** 376–81
[48] Soltani M, Chaker M, Haddad E and Kruzelesky R V 2006 Thermochromic vanadium dioxide smart coatings grown on kapton substrates by reactive pulsed laser deposition *J. Vac. Sci. Technol.* A **24** 612–7
[49] Jerominek H, Picard F and Vincent D 1993 Vanadium oxide films for optical switching and detection *Opt. Eng.* **32** 2092–100
[50] Burkhardt W, Christmann T, Franke S, Kriegseis W, Meister D, Meyer B K, Niessner, Schalch D and Scharmann A 2002 Tungsten and fluorine co-doping of VO_2 films *Thin Solid Films* **402** 226–31
[51] Kim D H and Kwok H S 1994 Pulsed laser deposition of VO_2 thin films *Appl. Phys. Lett.* **65** 3188–90
[52] Nagashima K, Yanagida T, Tanaka H and Kawai T 2006 Influence of ambient atmosphere on metal-insulator transition of strained vanadium dioxide ultrathin films *J. Appl. Phys.* **100** 063714
[53] Vasquez R A, Ewbank M D and Neugaonkar R R 1991 Photorefractive properties of doped strontium-barium niobate *Opt. Commun.* **80** 253–8
[54] Zhang Y L, Xie X S, Mo D, Pun E Y B and Shi L P 1996 Growth and photorefractive properties of Mn-doped $(Kna)_{0.1}(Sr_{0.6}Ba_{0.4})_{0.9}Nb_2O_6$ crystals *J. Appl. Phys.* **79** 8835–7
[55] Vre R D and Hesselink L 1994 Analysis of photorefractive stratified volume holographic optical elements *J. Opt. Soc. Am.* **11** 1800–8
[56] Aronson L B and Hesselink L 1990 Photorefractive integrated-optical switch arrays in $LiNbO_3$ *Opt. Left* **15** 30–2
[57] Hesselink L and Bashaw M C 1993 Optical memories implemented with photorefractive media *Opt. Quantum Electron* **25** S611–61
[58] Sakamoto S and Yazaki T 1973 Anomalous electro-optic properties of ferroelectric strontium barium niobate and their device applications *Appl. Phys. Lett.* **22** 429–31
[59] Nystrom M J, Wessels B W, Lin W P, Wong G K, Neumayer D A and Marks T 1995 Nonlinear optical properties of textured strontium barium niobate thin films prepared by metalorganic chemical vapor deposition *J. Appl. Phys. Lett.* **66** 1726–8
[60] Tanaka K, Nakagawa O, Nakano M, Shimuta T, Tabata H and Kawai T 1998 Epitaxial growth of (Sr, Ba) Nb_2O_6 thin films by pulsed laser deposition *Japan J. Appl. Phys.* **37** 6142
[61] Chiu T W, Wakiya N, Shinozaki K and Mizutani N 2002 Effects of ambient gas and and surface morphology of $Sr_{0.5}Ba_{0.5}Nb_2O_6$ thin films prepared by pulsed laser deposition *J. Ceram. Soc. Jpn.* **110** 368–72
[62] Nakaiso T, Sugiyama H, Noda M and Okuyama M 2000 Low-temperature preparation of $Sr_2(Ta_{1-x},Nb_x)2O_7$ thin films by pulsed laser deposition and its electrical properties *Japan J. Appl. Phys.* **39** 5517

[63] Song Z, Lin C, Wang L, Huang J, Hesse D, Zakharov N D, Xu H and Okuyama M 2000 Microstructure and electrical properties of textured $Sr_{0.51}Ba_{0.48}La_{0.01}Nb_2O_6$ thin films *Appl. Phys.* A **70** 355–8
[64] Li M, Chen X, Wang J, Yang J, Wu G and Zhao L 2000 Platinum silicide formation during pulsed laser annealing prepared by pulsed laser deposition *4th Int. Conf. Thin Film Phys. Appl.* vol 4086 *(Shanghai, China)* pp 823–6
[65] Neimash V, Shepelyavyi P, Dovbeshko G, Goushcha A O, Isaiev M, Melnyk V, Didukh O and Kuzmich A 2016 Nanocrystals growth control during laser annealing of Sn:(α-Si) composites *J. Nanomater.* **8** 7920238
[66] Chiu T W, Tonooka K and Kikuchi N 2008 Fabrication of ZnO and $CuCrO_2$:Mg thin films by pulsed laser deposition with *in situ* laser annealing and its application to oxide diodes *Thin Solid Films* **516** 5941–7
[67] Carey P G, Smith P M, Theiss S D and Wickboldt P 1999 Polysilicon thin film transistors fabricated on low temperature plastic substrates *J. Vac. Sci. Technol.* **17** 1946–9
[68] Peng Y C, Fu G S, Yu W, Li S Q and Wang Y L 2004 Crystallization of amorphous Si films by pulsed laser annealing and their structural characteristics *Semicond. Sci. Technol.* **19** 759
[69] Benko F A and Koffyberg F B 1986 Preparation and opto-electronic properties of semiconducting $CuCrO_2$ *Mater. Res. Bull.* **21** 753–7
[70] Nagarajan R, Draeseke A D, Sleight A W and Tate J 2001 p-Type conductivity in $CuCr_{1-x}Mg_xO_2$ films and powders *J. Appl. Phys.* **89** 8022–5
[71] Ozen I, Gulgün M A and Ozcan M 2004 Self-induced crystallinity in RF magnetron sputtered zno thin films *Key Eng. Mater.* **264** 1225–8
[72] Zhang Y, Du G, Zhang B, Cui Y, Zhu H and Chang Y 2005 Properties of ZnO thin films grown on Si substrates by MOCVD and ZnO/Si heterojunctions *Semicond. Sci. Technol.* **20** 1132
[73] Chiu T, Wakiya N, Shinozaki K and Mizutani N 2004 Effects of heating process on crystalline orientation and electrical properties of $(Bi,La)_4Ti_3O_{12}$ thin films derived by chemical solution deposition method *Integr. Ferroelectr.* **62** 189–92
[74] Tonooka K and Kikuchi N 2006 Preparation of transparent $CuCrO_2$:Mg/ZnO p–n junctions by pulsed laser deposition *Thin Solid Films* **515** 2415–8
[75] Debnath N, Kawaguchi T, Kumasaka W, Das H, Shinozaki K, Sakamoto N, Suzuki H and Wakiya N 2017 As-grown enhancement of spinodal decomposition in spinel cobalt ferrite thin films by dynamic aurora pulsed laser deposition *J. Magn. Magn. Mater.* **432** 391–5
[76] Dirnberger L, Dyer P E, Farrar S R and Key P H 1993 Proceeding of the second international conference on laser ablation *AIP Conf. Proc.* vol 288 (New York: AIP) 349
[77] Wakiya N, Nagamune T, Shinozaki K and Mizutani N 2004 *In-situ* magnetic field induced structure and properties of epitaxial spinel ferrite thin films prepared by pulsed laser deposition (PLD) (dynamic aurora pld method) *Mater. Res. Soc. Symp. Proc.* **853** I4.34
[78] Dirnberger L, Dyer P E, Farrar S R and Key P H 1994 Observation of magnetic-field-enhanced excitation and ionization in the plume of krf-laser-ablated magnesium *Appl. Phys.* A **59** 311–6
[79] Lash J S, Gilgenbach R M and Ching C H 1994 Laser-ablation-assisted-plasma discharges of aluminum in a transverse-magnetic field *Appl. Phys. Lett.* **65** 531–3
[80] Tachiki M and Kobayashi T 1999 Manipulation of laser ablation plume by magnetic field application *Japan J. Appl. Phys.* **38** 3642
[81] Kobayashi T, Akiyoshi H and Tachiki M 2002 Development of prominent PLD (Aurora method) suitable for high-quality and low-temperature film growth *Appl. Surf. Sci.* **197** 294–303

Chapter 4

ZnO-based ultraviolet light-emitting materials and devices

Weizhen Liu, Cen Zhang, Haiyang Xu and Yichun Liu

In the field of modern optoelectronics, short-wavelength light-emitting devices occupy a large market share in a wide range of applications, such as flat panel displays, solid-state lighting, information storage, environment protection, and life sciences, and they have attracted tremendous attention in recent years. The II–VI direct bandgap semiconductor zinc oxide (ZnO), which has a wide bandgap of 3.37 eV and a high exciton-binding energy of 60 meV, is regarded as a promising candidate for high-efficiency ultraviolet detectors, emitters and lasers. After nearly 20 years of development, great progress has been performed from ZnO-based semiconductors in the fields of material growth, impurity regulation, and device application. Considering the difficulties and challenges in high-quality p-type ZnO materials and their homojunction devices, in this chapter, we present a brief overview of ZnO light-emitting diodes (LEDs) and lasing diodes (LDs) based on P–N and metal-insulator-semiconductor heterostructures. Regarding the issues of the device efficiency and stability in heterojunction LEDs, some improvement methods are also introduced and then discussed in detail, including surface/interfacial modification, surface plasmons, single-crystalline nanostructures and emerging composites, which are employed in ZnO-based LEDs/LDs. Finally, the challenges and prospects for further development and application of ZnO-based light-emitting materials and devices are summarized.

4.1 Introduction

As third-generation semiconductors, wide bandgap materials have obvious advantages in solid-state lighting, short-wavelength semiconductor lasers, and ultraviolet (UV) optoelectronic detectors, and their further development will greatly improve people's productive lives. After decades of rapid development, group III nitrides, such as gallium nitride (GaN), have become the main representative of the third

doi:10.1088/978-0-7503-2556-1ch4

generation of wide bandgap semiconductors. Compared with group III nitrides, group II oxide semiconductors, represented by zinc oxide (ZnO) and their alloys, are expected to achieve exciton-type light-emitting devices and low-threshold lasing devices due to their higher exciton-binding energy. The bandgap of ZnO is 3.37 eV, which is similar to that of GaN, it has an emission wavelength in the blue-violet region, and its exciton-binding energy of 60 meV is second to none among common semiconductors. This value is much higher than the thermal ionization energy at room temperature (26 meV) and the excitons are therefore stable at room temperature. Furthermore, in many respects, ZnO is regarded as an alternate to GaN for optoelectronic device applications because of its relatively low production cost and superior photoelectric properties, which is widely applied in many fields (figure 4.1), such as light-emitting devices [1–15], photodetectors [16–23], photosensors [24–28], photovoltaics [29–35] and transistors [36–41]. In 1997, an excitonic collision induced stimulated emission with moderate optical pumping intensities was observed, which was the first time that ZnO optical pumping stimulated emission was achieved [42]. Soon afterwards, the above result was reported under the title 'Will UV lasers beat the blues?' in the *American Journal of Science* [43]. This opens a new era in the development of ZnO-based materials and devices. Since then, some milestone-like works have successively emerged based on ZnO spontaneous and stimulated emission in homojunction and heterojunction light-emitting devices. In 2005, a Japanese team first observed the blue–violet luminescence with electric injection from a ZnO homojunction at room temperature [44]. In the following year, Shen's team became the first research group in China to observe the electric injection emission in ZnO homojunction [45]. Later, electrically pumped waveguide lasing from ZnO nanowires (NWs) was first realized in homojunction devices [46]. However, stable and reproducible high-quality p-type doping remains a daunting challenge to produce ZnO homojunction light-emitting diodes (LEDs), lasing diodes (LDs). This challenge remains the most important topic and basic question in ZnO research today, and many research efforts are attempting to solve this problem. Our research group has made a contribution to overcome this bottleneck by developing a strategy for thermally oxidizing Zn_3N_2 thin films to achieve p-type N-doped ZnO [47, 48], and the thermal and optical stability was investigated in detail. Unfortunately, the results show that the stability of p-type ZnO still needs to be improved. Currently, the stability, reproducibility and controllability of p-type ZnO materials still do not satisfy the requirements for practical applications. Logically, an alternative strategy of investigation and development of ZnO-based heterojunction devices using other available p-type materials or design and construction of novel device configurations should be considered. A new round of research upsurge gradually emerged and promoted the continuous development of ZnO-based materials and devices. Actually, a 430-nm blue-violet electroluminescence (EL) was first already obtained in p-GaN/n-ZnO heterojunction LED in 2003 [49], and this kind of heterojunction device has attracted wide attention in the last 20 years. Much meaningful work has been done around this heterojunction [50–62]. However, a fatal flaw in this kind of heterojunction is that the source of luminescence has long been controversial. This led the researchers to think of another heterostructure that

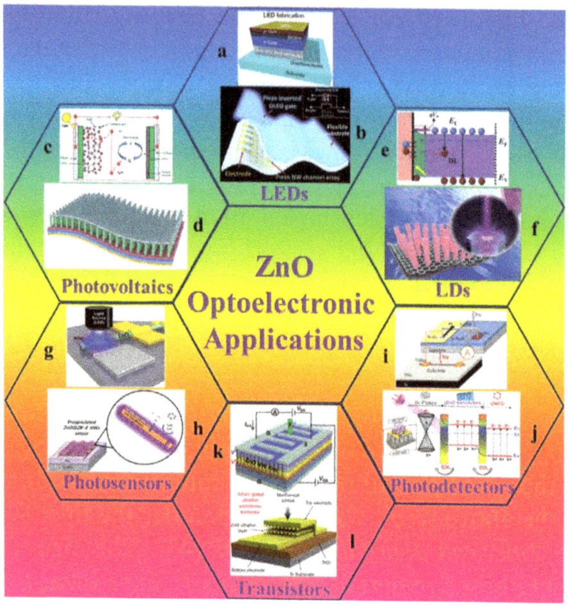

Figure 4.1. (a) Schematic of thin-film LED grown on graphene-layer substrate. [2] Reproduced with permission © Science. (b) Schematic illustration of inverted structural OLED based on ZnO NW channel array. Reprinted from [1], Copyright (2019), with permission from Elsevier. (c) The mechanism of dye sensitized solar cell for ZnO thin film as photoanode. Reprinted from [29], Copyright (2017), with permission from Elsevier. (d) The flexible device structural illustration of ZnO nanowire CIGS solar cell. Reprinted from [30], Copyright (2018), with permission from Elsevier. (e) Energy band diagram of a ZnO/MgO nanowire heterojunction. Reproduced from [7] with permission of The Royal Society of Chemistry. (f) Schematic of vertical ZnO nanowire micro-laser arrays. Reproduced from [8] with permission of The Royal Society of Chemistry. (g) Schematic illustration of ZnO-WSe_2 heterojunction photodiode operation under illumination. [24] John Wiley & Sons. © 2017 WILEY-VCH Verlag GmbH © Co., KGaA, Weinheim. (h) Structural diagram of MOF-based membrane encapsulated ZnO nanowire gas sensor. Reproduced with permission from [25]. Copyright (2016) American Chemical Society. (i) Schematic illustration of the integrated photodetector. [16] John Wiley & Sons. © 2011 WILEY-VCH Verlag GmbH & Co., KGaA, Weinheim. (j) The mechanism diagram of ZnO nanotube heterostructure photodetector. Reproduced with permission from [17]. Copyright (2020) Wiley-VCH. (k) Device structural schematic of ZnO nanowire thin film transistor. Reprinted from [37], Copyright (2018), with permission from Elsevier. (l) Schematic illustration of an ultrathin piezotronics transistor with two-terminal configuration. Reproduced with permission from [36]. Copyright (2018) American Chemical Society.

would try to overcome this factor. Metal-insulator-semiconductor (MIS) junction is also a common heterostructure, which is regarded as a perfect candidate for ZnO excitonic emission. Early in 1973, a blue emission (420 nm) from ZnO-based MIS junction was observed [63]. From then on, many improvements have been made to promote the development of ZnO-based MIS devices [64–67]. Until 2006, a pure UV EL was realized from Au/SiO_2/ZnO MIS LEDs [68], and electrically pumped ZnO lasing in the same structure was successfully achieving soon [69]. Later, the researchers obtained lower lasing threshold in ZnO-based MIS LDs using MgO as an insulating layer [70]. Since then, MIS heterojunction has ushered in a new era

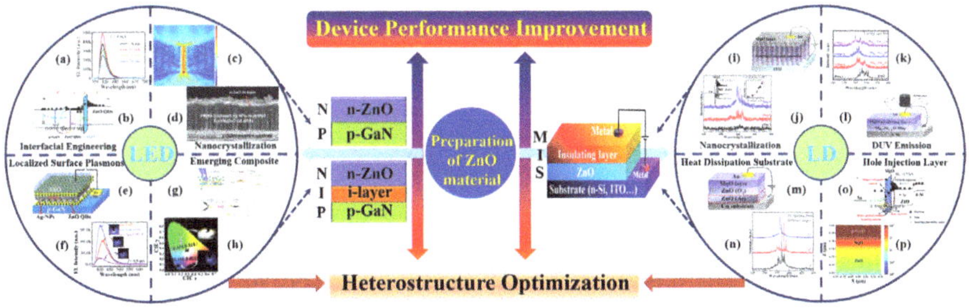

Figure 4.2. (a) EL spectra of the p-GaN/i-Al_2O_3/n-ZnO heterojunction LED device with different i-Al_2O_3 layer thicknesses. Reproduced from [86] with permission of The Royal Society of Chemistry. (b) Energy band alignment and carrier dynamic behavior of p-GaN/ZnO QDs/i-Al_2O_3/ZnO QDs (PAIT) heterojunction under forward bias. [87] John Wiley & Sons. © 2017 WILEY-VCH Verlag GmbH & Co., KGaA, Weinheim. (c) FDTD simulation of the EL intensity spatial distribution from a single nanorod optical waveguide. (d) A typical cross-sectional SEM image of the Ag-NPs decorated ZnO/MgZnO core/shell nanorod arrays LED device. Reproduced from [81] with permission of The Royal Society of Chemistry. (e) Structure schematic diagram of the Ag@PAIT-R Q-LED. (f) EL spectra of the three types of original PAIT, Ag@PAIT and Ag@PAIT-R Q-LEDs. [87] John Wiley & Sons. © 2017 WILEY-VCH Verlag GmbH & Co., KGaA, Weinheim. (g) Schematic structural diagram and energy band alignment of the 1L-MoS_2@ZnO QDs heterostructure. Reproduced from [101] with permission of The Royal Society of Chemistry. (h) CIE chromaticity coordinates of both LEDs and EL photos. Reprinted from [100], Copyright (2018), with permission from Elsevier. (i), (j) Structure schematic diagram of ZnO/MgO core/shell nanowire MIS heterojunction and EL spectra under different injection currents. Reprinted from [102], with the permission of AIP Publishing. (k), (l) EL spectra of Au/MgO/$Mg_xZn_{1-x}O$ MIS diodes with different Mg compositions and device structure diagram. Reprinted from [103], Copyright (2014), with permission from Elsevier. (m), (n) A structure diagram of Au/MgO/ZnO MIS-heterojunction LDs on the Cu substrate and the lasing spectra taken from different angles. Reproduced from [104]. © 2014 Optical Society of America. (o), (p) Energy band alignment under forward bias of Au/p-Cu_2O/MgO/ZnO heterostructure and simulated distribution profiles of hole concentration in p-Cu_2O decorated LED. Reproduced with permission from [105]. Copyright (2016) American Chemical Society.

in the development of ZnO LDs [71–77]. However, the device efficiency and stability are still not satisfactory in the above two heterostructures.

In this chapter, we present an overview of the recent advances in ZnO LEDs/LDs based on P–N and MIS heterostructures (figure 4.2), which focus on some contributions to improving device efficiency and stability. This chapter is organized as follows. In the first section, some traditional methods for preparing ZnO materials used in most devices are presented, and the quality and properties of these as-prepared ZnO materials, including thin films, NWs/nanorods (NRs), and quantum dots (QDs), are discussed in detail. The next section covers some ways to improve the UV emission efficiency in P–N or P–I–N heterojunction devices, including surface/interfacial engineering and modification (insertion of barrier, passivation, stress and strain, and retarding layers), surface plasmons (SPs, incorporation of Ag nanoparticles (NPs)/NWs, carbon QDs, and graphene layers), single-crystalline nanostructures (active layer substitution of ZnO NWs/NRs, and QDs), and emerging composites (combination with two-dimensional (2D) materials or perovskites). Some strategies for improving the lasing behavior and reducing the lasing threshold in ZnO MIS heterojunction devices are then discussed, including incorporation of core/shell (CS) NWs, interposition of $Mg_xZn_{1-x}O$ alloys, implantation of a heat-dissipating metal substrate, and insertion of a p-Cu_2O hole-injection layer. In the final section, the prospects and challenges for further development of ZnO-based heterojunction light-emitting devices and their multifunctional photoelectric device applications are briefly discussed.

4.2 Preparation and properties of ZnO materials

Preparation of high-quality ZnO materials is the fundamental problem and premise for realization of ZnO-based optoelectronic devices. Therefore, preparation of ZnO was first performed by our research group. Pulsed laser deposition (PLD) is a popular technique and an effective growth method for fabricating metastable phase films or multiple-quantum-well (MQW) films owing to the high peak energy of the laser light. The MgZnO ternary alloy has some particular advantages, including a continuous tunable bandgap from 3.37 (ZnO) to 7.8 eV (MgO) and tailored physicochemical properties, which facilitate bandgap engineering, deep-UV (DUV) detection, and emission. In 2009, we grew high Mg-content wurtzite MgZnO films on sapphire substrates by the PLD technique [78]. We found that the compositions, defect states, and electro-optical properties of the MgZnO film critically depended on the substrate temperature and oxygen pressure. In the temperature range from 300 °C to 800 °C (figures 4.3(a) and (b)), only the (002) diffraction peak was present. The decreasing *c*-axis lattice constant showed that as the Mg content in the single-phase wurtzite MgZnO film gradually increased, the bandgap widened. The effect of oxygen pressure on the structural properties was also investigated, as shown in figure 4.3(c). There was no evidence of other phases or impurities, confirming formation of single-phase wurtzite MgZnO. This work opened numerous possibilities for construction of DUV optoelectronic devices based on MgZnO alloys.

The PLD technique has also attracted interest for fabrication of semiconductor NWs and related nanodevices because of its good controllability, reproducibility,

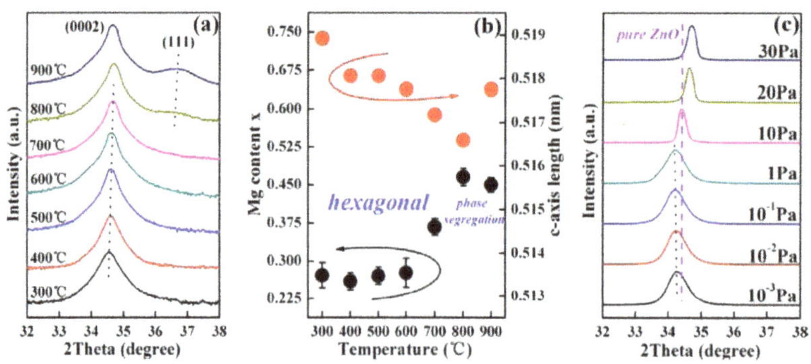

Figure 4.3. (a) XRD patterns of MgZnO films grown at different substrate temperatures. (b) Mg content and *c*-axis length as functions of substrate temperatures. (c) XRD patterns of MgZnO films deposited at various oxygen pressures. Reprinted from [78], with the permission of AIP Publishing.

and effective doping. In 2011, our group fabricated ZnO NWs by catalyst-free high-pressure PLD, and their size could be controlled by simply adjusting the growth conditions [79]. Experimental results showed that the NWs started to form when the substrate temperature and growth pressure exceeded the critical values of 700 °C and 700 Pa. The aspect ratio of the NWs decreased with increasing temperature or decreasing pressure, which is related to surface migration and scattering of ablated atoms. The variations of the ZnO NW diameter, length, and aspect ratio with temperature and pressure are shown in figure 4.4, and their sizes in scanning electron microscope images strongly depended on these growth conditions. The morphology, microstructure and temperature-dependent photoluminescence (PL) properties also demonstrated formation of single crystalline ZnO NWs. This work provided a new method for preparing high-quality ZnO nanostructures.

Compared with PLD-growth of ZnO NWs, the hydrothermal synthesis method was used to prepare ZnO NWs earlier because it has the advantages of a simple

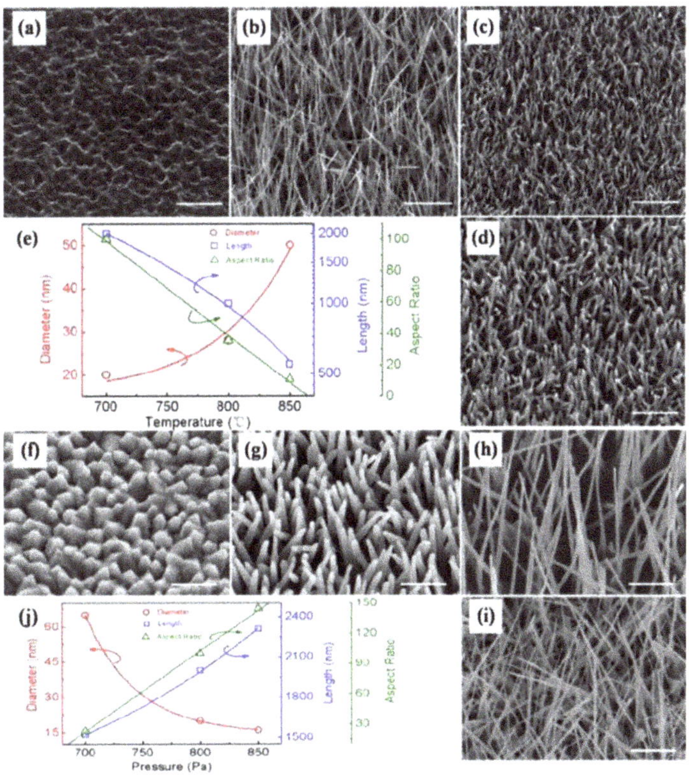

Figure 4.4. Tilted-view SEM images of series I samples grown at (a) 600 °C, (b) 700 °C, (c) 800 °C, (d) 850 °C and the fixed pressure of 800 Pa. The scale bars in all images are 1 μm. Tilted-view SEM images of series II samples grown at (f) 600, (g) 700, (h) 800, (i) 850 Pa and the same substrate temperature of 700 °C. The scale bars in all images are 500 nm. (e) and (j) show the variations of diameter, length, and aspect ratio of ZnO NWs with the substrate temperature and growth pressure. Reprinted from [79], with the permission of AIP Publishing.

preparation process and mild growth conditions. We designed ZnO/MgZnO coaxial NR arrays (NRAs) to investigate their microscopic structures, heteroepitaxial growth mechanisms, and luminescence characters [80]. Vertically aligned, single-crystalline ZnO/MgZnO coaxial NRAs with high-quality interfaces and a coherent epitaxial relationship, confirmed by Z-contrast scanning transmission electron microscopy (TEM) and line-scan compositional analysis (figures 4.5(a)–(e)), were successfully fabricated on substrates. Because of the shadow effect, this type of nonuniform coating gave us the opportunity to investigate the carrier confinement and surface passivation on a single NR (figure 4.5(f)). A novel model was established to understand the phenomenon that the near-band-edge emission of ZnO cores was variable along the length, and strongly depended on the shell thickness (figure 4.5(g)). The ZnO/MgZnO coaxial NRAs with high UV emission efficiency are suitable for developing devices, and their highly aligned orientation will facilitate 'bottom-up' integration of nanodevices.

To realize EL application, a similar ZnO/MgZnO CS structure prepared by the same synthetic method was used in LED devices [81]. It was confirmed that the

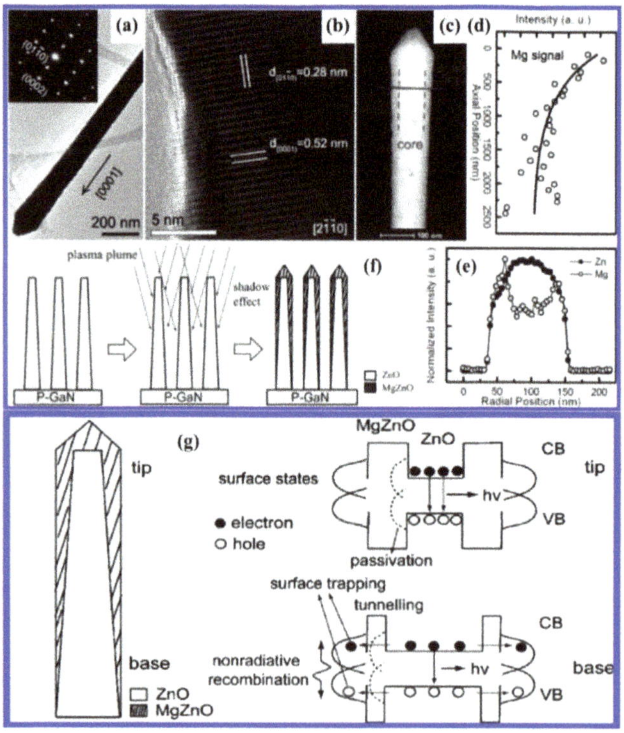

Figure 4.5. (a) Typical TEM image of a single coaxial NR. (b) Lattice-resolved HRTEM image of the MgZnO shell. (c) HAADF-STEM image of a single coaxial NR. (d), (e) The line-scan composition profiles of Mg and Zn elements along the axial and radial direction of the NR. (f) A schematic diagram of the growth process and structure of coaxial NRAs. (g) A schematic diagram showing the energy band structure, surface trapping, and radiative/nonradiative recombination processes in the upper and lower regions of the coaxial NR. Reproduced with permission from [80]. Copyright (2010) American Chemical Society.

improved UV emission was derived from optical waveguide-type EL transmission, and indirectly indicated that our prepared nanostructures were high-quality and of great application value. Moreover, for ZnO nanostructures, the QD system has been studied in many fields. We synthesized high-quality monodisperse uniformly distributed ZnO QDs by a solution-processed method [82]. The optimized experimental scheme and clear microstructural characterization demonstrated that these ZnO QDs can be applied to light-emitting devices to further explore the luminescence mechanism, source and process. Related studies of ZnO nanostructures will be discussed in detail in the following chapter.

Overall, we focus on high-quality thin films and nanostructures to prepare ZnO materials, and we use different methods to investigate their material properties and some valuable applications. These efforts provide a strong material basis for subsequent device fabrication and luminescence property research.

4.3 P–N and P–I–N heterojunction LEDs

Because of the difficulty in p-type doping ZnO, an alternative strategy is to find a p-type material to construct heterojunction devices. Among the mature p-type semiconductors such as GaN, Si, $CuAlO_2$, NiO, SiC, and even some organics, GaN is taken for the most suitable due to its identical crystal structure, similar energy band alignment and small lattice mismatch, which favor heteroepitaxial growth and form a high-quality heterointerface. However, the EL origin in this heterostructure remains controversial. This issue has been studied and discussed by our group [83]. For comparison, a series of heterojunctions were designed and fabricated (figure 4.6(a)). The current–voltage (*I*–*V*) characteristics (figure 4.6(b)), energy band diagram (figure 4.6(d)), and carrier dynamics mechanism of these devices were discussed in detail. For the pristine heterojunction LED, the near-UV luminescence was attributed to the interfacial transition from the ZnO conduction band minimum to the GaN acceptor level (figures 4.6(c) and (d)). For n-MgZnO/p-GaN devices, UV emission was independent of the Mg composition, and it originated from donor–acceptor pair emission in the GaN layer. Our experimental results indicated that ZnO or MgZnO related exitonic emission can hardly be obtained in these heterojunctions. Therefore, we believe that rational device design relying on further energy band analysis towards ZnO excitonic emission is more significant. In brief, feasibility analysis of the energy band in the P–N heterojunction revealed that the EL may, in principle, consist of several recombinations, and electrons and holes overcome almost the same barriers to realize carrier injection. Therefore, the origin of the EL will be mainly determined by the difference in the carrier mobility between n-ZnO and p-GaN. Generally, the electron mobility is higher than the hole mobility. Hence, electron injection prevails over hole injection under forward bias, that is, electron injection dominates the whole carrier transport process. Thereby, the blue emission usually occupies a large proportion in the EL spectrum, whereas radiative recombination in the n-ZnO side is always suppressed. This disadvantage has long plagued the ZnO UV emission efficiency and further development in this type of P–N heterojunction LEDs. Although some reports have focused on

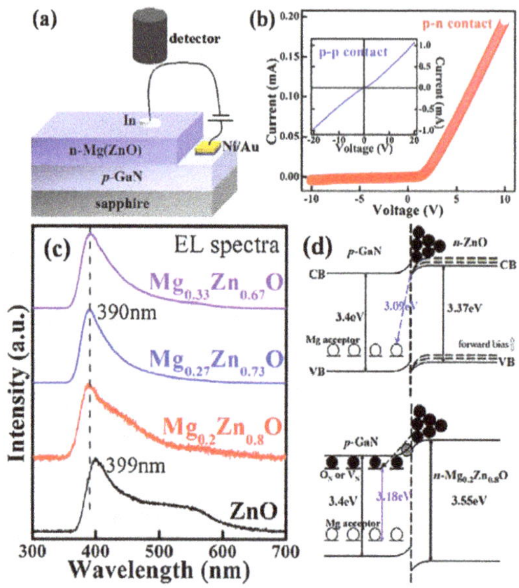

Figure 4.6. (a) The schematic of device structure. (b) *I–V* characteristic of n-MgZnO/p-GaN heterojunction. (c) EL spectra of different n-$Mg_xZn_{1-x}O$/p-GaN heterojunction diodes. (d) The energy band diagram of both heterojunctions. [83] John Wiley & Sons. © 2013 WILEY-VCH Verlag GmbH & Co., KGaA, Weinheim.

solving this problem, deep insight into the carrier dynamics mechanism is still lacking to the point of not being able to deeply understand this luminescence process.

Considering the above reasons, there are two inevitable problems of low emission efficiency and poor stability in such P–N heterojunctions. Therefore, it is still necessary and important to reconsider and redesign device structures to realize high-efficiency and stable ZnO-based heterojunction LEDs. To activate excitonic emission from ZnO, some alternatives to improve the device efficiency and stability in P–N or P–I–N heterojunction devices are implemented into light-emitting devices, including surface/interfacial engineering, SPs, single-crystalline nanostructures, and emerging composites. These novel and interesting alternatives will be discussed in greater detail in the following subsection.

4.3.1 Surface/interfacial engineering and modification

To compensate for the difference in the mobility, in 2005, our research group proposed the idea of introducing a thin semi-insulating layer into the n-ZnO/p-GaN heterojunction interface [84]. An i-ZnO layer was first grown on p-GaN substrate, and then a n-ZnO layer was deposited. The purpose of inserting this layer is to confine the parts of the carriers to recombine in i-ZnO region. The results indicated that the bandgaps indirectly calculated by the PL spectra of p-GaN, i-ZnO and n-ZnO were 3.10, 3.24, and 3.31 eV, respectively (figure 4.7(a)). More importantly, EL in reconstructed P–I–N heterojunction at 3.21 eV regarded as ZnO excitonic emission with an obvious blue-shift compared to the P–N heterojunction was obtained (figure 4.7(b)). That is, by constructing a P–I–N heterojunction, radiative

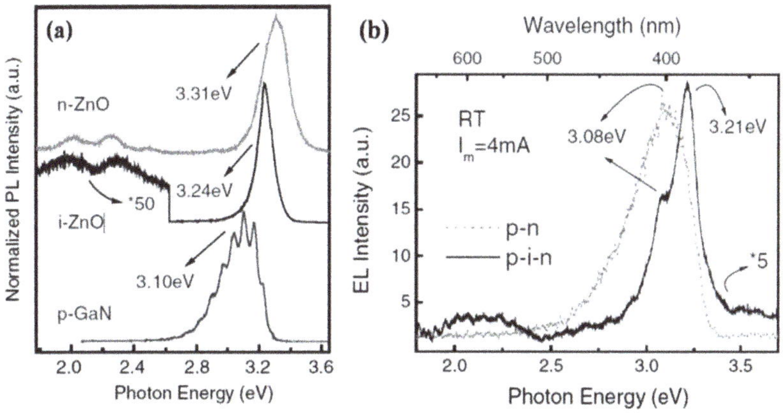

Figure 4.7. (a) PL spectra of p-GaN, i-ZnO and n-ZnO. (b) EL spectra of P–N and P–I–N heterojunction LEDs. [84] (2004) Copyright. With permission of Springer.

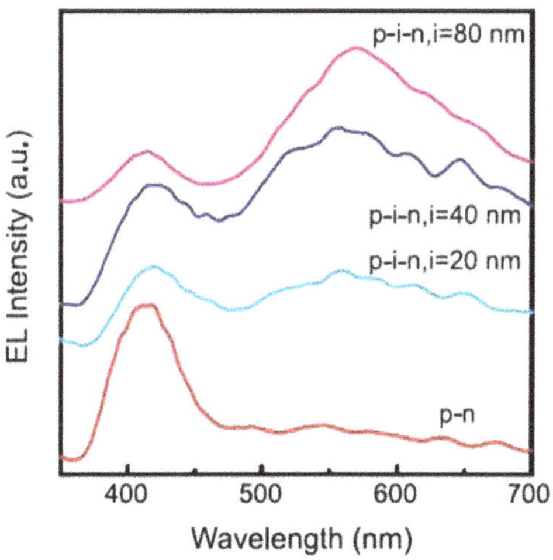

Figure 4.8. (a) EL spectra of p-GaN/n-ZnO and p-GaN/i-ZnO/n-ZnO heterojunctions with i-layer thickness of 20, 40, and 80 nm. [85] (2008) Copyright. With permission of Springer.

recombination was partly transferred from the p-GaN region to the i-ZnO region. Therefore, ZnO excitonic emission was activated and realized by inserting a barrier layer (i-ZnO) at the interface of the active region. This P–I–N heterojunction has potential applications that need to be durable under high-temperature operation. On the basis of the above work, a series of similar heterojunctions with different i-layer thickness showed simultaneous acceptor and deep-level emission [85]. By further changing the experimental conditions and analyzing the EL spectra, the yellow luminescence intensity, that is, the ratio of yellow to blue–violet emission, was tunable by precise control of the insulating layer thickness. The optimized heterojunction with a 20 nm-thick i-ZnO layer exhibited better white EL (figure 4.8). We proposed a new and simple way to prepare one-chip white LEDs, and considered that this diode is promising for realizing multi-color display and wavelength adjustable luminescence.

In the above P–I–N heterostructure, the incorporation of a thin semi-insulating ZnO layer could improve the UV emission, but the efficiency was still not high. From the viewpoint of energy band, this i-layer can be replaced with some other materials with a wider bandgap, such as MgO, Al_2O_3 or AlN. By appropriately adjusting the properties of this layer, the electrons can be better confined to the ZnO side in the P–I–N heterojunction. Based on the typical n-ZnO/p-GaN structure, in 2009, Shan's group developed a kind of heterojunction based on n-ZnO/i-MgO/p-GaN structure, which managed to keep most of the electrons on the ZnO side and promoted majority holes in p-GaN to be effectively injected into n-ZnO (figure 4.9) [51]. Furthermore, a laser behavior operating in the near-UV region with the threshold of 0.8 mA, which was considered the smallest value reported at the time, had been realized in ZnO. This work opens a new research direction of ZnO-based P–N or P–I–N heterojunction LDs, and predicts that similar results will be replicated in other feasible structures.

In our case, a thin i-Al_2O_3 layer, which is regarded as an ideal candidate because of its mature film preparation technique and similar crystal structure to GaN and ZnO, was introduced into the interface between p-GaN and n-ZnO [86]. The EL spectra in heterojunction devices without and with the Al_2O_3 interlayer are shown in figures 4.10(a) and (c). Analysis of the energy bands in figures 4.10(b) and (d)

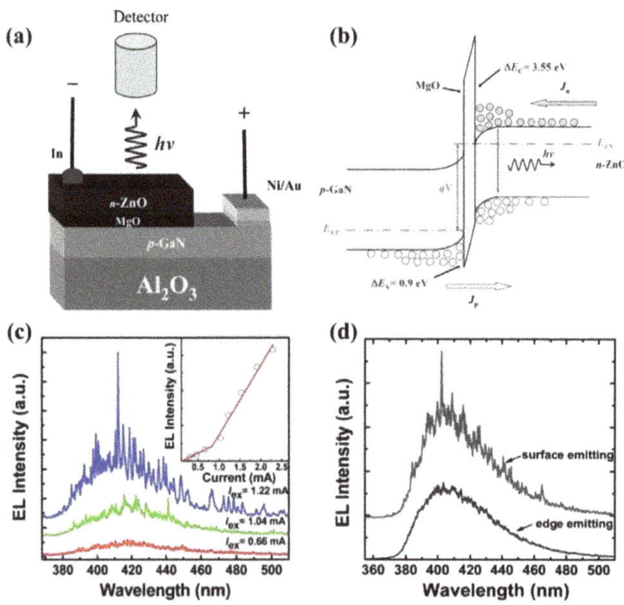

Figure 4.9. (a) Structural diagram of n-ZnO/i-MgO/p-GaN (P–I–N) heterojunction. (b) Schematic illustration of energy band alignment under forward bias. (c) EL spectra of the device under different currents. (d) The upper surface and edge EL spectra of LED device above the threshold, indicating a directional random emission behavior. [51] John Wiley & Sons. © 2009 WILEY-VCH Verlag GmbH & Co., KGaA, Weinheim.

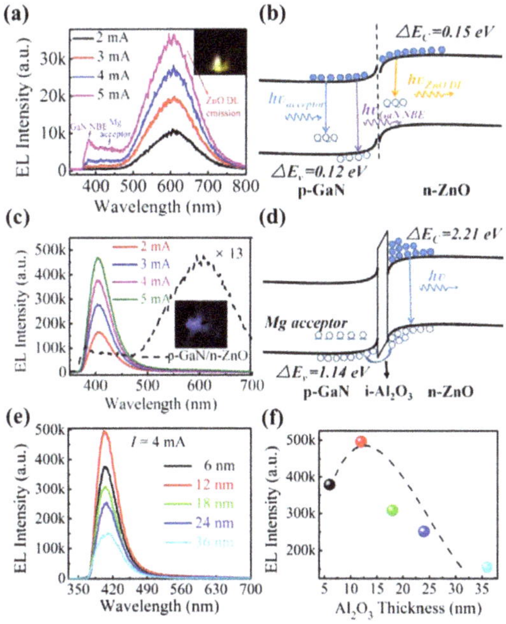

Figure 4.10. (a), (c) EL spectra of the p-GaN/n-ZnO and p-GaN/i-Al_2O_3/n-ZnO heterojunction LEDs. (b), (d) Energy band alignments and carrier dynamic processes of the p-GaN/n-ZnO heterostructure without and with the Al_2O_3 blocking layer under forward bias. (e) EL spectra of the p-GaN/i-Al_2O_3/n-ZnO heterojunction LED devices with different i-Al_2O_3 layer thicknesses. (f) EL intensity of such a P–I–N heterojunction as a function of the Al_2O_3 layer thickness. Reproduced from [86] with permission of The Royal Society of Chemistry.

allowed us to understand the reason for improved UV emission. Moreover, by optimizing the electron blocking layer thickness (~12 nm) with precise control by the atomic layer deposition technique, the effective electron accumulation and hole injection were simultaneously achieved, resulting in the improvement of near-UV EL in this P–I–N device (figures 4.10(e) and (f)).

To further explore the role of the i-Al_2O_3 layer and modulate the transport behavior of injected carriers in the P–I–N heterojunction, a thin i-Al_2O_3 layer (~10 nm thick) was inserted into ZnO QDs region as a retarding layer (figure 4.11(b)), separating them into two discrete layers [87]. The entire device structure is described below. The first QD layer sandwiched between i-Al_2O_3 and p-GaN acts as an active layer (A layer), the second QD layer on top of i-Al_2O_3 layer can be regarded as an electron transport layer (T layer). This multilayer structure QD LED (Q-LED) called a 'PAIT' heterojunction for simplicity was constructed. The EL spectra of both Q-LEDs are shown in figure 4.11(c). There was stronger near-UV emission in the PAIT heterojunction. The normalized EL spectrum (figure 4.11(d)) of the PAIT heterojunction indicated that the near-UV emission peak position experienced a blue-shift, and the entire spectrum full width at half-maximum showed ~30 nm narrowing. By analyzing and comparing two energy band alignments, we believe that the injected electrons will be first transported in the T

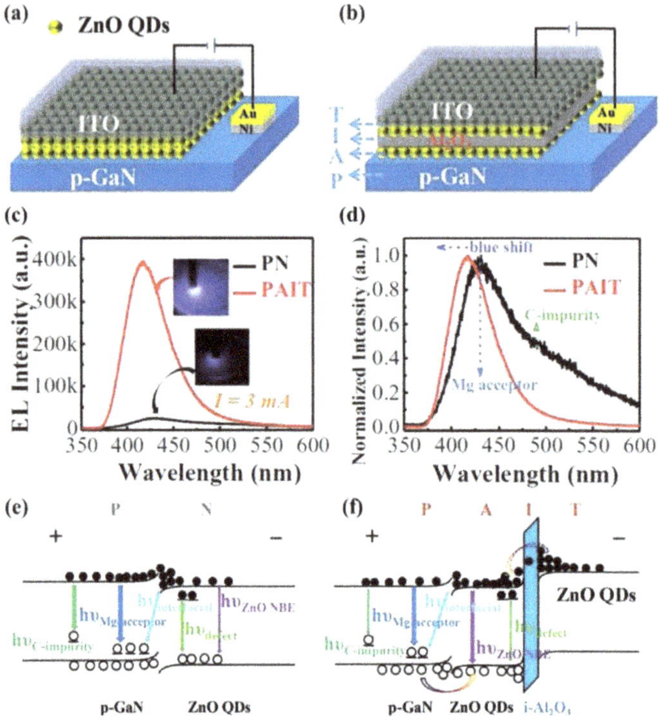

Figure 4.11. (a), (b) Device structure schematic diagrams of the P–N and PAIT Q-LEDs. (c) EL spectra of the P–N and PAIT Q-LEDs. (d) Normalized EL spectra of both Q-LEDs. (e), (f) Energy band alignments and carrier dynamic behaviors of the P–N and PAIT heterojunctions under forward bias. [87] John Wiley & Sons. © 2017 WILEY-VCH Verlag GmbH & Co., KGaA, Weinheim.

layer and then retarded (or even blocked) by i-Al_2O_3 layer. Actually, most of the electrons could still tunnel through this thin i-Al_2O_3 layer, but their migration ability is weakened because these electrons experience strong scattering. In this case, these electrons will be detained in the A layer. Meanwhile, holes will have sufficient time to enter the A layer, and radiative recombination results in an enhanced ZnO excitonic emission. In other words, the inserted thin i-Al_2O_3 layer can reflect the radiative recombination efficiency between electrons and holes. The electron migration rate can be effectively controlled by optimizing the thickness of the i-Al_2O_3 charge retarding layer.

In addition to the device efficiency, the stability is also used to evaluate the performance of the device. This leads us to think of some starting points for research [88]. It is known that surface defects and adsorption can greatly degrade the optical quality of ZnO nanostructures. Thus, surface coating and modifying may prevent undesirable negative effect. As a suitable research object, MgZnO-coated ZnO NRA heterostructures were mentioned again. The MgZnO-coated LED showed stronger EL emission than the uncoated LED (figures 4.12(a) and (b)). This is because that the effective surface passivation and carrier confinement were induced by introducing the MgZnO coating layer, resulting in an enhancement of radiative

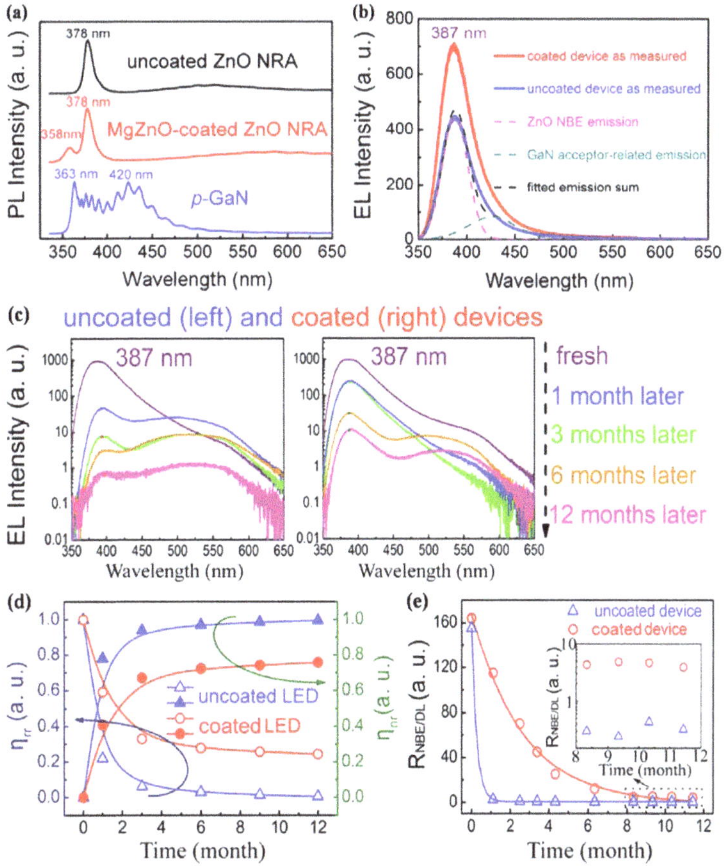

Figure 4.12. (a) PL spectra of MgZnO coated/uncoated ZnO NRs and GaN film. (b) EL spectra and Gaussian deconvolution analysis of uncoated and coated devices. (c) EL spectra of uncoated (left) and coated (right) devices obtained at different air-exposure periods. (d), (e) The variations of radiative recombination rate, nonradiative recombination rate, and the intensity ratio of near-band-edge to deep-level emission of both LEDs with air-exposure time. Reprinted from [88], with the permission of AIP Publishing.

recombination in the ZnO NRs. Further findings indicated that the nonradiative and deep level recombination processes decreased the device efficiency (figures 4.12(c)–(e)). However, suitable MgZnO coating can reduce the surface effect of ZnO NRs, thus improving the emission efficiency and stability of the device.

Confining the carriers at the interface of the heterostructure is considered to be an effective way to activate UV and DUV emission in such P–N heterojunctions. MQW structures have the natural advantage of type-I energy band alignment. The MgZnO ternary alloy can also be used as a potential barrier material for ZnO/MgZnO quantum wells (QWs). Therefore, in this case, to further shorten the luminescence wavelength, the MgZnO film should act as an active layer and wide bandgap MgO can be chosen as another barrier material. That is, the large energy-band offset between MgZnO and MgO will make the carrier confinement stronger.

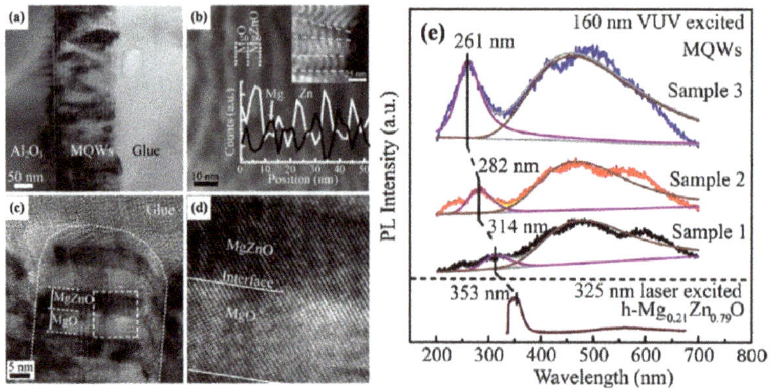

Figure 4.13. Cross-sectional TEM images of an MgZnO/MgO S-MQW nanocolumnar film with MgZnO layer thickness of ~5 nm. (a) A large-scale bright field image of the cross section of the above sample. (b) A magnified HAADF-STEM image taken on a single nanocolumn. (c) High resolution TEM image of an individual nanocolumn. (d) Lattice-resolved image of the rectangular region marked in (c). (e) The room-temperature PL spectra of MgZnO/MgO MQWs and pure $Mg_{0.21}Zn_{0.79}O$ film. Reprinted from [89], Copyright (2012), with permission from Elsevier.

Combined with the technical advantages of the aforementioned PLD method, closely arranged and vertically oriented MgZnO/MgO strained MQW (S-MQW) had been designed and prepared [89]. Alternating light/dark stripes in TEM images were clearly observed (figures 4.13(a)–(d)). The sharp interface between the MgZnO and MgO layers indicated that both layers suffered from in-plane stress. Wavelength-tunable DUV emission was obtained (figure 4.13(e)). This study on stress-induced structural transition and DUV emission developed a novel QW structure which will probably be applied to DUV detectors.

We further analyzed the above-mentioned S-MQW structures, and their growth dynamic processes and optical properties were investigated in detail. ZnO has a hexagonal wurtzite structure, whereas MgO has a cubic rock-salt structure. Compared with hexagonal wurtzite MgZnO, rock-salt MgZnO possesses higher structure symmetry. As a result, less phonon scattering and higher ballistic electron velocities are expected for rock-salt MgZnO, that is, rock-salt MgZnO is more favorable for DUV detector. Moreover, for DUV emitter application, the MQW structure is widely used as an active layer in light-emitting devices. From the above considerations, we grew highly biaxial rock-salt $Mg_{0.21}Zn_{0.79}O$/MgO MQW films on m-sapphire substrates [90]. A coherent epitaxial relationship with a high-quality interface was confirmed by cross-sectional TEM images (figures 4.14(a)–(d)). The room-temperature PL (figure 4.14(e)) showed a relatively broad DUV emission peak with a strong visible emission. The optical bandgap of the anomalous rock-salt phase was determined by taking into account both the quantum confinement effect and strain induced energy shift. The results indicated that rock-salt MQW structures had potential applications in DUV emitters.

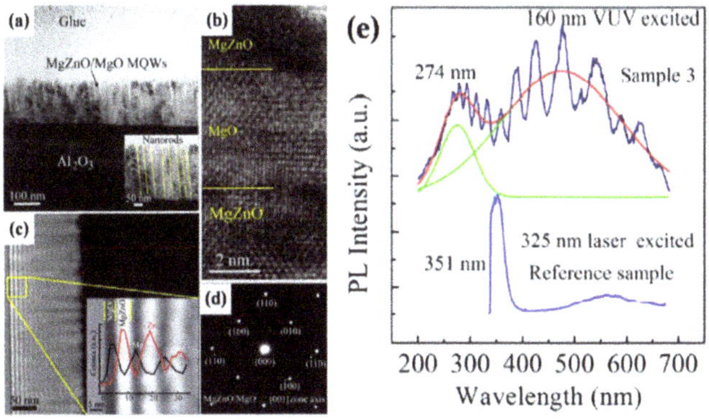

Figure 4.14. Cross-sectional TEM images of MgZnO/MgO MQWs with a well layer thickness of ~5 nm. (a) Large-scale cross-sectional bright field image. (b) High-resolution TEM image of MQW structure. (c) Magnified HAADF-STEM image. (d) SAED pattern for MgZnO/MgO MQWs. (e) Room-temperature PL spectra for the MgZnO/MgO MQWs and the reference sample. Reprinted from [90], with the permission of AIP Publishing.

4.3.2 Incorporation of localized surface plasmons

Although much progress has been performed in ZnO-based heterojunction LEDs, the energy barrier in such a P–N or P–I–N heterojunction will inevitably reduce the carrier injection efficiency. In recent years, localized surface plasmons (LSPs) have been found to be an effective strategy for improving device efficiency. Metal Ag nanostructures generally have a broad LSP absorption band, which overlaps with the energy of ZnO excitonic emission, suggesting the possibility of resonant coupling. In this subsection, we will review some related studies.

Actually, some works about LSP enhanced near-band-edge emission of ZnO in PL measurements were discovered and demonstrated earlier. However, how to realize the further improvement of ZnO UV EL is of more practical significance. In 2011, Zhang's group first demonstrated LSP-enhanced UV EL from n-ZnO/i-AlN/p-GaN heterojunction LED by inserting Ag NPs (figure 4.15(a)) [91]. It was found that the devices modified with Ag NPs exhibited significantly selective UV emission enhancement under the same injection current (figures 4.15(b) and (c)). A direct evidence, as well as the reduced PL decay lifetime of ZnO, suggested that the improved performance was attributed to the coupling interaction between ZnO excitons and Ag LSPs. Meanwhile, by tailoring properties of Ag LSPs and optimizing device structures, the improved EL performance of those LEDs with different sizes of Ag NPs decoration were also obtained [92], and a maximum enhancement in such a device with 10 nm-size Ag NPs decoration was observed (figure 4.15(d)). Their work demonstrated that Ag LSPs could enhance ZnO excitonic emission, and their sizes had an important effect on the regulation and improvement of ZnO UV EL.

In order to further study this physical process, we designed and prepared similar structural n-ZnO/i-ZnO/p-GaN film heterojunction LED. To suppress the undesired charge transfer and nonradiative Förster resonant energy transfer, a MgO layer and

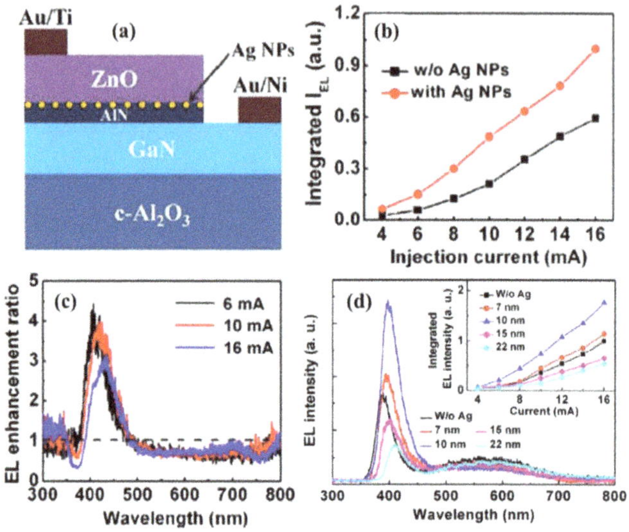

Figure 4.15. (a) Structural diagram of n-ZnO/i-AlN/p-GaN (P–I–N) heterojunction with Ag NPs decoration. (b) The EL integrated intensity of both LEDs with and without Ag NPs as a function of injection current. (c) The dependence of EL enhancement ratio on the emission wavelength under different injection currents. (d) EL spectra of n-ZnO/i-AlN/p-GaN heterojunction with and without different sizes of Ag NPs decoration. The inset presents the EL integrated intensity of both heterojunction devices with and without Ag NPs as a function of injection current. (a)–(c) Reprinted from [91], with the permission of AIP Publishing. (d) Reprinted from [92], with the permission permission of AIP Publishing.

Ag NPs were successively embedded on the surface of i-ZnO active layer (figure 4.16(a)) [93]. The PL spectra of those heterojunctions with different MgO thicknesses are shown in figure 4.16(b). When the MgO thickness was optimized to ~12 nm, 8-fold UV emission enhancement in PL measurement was observed. For Ag/MgO interlayer decorated LED device, the wavelength of the maximum EL enhancement (figures 4.16(c) and (d)) was close to that of the PL enhancement ratio, indicating that it originated from the same mechanism: the coupling between ZnO excitons and Ag LSPs. This was verified by the shortened PL lifetimes and increased internal quantum efficiency (IQE) from time-resolved and temperature-dependent PL measurements (figures 4.16(e) and (f)). The study indicated that LSPs can offer a promising way to improve device efficiency.

Afterwards, we attempted to incorporate Ag-NWs into p-GaN/i-Al_2O_3/n-ZnO heterojunction LEDs (figure 4.17(a)) [86]. Actually, the absorption of the Ag-NW LSPs showed a greater overlap with ZnO UV emission than that of Ag NPs. After optimizing i-Al_2O_3 electron blocking layer thickness, commercially available Ag-NWs were spin-coated onto ZnO active layer. Both PL and EL measurements of the devices with Ag-NWs showed significant enhancement in UV region (figures 4.17(b) and (d)). These phenomena are easily associated with a common physical mechanism: exciton-LSP resonant coupling (figures 4.17(e) and (f)). The current work demonstrated a feasible way to enhance ZnO excitonic emission in P–I–N type LEDs.

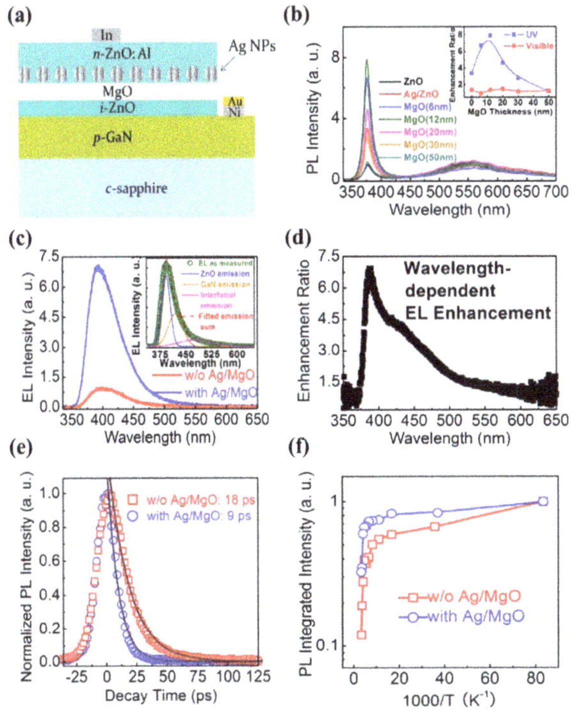

Figure 4.16. (a) Schematic diagram of the LSP-enhanced LED structure. (b) PL spectra of Ag/MgO/ZnO multilayer films with different spacer-layer thicknesses. The inset shows the variation of enhancement ratio of integrated UV/visible emission intensity with the MgO spacer-layer thickness. (c) EL spectra of the LEDs with and without Ag/MgO interlayer. (d) The dependence of EL enhancement ratio on emission wavelength. (e) 20 k TR-PL spectra of ZnO films with and without Ag/MgO layer. (f) Arrhenius plots of the normalized integrated UV-PL intensity of ZnO films with and without Ag/MgO layer. Reprinted from [93], with the permission of AIP Publishing.

Next, we carried out a detailed study on the luminescence improvement in a p-GaN/n-ZnO heterojunction LED with Ag-NW decoration (figure 4.18(a)) [94]. Four different surface densities of Ag-NWs were spin-coated on the n-ZnO layer. In PL measurements, the Ag-NWs covered ZnO films showed markedly improved UV emission (figure 4.18(b)). More importantly, the simultaneous appearance of blue–violet EL enhancement and yellow–green emission weakening with increasing coverage density of Ag-NWs confirmed that the coupling interaction mainly occurred between ZnO excitons and Ag-NW LSPs (figures 4.18(c) and (d)). This interesting behavior made a significant contribution to the color-tunable EL spectra (figures 4.18(d)–(f)). A simulation demonstrated the field distribution of Ag-NW LSPs reached the maximum at an excitation wavelength of 382 nm, which is close to ZnO excitonic emission (figure 4.18(g)), proving that the coupling effect in UV region was the most effective. The chromatic properties, chromaticity coordinates and color temperature of these P–N heterojunction LEDs were further investigated (figures 4.18(h) and (i)). With increasing surface coverage density of Ag-NWs, the

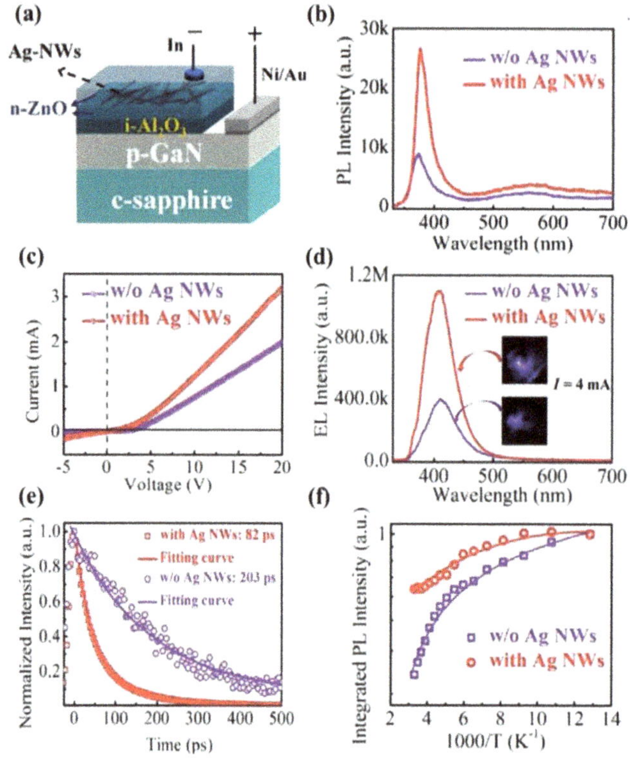

Figure 4.17. (a) Structural diagram of the Ag-NW decorated P–I–N heterojunction LED. (b) PL spectra of the ZnO films with and without Ag-NW decoration. (c) *I–V* curves of both LED devices with and without Ag-NWs. (d) EL spectra of both LEDs with and without Ag-NW decoration. (e) TR-PL spectra of the ZnO films with and without Ag-NWs. (f) Arrhenius plots of the normalized integrated UV-PL intensity of the ZnO films with and without Ag-NWs. Reproduced from [86] with permission of The Royal Society of Chemistry.

apparent color coordinates shifted from yellow–green to blue–violet region and the correlated color temperature obviously exhibited a broad range, indicating that these Ag-NW decorated LEDs have potential applications in a variety of situations. The current proposed strategy represents a reasonable strategy for adjusting the EL spectral profile and LED color.

4.3.3 Cooperative effect of nanocrystallization and surface plasmons

The previous results show that a P–N or P–I–N heterojunction still suffers from some inevitable interfacial problems, resulting in unsatisfactory device efficiency. However, the problems can be mitigated to some extent. An effective method is to introduce one-dimensional single-crystalline ZnO nanostructures as an active layer, and it is theoretically predicted that their incorporation into a P–N or P–I–N LED can increase carrier injection rate for nano-sized heterojunctions. In 2009, Wang's group developed a high-brightness blue LED based on ZnO NWs/p-GaN nano-heterostructure (figure 4.19(f)) [54]. It was found that the EL intensity increased and

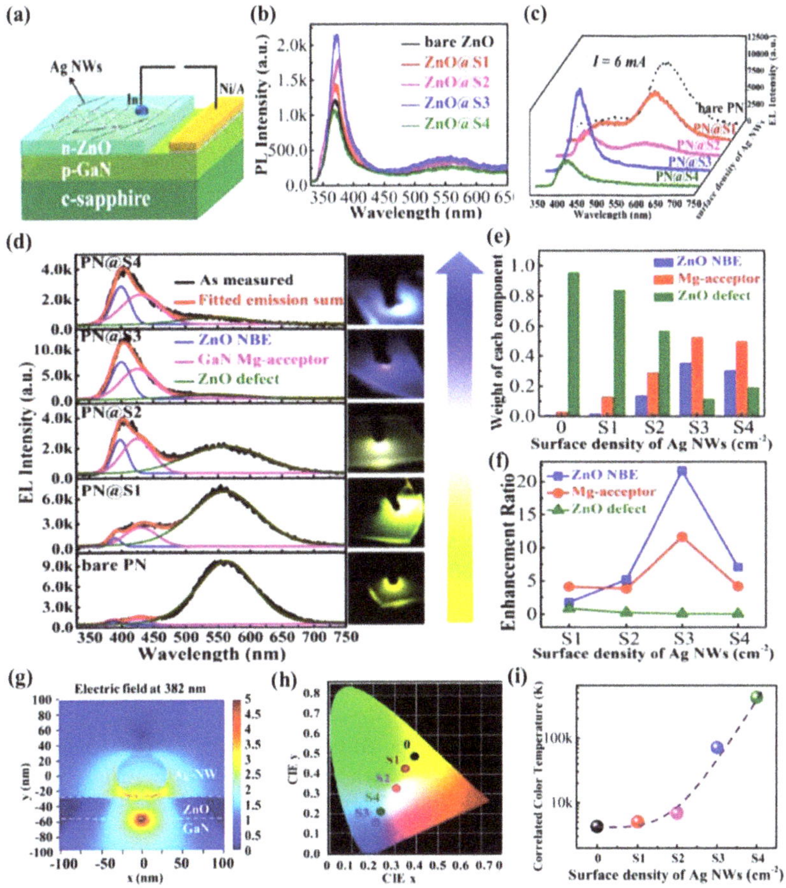

Figure 4.18. (a) Structure diagram of the Ag-NW decorated P–N heterojunction LED. (b) PL spectra of the ZnO films with and without Ag-NWs. (c) EL spectra of the Ag-NW decorated devices. (d) Gaussian deconvolution fittings of the p-GaN/n-ZnO LEDs with different surface densities of Ag-NWs. Right panel shows the EL photograph of these LEDs. (e) Relative spectral weight of three EL components. (f) Variation trend of the EL enhancement ratio for each luminescence component at different Ag-NW surface coverage densities. (g) FDTD simulations of the electric field intensity distributions of a single Ag-NW decorated ZnO/GaN heterojunction, and the excitation wavelength employed for simulation is 382 nm. (h) CIE 1931 chromaticity diagram of the Ag-NW decorated LEDs. (i) CCT of the Ag-NW decorated LEDs as a function of the Ag-NW surface density. Reproduced with permission from [94]. Copyright (2018) American Chemical Society.

the blue shift occurred with the increase of the external voltage, which is caused by the change of the energy band of the heterojunction in the dissipated region (figure 4.19(g)). Meanwhile, this LED device demonstrated a high sensitivity to UV irradiation because of the re-excited residual carriers and the density of charge carriers in ZnO (figure 4.19(h)). This study proved that a blue-violet LED can be achieved by constructing a nanoscale P–N heterojunction. Later, the ordered arrays of ZnO NWs were fabricated on p-GaN substrate by wet chemical method and electron beam lithography (figures 4.19(a)–(c)) [52]. The whole arrays consisted of

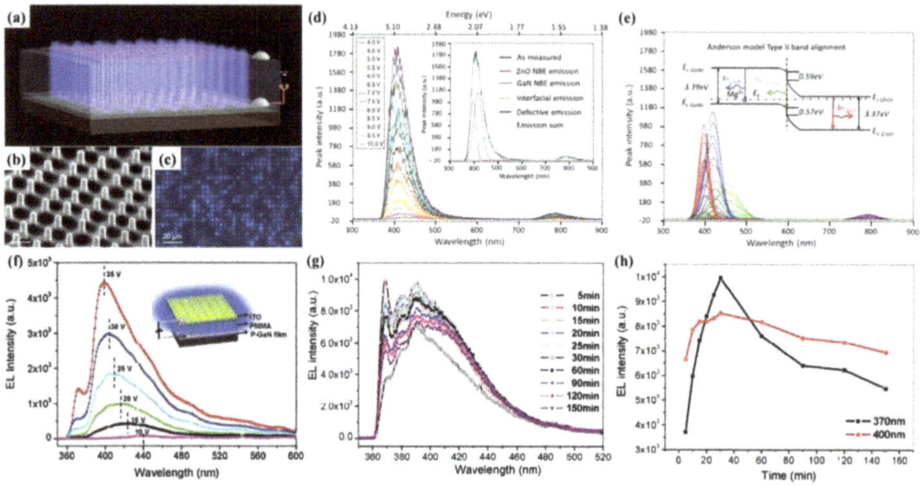

Figure 4.19. (a) Device structural design overview. (b) Tilted-view SEM image of the as-grown patterned vertical ZnO NWs coated with SiO_2 and wrapped with PMMA. (c) The LED optical image with a drive voltage applied. (d) EL spectra as a function of forward bias voltage. The inset shows Gaussian deconvolution analysis with the emission decomposed into three distinct bands. (e) The dependence of four emission bands (band width, height, and intensity) on applied forward bias voltage. The inset shows the energy band diagram of the n-ZnO/p-GaN heterojunction under no or small forward bias voltage. (f) The EL spectra of the n-ZnO NWs/p-GaN heterojunction LED under various forward bias voltages. The inset presents schematic illustration of n-ZnO NWs/p-GaN heterojunction LED device. (g) Series of EL emission spectra from the P–N heterojunction as a function of time after turning off the UV illumination. (h) The dependence of EL intensity at wavelength of 370 and 400 nm on the time when UV irradiation was turned off. [52] John Wiley & Sons. © 2010 WILEY-VCH Verlag GmbH & Co., KGaA, Weinheim. [54] John Wiley & Sons. © 2009 WILEY-VCH Verlag GmbH & Co., KGaA, Weinheim.

each single NW as a light emitter, and the obvious near-UV emission assigned to three distinct recombination processes was obtained (figures 4.19(d) and (e)). This heterojunction device demonstrated a considerable external quantum efficiency of 2.5%, which would promote some potential applications in high-resolution electronic display and high-density data storage.

Another strategy is to incorporate SPs, including metal nanostructures (Ag NPs or NWs) and nonmetallic materials (carbon nanodots (CDs) or graphene), into the heterojunction LED to improve the device efficiency. Since the realization of Ag LSP-enhanced EL from ZnO-based heterojunction LEDs, the research of using SPs to improve ZnO emission efficiency has aroused great interest. In 2012, Shan's group developed LSP-enhanced LEDs with different types of metals decoration (figure 4.20) [95]. It was found that the EL improvement strongly depended on the size and type of metals, and the selective enhancement at specific wavelengths which matched the absorption of metal nanostructures was generally observed. In earlier study, the researchers found graphene sheets presented a strong abnormal absorption in the violet-frequency region, probably resulting in a coupling between ZnO excitons and graphene SPs. Moreover, they observed substantially enhanced PL from hybrid structures of graphene/ZnO films with single-layer, bilayer and

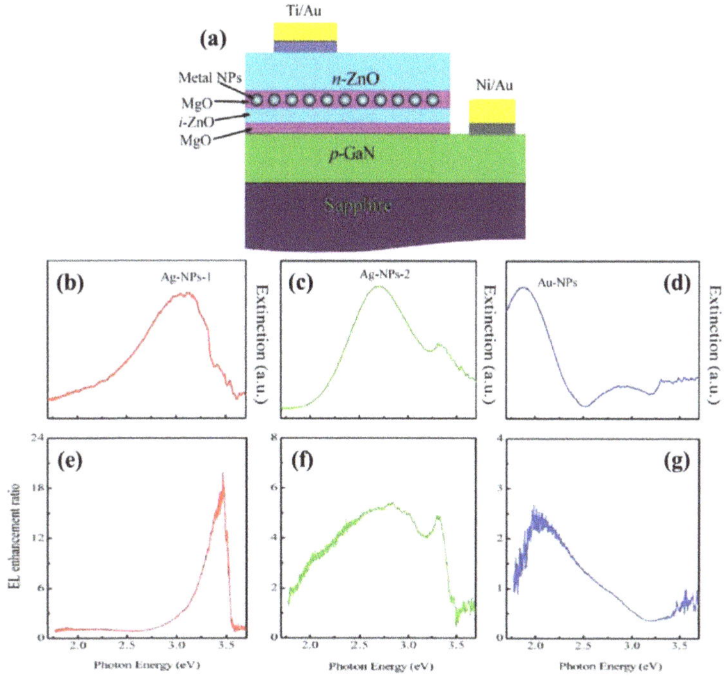

Figure 4.20. (a) Schematic diagram of n-ZnO/i-ZnO/MgO/p-GaN heterojunction device with metal NPs decoration. The extinction spectra of different metal NPs, including Ag NPs-1 (b), Ag NPs-2 (c), and Au NPs (d). The EL enhancement ratio of different metal NPs modified devices, including Ag NPs-1 (e), Ag NPs-2 (f) and Au NPs (g). Reproduced from [95] with permission of The Royal Society of Chemistry.

few-layer graphene incorporation [96]. How to reach an agreement between ZnO surface corrugation and dispersion relation of graphene SPs was the crucial factor for achieving UV luminescence enhancement. Furthermore, this PL enhancement depended on the thickness of graphene layer, annealing temperature of ZnO film, and PL measurement temperature (figure 4.21). The above results demonstrate that either metallic or nonmetallic SPs can effectively improve ZnO UV luminescence.

If both strategies are combined into device design and construction, more desirable improvement of ZnO UV emission and the device efficiency can be expected. Therefore, the cooperative effect of nanocrystallization and SPs will be discussed in this subsection. Based on the typical p-GaN/n-ZnO film heterojunction, we replaced the ZnO film with ZnO NRAs and redesigned the device structure to fabricate a nano-size LED device. Next, the Ag-NPs/PMMA composite prepared by solution synthesis was incorporated into the above heterojunction [97]. Four samples with different concentration of Ag-NP solution were prepared to investigate the effect of Ag-NP concentration on the luminescence performance. The PL measurements of these heterostructures (figure 4.22(b)), which consist of an excitonic emission and broad deep-level emission from ZnO, indicated enhanced ZnO UV emission after embedding the Ag-NPs/PMMA composites. The EL spectra, enhancement ratio and weight of each component analysis demonstrated that the

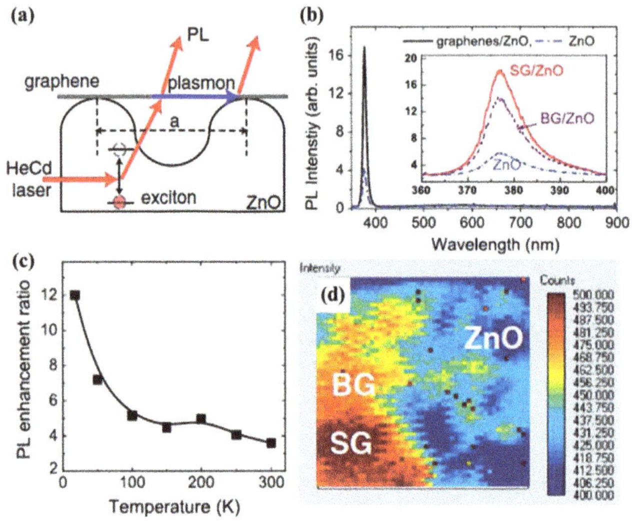

Figure 4.21. (a) Schematically described PL enhancement from a graphene/ZnO heterostructure. (b) PL spectra of graphene/ZnO hybrid structure and pure ZnO film. The inset shows PL spectra of single-layer and bilayer graphene-decorated ZnO films. (c) PL enhancement ratio as a function of measurement temperature. (d) The corresponding micro-PL mapping image. Reprinted figure with permission from [96]. Copyright (2010) by the American Physical Society.

UV EL component was selectively improved and the entire spectral line-shape changed (figures 4.22(c) and (d)). The spatial uniformity also improved and the line-width increased (figures 4.22(e) and (f)). These observations can be attributed to near-field optical coupling between Ag LSPs and ZnO excitons. This work provided a feasible way to improve device efficiency, and it is helpful for designing of lighting display.

The above work demonstrated the advantages of combining the two strategies. However, ZnO NRA LEDs usually suffer from efficiency and stability degradation. Coating the ZnO NRs with other materials, to form so-called 'core/shell' structures, can effectively solve this problem. Low Mg-content MgZnO possesses the same wurtzite structure as ZnO with small lattice mismatch, besides, good heteroepitaxial relationship and effective surface passivation is favorable for device design. MgZnO has a wider bandgap, which is beneficial for confinement of carriers in the core of the NRs. Moreover, the MgZnO shell can act as a spacer layer, which is favorable for producing the LSP-induced field enhancement effect. Therefore, LSP-enhanced waveguide-type UV LED was designed and fabricated [81]. Two dynamical processes were considered to understand significantly selective enhancement of ZnO UV EL (figure 4.23). Besides, an ultrathin SiO_2 coating layer can be easily prepared under mild conditions. This process allows a layer of dense silica to grow on the surface of the coated material, and the thickness of this layer can be precisely controlled. This method inspired us to grow a conformal silica shell on ZnO NRs in a stable solution system in order to develop a type of UV LED device. We constructed LSP-improved LED by spin-coating Ag-NPs onto ZnO/SiO_2 CS

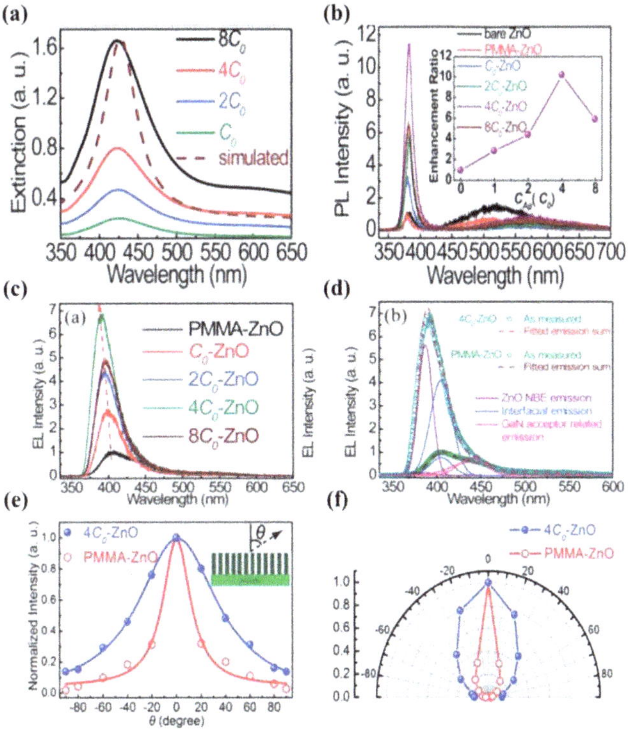

Figure 4.22. (a) Experimentally measured and theoretically simulated extinction spectra of the Ag-NPs/PMMA composites. (b) PL spectra of bare and embedded ZnO NRAs with different Ag concentration. (c) EL spectra of the NRA LEDs with different Ag concentration. (d) Gaussian deconvolution analyses of the near-UV EL spectra for both LED devices embedded in pure PMMA and the Ag-NPs/PMMA composite. (e), (f) The angle-dependence of the integrated EL intensity for LEDs with and without Ag-NPs in rectangular coordinate and polar coordinate systems. Reproduced from [97] with permission of The Royal Society of Chemistry.

NRAs/p-GaN structures with a low-cost method for performing the whole process in solution (figure 4.24(a)) [98]. Varying degrees of PL and EL enhancement was observed (figures 4.24(b) and (c)). The simulated results revealed that the higher EL enhancement originated from two contributions: IQE improvement induced by exciton-LSP coupling and light-extraction-efficiency (LEE) improvement aroused by photon-LSP coupling (figures 4.24(d)–(f)).

The other work, described previously in the subsection of surface/interfacial engineering and modification, was performed in the ZnO-based Q-LED system [87]. Ag LSPs were employed by spin-coating Ag@ZnO onto a PAIT Q-LED (figure 4.25(d)). This layer, whose extinction spectra were regulated by controlling the reaction time with good electrical conductivity (figures 4.25(a)–(c)). Compared with the pristine PAIT Q-LED, the Ag@PAIT Q-LED showed greater UV EL enhancement confirmed by exciton-LSP resonant coupling effect as shown in figure 4.25(e). This work demonstrated that the incorporation of ZnO

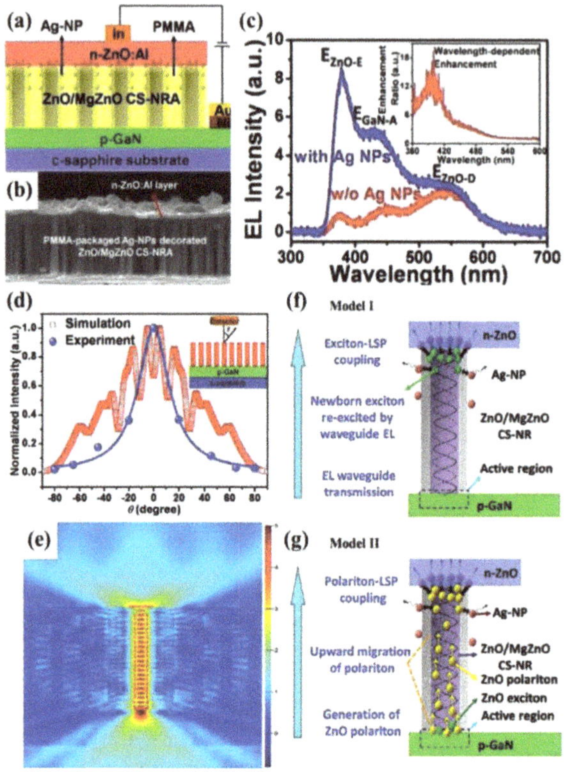

Figure 4.23. (a), (b) Structural schematic diagram and a typical cross-sectional SEM image of the Ag-NPs decorated CS-NRA LED device. (c) EL spectra of the devices with and without Ag-NPs. The inset shows the dependence of the EL enhancement ratio on the emission wavelength. (d) The measured and simulated angle-dependence of the EL intensity for a CS-NRA LED. (e) FDTD simulation of the EL intensity spatial distribution from a single NR optical waveguide. (f), (g) Two models showing the dynamic process of interactions among excitons, photons, and LSPs in an Ag-NPs decorated CS-NR. Reproduced from [81] with permission of The Royal Society of Chemistry.

nanostructures and metal LSPs is an effective strategy to further improve the device efficiency, which can also be used for other types of devices for performance improvement.

Recently, graphitic carbon materials have been found to present clear SP resonance behavior owing to the induced collective oscillations of π and $\pi + \sigma$ electrons in these carbon materials with Dirac-cone-like energy band alignments. Furthermore, graphitic carbon SPs have the advantages of low optical loss and continuously adjustable resonant absorption wavelength, which are of great scientific significance and economic value for developing carbon-based SP improved LEDs. In this regard, we have performed two types of studies. First, a SP-enhanced near-UV light-emitting device was fabricated by incorporating CDs into p-GaN/i-Al_2O_3/n-ZnO Q-LEDs and optimizing the molar ratio of ZnO QDs to CDs (figure 4.26(a)) [82]. The corresponding SP absorption wavelength varied in different experimental dielectric environments (figure 4.26(b)). These broad-range spectra

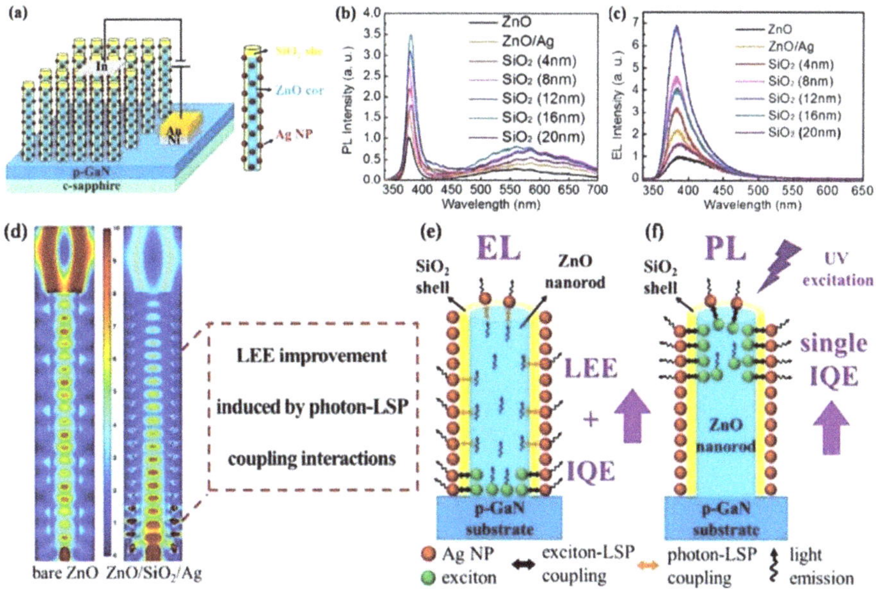

Figure 4.24. (a) Structural diagram of LSP-enhanced ZnO/SiO_2 core/shell NRAs LED. (b) PL spectra of bare ZnO NRAs and Ag-NPs decorated ZnO/SiO_2 core/shell NRAs with different SiO_2 layer thicknesses. (c) EL spectra of bare ZnO NRAs and Ag-NPs decorated core/shell NRAs with different SiO_2 shell thicknesses. (d) FDTD simulations of the EL propagation behaviors in bare ZnO nanorod and Ag-NPs decorated core/shell nanorod. The latter clearly shows the LEE improvement induced by photon-LSP coupling interaction. (e) Model illustrating the EL (IQE improvement + LEE improvement) enhancement mechanisms. (f) Model illustrating the PL enhancement mechanisms (IQE improvement). Reproduced with permission from [98]. Copyright (2016) American Chemical Society.

overlapped well with the excitonic emission of ZnO QDs, indicating the possibility of luminescence improvement. More importantly, a significant enhancement of the near-UV PL (figure 4.26(c)) and EL (figure 4.26(d)) intensities from the ZnO-based Q-LEDs was achieved. Time-resolved spectroscopy revealed that the luminescence enhancement resulted from resonant coupling interaction (figures 4.26(e) and (f)). Second, some favorable factors have shown the potential of graphene to generate SP resonant coupling, which is now considered to be a more promising plasmonic material than metals. A hybrid three-dimensional structure of one-dimensional ZnO NRAs and 2D graphene plasmon layer were fabricated on a p-GaN substrate (figure 4.27(a)) [99]. By an effective method, the graphene layer was directly embedded into the active region, as well as the heterointerface. Similar near-UV improvement of PL and EL was observed as shown in figures 4.27(b) and (c). Much direct and indirect evidence also suggested that the coupling interaction between ZnO excitons and graphene SPs was the cause of the emission enhancement (figures 4.27(d)–(f)).

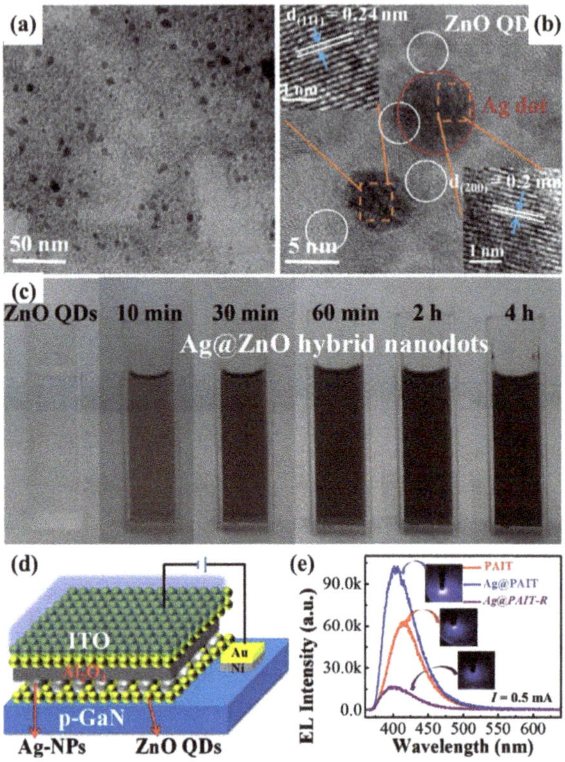

Figure 4.25. (a), (b) Typical TEM and HRTEM images of the prepared Ag@ZnO hybrid nanodots. (c) Digital photographs of the pure ZnO QDs and Ag@ZnO hybrid nanodots solution with different reaction time. (d) Structure schematic diagram of the Ag@PAIT-R Q-LED. (e) EL spectra of the three types of PAIT Q-LEDs. [87] John Wiley & Sons. © 2017 WILEY-VCH Verlag GmbH & Co., KGaA, Weinheim.

4.3.4 Emerging composite heterojunctions

In recent years, the emerging materials have attracted great interest. As a representative of emerging semiconductor material, lead halide perovskite (LHP) materials have attracted considerable attention because of their excellent material properties and fascinating electro-optical properties. There are two main groups of LHPs. The first is organic-LHP $APbX_3$ ($A = CH_3NH_3$ or CHN_2H_4, X = Cl, Br and I). The second is all-inorganic LHP materials, such as $CsPbX_3$ (X = Cl, Br and I). Both material systems have rapidly developed and frequently used to prepare solar cells with high power-conversion-efficiency over 20% and LED devices with high external-quantum-efficiency (EQE) over 14%. Another representative type of emerging materials is transition metal dichalcogenides (TMDs), which are a class of 2D layered semiconductors, have stimulated research upsurge around the world. As the thickness decreases to a monolayer, some indirect-band-gap TMDs become direct-band-gap semiconductors. Thus, various optoelectronic devices can be fabricated based on these monolayer TMDs for multiple purposes. In combination with the advantages of ZnO-based UV light-emitting materials and devices,

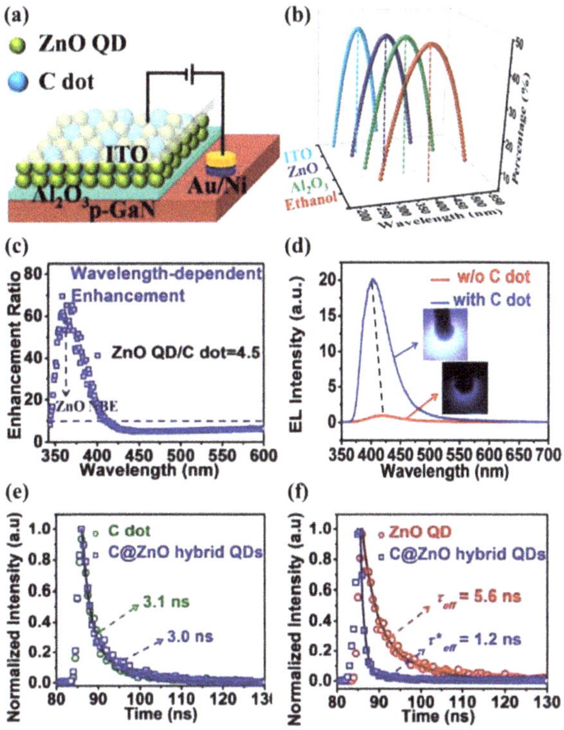

Figure 4.26. (a) Structure schematic diagram of the carbon SP-enhanced ZnO Q-LED. (b) Calculated SP resonance curves of these CDs in four different dielectric environments. (c) The dependence of the PL enhancement ratio on emission wavelength. (d) EL spectra of the two devices with and without CD decoration. (e) TR-PL spectra of the pure CD solution and CD/ZnO QD mixed solution. The monitoring wavelength is set at the CD fluorescence wavelength of 440 nm. (f) TR-PL spectra of the pure ZnO QDs and C@ZnO hybrid QDs. The monitoring wavelength is 360 nm. Reproduced from [82] with permission of The Royal Society of Chemistry.

ZnO-based emerging composite heterostructures are expected to generate interestingly novel phenomena and promote more meaningful applications.

All-inorganic $CsPbX_3$ perovskites have excellent environmental and thermal stability. This means that they are suitable for carrier-injection type device. In particular, their PL emissions show a regular blueshift (or redshift) as $CsPbBr_3$ gradually changes to $CsPbCl_3$ (or $CsPbI_3$). A LED with a p-GaN/n-ZnO heterojunction as a blue–violet excitation chip and $CsPbBr_3$ nanocrystals as green phosphors has been fabricated [100]. First, a simple and effective strategy of solution synthesis was performed to prepare all-inorganic $CsPbX_3$ perovskites. Characterization and luminescence mechanism of as-prepared all-inorganic $CsPbX_3$ indicated that the stable green emission originated from excitonic radiative recombination of the $CsPbBr_3$ nanocrystals compared with commercially available perovskite (figures 4.28(a)–(c)). Next, to demonstrate the application prospects of these $CsPbBr_3$ nanocrystals, a proof-of-concept LED was fabricated by covering the synthesized perovskite nanocrystals on the excitation chip. Near-white light

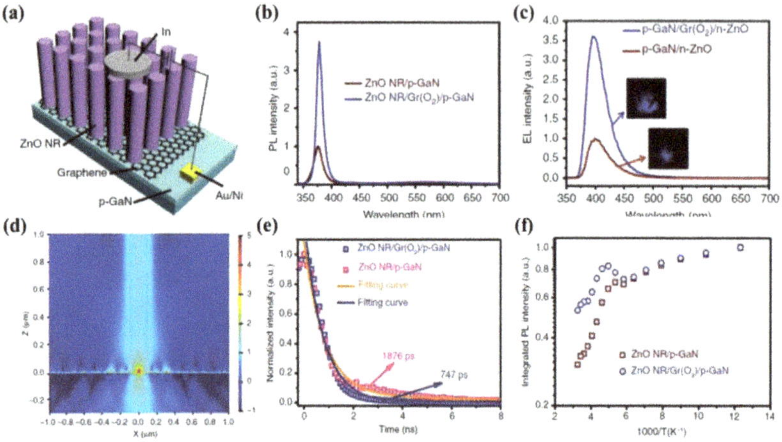

Figure 4.27. (a) Device structure diagram of graphene-decorated ZnO NRAs/p-GaN heterojunction. (b) PL measurement of ZnO NRAs/p-GaN without and with graphene plasmon layer decoration. (c) EL measurement of light-emitting devices without and with a graphene plasmon layer. (d) FDTD simulations of electric field intensity spatial distributions at a ZnO NRAs/p-GaN interface with graphene plasmon layer decoration. (e) TR-PL measurements of both heterojunctions. (f) Arrhenius plots of the normalized intensity from ZnO NRAs without and with graphene plasmon layer decoration. [99] Reproduced with permission © De Gruyter.

emission with chromaticity coordinates of (0.309, 0.323), luminance of 267 cd m^{-2}, current efficiency of 0.17 cd A^{-1}, and a maximum EQE close to 0.042% were achieved (figures 4.28(d)–(g)). Although further optimization of the device structure and performance is still needed, the present work demonstrated a feasible strategy to prepare all-inorganic perovskite and their potential application in light-emitting devices.

For other emerging materials to form better heterostructure properties with ZnO, the critical factor for developing practical and reliable optoelectronic devices is achieving effective control of their charge carrier polarity and doping in TMDs. Doping is the most effective method for changing the properties of traditional semiconductors, and it has also been employed to adjust the device performance of 2D TMDs. By forming heterostructures with different chemical compositions or physical dimensionalities, charge transfer can occur, which causes the so-called 'n-type' or 'p-type' doping effect and alters the optoelectronic properties of these monolayer TMDs. A hybrid heterostructure of 2D and zero-dimensional (0D) materials can act as an excellent platform for discovering novel physics phenomena or exploring the intrinsic mechanisms of those studied materials. Therefore, the novel 2D/0D hybrid heterostructures with easily adjustable fluorescence intensity and carrier concentration have been produced by decorating monolayer MoS_2 (1L-MoS_2) flakes with semiconductor-dots (carbon-dots or ZnO-dots) [101]. By carbon-dot decoration, the fluorescence intensity of 1L-MoS_2 was significantly suppressed owing to the n-type doping effect of electron transfer from the carbon-dots to 1L-MoS_2 (see the energy band alignment and PL measurements in figures 4.29(a)–(g)). In contrast, 1L-MoS_2 decorated with ZnO-dots exhibited

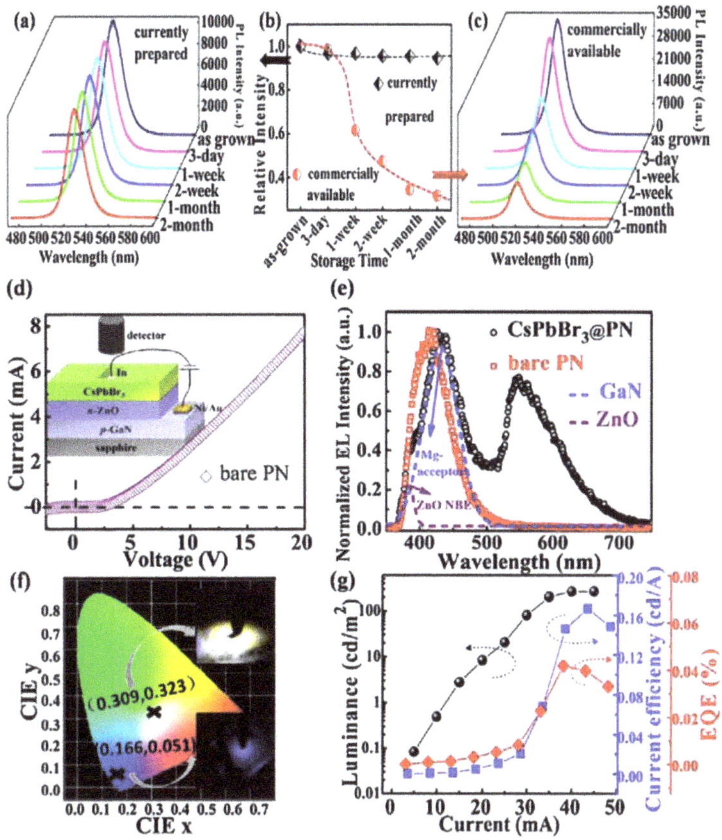

Figure 4.28. PL spectra of the currently prepared (a) and commercially available (c) $CsPbBr_3$ nanocrystals collected during a two-month storage period. (b) Extracted relative intensity of the two comparative samples for a better clarity. (d) *I–V* curve of the p-GaN/n-ZnO excitation chip. (e) EL spectra of the bare P–N chip and $CsPbBr_3$@P–N LED. (f) CIE chromaticity coordinates of the two EL spectra. (g) Plots of luminance, current efficiency and EQE as a function of injection current. Reprinted from [100], Copyright (2018), with permission from Elsevier.

remarkably improved PL because of the effective p-type doping modulation of electron transfer from 1L-MoS_2 to the ZnO-dots (see the energy band alignment and PL measurements in figures 4.29(h)–(n)). This study proposed an effective strategy for adjusting the electronic doping and optical properties of 2D semiconductors, which will pave the way for developing new optoelectronic devices based on these new emerging 2D materials and heterostructures.

4.4 MIS heterojunction LDs

Considering the difficulty in p-type doping of ZnO and its homojunction devices, our research group has investigated the efficiency and stability of ZnO-based P–N and P–I–N heterojunction LED devices, and proposed some improved strategies and methods. Among the heterostructures, MIS junctions are widely used to

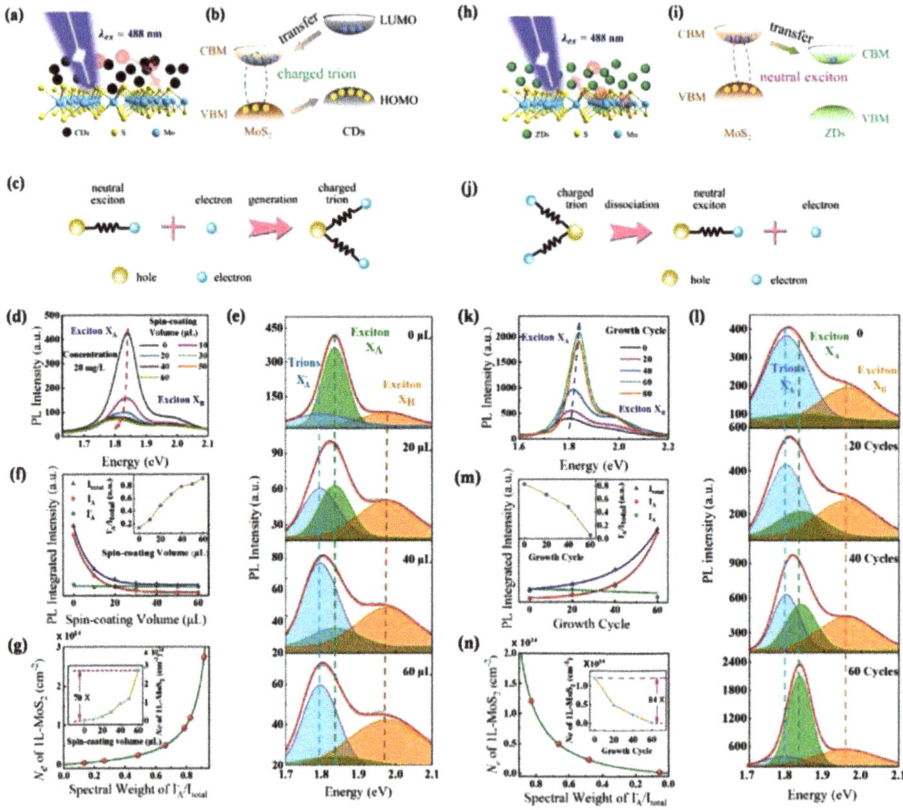

Figure 4.29. Schematic structural diagram (a) and energy band alignment (b) of the 1L-MoS_2@CD heterostructure. (c) Sketch illustrating the formation process of a negatively-charged trion. (d) PL spectra of the bare 1L-MoS_2 and 1L-MoS_2@CD heterostructures with varied spin-coating volumes of CD solution. (e) Gaussian deconvolution fittings of these PL spectra. (f) PL integrated intensity of each luminescence component at different loading amounts of CDs. (g) Calculated Ne data as a function of the I_{A-}/I_{total} value. The inset shows the variation tendency of Ne with the spin-coating volume of CDs. Schematic structural diagram (h) and energy band alignment (i) of the 1L-MoS_2@ZD heterostructure. (j) Sketch illustrating the dissociation process of a negatively-charged trion. (k) PL spectra of the bare 1L-MoS_2 and 1L-MoS_2@ZD heterostructures with varied growth cycles of the ZD layer. (l) Gaussian deconvolution fittings of these PL spectra. (m) PL integrated intensity of each luminescence component at different ZnO growth cycles. (n) Calculated Ne data as a function of the I_{A-}/I_{total} value. The inset shows the variation tendency of Ne with the ZnO growth cycle number. Reproduced from [101] with permission of The Royal Society of Chemistry.

construct LEDs and LDs for wide-band gap semiconductors because their EL mechanism is based on the so-called 'impact-ionization' effect, which has no rigid requirements for the p-type counterparts. Besides, the emission wavelength of this type of heterostructure does not suffer from a redshift caused by p-type materials. In other words, the MIS structure can more easily obtain the intrinsic emission of these active luminescent materials. For the working principle of MIS heterojunction, the typical Au/MgO/ZnO MIS structure is taken as an example. Because of the large conduction band offset between ZnO and MgO in the energy band alignment (~3.55 eV), electrons will be blocked and accumulated at the heterointerface. Most applied voltages will drop on the MgO layer, and the local electric field strength could be as high as 10^7 to 10^8 V m^{-1}. T^{-1} h^{-1} u^{-1} s^{-1}, electrons and holes can be generated by the 'impact-ionization' process in the insulating MgO layer. The generated holes will be driven into the ZnO active layer and radiatively recombine with the electrons accumulated at the ZnO/MgO interface, resulting in the UV emission of ZnO. Furthermore, when the population inversion condition is satisfied and the optical gain is larger than the loss through multiple scatterings, the lasing behavior dominates EL emission.

In 2006, Yang's group developed ZnO-based MIS heterojunction LEDs, and pure UV EL was first realized in Au/SiO_2/ZnO structure (figures 4.30(a) and (b)) [68].

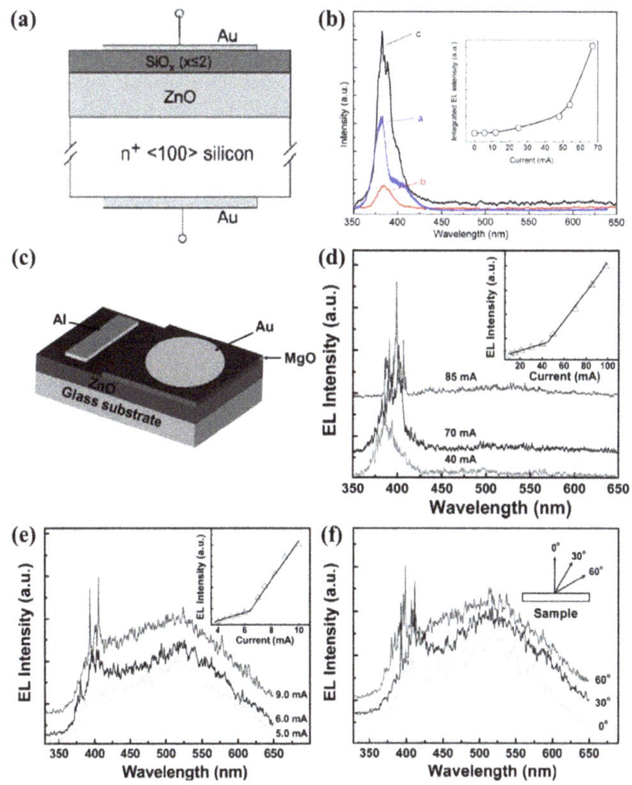

Figure 4.30. (a) Schematic diagram of Au/SiO_X/ZnO MIS heterojunction. (b) PL spectrum of ZnO film (curve a) and EL spectra of the ZnO-based MIS heterojunction LED under different injection currents (curves b and c). The inset shows EL integrated intensity as a function of injection current. Reprinted from [68], with the permission of AIP Publishing. (c) Schematic illustration of Au/MgO/ZnO heterojunction device. (d) EL spectra of MIS heterojunction under different injection current. The inset shows the dependence of EL integrated intensity on injection current. (e) EL spectra of Au/i-ZnO/MgO/ZnO heterojunction device under different injection current. The inset shows the dependence of EL integrated intensity on injection current. (f) Lasing spectra of Au/i-ZnO/MgO/ZnO heterojunction at different measurement angle. [70] John Wiley & Sons. © 2010 WILEY-VCH Verlag GmbH & Co., KGaA, Weinheim.

Study have shown that the thickness of SiO_2 can affect the performance of carrier injection and electron blocking. In addition, they predicted that the EL performance of this MIS junction was likely to depend on the optical and electrical properties of the ZnO film. Since then, many projects have realized UV spontaneous and stimulated emission, but the low emission efficiency and high laser threshold have hindered the further development of ZnO-based MIS junction. Until 2010, Shan's group fabricated Au/MgO/ZnO MIS heterostructure (figure 4.30(c)) [70]. The pure ZnO excitonic spontaneous emission and electrically pumped random lasers had been demonstrated, and a threshold current of 43 mA could be derived (figure 4.30(d)). Actually, relatively wide bandgap of MgO is not conducive to increase the generation efficiency of holes. Therefore, they believed that introducing narrow bandgap of an insulating ZnO layer is helpful to the generation of holes. Understandably, the Au/i-ZnO/MgO/ZnO structure was used for further study. It was found that the threshold current of this LD device could be reduced to 6.5 mA and the laser mode is still random (figures 4.30(e) and (f)). This study provides a simple route to achieve low threshold lasers, and promises a significant step towards future applications of this type of lasers. Thus, in current research of ZnO-based LEDs/LDs at the present stage, MIS heterojunctions are still of great importance and practical significance. In this subsection, we will summarize our recent efforts on the characteristics of ZnO-based MIS heterojunction devices.

We first fabricated a device based on ZnO/MgO CS NWs [102]. Microscopic images (figures 4.31(a)–(e)) revealed that the MgO layer was not only covered on the top of ZnO NWs to form a quasi-continuous film, but it was also coated on their side wall to form a CS structure. A set of EL spectra of the CS NW LD demonstrate a process from spontaneous to stimulated radiation in ZnO, and the lasing threshold current density was fixed at 2.3 A cm^{-2} (figure 4.31(f)). It is worth noting that compared with the planar device, the CS NW LD showed higher emission efficiency (figure 4.31(g)). This type of lasing behavior was random, which was demonstrated by the random changes about the position and intensity of those sharp peaks, and the mode spacing was not uniform, which supports the intrinsic properties of random laser action (figures 4.31(h) and (i)). The improved device performance suggested that this MIS junction can act as good building blocks for high-efficiency light-emitting devices.

We further improved and optimized the performance of the typical MIS heterostructure in three aspects: DUV emission, heat dissipation and the hole injection efficiency. First, a kind of Au/MgO/$Mg_xZn_{1-x}O$ MIS heterojunctions for DUV emission was fabricated [103]. Based on the energy band, impact-ionization process and disordered optical structure, the EL mechanism and laser forming process were demonstrated. The EL spectra of pure ZnO and low Mg-content MgZnO devices, showed single excitonic emission. By contrast, in the MgZnO devices, the emission came from self-formed Zn-rich MgZnO clusters, which were verified to be thermally stable luminescence centers (figures 4.32(a)–(d)). Such a MIS junction is usually fabricated on Si, Al_2O_3, SiC and GaN substrates. However, their low thermal conductivity and high transparency result in some negative effects, such as poor heat dissipation and large light loss in the direction of the substrate. Thus, we transferred the Au/MgO/ZnO MIS heterojunction to a metal Cu substrate, and

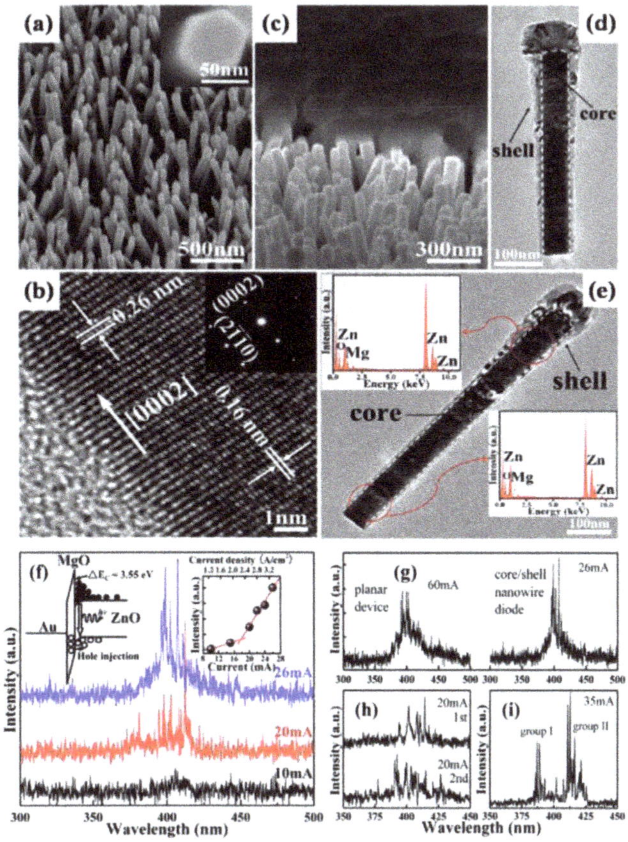

Figure 4.31. (a), (c) Tilted-view SEM images of as-grown ZnO NWs and NWs covered by MgO. (b) HRTEM image and SAED pattern of an individual NW. (d), (e) TEM images of ZnO/MgO CS-NWs and two EDX spectra obtained from upper and lower regions of a single NW. (f) EL spectra of the CS-NW LD under different injection currents. The left and right insets exhibit an energy band alignment under forward bias and the super-linear dependence of integrated emission intensity on injection current. (g) Two lasing spectra from our NW and film LDs. (h) The lasing spectra of the CS-NW LD obtained in two successive measurements. (i) A representative lasing spectrum showing that the emission wavelength is separated into two different regions. Reprinted from [102], with the permission of AIP Publishing.

demonstrated the feasibility of directly preparing ZnO UV LDs [104]. The high thermal and electrical conductivity of the Cu substrate increased the heat dissipation and decreased the series resistance, resulting in lower thermal effect and higher emission wavelength stability (figures 4.32(e)–(h)). Finally, considering the problem of the low intrinsic hole–carrier concentration in such a MIS junction, a thermally oxidized Cu_2O hole injection layer was introduced into the Au/MgO/ZnO structure, and improved UV EL was achieved [105]. The improvement mechanism was ascribed to effective hole transfer from p-Cu_2O to i-MgO, which increases the initial carrier concentration and further promotes the 'impact-ionization' effect to induce carrier generation and injection (figures 4.32(i)–(m)). This study proposed a new route to improve device performance of MIS junction by introducing extrinsic hole suppliers.

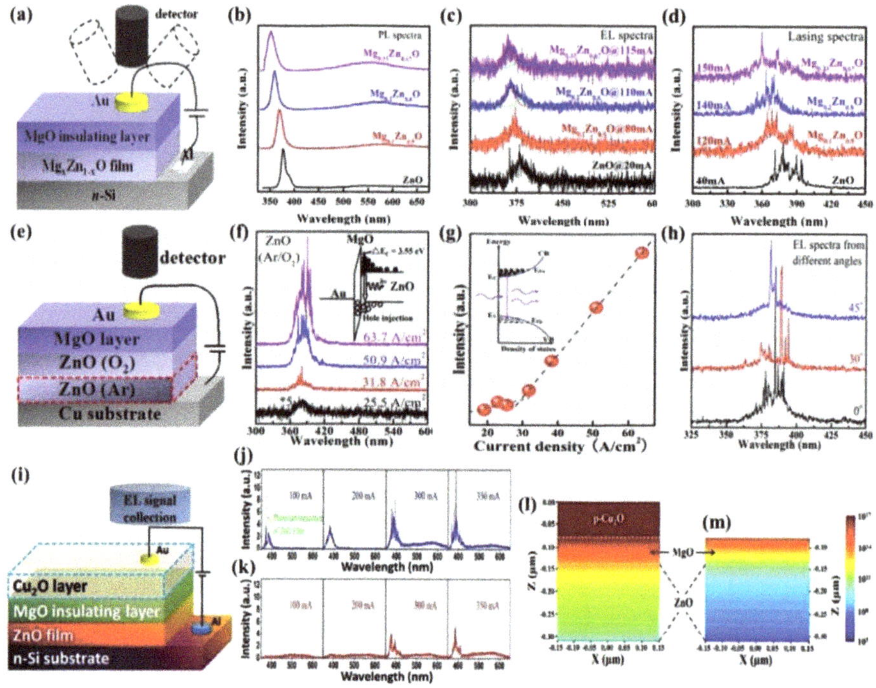

Figure 4.32. (a) Structure diagram of Au/MgO/$Mg_xZn_{1-x}O$ MIS diodes. (b) PL spectra of $Mg_xZn_{1-x}O$ films. (c) and (d) EL spectra and lasing spectra of $Mg_xZn_{1-x}O$ MIS diodes. Reprinted from [103], Copyright (2014), with permission from Elsevier. Elsevier. (e) A structure diagram of Au/MgO/ZnO MIS-heterojunction LDs on the Cu substrate. (f) EL spectra of ZnO LDs with an anti-oxidation layer under different injection currents. (g) Super-linear dependence of integrated EL intensity on the injection current density, the inset illustrates the process of population inversion and stimulated emission. (h) The lasing spectra of ZnO LDs on the Cu substrate taken from different angles. Reproduced from [104]. © 2014 Optical Society of America. (i) Structural schematic diagram of Au/p-Cu_2O/MgO/ZnO heterostructure. (j), (k) EL spectra of the Au/MgO/ZnO heterostructures with and without p-Cu_2O layer under different injection currents. (l), (m) Simulated distribution profiles of hole concentration in the two LEDs with and without p-Cu_2O layer respectively, which clearly show that the hole concentrations in MgO and ZnO layers are remarkably improved by introducing p-Cu_2O hole injection layer. Reproduced with permission from [105]. Copyright (2016) American Chemical Society.

4.5 Prospects and challenges

Like the initial analysis, some effective improvement strategies and methods have been proposed for the critical issues of the device efficiency and stability in ZnO-based heterojunction LEDs. Some interesting physical phenomena and fundamental applications of such P–N, P–I–N, and MIS heterojunctions have demonstrated the great advantages of ZnO in the field of UV light-emitting materials and devices. However, the following challenges still remain to realize further applications of short-wavelength optoelectronic devices. (1) Although a lot of efforts have been devoted to regulate and improve these P–N and P–I–N heterojunctions, the structure and performance of the device can be further optimized. (2) In MIS heterojunction devices, three aspects of the performance improvement show a positive trend to develop UV LEDs and LDs, but the low device efficiency and high threshold still restrict their applications of UV emitter, detector and laser. (3) In terms of light-emitting applications, the EQE of ZnO-based heterojunction LEDs and LDs is still low, resulting in great difficulty in connecting with real life.

Overall, our efforts on optimization and improvement of ZnO-based heterojunction devices are worthy of recognition and deliberation. That is, ZnO-based materials and devices still show considerable promise for achieving UV detection, emission, and storage. We believe that short-wavelength light-emitting devices and photodetectors will continue to attract attention for a wide range of optoelectronic applications. In particular, some recent efforts to achieve multifunctional expansion of ZnO-based materials and devices are indeed promising (figure 4.33), including electrically driven random laser memory [9], 3D-printed flexible photodetector [20],

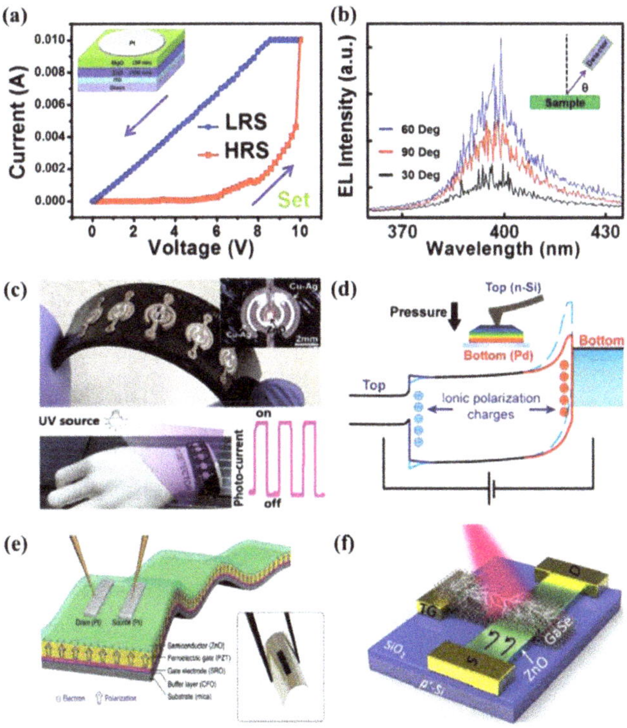

Figure 4.33. (a) The high and low resistance characteristics of Pt/MgO/ZnO MIS heterojunction. The inset shows schematic illustration of such heterojunction device. (b) Lasing spectra of Pt/MgO/ZnO heterojunction at different measurement angle. [9] John Wiley & Sons. © 2015 WILEY-VCH Verlag GmbH & Co., KGaA, Weinheim. (c) Top view and bending status images of 3D-printed UV photodetector (upper), and photosensitivity under UV irradiation (lower). Reproduced with permission from [20]. Copyright (2020) American Chemical Society. (d) Band diagram of piezotronic behavior in a 2D piezotronic transistor device. (e) The configuration and microstructure of flexible all-inorganic ferroelectric field effect transistors. (f) Structural diagram of light-driven WSe_2-ZnO heterojunction field-effect transistors. [41] John Wiley & Sons. © 2017 WILEY-VCH Verlag GmbH & Co., KGaA, Weinheim. [106, 107] John Wiley & Sons. © 2017 WILEY-VCH Verlag GmbH & Co., KGaA, Weinheim.

ultrasensitive 2D piezotronic transistor [41], flexible ferroelectric thin film transistor [106], and light-driven field-effect transistors for high-performance photodetection [107]. The cross integration of these emerging fields will certainly bring a new upsurge to further development of ZnO-based materials and devices.

References

[1] Bia S, Lia Q, He Z, Guo Q, Yeboah K A-, Liu Y and Jiang C 2019 Highly enhanced performance of integrated piezo photo-transistor with dual inverted OLED gate and nanowire array channel *Nano Energy* **66** 104101

[2] Chung K, Lee C-H and Yi G-C 2010 Transferable GaN layers grown on ZnO-coated graphene layers for optoelectronic devices *Science* **330** 655–7

[3] Zhao W, Xiong X, Han Y, Wen L, Zou Z, Luo S, Li H, Su J, Zhai T and Gao Y 2017 Fe-doped p-ZnO nanostructures/n-GaN heterojunction for 'Blue-Free' orange light-emitting diodes *Adv. Opt. Mater.* **5** 1700146

[4] Son D I, Kwon B W, Park D H, Seo W-S, Yi Y, Angadi B, Lee C-L and Choi W K 2012 Emissive ZnO-graphene quantum dots for white-light-emitting diodes *Nat. Nanotechnol.* **7** 465–71

[5] Lee C-H, Kim Y-J, Hong Y J, Jeon S-R, Bae S, Hong B H and Yi G-C 2011 Flexible inorganic nanostructure light-emitting diodes fabricated on graphene films *Adv. Mater.* **23** 4614–9

[6] Qian L, Zheng Y, Xue J and Holloway P H 2011 Stable and efficient quantum-dot light-emitting diodes based on solution-processed multilayer structures *Nat. Photon.* **5** 543–8

[7] Shi Z-F, Zhang Y-T, Cui X-J, Zhuang S-W, Wu B, Chu X-W, Dong X, Zhang B-L and Du G-T 2015 Photoluminescence performance enhancement of ZnO/MgO heterostructured nanowires and their applications in ultraviolet laser diodes *Phys. Chem. Chem. Phys.* **17** 13813–20

[8] Qin F F, Xu C X, Zhu Q X, Lu J F, Chen F, You D T, Zhua Z and Manohari A G 2018 Optical performance improvement in hydrothermal ZnO/graphene structures for ultraviolet lasing *J. Mater. Chem.* C **6** 3240–4

[9] Wang C-S, Nieh C-H, Lin T-Y and Chen Y-F 2015 Electrically driven random laser memory *Adv. Funct. Mater.* **25** 4058–63

[10] Fallert J, Dietz R J B, Sartor J, Schneider D, Klingshirn C and Kalt H 2009 Co-existence of strongly and weakly localized random laser modes *Nat. Photon.* **3** 279–82

[11] Gargas D J, Moore M C, Ni A, Chang S-W, Zhang Z, Chuang S-L and Yang P 2010 Whispering gallery mode lasing from zinc oxide hexagonal nanodisks *ACS Nano* **4** 3270–6

[12] Lu Y-J, Shan C-X, Jiang M-M, Hu G-C, Zhang N, Wang S-P, Li B-H and Shen D-Z 2015 Random lasing realized in n-ZnO/p-MgZnO core–shell nanowire heterostructures *CrystEngComm* **17** 3917–22

[13] Chen R, Ling B, Sun X W and Sun H D 2011 Room temperature excitonic whispering gallery mode lasing from high-quality hexagonal ZnO microdisks *Adv. Mater.* **23** 2199–204

[14] Huang M H, Mao S, Feick H, Yan H, Wu Y, Kind H, Weber E, Russo R and Yang P 2001 Room-temperature ultraviolet nanowire nanolasers *Science* **292** 1897–9

[15] Huang J, Chu S, Kong J, Zhang L, Schwarz C M, Wang G, Chernyak L, Chen Z and Liu J 2013 ZnO p-n homojunction random laser diode based on nitrogen-doped p-type nanowires *Adv. Opt. Mater.* **1** 179–85

[16] Bie Y-Q, Liao Z-M, Zhang H-Z, Li G-R, Ye Y, Zhou Y-B, Xu J, Qin Z-X, Dai L and Yu D-P 2011 Self-powered, ultrafast, visible-blind UV detection and optical logical operation based on ZnO/GaN nanoscale p-n junctions *Adv. Mater.* **23** 649–53
[17] Huang B-R, Saravanan A and Lu H-C 2020 Structural engineering of dispersed graphene flakes into ZnO nanotubes on discontinues ultra-nanocrystalline diamond substrates for high-performance photodetector with excellent UV light to dark current ratios *Adv. Mater. Interfaces* **7** 1901694
[18] Zhang Y, Hu M and Wang Z 2020 Enhanced performances of p-si/n-ZnO self-powered photodetector by interface state modification and pyro-phototronic effect *Nano Energy* **71** 104630
[19] Zhang Z, Wang K, Zheng K, Deng S, Xu N and Chen J 2018 Electron bombardment induced photoconductivity and high gain in a flat panel photodetector based on a ZnS photoconductor and ZnO nanowire field emitters *ACS Photonics* **5** 4147–55
[20] Lee D, Seol M-L, Motilal G, Kim B, Moon D-I, Han J-W and Meyyappan M 2020 All 3D-printed flexible ZnO UV photodetector on an ultraflat substrate *ACS Sens.* **5** 1028–32
[21] Hu M, Teng F, Chen H, Jiang M, Gu Y, Lu H, Hu L and Fang X 2017 Novel Ω-shaped core–shell photodetector with high ultraviolet selectivity and enhanced responsivity *Adv. Funct. Mater.* **27** 1704477
[22] Nasiri N, Bo R, Wang F, Fu L and Tricoli A 2015 Ultraporous electron-depleted ZnO nanoparticle networks for highly sensitive portable visible-blind UV photodetectors *Adv. Mater.* **27** 4336–43
[23] Ahn S, Chen W, M-Gonzalez M A, Lockett M, Wang J and Mena O V 2020 Enhanced charge transfer and responsivity in hybrid quantum dot/graphene photodetectors using ZnO as intermediate electron-collecting layer *Adv. Electron. Mater.* **6** 2000014
[24] Lee Y T *et al* 2017 Mixed-dimensional 1D ZnO-2D WSe_2 van der Waals heterojunction device for photosensors *Adv. Funct. Mater.* **27** 1703822
[25] Drobek M, Kim J-H, Bechelany M, Vallicari C, Julbe A and Kim S S 2016 MOF-based membrane encapsulated ZnO nanowires for enhanced gas sensor selectivity *ACS Appl. Mater. Interfaces* **8** 8323–28
[26] Thongma S, Tantisantisom K, Grisdanurak N and Boonkoom T 2019 UV enhanced white-light response based on p-Si/n-ZnO nanorod heterojunction photosensor *Sens. Actuators A: Phys.* **296** 324–30
[27] Kim Y, Kim J, Kim H-M and Jang J 2020 Quantum-dots photosensor with wide bandgap P-type and N-type oxide semiconductors for high detectivity and responsivity *Adv. Mater. Technol.* **5** 1900857
[28] Tian H, Fan H, Li M and Ma L 2016 Zeolitic imidazolate framework coated ZnO nanorods as molecular sieving to improve selectivity of formaldehyde gas sensor *ACS Sens.* **1** 243–50
[29] Dhamodharan P, Manoharan C, Bououdina M, Venkadachalapathy R and Ramalingam S 2017 Al-doped ZnO thin films grown onto ITO substrates as photoanode in dye sensitized solar cell *Sol. Energy* **141** 127–44
[30] Qiao S, Liu J, Fu G, Ren K, Li Z, Wang S and Pan C 2018 ZnO nanowire based CIGS solar cell and its efficiency enhancement by the piezo-phototronic effect *Nano Energy* **49** 508–14
[31] Chuang C-H M, Maurano A, Brandt R E, Hwang G W, Jean J, Buonassisi T, Bulovic V and Bawendi M G 2015 Open-circuit voltage deficit, radiative sub-bandgap states, and prospects in quantum dot solar cells *Nano Lett.* **15** 3286–94

[32] Kim Y *et al* 2019 A facet-specific quantum dot passivation strategy for colloid management and efficient infrared photovoltaics *Adv. Mater.* **31** 1805580
[33] Chuang C-H M, Brown P R, Bulović V and Bawendi M G 2014 Improved performance and stability in quantum dot solar cells through band alignment engineering *Nat. Mater.* **13** 796–801
[34] Xu J *et al* 2018 2D matrix engineering for homogeneous quantum dot coupling in photovoltaic solids *Nat. Nanotechnol.* **13** 456–63
[35] Eisner F *et al* 2018 Solution-processed In_2O_3/ZnO heterojunction electron transport layers for efficient organic bulk heterojunction and inorganic colloidal quantum-dot solar cells *Sol. RRL* **2** 1800076
[36] Wang L *et al* 2018 Ultrathin piezotronic transistors with 2 nm channel lengths *ACS Nano* **12** 4903–08
[37] Pan Z, Peng W, Li F and He Y 2018 Carrier concentration-dependent piezotronic and piezo-phototronic effects in ZnO thin-film transistor *Nano Energy* **49** 529–37
[38] Yang X *et al* 2019 Coupled ion-gel channel-width gating and piezotronic interface gating in ZnO nanowire devices *Adv. Funct. Mater.* **29** 1807837
[39] Keil P, Trapp M, Novak N, Frömling T, Kleebe H-J and Rödel J 2018 Piezotronic tuning of potential barriers in ZnO bicrystals *Adv. Mater.* **30** 1705573
[40] Afouxenidis D, Halcovitch N R, Milne W I, Nathan A and Adamopoulos G 2020 Films stoichiometry effects on the electronic transport properties of solution-processed yttrium doped indium-zinc oxide crystalline semiconductors for thin film transistor applications *Adv. Electron. Mater.* **6** 1900976
[41] Liu S, Wang L, Feng X, Wang Z, Xu Q, Bai S, Qin Y and Wang Z L 2017 Ultrasensitive 2D ZnO piezotronic transistor array for high resolution tactile imaging *Adv. Mater.* **29** 1606346
[42] Zu P, Tang Z K, Wong G K L, Kawasaki M, Ohtomo A, Koinuma H and Segawa Y 1997 Ultraviolet spontaneous and stimulated emissions from ZnO microcrystallite thin films at room temperature *Solid State Commun.* **103** 459–63
[43] R F Service 1997 Will UV lasers beat the blues? *Science* **276** 895
[44] Tsukazaki A *et al* 2005 Repeated temperature modulation epitaxy for p-type doping and light-emitting diode based on ZnO *Nat. Mater.* **4** 42–6
[45] Jiao S J, Zhang Z Z, Lu Y M, Shen D Z, Yao B, Zhang J Y, Li B H, Zhao D X, Fan X W and Tang Z K 2006 ZnO p-n junction light-emitting diodes fabricated on sapphire substrates *Appl. Phys. Lett.* **88** 031911
[46] Chu S, Wang G, Zhou W, Lin Y, Chernyak L, Zhao J, Kong J, Li L, Ren J and Liu J 2011 Electrically pumped waveguide lasing from ZnO nanowires *Nat. Nanotechnol.* **6** 506–10
[47] Xiao Z Y, Liu Y C, Mu R, Zhao D X and Zhang J Y 2008 Stability of p-type conductivity in nitrogen-doped ZnO thin film *Appl. Phys. Lett.* **92** 052106
[48] Li X H, Xu H Y, Zhang X T, Liu Y C, Sun J W and Lu Y M 2009 Local chemical states and thermal stabilities of nitrogen dopants in ZnO film studied by temperature-dependent x-ray photoelectron spectroscopy *Appl. Phys. Lett.* **95** 191903
[49] Alivov Y I, Nostrand J E V, Look D C, Chukichev M V and Ataev B M 2003 Observation of 430 nm electroluminescence from ZnO/GaN heterojunction light-emitting diodes *Appl. Phys. Lett.* **83** 2943–45
[50] Fu H-K, Cheng C-L, Wang C-H, Lin T-Y and Chen Y-F 2009 Selective angle electroluminescence of light-emitting diodes based on nanostructured ZnO/GaN heterojunctions *Adv. Funct. Mater.* **19** 3471–75

[51] Zhu H *et al* 2009 Ultralow-threshold laser realized in zinc oxide *Adv. Mater.* **21** 1613–17
[52] Xu S *et al* 2010 Ordered nanowire array blue/near-UV light emitting diodes *Adv. Mater.* **22** 4749–53
[53] Park W I and Yi G-C 2004 Electroluminescence in n-ZnO nanorod arrays vertically grown on p-GaN *Adv. Mater.* **16** 87–90
[54] Zhang X-M, Lu M-Y, Zhang Y, Chen L-J and Wang Z L 2009 Fabrication of a high-brightness blue-light-emitting diode using a ZnO-nanowire array grown on p-GaN thin film *Adv. Mater.* **21** 2767–70
[55] Shi Z *et al* 2013 High-performance ultraviolet-blue light-emitting diodes based on an n-ZnO nanowall networks/p-GaN heterojunction *Appl. Phys. Lett.* **103** 021109
[56] Wang D, Wang F, Wang Y, Fan Y, Zhao B and Zhao D 2015 Interfacial emission adjustment in ZnO quantum dots/p-GaN heterojunction light-emitting diodes *J. Phys. Chem.* C **119** 2798–803
[57] Tang X, Li G and Zhou S 2013 Ultraviolet electroluminescence of light-emitting diodes based on single n-ZnO/p-AlGaN heterojunction nanowires *Nano Lett.* **13** 5046–50
[58] Yang Z-P, Xie Z-H, Lin C-C and Lee Y-J 2015 Slanted n-ZnO nanorod arrays/p-GaN light-emitting diodes with strong ultraviolet emissions *Opt. Mater. Express* **5** 399–407
[59] Zhu G Y, Xu C X, Lin Y, Shi Z L, Li J T, Ding T, Tian Z S and Chen G F 2012 Ultraviolet electroluminescence from horizontal ZnO microrods/GaN heterojunction light-emitting diode array *Appl. Phys. Lett.* **101** 041110
[60] Fang X *et al* 2016 Ultraviolet electroluminescence from ZnS@ZnO core–shell nanowires/p-GaN introduced by exciton localization *ACS Appl. Mater. Interfaces* **8** 1661–6
[61] Dai J, Xu C X and Sun X W 2011 ZnO-microrod/p-GaN heterostructured whispering-gallery-mode microlaser diodes *Adv. Mater.* **23** 4115–9
[62] Jeong M-C, Oh B-Y, Ham M-H, Lee S-W and Myoung J-M 2007 ZnO-nanowire-inserted GaN/ZnO heterojunction light-emitting diodes *Small* **3** 568–72
[63] Thomas B W and Walsh D 1973 Metal-insulator-semiconductor electroluminescent diodes in single crystal zinc oxide *Electron. Lett.* **9** 362–3
[64] Minami T, Tanigawa M, Yamanishi M and Kawamura T 1974 Observation of ultraviolet luminescence from the ZnO MIS diodes *Japan J. Appl. Phys.* **13** 1475–6
[65] Shimizu A, Kanbara M, Hada M and Kasuga M 1978 ZnO green light emitting diode *Jpn. J. Appl. Phys.* **17** 1435–6
[66] Tang Z K, Wong G K L, Yu P, Kawasaki M, Ohtomo A, Koinuma H and Segawa Y 1998 Room-temperature ultraviolet laser emission from self-assembled ZnO microcrystallite thin films *Appl. Phys. Lett.* **72** 3270–2
[67] Ya I, Alivov D C, Look B M, Ataev M V, Chukichev V V, Mamedov V I, Zinenko Y A and Agafonov A N 2004 Pustovit, fabrication of ZnO-based metal-insulator-semiconductor diodes by ion implantation *Solid-State Electron.* **48** 2343–6
[68] Chen P, Ma X and Yang D 2006 Fairly pure ultraviolet electroluminescence from ZnO-based light-emitting devices *Appl. Phys. Lett.* **89** 111112
[69] Ma X, Chen P, Li D, Zhang Y and Yang D 2007 Electrically pumped ZnO film ultraviolet random lasers on silicon substrate *Appl. Phys. Lett.* **91** 251109
[70] Zhu H *et al* 2010 Low-threshold electrically pumped random lasers *Adv. Mater.* **22** 1877–81
[71] Tian Y, Ma X, Jin L and Yang D 2010 Electrically pumped ultraviolet random lasing from ZnO films: compensation between optical gain and light scattering *Appl. Phys. Lett.* **97** 251115

[72] Liu X-Y, Shan C-X, Wang S-P, Zhang Z-Z and Shen D-Z 2012 Electrically pumped random lasers fabricated from ZnO nanowire arrays *Nanoscale* **4** 2843–6

[73] Li Y, Wang C, Jin L, Ma X and Yang D 2013 Electrically pumped random lasing in ZnO-based metal-insulator-semiconductor structured devices: effect of ZnO film thickness *J. Appl. Phys.* **113** 213103

[74] Ni P-N, Shan C-X, Li B-H, Wang S-P and Shen D-Z 2014 Bias-polarity dependent ultraviolet/visible switchable light-emitting devices *ACS Appl. Mater. Interfaces* **6** 8257–62

[75] Singh S 2016 Al doped ZnO based metal-semiconductor-metal and metal-insulator-semiconductor-insulator-metal UV sensors *Optik* **127** 3523–6

[76] Yang X, Ni P-N, Jing P-T, Zhang L-G, Ma R-M, Shan C-X, Shen D-Z and Genevet P 2019 Room temperature electrically driven ultraviolet plasmonic lasers *Adv. Opt. Mater.* **7** 1801681

[77] Han S, Zhang H, Lu Y, Xu W, Fang M, Liu W, Cao P and Zhu D 2020 Self-powered Au/MgZnO/nanolayered Ga-doped ZnO/In metal-insulator-semiconductor UV detector with high internal gain at deep UV light under low voltage *ACS Appl. Nano Mater.* **3** 120–30

[78] Liu C Y, Xu H Y, Wang L, Li X H and Liu Y C 2009 Pulsed laser deposition of high Mg-content MgZnO films: effects of substrate temperature and oxygen pressure, *J. Appl. Phys.* **106** 073518

[79] Liu W Z, Xu H Y, Wang L, Li X H and Liu Y C 2011 Size-controlled growth of ZnO nanowires by catalyst-free high-pressure pulsed laser deposition and their optical properties *AIP Adv.* **1** 022145

[80] Liu W, Liang Y, Xu H, Wang L, Zhang X, Liu Y and Hark S 2010 Heteroepitaxial growth and spatially resolved cathodoluminescence of ZnO/MgZnO coaxial nanorod arrays *J. Phys. Chem.* C **114** 16148–52

[81] Zhang C, Marvinney C E, Xu H Y, Liu W Z, Wang C L, Zhang L X, Wang J N, Ma J G and Liu Y C 2015 Enhanced waveguide-type ultraviolet electroluminescence from ZnO/MgZnO core/shell nanorod array light-emitting diodes via coupling with Ag nanoparticles localized surface plasmons *Nanoscale* **7** 1073–80

[82] Zhang C, Zhu F, Xu H, Liu W, Yang L, Wang Z, Ma J, Kang Z and Liu Y 2017 Significant improvement of near-UV electroluminescence from ZnO quantum dot LEDs via coupling with carbon nanodot surface plasmons *Nanoscale* **9** 14592–601

[83] Liu C, Xu H, Ma J and Liu Y 2013 Origin of ultraviolet electroluminescence in n-ZnO/p-GaN and n-MgZnO/p-GaN heterojunction light-emitting diodes *Phys. Status Solidi* a **210** 2751–5

[84] Xu H Y, Liu Y C, Liu Y X, Xu C S, Shao C L and Mu R 2005 Ultraviolet electroluminescence from p-GaN/i-ZnO/n-ZnO heterojunction light-emitting diodes *Appl. Phys.* B **80** 871–4

[85] Zhao L, Xu C S, Liu Y X, Shao C L, Li X H and Liu Y C 2008 A new approach to white light emitting diodes of p-GaN/i-ZnO/n-ZnO heterojunctions *Appl. Phys.* B **92** 185–8

[86] Yang L, Liu W, Xu H, Ma J, Zhang C, Liu C, Wang Z and Liu Y 2017 Enhanced near-UV electroluminescence from p-GaN/i-Al_2O_3/n-ZnO heterojunction LEDs by optimizing the insulator thickness and introducing surface plasmons of Ag nanowires *J. Mater. Chem.* C **5** 3288–95

[87] Yang L, Liu K, Xu H, Liu W, Ma J, Zhang C, Liu C, Wang Z, Yang G and Liu Y 2017 Enhanced electroluminescence from ZnO quantum dot light-emitting diodes via introducing Al_2O_3 retarding layer and Ag@ZnO hybrid nanodots *Adv. Opt. Mater.* **5** 1700493

[88] Liu W Z, Xu H Y, Ma J G, Liu C Y, Liu Y X and Liu Y C 2012 Effect of oxygen-related surface adsorption on the efficiency and stability of ZnO nanorod array ultraviolet light-emitting diodes, *Appl. Phys. Lett.* **100** 203101
[89] Wang L, Xu H Y, Zhang C, Li X H, Liu Y C, Zhang X T, Tao Y, Huang Y and Chen D L 2012 MgZnO/MgO strained multiple-quantum-well nanocolumnar films: stress-induced structural transition and deep ultraviolet emission *J. Alloys Compd.* **513** 399–403
[90] Wang L, Ma J, Xu H, Zhang C, Li X and Liu Y 2013 Anisotropic strained cubic MgZnO/MgO multiple-quantum-well nanorods: growths and optical properties, *Appl. Phys. Lett.* **102** 031905
[91] Zhang S G, Zhang X W, Yin Z G, Wang J X, Dong J J, Gao H L, Si F T, Sun S S and Tao Y 2011 Localized surface plasmon-enhanced electroluminescence from ZnO-based heterojunction light-emitting diodes *Appl. Phys. Lett.* **99** 181116
[92] Zhang S G, Zhang X W, Yin Z G, Wang J X, Si F T, Gao H L, Dong J J and Liu X 2012 Optimization of electroluminescence from n-ZnO/AlN/p-GaN light-emitting diodes by tailoring Ag localized surface plasmon *J. Appl. Phys.* **112** 013112
[93] Liu W Z, Xu H Y, Zhang L X, Zhang C, Ma J G, Wang J N and Liu Y C 2012 Localized surface plasmon-enhanced ultraviolet electroluminescence from n-ZnO/i-ZnO/p-GaN heterojunction light-emitting diodes via optimizing the thickness of MgO spacer layer, *Appl. Phys. Lett.* **102** 142101
[94] Yang L, Wang Y, Xu H, Liu W, Zhang C, Wang C, Wang Z, Ma J and Liu Y 2018 Color-tunable ZnO/GaN heterojunction LEDs achieved by coupling with Ag nanowire surface plasmons *ACS Appl. Mater. Interfaces* **10** 15812–9
[95] Qiao Q, Shan C-X, Zheng J, Li B-H, Zhang Z-Z, Zhang L-G and Shen D-Z 2012 Localized surface plasmon enhanced light-emitting devices *J. Mater. Chem.* **22** 9481–4
[96] Hwang S W *et al* 2010 Plasmon-enhanced ultraviolet photoluminescence from hybrid structures of graphene/ZnO films *Phys. Rev. Lett.* **105** 127403
[97] Liu W Z, Xu H Y, Wang C L, Zhang L X, Zhang C, Sun S Y, Ma J G, Zhang X T, Wang J N and Liu Y C 2013 Enhanced ultraviolet emission and improved spatial distribution uniformity of ZnO nanorod array light-emitting diodes via Ag nanoparticles decoration *Nanoscale* **5** 8634–9
[98] Liu W *et al* 2016 Effect of SiO_2 spacer-layer thickness on localized surface plasmon enhanced ZnO nanorod array LEDs *ACS Appl. Mater. Interfaces* **8** 1653–60
[99] Zhang C, Qiu Y, Liu W, Xu H, Yang L, Wang C and Liu Y 2019 Improved near-UV electroluminescence of ZnO nanorod array LEDs by coupling with a graphene plasmon layer *Nanophotonics* **8** 2203–13
[100] Wang Y, Yang L, Chen H, Liu C, Liu W, Xu H, Zhang C, Wang Z and Liu Y 2018 White LED based on $CsPbBr_3$ nanocrystal phosphors via a facile two-step solution synthesis route *Mater. Res. Bull.* **104** 48–52
[101] Feng Q, Shi J, Yang W, Zhong W, Li Y, Chen H, Liu W, Xu H, Liu X and Liu Y 2019 Engineering fluorescence intensity and electron concentration of monolayer MoS_2 by forming heterostructures with semiconductor dots *Nanoscale* **11** 6544–51
[102] Liu C Y, Xu H Y, Ma J G, Li X H, Zhang X T, Liu Y C and Mu R 2011 Electrically pumped near-ultraviolet lasing from ZnO/MgO core/shell nanowires, *Appl. Phys. Lett.* **99** 063115
[103] Liu C Y, Xu H Y, Sun Y, Zhang C, Ma J G and Liu Y C 2014 Ultraviolet electroluminescence from Au/MgO/$Mg_xZn_{1-x}O$ heterojunction diodes and the observation of Zn-rich cluster emission *J. Lumin.* **148** 116–20

[104] Liu C Y, Xu H Y, Sun Y, Ma J G and Liu Y C 2014 ZnO ultraviolet random laser diode on metal copper substrate *Opt. Express* **22** 16731–7
[105] Zhang C, Zhang J, Liu W, Xu H, Hou S, Wang C, Yang L, Wang Z, Wang X and Liu Y 2016 Enhanced ultraviolet random lasing from Au/MgO/ZnO heterostructure by introducing p-Cu_2O hole-injection layer *ACS Appl. Mater. Interfaces* **8** 31485–90
[106] Ren C *et al* 2020 Highly robust flexible ferroelectric field effect transistors operable at high temperature with low-power consumption *Adv. Funct. Mater.* **30** 1906131
[107] Guo N *et al* 2020 Light-driven WSe_2-ZnO junction field-effect transistors for high-performance photodetection *Adv. Sci.* **7** 1901637

Chapter 5

Heterojunction oxide thin-film transistors

Hendrik Faber, Emre Yarali, Emre Yengel and Thomas D Anthopoulos

Metal oxide (MO) semiconductors are an important class of materials for a large range of (opto)electronic applications, including as active channel material in thin-film transistors (TFTs) for display technologies. Research effort in this field focuses on the improvement of device characteristics, with one of the main figures of merit being the electron mobility, as well as the stability against bias stress. While TFTs with a single layer MO as the channel material are limited by the inherent bulk properties of that MO, using heterojunctions (HJs) of bilayer or multilayer stacks instead offers ways to overcome those limitations. For example, interface effects such as the formation of a two-dimensional electron gas, can enhance the mobility beyond that of the single material components of the HJ. On the other hand, the combination of two layers with distinct functionalities (e.g. high mobility of material 1 with high stability and low off-currents for material 2) gives rise to well-balanced overall device characteristics that would not be achievable with either MO by itself. This chapter gives an overview of the materials, mechanisms and improvements involved with heterojunction metal oxide TFTs.

5.1 Introduction

MO semiconductors find widespread application in a variety of (opto)electronic devices due to a multitude of beneficial properties. Transparency and high electron mobility together with decent environmental stability, non-toxicity, and a range of possible deposition schemes have made them an excellent material class for TFT applications of all kinds, with specific emphasis on display technologies for (consumer) electronics [1–3]. Target requirements in this field are a high charge carrier mobility μ_{FE}, which allows small pixel sizes and fast refresh rates that can cope even with demanding applications, e.g. in virtual reality displays, as well as good stability during continuous use cases over extended time periods. Both requirements should ideally be met using materials and fabrication methods that are cost-effective and easily applicable over large areas. Aside from chasing this goal

doi:10.1088/978-0-7503-2556-1ch5

via optimizing individual materials, through composition and process control, another approach that has gained attention in recent years is to make use of active channels composed of layers of two or more different MO materials [4–6]. These HJ films can lead to remarkable device properties when integrated into TFTs, allowing the improvement of both targets, μ_{FE} and bias stability, simultaneously. In many cases, the HJ TFTs achieve better results than either of the individual MO films is capable of.

The proposed reasons and mechanisms for an improved performance that will be discussed in the course of this chapter may vary from case to case. Among them are the creation of a two-dimensional electron gas (2DEG) at the HJ [7], the combination of positive properties of individual MO layers while mitigating their drawbacks [8], the control of chemical composition, stoichiometry and morphology as well as interplays between them.

When referring to MO HJ films there are a number of terms that can be found in literature, e.g. bilayer MO (BMO) [9], dual active layer (DAL) [10, 11], double-active layer [12] stacked active layer (SAL) [13], or (quasi)-superlattice [14–16]. The latter two expressions in particular hint at the fact that the number of layers in an HJ film is not limited to two, but that multiple layers of different MOs can be stacked together. In addition to multilayers, this chapter also discusses reports of homo-junction films where the respective layers feature nominally the same material (e.g. IGZO) which were processed in a way to create distinct properties per layer [17–21].

HJ films are created by sequential deposition of the individual components. In the literature, different naming conventions can be found when describing the sequence of a bilayer film. While describing a film stack of In_2O_3/ZnO, for one research group this refers to the sequence starting from the substrate, i.e. In_2O_3 is deposited first onto the substrate, followed by a ZnO layer on top. Other authors, however, may use the same term (In_2O_3/ZnO) to refer to the opposite stacking direction, starting with the material on the very top. To avoid possible confusion, throughout this chapter we will be using the first naming convention that describes layers according to the sequence in which they are deposited, as depicted schematically in figure 5.1. Another way to refer to individual films is by the names front channel and back channel, whereas front channel is the film directly at the interface with the dielectric and the back channel is the film facing away from the dielectric, typically in contact

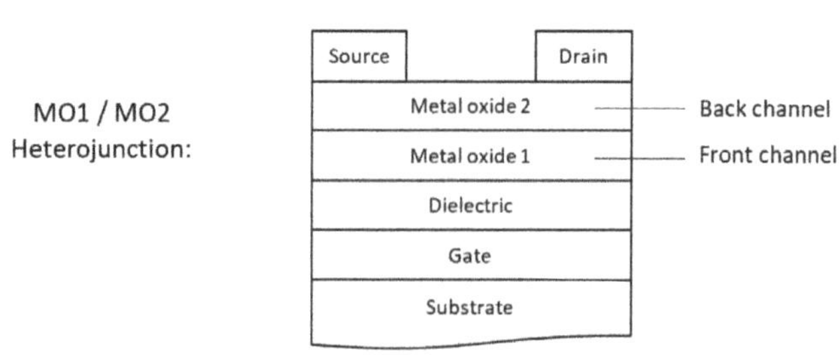

Figure 5.1. Naming convention for heterojunction film stacks as part of a TFT. The expression MO1/MO2 refers to metal oxide 1 being the first film to be deposited onto the substrate or dielectric and metal oxide 2 being the top layer.

with the environment or a passivation layer for the bottom-gate top-contact device architecture. Due to the multitude of different MOs and their combinations in ternary or higher compound materials, abbreviations such as IGZO for indium gallium zinc oxide will be used hereafter. In these abbreviations, the following letter-element combinations will be employed: I for indium, G for gallium, Z for zinc, T for tin, A for aluminium, H for hafnium, O for oxide. Among the material pairings within the HJs, there can be found junctions between insulator/insulator [22], semiconductor/semiconductor [14, 23, 24], or semiconductor/insulator [15, 25–28].

In the course of this chapter we will first take a look at the underlying mechanisms for improved device performances caused by HJ metal oxide (HJMO) films, as well as typical material combination and deposition techniques used to process the HJs (section 5.2). Next, section 5.3 provides a general overview of HJMO TFTs and their field-effect mobility μ_{FE} as the main figure of merit. It then goes on to describe examples how specific parameters, such as film thickness ratio or film stacking sequence in HJ and multilayer TFTs, can have an impact on the device characteristics. Finally, section 5.4 showcases the effects that HJ channels have to offer in regards to bias stress stability, followed by the conclusion and outlook section 5.5.

5.2 Mechanisms and materials for HJ TFTs

5.2.1 2DEG formation

A single semiconductor channel material will always be bound to its inherent limitations when it comes to charge carrier mobility, but the combination of two or more materials allows the exploitation of interface effects that enable new transport properties as well as mobility values that surpass those measured for the individual HJ components. Joining two materials with differently sized bandgaps and a large enough conduction band offset ΔE_C can lead to the formation of a potential well at the heterointerface in which electrons are confined and can form a 2DEG. This phenomenon has been shown in epitaxially grown HJs based on, for example, GaAs/AlGaAs [29] or AlGaN/GaN [30]. An additional electron transfer into the potential well can be initiated by a process called modulation doping (MOD) where the barrier layer is intentionally doped. This results in an increased electron density in the 2DEG and, importantly, it physically separates the ionized donors from the charge carriers. As a consequence, the effects of impurity scattering are all but eliminated, which allows the carrier mobility to reach larger value than that measured for the bulk semiconductors that form the HJ [29, 31].

For oxide based materials, 2DEG formation has been reported, e.g. for heterojunctions based on the insulating perovskite oxides $SrTiO_3$ and $LaAlO_3$ [22] as well as for ZnO/ZnMgO [32–34]. While the alternating AO and BO_2 planes that define the perovskite structure are charge neutral for the case of $SrTiO_3$ ($Sr^{2+}O^{2-}$ and $Ti^{4+}(O^{2-})_2$) they either carry one positive or negative charge for $LaAlO_3$ ($La^{3+}O^{2-}$ and $Al^{3+}(O^{2-})_2$). This leaves the HJ of $SrTiO_3/LaAlO_3$ with either half an extra electron or hole per unit cell that can be transferred across the interface where it is spatially confined [4]. Ohtomo and Hwang investigated $SrTiO_3/LaAlO_3$ HJ deposited by pulsed laser deposition (PLD) [22]. They reported that the junction formed by

$Al^{3+}(O^{2-})_2/Sr^{2+}O^{2-}$ remains an insulator, however, if the junction is formed by $La^{3+}O^{2-}$ and $Ti^{4+}(O^{2-})_2$ it shows n-type conductivity and a high Hall mobility μ_H above 10 000 $cm^2\ V^{-1}\ s^{-1}$ measured at 2 K. The polarity discontinuation at the interface was proposed as the main mechanism for the electron transfer and 2DEG creation.

The formation of a 2DEG in the ZnO/ZnMgO system was reported as early as 2003. Edahiro *et al* used PLD to grow a ZnMgO/ZnO HJ on a sapphire substrate and as a result found a high sheet carrier concentration of $8.8 \times 10^{12}\ cm^{-2}$ and a clear bump in carrier density at the location of interface, revealed by capacitance–voltage (C–V) carrier profiling [32]. A high room temperature (RT) Hall mobility μ_H of 130 $cm^2\ V^{-1}\ s^{-1}$ was determined, which increased further for reduced temperatures down to 100 K then remained stable and featured a value of 360 $cm^2\ V^{-1}\ s^{-1}$ at 1.85 K. This temperature dependant behaviour was indicative of suppressed ionized impurity scattering as compared to a single ZnO layer, whose mobility fell drastically for $T < 100$ K. Additionally, clear Shubnikov–de Haas (SdH) oscillations were observed in magnetoresistance measurements at 1.85 K. In 2004, Koike *et al* realized a double heterojunction of ZnMgO/ZnO/ZnMgO, grown by molecular beam epitaxy (MBE) on sapphire [35]. A deep potential well was formed as a result of the large conduction band discontinuity ΔE_c of 0.9 eV between both oxides. The lattice mismatch between ZnO and ZnMgO led to a polarization induced high sheet charge and the accumulation of free electrons to form the 2DEG at the interface. Hall effect measurements deduced an electron density of $1.2 \times 10^{13}\ cm^{-2}$ as well as μ_H of 170 and 400 $cm^2\ V^{-1}\ s^{-1}$ at RT and 77 K, respectively. Tampo *et al* altered the MBE conditions to successfully grow the HJ on Zn polar faces of ZnO where it was previously only possible to use the O polar surface [23, 36]. This resulted in a boost in μ_H with a value of 250 $cm^2\ V^{-1}\ s^{-1}$ at RT and up to 2750 $cm^2\ V^{-1}\ s^{-1}$ at 4 K. Further refinement of the Mg content later improved μ_H to 319 $cm^2\ V^{-1}\ s^{-1}$ at RT [7]. In work by Nakano *et al* [37], the authors report a ZnO/ZnMgO HJ grown by MBE on Zn faced (0001) ZnO substrates, where a conductive polymer is spin cast on top of the stack to act as a Schottky contact. This allowed field-controlled measurements of Hall mobility with a value up to 21 000 $cm^2\ V^{-1}\ s^{-1}$ at 2 K. This value was further increased in studies by Tzukasaki *et al* who were able to observe fractional quantum Hall effect in a ZnO/ZnMgO HJ and reached μ_H of 180 000 $cm^2\ V^{-1}\ s^{-1}$ at 0.06 K [33] as well as a report by Falson *et al*, who measured a maximum μ_H of 700 000 $cm^2\ V^{-1}\ s^{-1}$ at 0.5 K in films with a very dilute Mg concentration [38].

Collectively, the early investigations pointed out a number of properties that are exemplary for the formation of a 2DEG. Among those are a high conduction band offset ΔE_c to confine the electrons at the HJ (figure 5.2(a)), typically a high sheet carrier density on the order of 10^{12}–$10^{13}\ cm^{-2}$, which is largely independent of temperature, and a clear peak in the cross-sectional carrier profile at the location of the HJ (figures 5.2(b) and (c)). Additionally, the importance of a high quality and flat interface of the HJ was highlighted [7, 39]. Further, the mobility was found to be increasing with decreasing temperature in contrast to a mobility decline at low temperatures for a single layer ZnO. Finally, HJs that formed a 2DEG showed

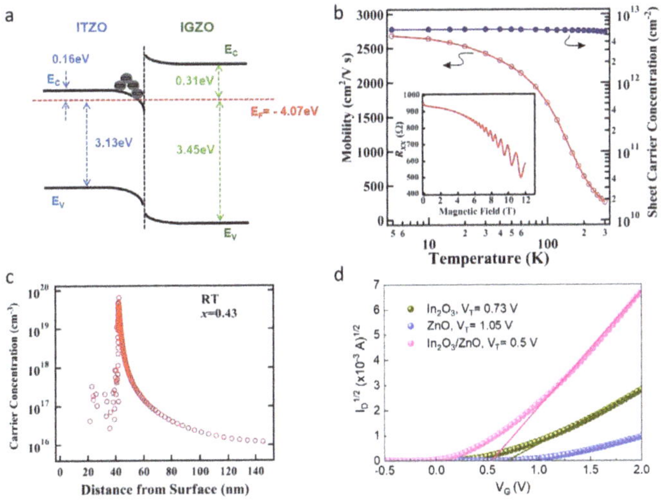

Figure 5.2. Properties observed for 2DEGs HJ films. (a) Band diagram of ITZO/IGZO HJ with proposed 2DEG formation. The band offset makes it favourable for electrons to accumulate at the HJ interface. [43] John Wiley & Sons. © 2018 WILEY-VCH Verlag GmbH & Co., KGaA, Weinheim. (b) Electron carrier concentration across a ZnO/ZnMgO HJ cross-section in the direction from the top surface towards the substrate. A clear spike in carrier concentration indicates the location of the 2DEG at the interface of the HJ. Reprinted from [7], with the permission of AIP Publishing. (c) Temperature dependent mobility and sheet carrier concentration for a ZnO/ZnMgO system. The inset shows SdH oscillations in magnetoresistance measurements at 0.5 K. Reprinted from [7], with the permission of AIP Publishing. (d) Plot of square root of drain current versus gate voltage for solution processed ZnO, In_2O_3 and bilayer In_2O_3/ZnO TFTs. The HJ device features improved performance over both individual MO films. [44] John Wiley & Sons. © 2020 WILEY-VCH Verlag GmbH & Co., KGaA, Weinheim.

SdH oscillations at low temperature magnetoresistance measurements (inset in figure 5.2(c)).

The first transistors using a MO HJ with a 2DEG in the ZnO/ZnMgO system were demonstrated between 2005 and 2007 by Koike *et al* [40] and Sasa *et al* [41, 42]. All cases used MBE to grow the channel materials on top of sapphire substrates and employed a top-gate device architecture with MgO, Al_2O_3 or HfO_2 as gate dielectrics and mobilities μ_{FE} between 62 and 240 $cm^2\ V^{-1}\ s^{-1}$ were achieved. Despite the exceptionally high mobility values, a drawback for practical applications of these structures was the low drain current on/off ratio that stayed below 10^3.

In 2010, the first experiments were carried out to confirm that 2DEGs could be obtained with other deposition methods that did not rely on epitaxial growth on special substrates. Chin *et al* used RF sputtering to fabricate HJs of ZnO/ZnMgO on glass substrates and despite the polycrystalline nature of the films observed 2DEG characteristics, such as a high sheet carrier density, which increased to a maximum around $1 \times 10^{14}\ cm^{-2}$ with increasing Mg content due to increased polarization [45]. Additionally, both the carrier density and the Hall mobility were found to be approximately temperature independent. It was argued that the high carrier density of the 2DEG could successfully screen the grain boundary potential [46] but that improvements in surface roughness were required to increase the only moderate μ_H of 39.4 cm^2 V^{-1} s^{-1}. The same approach of sputtered ZnO/ZnMgO was integrated by Cheng *et al* to produce top-gated HJ TFTs with Al_2O_3 dielectric and achieved a high μ_{FE} of 84.2 cm^2 V^{-1} s^{-1} while maintaining a large on/off ratio > 10^6, indicating the practical applicability of polycrystalline MO HJs forming 2DEGs [47]. Similarly, Remashan *et al* used metal organic chemical vapor deposition (MOCVD) to grow polycrystalline ZnO on top of a ZnMgO layer and reported an increased μ_{FE} of the HJ TFT of 9.1 cm^2 V^{-1} s^{-1} compared to a single layer of ZnO (2.3 cm^2 V^{-1} s^{-1}) [48].

The 2D nature of electrons was also described in materials systems other than crystalline ZnO/ZnMgO, for example for a thin polycrystalline ITO layer on top of amorphous IGZO or in superlattices based on ZnO and Al_2O_3 [15, 49]. Taniguchi *et al* described a type 2 band offset for their sputter deposited IGZO/ITO heterojunction that favors electron transport from the ITO conduction band to that of IGZO and the 2D nature of electrons was assigned based on temperature dependent Hall effect measurements [49]. Resulting HJ TFTs achieved a μ_{FE} of 20 cm^2 V^{-1} s^{-1} compared to 7.5 cm^2 V^{-1} s^{-1} for a single layer of IGZO. On the other hand, Ahn *et al* produced semiconductor/dielectric superlattices by atomic layer deposition (ALD) of ZnO/Al_2O_3 at moderate 200 °C on common Si/SiO_2 substrates [15]. ZnO films grown on alumina showed improved crystallinity and preferred *c*-axis orientation compared to ZnO grown on the bare substrate. Together with the 2D confinement of electrons in the ZnO layers due to the large Al_2O_3 band gap, this superlattice channel showed enhanced mobility (27.8 cm^2 V^{-1} s^{-1}) and on/off ratio (10^9) compared to ZnO only devices, as well as improved bias stress behavior.

After 2DEG formation had been observed even for imperfect, polycrystalline vacuum deposited films, the question arose whether similar effects could also be realized with solution processing methods. To this end, Lin *et al* developed In_2O_3/ZnO HJs based on ultrathin (<10 nm) films processed via spin coating of precursors and subsequent annealing at moderate temperatures below 200 °C [14]. Due to an offset in Fermi level and conduction band, an electron transfer from ZnO to In_2O_3 was proposed, where available energy levels are quantized because of the low film thickness, leading to a quasi 2D-electron-gas-like system at the MO HJ. The bilayer structure was extended by adding thin layers of equally solution processed Ga_2O_3 to form a quasi superlattice (QSL) system consisting of In_2O_3/Ga_2O_3/ZnO/Ga_2O_3/In_2O_3. The addition of Ga_2O_3 served the purpose to enhance confinement and interface planarity as well as for trap passivation. This QSL setup exhibited high μ_{FE} of 30 and 40 cm^2 V^{-1} s^{-1}, using a SiO_2 or a Al_2O_3/ZrO_2 bilayer dielectric, respectively. The concept around using discrete solution processed In_2O_3 and ZnO bilayers was further developed by the same research group, exploring a new solution-based deposition technique (spray pyrolysis) [50], the effects of high temperature annealing up to 600 °C [51], photonic curing as alternative method for the rapid precursor conversion [44, 52], the impact of modulation doping (MOD) through Li-doped ZnO films [53], the variation in stacking sequence within multi-layer junctions [16] and the incorporation of ZnO nanoparticles and an interlayer of polystyrene for additional trap passivation [54]. Several properties that resemble or hint at a 2DEG system were observed repeatedly in these studies, including a spike in electron concentration at the location of the heterojunction (via *C–V* carrier profiling), a change towards a temperature independent μ_{FE} and a shift in transport properties from a trap limited conduction (TLC) to a percolation conduction (PC) [14, 16, 44, 50]. Additionally, the mobility of the various HJ structures reached consistently high values between ≈20–50 $cm^2\ V^{-1}\ s^{-1}$, and in all cases was significantly enhanced when compared to TFTs based on either of the individual MO components (as depicted in figure 5.2(d)). This synergetic effect demonstrates the unique properties that can be achieved by 2D systems engineered at MO HJs.

Other solution processed MO bilayer systems for that 2DEG formation reported include e.g. the combination of ITZO/IGZO [43], In_2O_3/AlInO [55], or a homojunction of undoped and polymer doped In_2O_3 [56]. Lee *et al* first investigated the properties of planar junctions depending on the bottom ITZO layer thickness, and then created a corrugated stripe structure of ITZO/IGZO bilayers with alternating thin/thick IZO bottom layers in the stripes running perpendicular to the channel [43]. Here, the thick-IZO/IGZO parts were identified to enhance μ_{FE}, while the thin-IZO/IGZO parts were able to maintain a low off-current level. This combination resulted in TFTs with high mobility up to 51 $cm^2\ V^{-1}\ s^{-1}$ in combination with an on/off ration > 10^8. In the study by Chen *et al* the authors used the polymer polyethylenimine (PEI) to dope the back channel In_2O_3 on top of an undoped In_2O_3 front channel [56]. The resulting energy offsets gave rise to electron confinement and 2DEG formation. Transistor performance based on In_2O_3/In_2O_3:PEI surpassed that of single layer In_2O_3 devices and reached mobilities >10 $cm^2\ V^{-1}\ s^{-1}$ and as high as 30 $cm^2\ V^{-1}\ s^{-1}$ using SiO_2 or a high-k ZrO_2 dielectric, respectively.

5.2.2 Dual active layers and other mechanisms

While HJ layer stacks that form a 2DEG typically lead to TFTs in which the bilayer is showing improved characteristics (e.g. higher mobility and currents, improved bias stability) compared to either of the individual films, another widely used approach to work with layered MOs is to balance out properties of either components [6, 9, 21, 57]. In this case, a thin but highly conductive MO with high charge carrier density N_e and mobility μ_{FE}, typically deposited as the front channel, serves as the main transport layer. By itself, the drain current for this film may be difficult to modulate through the gate voltage resulting in either a negative V_{th} or even the device not being able to fully turn off. In order to achieve successful gate modulation, a second and more resistive MO with smaller mobility is used, mostly as the back channel material. Compared to devices based on the front channel material alone, this bilayer combination can have the drawback reducing the drain current in the on-state slightly but it comes with the substantial benefit that it is now possible to fully turn-off the device. For this combination, the back channel takes control over the turn-on behavior and V_{th}, while the front channel still delivers reasonably high currents and mobility values. Because both layers play their distinctive part in the assembly, the term dual active layer is often used to describe them. An example of this behavior is shown in the transfer characteristics in figure 5.3, where a rather

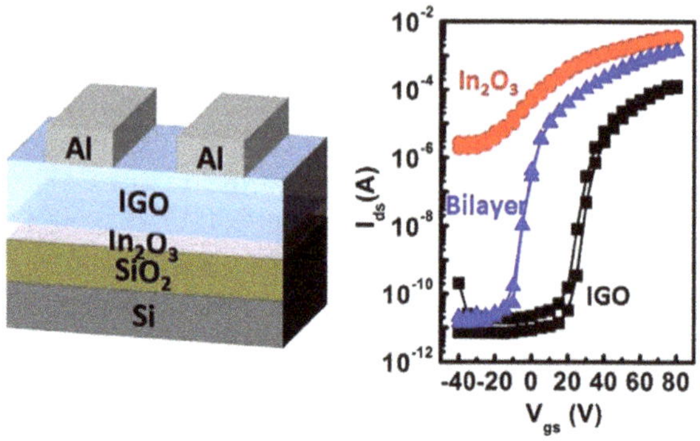

Figure 5.3. Schematic cross-section transfer characteristics of TFTs based on single layer In_2O_3 and IGO as well as their combination in an In_2O_3/IGO dual active layer. The bilayer combines beneficial attributes from both components, such as the higher mobility and on-current of the front channel and the low off-current of the back channel material. Reprinted with permission from [8]. Copyright (2013) American Chemical Society.

conductive In_2O_3 is combined with a layer of IGO and the combined solution processed In_2O_3/IGO bilayer film shares the off-current with IGO yet still delivers an increased I_D in the on-state [8].

The approach to use a high mobility transport/high N_e transport layer with a more stable film to control off-current and V_{th} has been pursued with a number of MO combinations and different deposition techniques. For sputter deposited HJs this includes, for example, ITO/ZTO [58], IZO/HIZO [59, 60], IZO/IGZO [61] or IGZO:H/IGZO [21]. Similarly, solution based dual layer HJ TFTs have been reported based on, for example, IZO/AIZO [62, 63], In_2O_3/IGO [8], or also indium-free composition of SnO/ZnO [64].

Other reasons to employ a bilayer HJ include the control and improvement of chemical composition, stoichiometry or morphology. For example, as reported for sputtered as well as solution processed back channels of tin containing materials, i.e. IZO/ZTO [65], IGZO/ITO [10] or IGZO/ZTO [11], the diffusion of Sn^{4+} into the respective bottom layer could lead to improvements in device performance. The tin cations served as n-dopants by substituting for In^{3+}, Ga^{3+} or Zn^{2+}, thus increasing the carrier concentration in the channel.

MO films deposited by various solution processes may still possess a considerable degree of porosity as well as pinholes which are detrimental to an efficient charge transport. Consequently, the subsequent application of a second MO layer has the chance to fill up gaps in the bottom layer and thus help to improve the channel properties. This beneficial effect was observed for spin cast ZTO films deposited on top of IGZO [10, 11], as well as for inkjet printed ZTO/In_2O_3 bilayers in the study by Lee and Choi, who reported film densification through pore filling and passivation as well as reduced surface roughness leading to improved μ_{FE} in devices [12].

In a rare example of p-type MO HJs, Al-Jawhari *et al* conducted a study on SnO/Cu_2O TFTs, where the Cu_2O capping layer could be used to help control the SnO stoichiometry [66]. Depending on the oxygen pressure during sputter deposition, the layer thickness ratios, and annealing temperatures between 140 °C and 200 °C, the resulting TFTs were either found to exhibit n-type behavior, p-type behavior or no functionality. With the correct parameter set, an improved p-type mobility of 0.66 $cm^2\ V^{-1}\ s^{-1}$ could be established compared to 0.23 $cm^2\ V^{-1}\ s^{-1}$ for a single film of SnO.

It is worth noting that the various possible origins for improved device characteristics of HJ TFTs described here in section 5.2 do not have to be seen as isolated mechanisms. Rather, a combination of several of them acting simultaneously may be responsible and the exact root cause determination for enhanced charge transport properties may not always be possible.

5.3 TFTs based on bilayer HJs and multilayer stacks

5.3.1 Overview

Before looking at a few specific examples of how the properties of HJMO TFTs can be tailored, we show in figure 5.4 an overview of mobility values that were reported to date, with respect to the deposition methods and maximum substrate temperature (figure 5.4(a)) as well as the year of publication (figure 5.4(b)). The data used to generate figure 5.4 is additionally listed in tables 5.1 and 5.2. In these plots, the distinction between deposition categories is as follows. Molecular beam epitaxy (MBE), pulsed laser deposition (PLD), and (plasma enhanced) atomic layer deposition (PEALD) are considered *advanced vacuum* techniques, all sputtering variations as well as MOCVD are grouped as *vacuum* techniques, whereas methods such as spin coating, spray pyrolysis [50, 67] or inkjet printing [12, 68] are in the *solution* processing category.

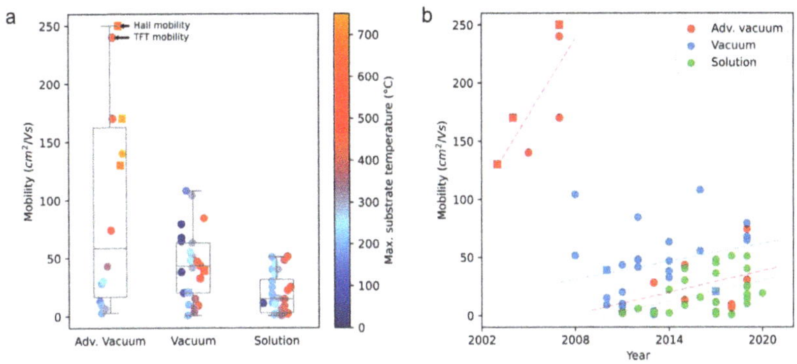

Figure 5.4. Overview of mobility values achieved in HJ MO structures, based on the values listed in tables 5.1 and 5.2. (a) Mobility broken down by deposition categories, i.e. advanced vacuum deposition techniques (MBE, ALD, PLD), vacuum techniques (sputtering and MOCVD) and various solution processing techniques (spin coating, spray pyrolysis, inkjet). The color of each marker is indicative of the maximum substrate temperature during fabrication as shown in the color bar on the right. Within each category, markers are additionally distributed from left to right with increasing maximum temperature. (b) Progression of mobility values with the year of publication. Here, the color of each marker denotes the deposition category for the MO semiconductors. The dashed lines are linear trend lines as guides for the eye. In both (a) and (b), square markers denote mobility values extracted via Hall effect measurements at RT, whereas circular markers correspond to device mobility values extracted from TFTs.

Table 5.1. Overview of selected publications on HJ MO films and TFTs processed by vacuum techniques. Entries with the added asterisk (*) in the mobility field denote Hall mobilities μ_H rather than TFT device mobilties μ_{FE}.

Year	Heterojunction structure	Deposition method	Mobility ($cm^2\ V^{-1}\ s^{-1}$)	Dielectric	Maximum temperature (°C)	Reference
		Deposition category: advanced vacuum				
2003	ZnMnO:Al/ZnO	PLD	130 (*)	n/a	700	[32]
2004	ZnMgO/ZnO/ZnMgO	MBE	170	n/a	750	[35]
2005	ZnMgO/ZnO/ZnMgO	MBE	140	MgO	750	[40]
2007	ZnO/ZnMgO	MBE	240170	$HfO_2Al_2O_3$	500	[42]
2007	ZnO/ZnMgO	MBE	250 (*)	n/a	650	[23]
2011	ZnO/Al_2O_3/ZnO	PEALD	2.8	Al_2O_3	200	[26]
2013	Al_2O_3/ZnO superlattice	ALD	27.8	SiO_2	200	[15]
2015	ZnO/SnO multilayers	PLD	43	Al_2O_3	350	[70]
2015	ZnO/HfO_2	ALD	13	HfO_2	160	[27]
2018	In_2O_3/ZnO supercycles	ALD	9.3	SiO_2	200	[71]
2018	In_2O_3/ZnO supercycles	ALD	6.5	SiO_2	300	[69]
2019	In_2O_3/ZnO supercycles	PEALD	30.3	Al_2O_3	250	[77]
2019	In_2O_3/ZnO/Ga_2O_3 supercycles	PEALD	74	SiO_2	450	[78]
		Deposition category: vacuum				
2008	IZO/GIZOor ITO/GIZO	Sputtering	51.3104	SiO_2	300	[72]
2010	MgZnO/ZnO	MOCVD	9.1	Si_3N_4	450	[48]
2010	ZnO/MgZnO	Sputtering	39 (*)	n/a	600	[79]
2010	HIZO/HIZO Hf content varied	Sputtering	15	Si_3N_4/SiO_2	370	[17]
2011	GZO/IGZO	Sputtering	10	SiO_2	200	[73]
2011	IGZO/IGZO:N	Sputtering	7.3	SiO_2	350	[18]
2011	ITO/ZTO	Sputtering	43.2	SiO_2	500	[58]
2011	IGZO/ITO	Sputtering	20	SiO_2	80	[49]
2012	IZO/HIZO	Sputtering	48	Si_3N_4/SiO_2	350	[60]
2012	ZnO/MgZnO	Sputtering	84.2	Al_2O_3	600	[47]
2012	IZO/HIZO	Sputtering	41.4	SiO_2	300	[59]

2012	IZO/IGZO	Sputtering	47.7	SiO_2	250	[61]
2013	SnO/Cu_2O	Sputtering	0.66	Al_2TiO_5	190	[66]
2013	IGZO (O poor)/IGZO (O rich)	Sputtering	3.65	SiO_2	350	[19]
2014	IGZO/IGZO:Ti	Sputtering	63	HfO_2	300	[74]
2014	IGZO/IZO	IGZO: spin coatingIZO: sputtering	46.7	Al_2O_3	430	[80]
2014	In_2O_3/InZnO	Sputtering	37.9	ZrO_2	25	[9]
2014	IZO/ZTO	Sputtering	32.3	SiO_2	500	[65]
2016	IGZO:H/IGZO	Sputtering	55.3	SiO_2	250	[21]
2016	ATZO/ATZO varied O pressure	Sputtering	108	SiO_2	150	[20]
2017	In_2O_3/Al_2O_3	Sputtering	20.5 (*)	n/a	200	[28]
2019	IGZO/In_2O_3	RF sputtering	64.467.579.1	SiO_2 HfO_2 Si_3N_4	25	[24]

Table 5.2. Overview of selected publications on HJ MO TFTs processed by solution processing techniques. Entries with the added asterisk (*) in the temperature field were treated by photonic curing and the stated values are the approximate substrate temperatures during the process, based on simulations.

Year	Heterojunction structure	Deposition method	Mobility ($cm^2\ V^{-1}\ s^{-1}$)	Dielectric	Maximum temperature (°C)	Reference
2011	IZO/AIZO	Spin coating	1.57	Tetraethyl-orthosilicate	350	[62]
2012	IZO/AIZO	Spin coating	5.62	SiO_2	350	[63]
2013	IGZO/ZTO	Spin coating	2.09	SiO_2	450	[10]
2013	IGZO/IGZO/IGZO	Spin coating	2.4	SiO_2	450	[13]
2013	In_2O_3/IGO	Spin coating	2.6	SiO_2	250	[8]
2014	IGZO/ZTO	Spin coating	1.97	SiO_2	450	[11]
2014	IZTO/IGZO	Spin coating	22.16	SiO_2	450	[57]
2015	$In_2O_3/Ga_2O_3/ZnO/Ga_2O_3/In_2O_3$	Spin coating	30 40	SiO_2 Al_2O_3/ZrO_2	200	[14]
2015	ZTO/In_2O_3	Inkjet	8.6	SiO_2	400	[12]
2016	SnO_2/ZnO	Spin coating	15.4	SiO_2	300	[64]
2017	ZTO/Al_2O_3	Spin coating	3.08	SiO_2	400	[25]
2017	In_2O_3/ZnO	Spray pyrolysis and spin coating	45	SiO_2	250	[50]
2017	In_2O_3/Li:ZnO	Spin coating	11.4	SiO_2	200	[53]
2017	In_2O_3/GaOx	Spin coating	11	SiO_2	50	[75]
2017	In_2O_3/ZnO	Spin coating	47.9	SiO_2	400	[51]
2017	In_2O_3/ZnO	Spin coating	36	Al_2O_3/ZrO_2	250 (*)	[52]
2017	ZnO/Ga_2O_3	Spray pyrolysis	1	SiO_2	233	[67]
2018	ITZO/IGZO (corrugated structure)	Spin coating	51	Al_2O_3	450	[43]
2018	ZnO/In_2O_3	Spin coating	0.5	SiO_2	350	[76]
2019	In_2O_3/In_2O_3:PEI	Spin coating	10.0524.8	SiO_2ZrO_2	250 500	[56]
2019	In_2O_3/ZnO:Li/In_2O_3/ZnO:Li	Spin coating	18.25	SiO_2 Al_2O_3/ZrO_2	200	[16]
2019	In_2O_3/ZnO NP:Al/PS/ZnO	Spin coating	50.7	SiO_2	200	[54]
2019	In_2O_3/AlInO	Spin coating	40	SiO_2	300	[55]
2019	In_2O_3/IGZO	Inkjet	14.5	SiO_2	400	[68]
2020	In_2O_3/ZnO	Spin coating	19	Al_2O_3/ZrO_2	200 (*)	[44]

The largest range of μ_{FE} is found within the advanced vacuum category with values between 6.5 for a In_2O_3/ZnO HJ and 240 $cm^2\ V^{-1}\ s^{-1}$ for ZnO/ZnMgO [42, 69], with the latter value being close to Hall effect mobility of 250 $cm^2\ V^{-1}\ s^{-1}$ measured in a comparable ZnO/ZnMgO system at RT [23]. This large variation can be explained mainly by two factors. First, the deposition temperature can have a major impact on the resulting films, where typically higher temperatures lead to films with improved structural quality. Second, processes such as MBE especially require specialized substrates (e.g. sapphire) to reach the full potential of epitaxially grown materials [7, 23, 35, 40, 42]. Common substrate materials, such as Si/SiO_2, glass, etc, are far more commercially viable but do not yield the same quality films [15, 69–71]. The range as well as maximum of reported mobility in the regular vacuum deposition and solution processing category is considerably smaller with values over 100 $cm^2\ V^{-1}\ s^{-1}$ [20, 72] and close to 50 $cm^2\ V^{-1}\ s^{-1}$ [43, 51, 54], respectively. The decline in performance is expected, as especially in the case of solution processing the films typically still contain a significant number of defects that hinder charge transport. Interestingly, there does not seem to be a strong correlation between the maximum temperature and the resulting mobilities for the HJ films in the vacuum and solution processing deposition categories.

As can be seen in figure 5.4(b), the initial studies on HJs were started from 2003 to approximately 2008 [7, 23, 32, 35, 40, 42]. Here, techniques such as MBE and PLD were used to grow high quality ZnO/ZnMgO HJ on sapphire substrates; an approach that lead to the highest reported mobility values. From 2008 onwards, it was attempted to reproduce the benefits of HJ films by more scalable and commercially viable vacuum deposition techniques, predominantly by sputtering on more common substrates [9, 17–21, 24, 28, 45, 47, 49, 58–61, 65, 66, 72–74]. As a result, exceptionally high mobility values over 80 $cm^2\ V^{-1}\ s^{-1}$ were achieved in the ZnO/ZnMgO [47] system but also for ITO/GIZO [72] and ATZO/ATZO [20] (front and back channel processed with different oxygen pressure). Similarly, for solution processing techniques, research into HJMO TFTs was intensified from 2011 onwards, using spin coating in the vast majority of cases [8, 10, 11, 13, 14, 16, 25, 43, 44, 51–57, 62–64, 75, 76], but also including spray pyrolysis [50, 67] and inkjet printing [12, 68]. This increasing shift in research efforts towards large scale and higher throughput fabrication techniques versus the ultrahigh quality epitaxial MO HJs of the initial phase can be seen as the reason why the overall average mobility actually appears to trend downwards over the years. However, if the advanced vacuum category is further divided into periods of epitaxial (before 2009) and polycrystalline growth (after 2009), within all categories the mobility trend-line is actually increasing over time (see dashed lines in figure 5.4(b)).

5.3.2 Distinct heterojunction layers versus mixed phase oxides

When investigating HJ stacks and their impact on device performance relative to single layers MO films, one important aspect is to verify whether any improvements are indeed the results of creating distinct layers and the subsequent interface effects taking place or if the same could have been achieved simply by creating a single

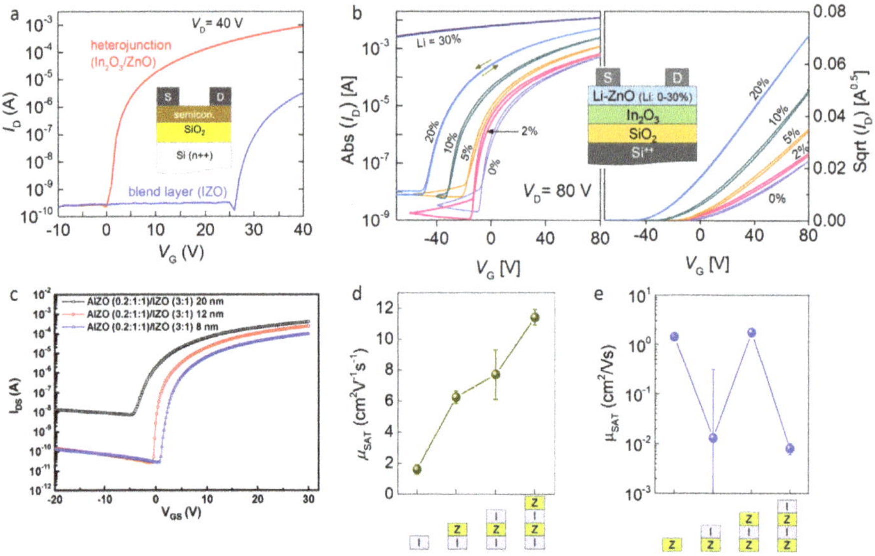

Figure 5.5. (a) Comparison between an HJ TFT with an active channel comprised of solution processed In_2O_3/ZnO and a device based on a single blend layer of IZO using the same precursor components. The distinct HJ device exhibits superior performance. [16] John Wiley & Sons. © 2019 WILEY-VCH Verlag GmbH & Co., KGaA, Weinheim. (b) Effect of MOD on the transfer characteristics in In_2O_3/ZnO HJ TFTs via Li-doping of the ZnO layer. [53] John Wiley & Sons. © 2017 WILEY-VCH Verlag GmbH & Co., KGaA, Weinheim. (c) Evolution of transfer curves for TFTs based on solution processed IZO/AIZO HJs with increasing IZO front channel thickness between 8 and 20 nm. © (2012) IEEE. Reprinted, with permission, from [63]. (d), (e) The effect of stacking sequence on μ_{FE} for In_2O_3 (I) and ZnO (Z) multilayer TFTs. [16] John Wiley & Sons. © 2019 WILEY-VCH Verlag GmbH & Co., KGaA, Weinheim.

multicomponent mixed phase oxide instead. Such a reference test was carried out for example by Khim *et al* who researched solution processed In_2O_3/ZnO HJs [16]. As shown in figure 5.5(a), it was found that the IZO based reference devices, created by mixing the respective In- and Zn-containing precursors, showed poor transport behavior with low on-currents on the order of 10^{-6} A and a high turn-on voltage of 26 V. In contrast, the In_2O_3/bilayer exhibited superior properties with increased on-current (10^{-3} A) and a turn on close to 0 V. It was argued that the 200 °C annealing temperature did not supply enough thermal energy for a complete IZO conversion, whereas electron confinement at the successfully formed In_2O_3/ZnO HJ was the reason for superior properties in the bilayer TFT. For a different material system of solution processed SnO_2/ZnO, Nam *et al* came to similar findings where TFTs based on control ZTO films showed inferior performance compared to the SnO_2/ZnO counterparts when annealed at a temperature of 300 °C [64].

5.3.3 Influence of layer thicknesses

The layer thicknesses of the individual MOs that form the HJ in the active TFT channel have a profound impact on the overall device characteristics. For example, the effect of the front channel thickness in solution processed IZO/AIZO HJs TFTs was investigated by Jeong *et al* [63]. As seen in figure 5.5(c), the on/off ratio decreases with increasing IZO front channel thickness between 8 and 20 nm. This is the result of increasing overall N_e for higher IZO thicknesses, which also lead to a negative V_{th} shift. The same trend was reported by Liu *et al* for sputtered In_2O_3/IZO HJ TFTs where μ_{FE} steadily increases with higher In_2O_3 thickness, albeit at the cost of an increasing subthreshold swing (SS), decreasing on/off ratio and negative V_{th} shift [9]. Consequently, a balanced compromise between high electron mobility and on-currents on one side and high on/off ratio, low SS and V_{th} close to 0 V on the other side, has to be made. While Lee *et al* found very similar properties for increasing ITZO film thicknesses in their ITZO/IGZO HJ devices, they made use of these findings by changing up the channel configuration [43]. They created corrugated bilayer films with stripes of thin ITZO/IGZO and thick ITZO/IGZO going perpendicular to the source-drain channel orientation. This procedure allowed for devices with high on-currents and mobility >50 $cm^2\ V^{-1}\ s^{-1}$ while simultaneously being able to maintain low off-currents and an on/off ratio $>10^8$.

An investigation by Lee and Choi about inkjet printed ZTO/In_2O_3 bilayer HJ TFTs examined how the increase in In_2O_3 back channel thickness impacts the device performance [12]. They found an initial improvement in μ_{FE}, on/off ratio, SS and V_{th} for thinner In_2O_3 films, which was attributed to the filling of pores in the ZTO front channel as well as a decrease in roughness. Increasing the back channel further, however, saw a deterioration of all quantities due to the introduction of increased contact resistance, trap density and elevated roughness.

In a study on PLD deposited multilayer stacks consisting of repeated ZnO/SnO_2 units, the thickness of the individual layers was simultaneously varied between 0.5 and 3 nm while keeping the total channel thickness to 30 nm [70]. The resulting transistors were found to exhibit a maximum μ_{FE} of 43 $cm^2\ V^{-1}\ s^{-1}$ for a thickness of 1.5 nm for both sub-units. Further increase in thickness only lead to a decline in performance. It was argued that high mobility electron channels formed at the ZnO/SnO_2 interfaces and that for higher sub-unit thicknesses the number of available interfaces decreased, thus reducing μ_{FE} in the device.

Krausmann *et al* produced HJ TFTs based on multilayers with repeating In_2O_3/ZnO units via ALD and changed the thickness ratio between In_2O_3 and ZnO in the repeat units [71]. They found that higher amounts of defects are introduced by a higher thickness of the ZnO layers, while opposite to that thicker In_2O_3 layers resulted in higher mobilities up to 12.2 $cm^2\ V^{-1}\ s^{-1}$. However, other device parameters (on/off, SS, V_{th}) began to deteriorate for the thickest In_2O_3 sub-units and a more balanced optimum performance was determined for a medium size In_2O_3 thickness. In addition, the study revealed a gradual change in the transport properties from an initially trap-limited conduction for a low In_2O_3/ZnO thickness ratio towards a percolation conduction for thicker In_2O_3 layers in the repeating units. This also manifested in nearly temperature independent μ_{FE} as opposed to a temperature activated behavior for devices with the lowest In_2O_3/ZnO ratios. For the same material system, with stacks of alternating In_2O_3 and ZnO processed by ALD, Sanctis *et al* varied the total number of stacks between 3 and 6 [69]. With a higher number of stacks an increase in channel conductivity (reduced on/off ratio, negative V_{th} shift) was observed and attributed to deep-donor-like states related to oxygen vacancies. An optimum number of four stacks was suggested based on the tradeoff between mobility, on/off ratio, SS and V_{th}.

5.3.4 Influence of stacking sequence

In many bilayer HJ TFTs the more conductive MO layer is typically employed as the front channel material in direct contact with the dielectric. In this position it can serve as charge transport layer with a high N_e that also can assist in trap passivation at the semiconductor/dielectric interface [8, 9, 21, 71]. This was evaluated and confirmed by Sanctis *et al* who fabricated multilayer stacks of repeated In_2O_3 and ZnO layers via ALD [69]. They tested the impact of starting the HJ sequence either with a layer of ZnO or In_2O_3. For ZnO as the starting material, the multilayer TFTs yielded a low mobility of 1 $cm^2 V^{-1} s^{-1}$, an on/off ratio of 5×10^6 and SS of 1.85 V/dec. All these quantities were found to be significantly improved when the sequence was reversed and In_2O_3 was the starting material, giving rise to a mobility, on/off ratio and SS of 6.5 $cm^2 V^{-1} s^{-1}$, 5×10^7, and 0.7 V/dec, respectively.

Another study on the influence of the stacking sequence for the same In_2O_3/ZnO material system was carried out by Khim *et al* [16]. Here, film stacks of up to four sequentially solution deposited single layers were formed, alternating between In_2O_3 and ZnO in consecutive deposition steps. Film as well as device properties were evaluated at each stage. Additionally, the films stack fabrication was either started by the deposition of In_2O_3 or ZnO as the initial layer. As shown in figures 5.5(d) and (e), the progression of device characteristics with each new layer was vastly different. For stacks beginning with In_2O_3, the mobility increases with each new step (from 1.6 to 11.4 $cm^2 V^{-1} s^{-1}$), making bigger jumps after a new ZnO film was deposited. In contrast to that, the devices using ZnO as the starting layer were limited by the mobility of a single ZnO film (1.4 $cm^2 V^{-1} s^{-1}$) when ZnO is the topmost layer and μ_{FE} even dropped about 2 orders of magnitude when In_2O_3 was the topmost layer. Based on AFM investigations, it was determined that the interface roughness plays a crucial role, whereas the In_2O_3/ZnO interface is favorable for charge transport but the ZnO/In_2O_3 interface is doing the opposite. Consequently, sequences starting with In_2O_3 inherently include more beneficial HJs that have an accumulative effect, while sequences starting with ZnO as the initial material contain more unfavorable ZnO/In_2O_3 junctions leading to more traps in the system and limited device performance.

5.3.5 Modulation doping

Modulation doping (MOD) is a technique to improve the device performance wherein two separate semiconductors are brought into close contact, one of them is undoped (intrinsic) while the other one is intentionally doped. Given the correct relative energy levels, an electron flow (in the case of n-type doping) from the doped to the undoped layer is initiated. Because the ionized dopant sites are physically separated from the undoped semiconductor the charge transport in the latter is not hampered by ionized impurity scattering, enabling higher carrier mobilities.

Modulation doping finds application for example in GaAs/(Al,Ga)As high electron mobility transistors [29].

The introduction of MOD for MO HJ structures was simulated for the ZnO/ZnMgO system [81] and also experimentally verified for sputtered films, where the ZnMgO layer was doped with a fixed concentration of Al [47, 79]. The addition of the doping layer led to an increased sheet carrier density in addition to the polarization induced charge carriers. When the Mg content in the doping layer was varied, the band gap increased with increasing Mg concentration and as a result the tendency of electron transfer to the 2DEG region increased, resulting in decreased sheet resistance and increased sheet carrier densities. According to their estimates, Chin *et al* suggested that up to 61% of the charge carriers in the heterostructure originated in the doping layer [79].

Regarding solution processed HJs, Khim *et al* used Li-doped ZnO films to apply MOD in In_2O_3/ZnO bilayer TFTs [53]. Increasing the Li concentration from 0 to 20 wt% (in the precursor solution) was found to move the ZnO Fermi level (E_F) closer to vacuum due to higher electron concentrations in the doped ZnO films. The resulting increased offset between the Fermi levels of In_2O_3 and ZnO:Li was suggested to facilitate electron transfer from ZnO:Li to In_2O_3. Modulation doping led to significantly reduced sheet resistance of bilayer films and the μ_{FE} of HJ TFTs saw an improvement from 3.2 to 11.4 cm^2 V^{-1} s^{-1} for 0 to 20 wt% Li doping. Figure 5.5(b) shows the evolution of transfer curves depending on the doping concentration. In a follow up study, MOD with Li was also used to boost the performance of MO multilayer TFTs based on a In_2O_3/ZnO/In_2O_3/ZnO channel architecture [16]. Here, an increased mobility of 18 cm^2 V^{-1} s^{-1} and improved uniformity could be demonstrated using a SiO_2 dielectric as well as μ_{FE} up to 25 cm^2 V^{-1} s^{-1} using a high-k dielectric based on AlO_x/ZrO_x. These results are a good indicator for the improvements that can be achieved using MOD.

5.4 Improvements in bias stress stability for HJ TFTs

For most of the targeted applications of HJMO TFTs, each single TFT is but one device out of a large group, working together to fulfill a specific system function. One simple example is an individual transistor switching on (or off) a single pixel as part of a display that comprises many thousands of pixels. To ensure the overall functionality is not compromised, it is therefore critical that each individual device stays within its operational parameters, even over prolonged times of activity and in different environments. Bias stress tests are one common way of testing a device's resistance against the application of constant biases over extended periods of time.

In a typical bias stress (BS) test, an initial transfer curve is recorded first, then a fixed set of drain voltage V_D and gate voltage V_G is applied over a long time frame, with intermittent transfer curve measurements at certain intervals. Depending on the polarity of V_G and ambient conditions, different types of bias stress tests are defined:

PBS: positive bias stress, i.e. V_G is set at a positive value [9–13, 15, 16, 19, 21, 25, 54, 57, 64, 68].

NBS: negative bias stress, i.e. V_G is set at a negative value [11, 19, 21, 25, 61, 68].

NBIS/PBIS: negative/positive bias stress under constant illumination [11, 25, 26, 61].

NBTS/PBTS: negative/positive bias stress including thermal stress at elevated temperatures [25, 64].

As it is the most commonly performed test, we will focus on PBS in the following examination. While other values, such as the SS or mobility μ_{FE} can be taken into consideration, most often the shift in threshold voltage V_{th} over time is used to assess the severity of bias stress on the devices. A bias stress comparison between different literature reports is challenging as the magnitude of V_{th} and its shift over time is influenced not only by the active semiconductor channel composition but also heavily by the choice of gate dielectric. Additionally, there is no clear and commonly agreed upon, framework regarding what voltage conditions should be applied during the testing and how long a bias stress test should be run for. While most investigations were carried out for durations below 3 h [9–13, 15, 19, 21, 25, 57, 68], only a few reports feature measurements up to 9–10 h [16, 64] or even up to 24 h [54]. In an effort to facilitate comparisons between bias stress tests, Lin *et al* [54] proposed to calculate the electric flux density D and electron trap density N_{tr} according to the equations below:

$$D = C_i V_{G,\,PBS} \tag{5.1}$$

$$N_{tr} = \frac{C_i\,|\Delta V_{th}|}{e} \tag{5.2}$$

with e being the electron charge. These quantities incorporate device and measurement parameters, namely the geometric gate capacitance C_i, the applied gate voltage $V_{G,PBS}$ as well as the shift in threshold voltages (ΔV_{th}) between the initial and final transfer curve during the bias stress procedure. They are therefore more meaningful to compare than the raw measure of ΔV_{th}. Note, however, that this approach is not taking into account the value of V_D during the bias stress procedure. Higher drain voltages can reduce the ΔV_{th} through a field-assisted de-trapping mechanism (Poole–Frenkel effect) [54].

The threshold voltage shift during PBS application is typically an effect of various sources of trap states, which capture electrons from the channel when they are filled. Consequently, as the stress time increases, fewer and fewer electrons can actively contribute to the drain current, unless the gate bias is increased as well in order to compensate for this effect. As a result, the transfer curve and V_{th} are shifting to more positive gate voltages in the course of a PBS experiment. Trap states can manifest as the result of various defects, such as oxygen vacancies V_O, imperfect metal–oxide–metal (M–O–M) network formation or impurities [25]. Another source of trap sites can be seen in the top-surface of the MO channel, which can interact with ambient gases in the environment. Specifically, the presence of oxygen during PBS causes adsorption on the channel surface, leading to the formation of O_2^- species as electrons are captured from the channel.

To improve the bias stress stability of a device, each of these possible trap sites should be addressed. Inherent defects caused by e.g. incomplete M–O–M formation, remaining film porosity or impurities are often a result simply of the deposition method, where low-temperature solution processed films typically show the highest trap numbers. This might be mitigated to some degree, either by novel approaches in precursor design [82, 83], higher process temperatures or new post-deposition treatments [44, 52], or simply changing to more advanced vacuum deposition techniques. However, for various reasons this is not always feasible, e.g. regarding process constraints (temperature sensitive substrates, large scale manufacturing), or cost considerations. Furthermore, as a means to reduce the number of V_O, it is a common strategy to introduce elements with a strong affinity towards oxygen into the MO matrix, such as Ga, Hf, or Al. Together with the desired V_O reduction, however, this often has the unwanted side effect to lower the charge carrier mobility [13, 21]. It is therefore challenging to simultaneously realize high μ_{eff} and a good bias stress stability (low trap density) within a single layer MO channel.

This is where HJ transistors come into play because they allow tackling both targets at the same time. A common approach is to combine a front channel material with a high mobility μ_{FE} and carrier density N_e together with a back channel with of lower mobility but increased stability. The higher carrier density helps to passivate large number of (interface) traps and the back channel additionally adds separation between the charge transporting channel near the dielectric and any gas molecule interactions at the top surface. Additionally, the back channel may be processed in a way to minimize the likelihood of oxygen adsorption, e.g. by creating an oxygen rich material during the deposition [19].

Bias stress investigations were carried out in this kind of HJ structures with different material combinations as front and back channel. For example, Abliz *et al* used a homojunction of IGZO/IGZO in which the front channel underwent a hydrogen plasma to increase N_e [21]. Similarly, Zhan *et al* varied the carrier concentrations in both components of an IGZO/IGZO bilayer by adjusting the oxygen pressure during deposition via RF sputtering, thus creating an oxygen-poor but conductive front channel and an oxygen-rich IGZO back channel [19]. Kim *et al* used solution processed IGZO film stacks with an increased Ga content for the topmost layer [13]. Other combinations rely on a front channel of In_2O_3 combined with IZO [9], ZnO [16], or IGZO [68] as the back channel. Exceptionally, In_2O_3 was only found as the back channel material in the study by Lee and Choi [12]. In a recent study, Lin *et al* employed a solution processed In_2O_3/ZnO bilayer and improved upon the trap passivation by inserting both ZnO nanoparticles (NPs) as well a thin, UV/ozone treated layer of polystyrene (PS) in between the front and back channel [54]. This hybrid organic–MO multilayer approach lead to exceptional improvements in PBS stability.

As non-In-containing HJ channel materials for bias stress tests, the combinations of SnO_2/ZnO [64] or ZTO/Al_2O_3 [25] were investigated. Al_2O_3 was additionally studied by Ahn *et al* as part of their ZnO/Al_2O_3 superlattices processed by ALD [15].

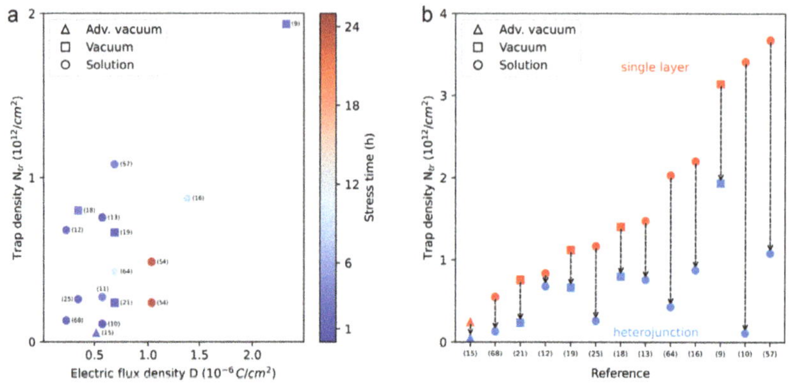

Figure 5.6. Overview of PBS literature results for MO HJ TFTs. (a) Trap density shifts for HJ TFTs over electric flux density applied during PBS application. The color coding of the marker indicates the total stress time up to 24 h. (b) Trap density shifts after PBS application for HJ TFTs (blue) and their single layer counterparts (red). In both panels, the shape of the marker indicates the deposition category, broken down to advanced and regular vacuum techniques as well as solution processing, as described earlier.

For all of the studies and HJ material combinations mentioned above, the effect was an increased resistance against bias stress. An overview comparing the results of various reports for PBS measurements of HJ TFTs is given in figure 5.6 and table 5.3. Figure 5.6(a) shows the trap density (as a measure for the shift in transfer characteristics throughout the BS measurements) depending on the electric flux density (as a quantity describing the magnitude of applied stress). Several aspects are immediately apparent. First of all, there are two distinct groupings, one at lower values of D between 0.23 to 1.4×10^{-6} C cm^{-2} and a single data point at around 2.3×10^{-6} C cm^{-2}. This reflects the fact that to date, most BS investigations on HJ-MO TFTs were carried out using bottom-gate top contact device architecture with a Si wafer as combined substrate and gate electrode, with a SiO_2 layer atop acting as the gate dielectric. The SiO_2 thickness ranges from 100 to 300 nm and the $V_{G,PBS}$ between 20 and 40 V, leading to the variation seen in the electric flux. In contrast, the investigation by Liu *et al* used a In_2O_3/InZnO HJ on top of solution processed ZrO_2 dielectric with a high C_i of 775 nF cm^{-2} which results in a high electric flux density over 2×10^{-6} C cm^{-2} even at low $V_{G,PBS}$ of 3 V [9]. While there are obviously differences between the HJ systems, as a rough trend, the resulting trap density increases with increasing electric flux, with the highest value being 1.9×10^{12} cm^{-2}. This intuitively makes sense as a more pronounced ΔV_{th} is expected when a higher gate bias is used during PBS. Finally, the color-coding in figure 5.6(a) refers to the total stress time, from which it can be seen that most PBS investigations to date are kept below 3 h while only a few are carried out for times >9 h [16, 64], with the longest one by Lin *et al* reaching up to 24 h [54].

As pointed out before, one of the benefits of using an HJ approach is in the improved bias stress stability. This is visualized in figure 5.6(b) where the resulting values of trap density for HJ TFTs are shown in comparison to a reference device based on a single layer channel material. The single layer films span over a wide

Table 5.3. Overview of PBS investigations on MO HJ structures.

Year	Heterojunction structure	Deposition method	Dielectric	C_i (nF cm^{-2})	$V_{G,PBS}$ (V)	ΔV_{th} (V)	Stress time (s)	D ($\times 10^{-6}$ C cm^{-2})	N_{tr} ($\times 10^{12}$ cm^{-2})	Reference
2011	IGZO/IGZO:N	Sputtering	SiO_2	34.5	10	3.7	10 000	0.34	0.8	[18]
2013	Al_2O_3/ZnO superlattice	ALD	SiO_2	17.3	30	0.5	7200	0.52	0.05	[15]
2013	IGZO/ZTO	Spin coating	SiO_2	28.8	20	0.6	1000	0.58	0.11	[10]
2013	IGZO/IGZO/IGZO	Spin coating	SiO_2	28.8	20	4.2	1000	0.58	0.75	[13]
2013	IGZO (O poor)/IGZO (O rich)	Sputtering	SiO_2	34.5	20	3.08	1500	0.69	0.66	[19]
2014	IGZO/ZTO	Spin coating	SiO_2	28.8	20	1.51	10 000	0.58	0.27	[11]
2014	ITZO/IGZO	Spin coating	SiO_2	34.5	20	5.01	10 000	0.69	1.08	[57]
2014	In_2O_3/InZnO	Sputtering	ZrOx	775.0	3	0.4	9000	2.33	1.93	[9]
2015	ZTO/In_2O_3	Inkjet	SiO_2	11.5	20	9.45	3600	0.23	0.68	[12]
2016	IGZO:H/IGZO	Sputtering	SiO_2	34.5	20	1.1	3600	0.69	0.24	[21]
2016	SnO_2/ZnO	Spin coating	SiO_2	34.5	20	1.98	36 000	0.69	0.43	[64]
2017	ZTO/Al_2O_3	Spin coating	SiO_2	17.3	20	2.4	4200	0.35	0.26	[25]
2019	In_2O_3/ZnO NP:Al/PS/ZnO In_2O_3/ZnO NP/PS/ZnO	Spin coating	SiO_2	34.5	30	2.251.1	86 400	1.04	0.480.24	[54]
2019	In_2O_3/IGZO	Inkjet	SiO_2	11.5	20	1.8	3600	0.23	0.13	[68]
2019	In_2O_3/ZnO	Spin coating	SiO_2	34.5	40	4.05	32 400	1.38	0.87	[16]

range of trap densities between 0.24 and $3.7 \times 10^{12}\ cm^{-2}$. Accordingly, the changes in trap density when going towards an HJ device vary over a similar range. Although not a clear correlation, the highest margins seem to be possible when the initial single layer TFT shows increased levels of N_{tr}, whereas the room for improvements is more limited with a low starting point. When comparing the different deposition methods (see marker shapes in figure 5.6(b)), the lowest trap density is achieved using atomic layer deposition [15] while higher values are the result of less sophisticated methods such as sputtering techniques or via solution based processes. In particular, solution-processed layers are often inherently more prone to defects due to, for example, non-fully converted precursors material, residual contaminants, or pinholes. As a result, those kinds of films can potential gain the most benefits of forming an HJ through the deposition of a second layer, which allows an effective defect compensation and thus more stable devices under bias stress.

5.5 Conclusions

The exploration of MO HJs and their implementations in TFTs is an exciting field of current research, ranging from basic research on 2DEG formation and transport physics on one side, to simultaneous improvements of mobility and bias stress stability for potential commercial applications in various display technologies, on the other. In particular, for deposition techniques that are geared more towards low-cost and high throughput fabrication on large area substrates, the formation of HJ structures is an approach that offers many benefits. Due to the multitude of possible combinations of HJ bilayers and even more so in connection with multilayer junctions, in terms of their composition and stacking sequences, many avenues are yet to be investigated. Advancements in interface and band gap engineering of MO materials for HJs promise superior device properties that are out of reach for single MO materials.

In addition to the improvements of current MO HJ concepts, the transfer of knowledge to further domains in the future is envisioned. Especially with regards to p-type oxides, the integration of organic materials into organic/inorganic hybrids or even considering sequences fully based on organic films, it will be exciting to see how far the idea of HJs can be extended. As a whole, an increased usage of HJ structures is expected to open new windows of opportunities for next generation (flexible) electronic applications.

Acknowledgements

This publication is based upon work supported by the King Abdullah University of Science and Technology (KAUST) Office of Sponsored Research (OSR) under Award No: OSR-2018-CARF/CCF-3079.

References

[1] Thomas S R, Pattanasattayavong P and Anthopoulos T D 2013 Solution-processable metal oxide semiconductors for thin-film transistor applications *Chem. Soc. Rev.* **42** 6910–23

[2] Petti L, Münzenrieder N, Vogt C, Faber H, Büthe L, Cantarella G, Bottacchi F, Anthopoulos T D and Tröster G 2016 Metal oxide semiconductor thin-film transistors for flexible electronics *Appl. Phys. Rev.* **3** 021303

[3] Yeon Kwon J and Kyeong Jeong J 2015 Recent progress in high performance and reliable n-type transition metal oxide-based thin film transistors *Semicond. Sci. Technol.* **30** 024002

[4] Labram J G, Lin Y-H and Anthopoulos T D 2015 Exploring two-dimensional transport phenomena in metal oxide heterointerfaces for next-generation, high-performance, thin-film transistor technologies *Small* **11** 5472–82

[5] Li Y, Zhao C, Zhu D, Cao P, Han S, Lu Y, Fang M, Liu W and Xu W 2020 Recent advances of solution-processed heterojunction oxide thin-film transistors *Nanomaterials* **10** 965

[6] Hong S, Park J W, Kim H J, Kim Y-g and Kim H J 2016 A review of multi-stacked active-layer structures for solution-processed oxide semiconductor thin-film transistors *J. Inf. Disp.* **17** 93–101

[7] Tampo H *et al* 2008 Polarization-induced two-dimensional electron gases in ZnMgO/ZnO heterostructures *Appl. Phys. Lett.* **93** 202104

[8] Yu X, Zhou N, Smith J, Lin H, Stallings K, Yu J, Marks T J and Facchetti A 2013 Synergistic approach to high-performance oxide thin film transistors using a bilayer channel architecture *ACS Appl. Mater. Interfaces* **5** 7983–8

[9] Liu G X, Liu A, Shan F K, Meng Y, Shin B C, Fortunato E and Martins R 2014 High-performance fully amorphous bilayer metal-oxide thin film transistors using ultra-thin solution-processed ZrO_x dielectric *Appl. Phys. Lett.* **105** 113509

[10] Kim C H, Rim Y S and Kim H J 2013 Chemical stability and electrical performance of dual-active-layered zinc-tin-oxide/indium-gallium-zinc-oxide thin-film transistors using a solution process *ACS Appl. Mater. Interfaces* **5** 6108–12

[11] Kim C H, Rim Y S and Kim H J 2014 The effect of a zinc–tin-oxide layer used as an etch-stopper layer on the bias stress stability of solution-processed indium–gallium–zinc-oxide thin-film transistors *J. Phys. D: Appl. Phys.* **47** 385104

[12] Lee S-H and Choi W-S 2015 Inkjet-printed oxide thin-film transistors using double-active layer structure *J. Disp. Technol.* **11** 698–702

[13] Kim D J, Rim Y S and Kim H J 2013 Enhanced electrical properties of thin-film transistor with self-passivated multistacked active layers *ACS Appl. Mater. Interfaces* **5** 4190–4

[14] Lin Y-H *et al* 2015 High electron mobility thin-film transistors based on solution-processed semiconducting metal oxide heterojunctions and quasi-superlattices *Adv. Sci.* **2** 1500058

[15] Ahn C H, Senthil K, Cho H K and Lee S Y 2013 Artificial semiconductor/insulator superlattice channel structure for high-performance oxide thin-film transistors *Sci. Rep.* **3** 2737

[16] Khim D, Lin Y-H and Anthopoulos T D 2019 Impact of layer configuration and doping on electron transport and bias stability in heterojunction and superlattice metal oxide transistors *Adv. Funct. Mater.* **29** 1902591

[17] Park J C, Kim S, Kim S, Kim C, Song I, Park Y, Jung U-I, Kim D H and Lee J-S 2010 Highly stable transparent amorphous oxide semiconductor thin-film transistors having double-stacked active layers *Adv. Mater.* **22** 5512–6

[18] Liu P-T, Chou Y-T, Teng L-F, Li F-H, Fuh C-S and Shieh H-P D 2011 Ambient stability enhancement of thin-film transistor with InGaZnO capped with InGaZnO:N bilayer stack channel layers *IEEE Electron Device Lett.* **32** 1397–9

[19] Zhan R, Dong C, Liu P-T and Shieh H-P D 2013 Influence of channel layer and passivation layer on the stability of amorphous InGaZnO thin film transistors *Microelectron. Reliab.* **53** 1879–85
[20] Cong Y *et al* 2016 High-performance Al–Sn–Zn–O thin-film transistor with a quasi-double-channel structure *IEEE Electron Device Lett.* **37** 53–6
[21] Abliz A *et al* 2016 Boost up the electrical performance of InGaZnO thin film transistors by inserting an ultrathin InGaZnO:H layer *Appl. Phys. Lett.* **108** 213501
[22] Ohtomo A and Hwang H Y 2004 A high-mobility electron gas at the $LaAlO_3$–$SrTiO_3$ heterointerface *Nature* **427** 423–6
[23] Tampo H, Matsubara K, Yamada A, Shibata H, Fons P, Yamagata M, Kanie H and Niki S 2007 High electron mobility Zn polar ZnMgO/ZnO heterostructures grown by molecular beam epitaxy *J. Cryst. Growth* **301–2** 358–61
[24] He J *et al* 2019 Defect self-compensation for high-mobility bilayer InGaZnO/In_2O_3 thin-film transistor *Adv. Electron. Mater.* **5** 1900125
[25] Cho S W, Kim D E, Kim K S, Jung S H and Cho H K 2017 Towards environmentally stable solution-processed oxide thin-film transistors: a rare-metal-free oxide-based semiconductor/insulator heterostructure and chemically stable multi-stacking *J. Mater. Chem.* C **5** 10498–508
[26] Oh H, Ko Park S-H, Hwang C-S, Yang S, Ki and Ryu M 2011 Enhanced bias illumination stability of oxide thin film transistor through insertion of ultrathin positive charge barrier into active material *Appl. Phys. Lett.* **99** 022105
[27] Nayak P K, Wang Z, Anjum D H, Hedhili M N and Alshareef H N 2015 Highly stable thin film transistors using multilayer channel structure *Appl. Phys. Lett.* **106** 103505
[28] Lee S Y, Kim J, Park A, Park J and Seo H 2017 Creation of a short-range ordered two-dimensional electron gas channel in Al_2O_3/In_2O_3 interfaces *ACS Nano* **11** 6040–7
[29] Drummond T J, Masselink W T and Morkoc H 1986 Modulation-doped GaAs/(Al, Ga)As heterojunction field-effect transistors: MODFETs *Proc. IEEE* **74** 773–822
[30] Ambacher O *et al* 1999 Two-dimensional electron gases induced by spontaneous and piezoelectric polarization charges in N- and Ga-face AlGaN/GaN heterostructures *J. Appl. Phys.* **85** 3222–33
[31] Schlom D G and Pfeiffer L N 2010 Oxide electronics: upward mobility rocks! *Nat. Mater.* **9** 881–3
[32] Edahiro T, Fujimura N and Ito T 2003 Formation of two-dimensional electron gas and the magnetotransport behavior of ZnMnO/ZnO heterostructure *J. Appl. Phys.* **93** 7673–5
[33] Tsukazaki A, Ohtomo A, Kita T, Ohno Y, Ohno H and Kawasaki M 2007 Quantum hall effect in polar oxide heterostructures *Science* **315** 1388–91
[34] Yano M, Hashimoto K, Fujimoto K, Koike K, Sasa S, Inoue M, Uetsuji Y, Ohnishi T and Inaba K 2007 Polarization-induced two-dimensional electron gas at $Zn_{1-x}Mg_xO$/ZnO heterointerface *J. Cryst. Growth* **301–2** 353–7
[35] Koike K, Hama K, Nakashima I, Takada G-Y, Ozaki M, Ogata K-I, Sasa S, Inoue M and Yano M 2004 Piezoelectric carrier confinement by lattice mismatch at ZnO/$Zn_{0.6}Mg_{0.4}O$ heterointerface *Japan J. Appl. Phys.* **43** L1372–5
[36] Tampo H, Shibata H, Matsubara K, Yamada A, Fons P, Niki S, Yamagata M and Kanie H 2006 Two-dimensional electron gas in Zn polar ZnMgO/ZnO heterostructures grown by radical source molecular beam epitaxy *Appl. Phys. Lett.* **89** 132113

[37] Nakano M, Tsukazaki A, Ohtomo A, Ueno K, Akasaka S, Yuji H, Nakahara K, Fukumura T and Kawasaki M 2010 Electric-field control of two-dimensional electrons in polymer-gated-oxide semiconductor heterostructures *Adv. Mater.* **22** 876–9

[38] Falson J, Maryenko D, Kozuka Y, Tsukazaki A and Kawasaki M 2011 Magnesium doping controlled density and mobility of two-dimensional electron gas in $Mg_xZn_{1-x}O/ZnO$ heterostructures *Appl. Phys. Express* **4** 091101

[39] Tsukazaki A, Akasaka S, Nakahara K, Ohno Y, Ohno H, Maryenko D, Ohtomo A and Kawasaki M 2010 Observation of the fractional quantum Hall effect in an oxide *Nat. Mater.* **9** 889–93

[40] Koike K, Nakashima I, Hashimoto K, Sasa S, Inoue M and Yano M 2005 Characteristics of a $Zn_{0.7}Mg_{0.3}O/ZnO$ heterostructure field-effect transistor grown on sapphire substrate by molecular-beam epitaxy *Appl. Phys. Lett.* **87** 112106

[41] Sasa S, Ozaki M, Koike K, Yano M and Inoue M 2006 High-performance ZnO/ZnMgO field-effect transistors using a hetero-metal-insulator-semiconductor structure *Appl. Phys. Lett.* **89** 053502

[42] Sasa S, Hayafuji T, Kawasaki M, Koike K, Yano M and Inoue M 2007 Improved stability of high-performance ZnO/ZnMgO hetero-MISFETs *IEEE Electron Device Lett.* **28** 543–5

[43] Lee M, Jo J W, Kim Y J, Choi S, Kwon S M, Jeon S P, Facchetti A, Kim Y H and Park S K 2018 Corrugated heterojunction metal-oxide thin-film transistors with high electron mobility via vertical interface manipulation *Adv. Mater.* **30** e1804120

[44] Yarali E *et al* 2020 Low-voltage heterojunction metal oxide transistors via rapid photonic processing *Adv. Electron. Mater.* **6** 2000028

[45] Chin H-A, Cheng I C, Huang C-I, Wu Y-R, Lu W-S, Lee W-L, Chen J Z, Chiu K-C and Lin T-S 2010 Two dimensional electron gases in polycrystalline MgZnO/ZnO heterostructures grown by rf-sputtering process *J. Appl. Phys.* **108** 054503

[46] Huang C-I, Chin H-A, Wu Y-R, Cheng I C, Chen J Z, Chiu K-C and Lin T-S 2010 Mobility enhancement of polycrystalline MgZnO/ZnO thin film layers with modulation doping and polarization effects *IEEE Trans. Electron Devices* **57** 696–703

[47] Cheng I-C, Wang B-S, Hou H-H and Chen J-Z 2012 MgZnOZnO heterostructure field-effect transistors fabricated by RF-sputtering *ECTS Trans.* **50** 83–93

[48] Remashan K, Choi Y S, Park S J and Jang J H 2010 High performance MOCVD-grown ZnO thin-film transistor with a thin MgZnO layer at channel/gate insulator interface *J. Electrochem. Soc.* **157** H1121

[49] Taniguchi S, Yokozeki M, Ikeda M and Suzuki T-k 2011 Transparent oxide thin-film transistors using n-$(In_2O_3)_{0.9}(SnO_2)_{0.1}/InGaZnO_4$ modulation-doped heterostructures *Japan J. Appl. Phys.* **50** 04df11

[50] Faber H *et al* 2017 Heterojunction oxide thin-film transistors with unprecedented electron mobility grown from solution *Sci. Adv.* **3** e1602640

[51] Tetzner K, Isakov I, Regoutz A, Payne D J and Anthopoulos T D 2017 The impact of post-deposition annealing on the performance of solution-processed single layer In_2O_3 and isotype In_2O_3/ZnO heterojunction transistors *J. Mater. Chem.* C **5** 59–64

[52] Tetzner K, Lin Y-H, Regoutz A, Seitkhan A, Payne D J and Anthopoulos T D 2017 Sub-second photonic processing of solution-deposited single layer and heterojunction metal oxide thin-film transistors using a high-power xenon flash lamp *J. Mater. Chem.* C **5** 11724–32

[53] Khim D, Lin Y-H, Nam S, Faber H, Tetzner K, Li R, Zhang Q, Zhang X and Anthopoulos T D 2017 Modulation-doped In_2O_3/ZnO heterojunction transistors processed from solution *Adv. Mater.* **29** 1605837

[54] Lin Y-H *et al* 2019 Hybrid organic–metal oxide multilayer channel transistors with high operational stability *Nat. Electron.* **2** 587–95
[55] Liu L, Chen S, Liang X and Pei Y 2019 Solution processed AlInO/In_2O_3 heterostructure channel thin film transistor with enhanced performance *Adv. Electron. Mater.* **5** 1900550
[56] Chen Y *et al* 2019 Polymer doping enables a two-dimensional electron gas for high-performance homojunction oxide thin-film transistors *Adv. Mater.* **31** e1805082
[57] Rim Y S, Chen H, Kou X, Duan H S, Zhou H, Cai M, Kim H Y and Yang Y 2014 Boost up mobility of solution-processed metal oxide thin-film transistors via confining structure on electron pathways *Adv. Mater.* **26** 4273–8
[58] Kim J-I *et al* 2011 Improvement in both mobility and bias stability of ZnSnO transistors by inserting ultra-thin InSnO layer at the gate insulator/channel interface *Appl. Phys. Lett.* **99** 122102
[59] Chong E and Lee S Y 2012 Influence of a highly doped buried layer for HfInZnO thin-film transistors *Semicond. Sci. Technol.* **27** 012001
[60] Kim H S *et al* 2012 Density of states-based design of metal oxide thin-film transistors for high mobility and superior photostability *ACS Appl. Mater. Interfaces* **4** 5416–21
[61] Park J C and Lee H-N 2012 Improvement of the performance and stability of oxide semiconductor thin-film transistors using double-stacked active layers *IEEE Electron Device Lett.* **33** 818–20
[62] Kim K M, Jeong W H, Kim D L, Rim Y S, Choi Y, Ryu M-K, Park K-B and Kim H J 2011 Low-temperature solution processing of AlInZnO/InZnO dual-channel thin-film transistors *IEEE Electron Device Lett.* **32** 1242–4
[63] Jeong W H, Kim K M, Kim D L, Rim Y S and Kim H J 2012 The effects of dual-active-layer modulation on a low-temperature solution-processed oxide thin-film transistor *IEEE Trans. Electron Devices* **59** 2149–52
[64] Nam S *et al* 2016 Solution-processed indium-free ZnO/SnO_2 bilayer heterostructures as a low-temperature route to high-performance metal oxide thin-film transistors with excellent stabilities *J. Mater. Chem.* C **4** 11298–304
[65] Jung H Y, Kang Y, Hwang A Y, Lee C K, Han S, Kim D H, Bae J-U, Shin W-S and Jeong J K 2014 Origin of the improved mobility and photo-bias stability in a double-channel metal oxide transistor *Sci. Rep.* **4** 3765
[66] Al-Jawhari H A, Caraveo-Frescas J A, Hedhili M N and Alshareef H N 2013 P-type Cu_2O/SnO bilayer thin film transistors processed at low temperatures *ACS Appl. Mater. Interfaces* **5** 9615–9
[67] Zeumault A and Subramanian V 2017 Use of high-k encapsulation to improve mobility in trap-limited metal-oxide semiconductors *Phys. Status Solidi* b **254** 1700124
[68] Liang K *et al* 2019 High-performance metal-oxide thin-film transistors based on inkjet-printed self-confined bilayer heterojunction channels *J. Mater. Chem.* C **7** 6169–77
[69] Sanctis S, Krausmann J, Guhl C and Schneider J J 2018 Stacked indium oxide/zinc oxide heterostructures as semiconductors in thin film transistor devices: a case study using atomic layer deposition *J. Mater. Chem.* C **6** 464–72
[70] Lee S-J, Hwang C-S, Pi J-E, Yang J-H, Byun C-W, Chu H Y, Cho K-I and Cho S-H 2015 High performance amorphous multilayered ZnO-SnO_2 heterostructure thin-film transistors: fabrication and characteristics *ETRI J.* **37** 1135–42

[71] Krausmann J, Sanctis S, Engstler J, Luysberg M, Bruns M and Schneider J J 2018 Charge transport in low-temperature processed thin-film transistors based on indium oxide/zinc oxide heterostructures *ACS Appl. Mater. Interfaces* **10** 20661–71
[72] Kim S I, Kim C J, Song I, Kim S W, Yin H and Lee E *et al* 2008 High performance oxide thin film transistors with double active layers *2008 IEEE Int. Electron Dev. Meeting.* pp 1–4
[73] Chong E, Jeon Y W, Chun Y S, Kim D H and Lee S Y 2011 Localization effect of a current-path in amorphous In–Ga–Zn–O thin film transistors with a highly doped buried-layer *Thin Solid Films* **519** 4347–50
[74] Hsu H-H, Chang C-Y, Cheng C-H, Chiou S-H and Huang C-H 2014 High mobility bilayer metal–oxide thin film transistors using titanium-doped InGaZnO *IEEE Electron Device Lett.* **35** 87–9
[75] Lee W, Kim J and Kim Y-H 2017 High-performance InO_x/GaO_x bilayer channel thin-film transistors made using persistent high-surface-energy induced by photochemical activation *J. Alloys Compd.* **723** 627–32.
[76] Lee H, Zhang X, Kim J W, Kim E J and Park J 2018 Investigation of the electrical characteristics of bilayer ZnO/In_2O_3 thin-film transistors fabricated by solution processing *Materials* **11** 2103
[77] Lee J-M, Lee H-J, Pi J-E, Yang J-H, Lee J H, Ahn S-D, Kang S-Y and Moon J 2019 All-oxide thin-film transistors with channels of mixed InO_x–ZnO_y formed by plasma-enhanced atomic layer deposition process *J. Vac. Sci. Technol.* A **37** 060910
[78] Sheng J, Hong T, Lee H M, Kim K, Sasase M, Kim J, Hosono H and Park J-S 2019 Amorphous IGZO TFT with high mobility of $\sim$ 70 $cm^2/(V\ s)$ via vertical dimension control using pEALD *ACS Appl. Mater. Interfaces* **11** 40300–9
[79] Chin H A, Cheng I C, Li C K, Wu Y R, Chen J Z, Lu W S and Lee W-L 2011 Electrical properties of modulation-doped rf-sputtered polycrystalline MgZnO/ZnO heterostructures *J. Phys. D: Appl. Phys.* **44** 455101
[80] Kim K T, Kim J, Kim Y-H and Park S K 2014 In-situ metallic oxide capping for high mobility solution-processed metal-oxide TFTs *IEEE Electron Device Lett.* **35** 850–2
[81] Cohen D J and Barnett S A 2005 Predicted electrical properties of modulation-doped ZnO-based transparent conducting oxides *J. Appl. Phys.* **98** 053705
[82] Kim M G, Kanatzidis M G, Facchetti A and Marks T J 2011 Low-temperature fabrication of high-performance metal oxide thin-film electronics via combustion processing *Nat. Mater.* **10** 382–8
[83] Lin Y-H, Faber H, Zhao K, Wang Q, Amassian A, McLachlan M and Anthopoulos T D 2013 High-performance ZnO transistors processed via an aqueous carbon-free metal oxide precursor route at temperatures between 80-180 °C *Adv. Mater.* **25** 4340–6

Chapter 6

Application of one-dimensional metal oxide semiconductor in field effect transistor

Fengyun Wang and Hongliang Zu

One-dimensional metal oxide nanostructures, especially nanowires (NWs) and nanofibers (NFs), due to their excellent physical and chemical properties, exhibit unique optical, electrical, chemical, thermal, and magnetic properties, making them suitable for the next-generation, high-performance electronic products. In this article, we will introduce the development and application of one-dimensional metal oxide field effect transistors (FETs), summarize the characteristics of different kinds of metal oxide FET and the methods to improve their performance, and discuss various application areas. Finally, the challenges and development prospects of metal oxide (NW) transistors are discussed, including the investigation of material properties, and the design and application of device structures.

6.1 Introduction

Two-dimensional metal oxide thin film materials have great value in scientific research and practical applications and have been successfully applied to thin film transistor (TFT) devices. However, as science and technology applications have progressed, two-dimensional thin-film materials have been found to have some shortcomings. The electron transmission path of two-dimensional thin-film semiconductor materials is not fixed [1], which leads to the scattering of electrons in the channel and reduces the carrier mobility of the device. Additionally, the specific surface area of the dimensional thin film material is small, which severely limits the application of any device it constitutes to detect gas and photoelectrons. One-dimensional metal oxides with unique, short electron transport channels and large specific areas have become the focus of research, and are being widely applied in the areas of logic circuits, photoelectric detectors, gas detectors and other fields [2–21].

In this article, we will summarize the latest progress in high-performance metal oxide nanowire field-effect transistors (NWFETs) [22], covering all aspects from

doi:10.1088/978-0-7503-2556-1ch6

basic materials to important applications. In section 6.2, we will briefly introduce the preparation of metal oxide NWs, and in section 6.3, we will introduce the basic physical principles of one-dimensional metal oxide semiconductor FETs. After that, in section 6.4, various FETs, manufactured using different metal oxide NWs or NFs, are introduced. In section 6.5, several examples of metal oxide NWFETs used in different fields will be discussed. Finally, at the end of the article, we summarize the existing problems and future developmental directions of metal oxide NWFETs.

6.2 Preparation of one-dimensional metal oxide materials

Since the preparation of carbon nanotubes by Dr Iijima in Japan in 1991, the study of one-dimensional materials has attracted great attention from experts and scholars in the fields of materials, physics, and chemistry. One-dimensional nanostructures have the characteristics of surface and interface effects, small size effects, quantum size effects, and macroscopic quantum tunneling effects, and they exhibit unique properties applicable in the fields of optics, electricity, thermodynamics, and magnetism. Compared with two-dimensional thin-film materials, one-dimensional metal oxide semiconductor nanomaterials have a directional electron transmission path and a large specific surface area. Therefore, one-dimensional metal oxide semiconductor nanomaterials were prepared with suitable forbidden band widths, large carrier concentrations, and good stability. Metal oxide semiconductor nanomaterials have become a research hot topic. The current methods for preparing one-dimensional metal oxide semiconductor nanomaterials are chemical vapor deposition, the sol–gel method, the solvo-thermal method, molecular beam epitaxy, and electrospinning [23].

6.2.1 Chemical vapor deposition

Chemical vapor deposition (CVD) is a relatively new technology for preparing inorganic materials [24]. Chemical vapor deposition has been widely used to purify substances, develop new crystals, and deposit various single crystal, polycrystalline, or glassy inorganic materials. These materials can be oxides, sulfides, nitrides, or carbides and binary or multi-element compounds in the groups III–V, II–IV, IV–VI. The doping process can also be precisely controlled. CVD has become a new field of inorganic synthetic chemistry [23, 25–28].

CVD technology is a process that uses gaseous substances to produce chemical and transport reactions on solids, thus producing the solid deposits [2–4], as shown in figure 6.1. It generally includes three steps:

1. Formation of volatile substances;
2. Transfer of these substances to the deposition area;
3. Production of chemical reactions on solids to produce solid deposits.

Metal oxides are one of the synthesized compounds. One-dimensional metal oxide nanostructures can be easily synthesized using CVD technology. Nano-sized metal catalysts are used in CVD technology to grow one-dimensional metal oxide nanostructures. The diameter or width of the one-dimensional metal oxide

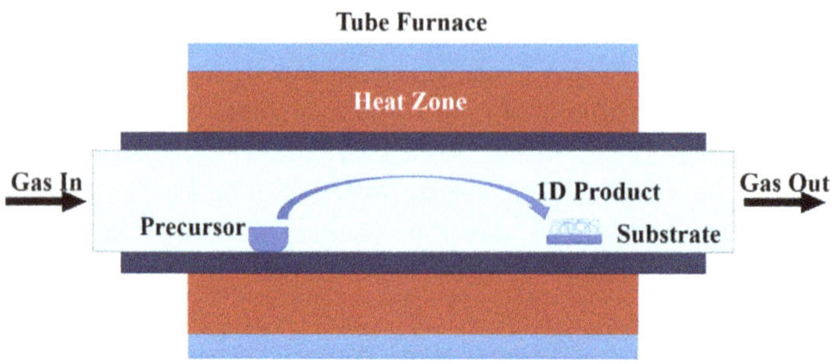

Figure 6.1. Schematic of a tube furnace for CVD synthesis of 1D nanostructures.

nanostructure depends largely on the size of the nanoscale metal catalyst. The structural morphology of the one-dimensional metal oxide nanostructures is related to the substrate temperature, molecular surface diffusion rate, and the concentration of metal and oxygen vapor.

Based on this knowledge of CVD, metal organic chemical vapor deposition (MOCVD) technology has been developed. MOCVD can reduce the deposition temperature for the growth of one-dimensional metal oxide nanostructures. MOCVD technology has the advantages of thickness control, precise doping, large area deposition and morphological diversity. Most CVD synthesis techniques can only provide autocatalytic growth or layer-by-layer growth of one-dimensional metal oxide nanostructures.

6.2.2 Sol–gel technique

The sol–gel method uses a compound containing a highly chemically active component as a precursor. These raw materials are uniformly mixed in the liquid phase. Then hydrolysis and condensation chemical reactions are preformed to stabilize the transparent sol system in solution [2–4, 23]. The colloidal particles slowly polymerize to form a three-dimensional network structure, and the solvent loses fluidity to form a gel. The gel is dried, sintered and solidified to prepare molecular and even nano-substructure materials. Sol–gel technology can be used to synthesize one-dimensional metal oxide nanostructures. The precursors are alkali metal oxides or metal salts, such as chlorides, nitrates and acetates, usually surrounded by various reactive ligands. Therefore, the precursor can undergo various hydrolysis and condensation reaction processes. The raw materials in the sol are processed to form dispersed metal oxides in contact with liquid water or dilute acid. Removal of the liquid water or dilute acid from the sol produces a gel, and the sol/gel transition controls the particle size and shape. For example, the calcination of a gel produces metal oxides. Figure 6.2 shows how a colloidal suspension (sol) of an alkali metal oxide or metal salt is dissolved in a solute. Hydrolysis and condensation occur in the sol when poured onto the rotating substrate. As the sol rotates, hydrolysis and condensation proceed outward. As the sol gradually turns into a gel, the colloidal suspension coalesces to form a 1D nanostructure by self-assembly. The gel can be heated to

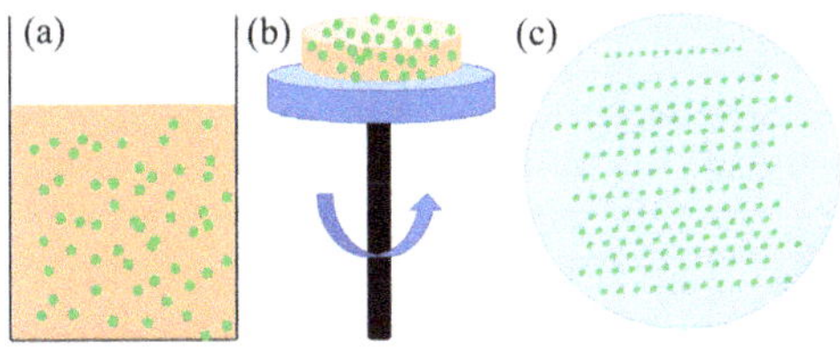

Figure 6.2. Schematic diagrams of the sol–gel technique: (a) the colloidal suspension in the sol is represented by green spots; (b) sol is spun to form a gel; and (c) after the heat treatment 1D metal-oxide nanostructures grow on the substrate.

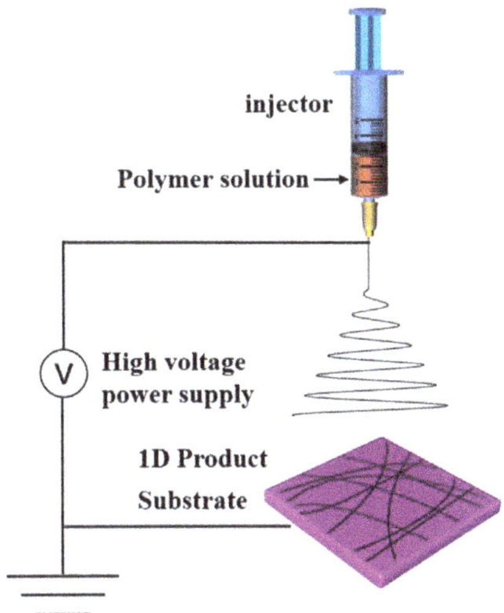

Figure 6.3. Schematic diagram of one-dimensional nanostructure prepared by electrospinning.

completely remove the liquid water or dilute acid, thus forming a one-dimensional metal oxide nanostructure.

6.2.3 Electrospinning technology

Electrostatic spinning equipment is mainly composed of three parts: a propulsion pump, a high-voltage power supply, and a receiving substrate, as shown in figure 6.3. The propulsion pump pushes the liquid out of the syringe, the droplet is stretched into a wire under the action of a high-voltage power supply, and then received by the receiving substrate. Electrospinning technology can not only be used to prepare nanofibers, but also nanotubes, core–shell structures, multi-wall nanofibers, nanoflowers and other similar nanostructures [29–35]. Electrospinning technology has the advantages of simple operation, low cost, high efficiency and large output. The prepared nanofibers are widely used in the fields of catalysis, energy, and electronic devices due to the advantages of larger specific surface areas, better stability and uniform morphology [35–41].

The configuration of the precursor solution determines the spinning quality. In general, *N*,*N*-dimethylformamide (DMF), alcohol, or deionized water, are usually used. In addition, it is necessary to add a polymer as a binder to the precursor to control the viscosity. The proper viscosity is a necessary condition for fiber formation and is very important for the formation of nanofibers. In the preparation of inorganic nanofibers, the following polymers are commonly used: polyvinylpyrrolidone (PVP), polylactic acid (PLA), polyvinyl alcohol (PVA), etc.

Compared with other technologies, electrospinning has the following advantages: (1) large area nanofiber thin layers can be prepared. (2) The prepared fiber has a very large specific surface area. (3) The method is simple, only a high-voltage power supply, propulsion pump and receiving substrate are necessary. (4) The cost is low. No precious metal is needed as a catalyst, and no subsequent photolithography technology is needed to prepare the electrode. (5) It is easy to uniformly incorporate other elements. Just add the doping metal salt to the precursor solution. These advantages make electrostatic spinning technology very suitable for the preparation of electronic devices and optoelectronic devices.

6.3 Basic principle of one-dimensional metal oxide semiconductor FET

The one-dimensional metal oxide semiconductor FET is a field effect transistor that uses one-dimensional metal oxide as a conductive channel, and controls the current of the NW by controlling the voltage of the gate [42, 43]. The basic structure is shown in figure 6.4.

Laid out from top to bottom are the source and drain electrodes, the semiconductor layer, the dielectric layer and the gate. The FET is a three-terminal device, that is, the opening and closing of the device is controlled by three electrodes, the source electrode, the drain electrode, and the gate electrode [44]. For enhanced n-channel FETs, where only a certain voltage value V_{DS} is added between the source and drain electrodes, the current in the transistor is very small or nonexistent. The concentration of freely moving carriers in the semiconductor layer is very low, so the current generated by the voltage is very low. The lowest current generated at this

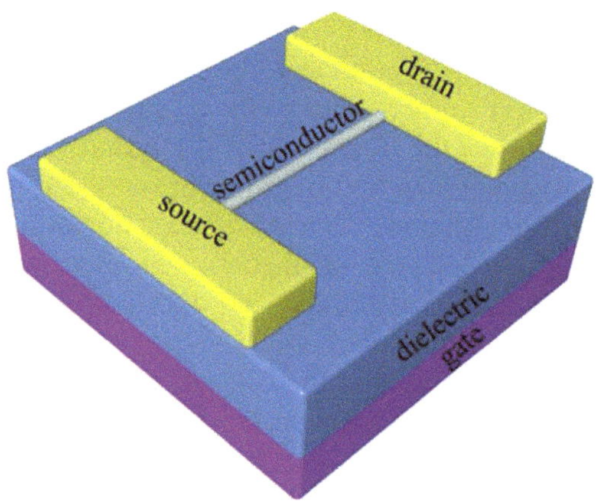

Figure 6.4. Schematic diagram of NW-FET structure.

time is called the off-state current of the FET. Therefore, to make the transistor work, a voltage V_{GS} is added between G and S and a voltage V_{DS} is added between D and S. When a bias is applied to the gate, the dielectric layer will form a capacitor, and carriers will be induced between the semiconductor layer and the node dielectric layer by the applied voltage. In this way, carriers will now move from the source to the drain, forming a current between the source and drain electrodes. The current channel is close to the interface between the semiconductor layer and the dielectric layer. The channel length (L) is defined as the distance between the source electrode and the drain electrode, and the channel width (W) is defined as the total length of the source electrode and the drain electrode in parallel. The channel is generally located a few nanometers above the dielectric layer, and its thickness is extremely narrow, which plays a crucial role in the performance of the transistor. The interface between the semiconductor layer and the dielectric layer is very important [45, 46].

There are six main evaluation parameters of FET performance: threshold voltage (V_{TH}), off-state current (I_{off}), on-state current (I_{on}), switching ratio (I_{on}/I_{off}), carrier mobility (μ), and sub-threshold swing (SS). The V_{TH} usually refers to the input voltage corresponding to the midpoint of the transition area where the output current in the transfer characteristic curve changes abruptly as the input voltage changes. The V_{TH} can be the value of V_{TH} at the intersection of the tangent and the V_G axis in the relationship curve between the square root of the current ${I_{DS}}^{1/2}$ and V_G. The V_{TH} of an ideal FET should be 0 V, but in there will always be deviations. In practical applications, there will be a high carrier concentration in the semiconductor or a current between the semiconductor and the interface of the dielectric layer sub-traps which will cause the V_{TH} to be positive or negative from time to time. When $V_{TH} < 0$ V, the device is in depletion mode (D-mode) and when $V_{TH} > 0$ V, the device is in enhancement mode (E-mode). Turning off a D-mode device requires an application of a negative pressure. Thus, a D-mode device consumes more energy than the E-mode device. The current switching ratio (I_{on}/I_{off}) represents the ratio of the on-state current to the off-state current. Since the transistor has a switching function in the circuit, the current switching ratio of an excellent transistor should be $I_{on}/I_{off} = 10^6$ or higher. Carrier mobility refers to the average drift velocity of carriers (electrons and holes) under the action of a unit electric field. In the different semiconductor material, carrier types are different, and mobility is different. Generally, the mobility of electrons is higher than that of holes. Mobility mainly affects two properties of transistors. First, the carrier concentration determines the conductivity (reciprocal of resistivity) of semiconductor materials. The greater the mobility, the smaller the resistivity. When passing the same current, the power consumption is smaller, and the current carrying capacity is greater. Second, the rate of carrier mobility also affects the refresh rate of the device. When the mobility is small, it can be used in devices with low refresh frequency requirements, such as e-books. High mobility is useful in some high-end devices, such as TVs, computers, and some screens used in eSports. The SS value is equal to the gate voltage increment ΔV_G required to change the current I_{DS} by an order of magnitude. Note that SS is extracted from the maximum slope on the V_G–I_D curve. It indicates the rise rate of the $\tilde{I_D}V_G$ relationship curve. In order to increase the operating speed of the subthreshold region of the transistor, the smaller the SS value, the better.

6.4 Individual one-dimensional metal oxide NW transistors

6.4.1 Binary metal oxide NW transistor

6.4.1.1 ZnO NW transistors

ZnO has excellent optical and electrical properties, has received great attention in the fields of electronics, optoelectronics and piezoelectric electronics, and is one of the most outstanding materials in the metal oxide family. ZnO is a direct gap wide band gap compound semiconductor material with a band gap width of 3.37 eV at room temperature [47]. It has a wurtzite structure and belongs to a hexagonal system of point groups. ZnO can produce high-efficiency ultraviolet laser emissions at room temperature, and has broad application prospects in the field of short-wavelength optical devices [26, 48–52]. ZnO is also a piezoelectric and thermoelectric semiconductor material with good performance. Because of its piezoelectric and photoelectric effects, ZnO nanofibers are widely used in electrical, optical, and detector devices, and are further made into photodetectors, gas detectors, and filters [13, 15, 17, 29, 31, 43, 47, 53–55].

Although significant progress has been made in the use of ZnO NWs in high-performance and low-cost electronic products, due to the large surface roughness and granular structure of the obtained NFs, they still suffer from insufficient device performance. It should be noted, Wang *et al* proposed a simple one-step electrospinning process to obtain a controllable ZnO NW network of high-performance, large-scale, low-power TFTs [31]. By precisely controlling the annealing temperature during the manufacturing process, the crystallinity, grain size distribution, surface morphology, etc can be reliably adjusted to enhance the performance of the transistor. An annealing temperature of 500 °C produces a device with a positive threshold voltage (V_{TH}) of approximately 0.9 V, a low leakage current of approximately 10^{-12} A and an excellent on/off current ratio 10^6. This is one of the best performing ZnO NF devices reported so far. When high-κ AlO_x films are used as the gate dielectric, the source/drain voltage (V_{DS}) can be reduced by a factor of 10 to only 0.3 V. These results indicate the potential of these nanofibers in next-generation low-power devices (figure 6.5) [31].

6.4.1.2 In_2O_3 NW transistors

In_2O_3 is a new n-type transparent semiconductor material with a wide band gap (direct band gap of about 3.6 eV and indirect band gap of about 2.5 eV), small resistivity, and high catalytic activity. It is widely used in the field of optoelectronics, gas sensors, catalysts, solar cells, and diodes. In_2O_3 has higher electron mobility and carrier concentration than other semiconductor materials and has better conductivity. When In_2O_3 nanofibers are built into transistors, the carrier mobility range is about 70–250 $cm^2\ V^{-1}\ s^{-1}$. However, the current switching ratio is one order of magnitude lower than that of ZnO semiconductor. Because In_2O_3 semiconductors

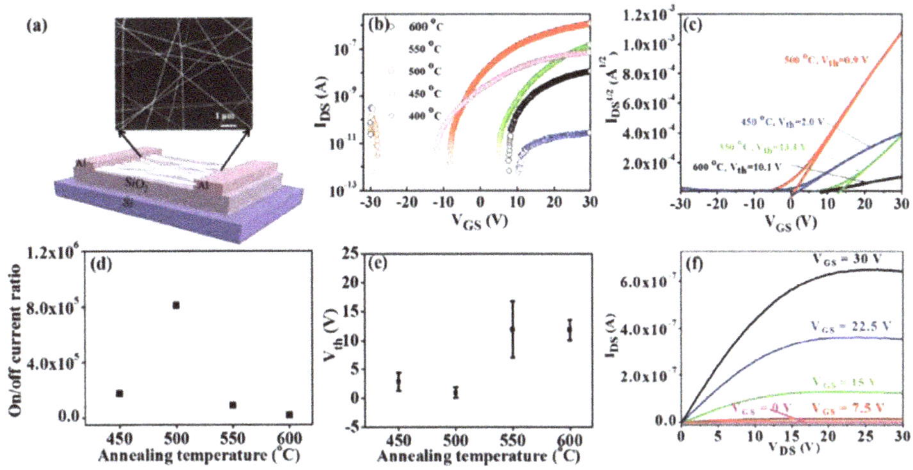

Figure 6.5. (a) Top: SEM image and bottom: model of a typical back-gated ZnO NFTFT fabricated with Al electrodes and heavily doped silicon substrates covered with 300 nm thick SiO_2. (b) The transfer curves, (c) V_{TH}, (d) average on/off current ratio, and (e) average V_{TH} of ZnO NFTFTs annealed at different temperatures. (f) The output curves of ZnO NFTFT annealed at 500 °C. [31] John Wiley & Sons. © 2018 WILEY-VCH Verlag GmbH & Co., KGaA, Weinheim.

have lower off-state currents due to lower resistivity, the switching ratio is about 10^3 to 10^6 [56, 57].

At the same time, In_2O_3 semiconductor nanomaterials also have good optical transmittance, and their transmittance in the visible light region generally exceeds 80% and 90%. Thus In_2O_3 is an ideal material for preparing transistors [13, 29–31]. Park *et al* prepared In_2O_3 nanofibers by electrospinning [58]. After annealing to remove the polymer, a bottom-gate top-contact FET was prepared. The transistor showed typical n-channel properties. In order to improve the performance of the transistor, insulating zirconia (ZrO_2) was used in a double-layer structure of a NW network. ZrO_2 has a high binding energy with oxygen which is helpful in reducing the oxygen-related defects in the NW. The resulting transistor has a high current switching ratio, high carrier mobility, small sub-threshold swing (figure 6.6) [58].

6.4.1.3 SnO_2 NW transistors

SnO_2 is also an n-type semiconductor material with a forbidden band width of 3.6 eV, a square rutile structure, a transmittance of 80% in the visible and infrared regions, and good chemical stability [48]. It was the first commercially available transparent conductive material. Due to the small size effect, quantum size effect, surface effect and macroscopic quantum tunneling effect, SnO_2 nanomaterials show significant changes to its physical properties in regards to light [59], heat, electricity, sound, magnetism, and other macroscopic properties compared with traditional SnO_2. The carriers of a SnO_2 semiconductor mainly comes from crystal defects, such as oxygen vacancies and electrons provided by the doping impurities [15, 16, 48, 59, 60].

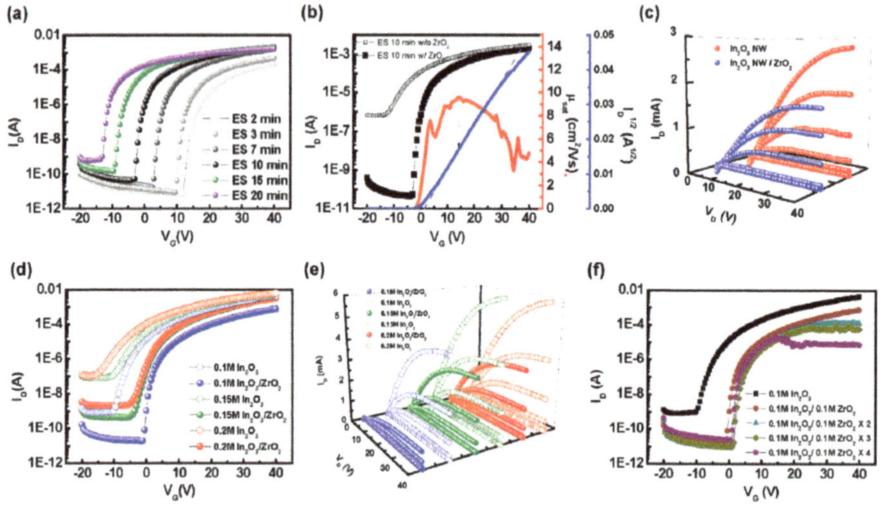

Figure 6.6. (a) Transfer characteristics of the hybrid In_2O_3 nanowire/ZrO_2 thin film FET as a function of electrospinning time. (b) Transfer characteristics and (c) output characteristics of bare In_2O_3 nanowire FET and nanowire/ZrO_2 film FET after the optimized collecting time (10 min). (d) Transfer and (e) output characteristics of single In_2O_3 channel and In_2O_3/ZrO_2 bilayer channel TFTs as a function of the In_2O_3 thickness. (f) Transfer characteristics of 0.1 m In_2O_3 TFTs with different thickness of ZrO_2 films. [58] John Wiley & Sons. © 2018 WILEY-VCH Verlag GmbH & Co., KGaA, Weinheim.

Wang *et al* have grown SnO_2 NWs on sapphire substrates by using a vapor–liquid–solid (VLS) process in a low-pressure chemical vapor deposition system [61]. The length of these parallel NWs can reach 100 μm. The fabricated transistor had an on/off current ratio of 10^6, a carrier mobility of about 71.68 $cm^2\ V^{-1}\ s^{-1}$ and a sufficiently high current to drive an external organic light-emitting diode display. Not only is the device sensitive to ultraviolet light, it can distinguish between 254 and 365 nm wavelengths of light, and detect NO_2 with a concentration of 0.2 ppb (figure 6.7) [61].

6.4.2 Ternary metal oxide NW transistor

With the continuous development of NW transistors, the disadvantages of binary metal oxides have been gradually revealed. In_2O_3 semiconductors are a good example. Although they have excellent electrical and optical properties, transistors based on In_2O_3 semiconductor nanofibers have many disadvantages: high leakage current, low current switching ratio, and negative threshold voltage. This leads to poor device performance and energy consumption. At this time, the electrical properties of NWs can be adjusted by introducing other metal elements to improve the performance of the transistor [15].

In order to improve the performance of In_2O_3 NW transistors, Zhang *et al* prepared Mg-doped In_2O_3 NWs by electrospinning. The produced a high-performance enhanced FET [29]. The performance of the transistor can be controlled by simply adjusting the Mg content in the NW. At the optimal Mg concentration of

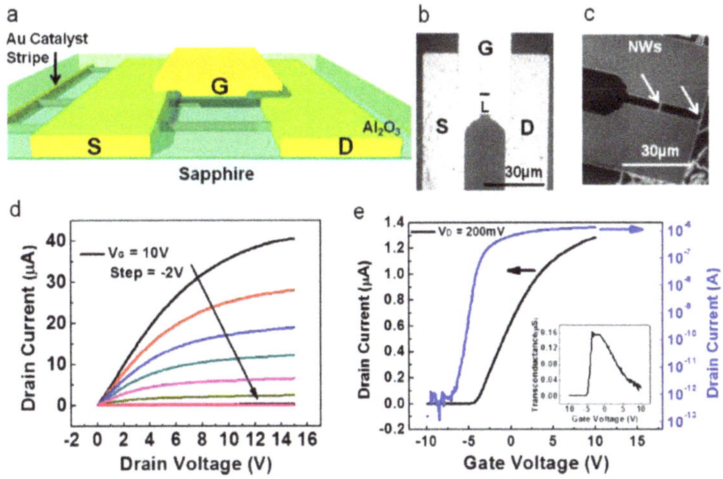

Figure 6.7. Aligned SnO_2 nanowire transistor study. (a) Diagram of device fabrication. (b) Top-view optical image of an aligned SnO_2 nanowire transistor. (c) SEM image of two aligned SnO_2 nanowires bridging the source and drain electrodes of a nanowire transistor. (d) I_D-V_D family plot of the transistor shown in (c). (e) I_D-V_G plot of the transistor in (c) plotted with standard scale in black and logarithmic scale in blue. The subplot shows the transconductance of the same device. Reprinted with permission from [61]. Copyright (2014) American Chemical Society.

2 mol%, the device carrier mobility is 2.04 $cm^2\ V^{-1}\ s^{-1}$, the current switching ratio is greater than 10^8 and the threshold voltage is 3.2 V (figure 6.8) [29].

Due to various shortcomings of single metal element oxides, Song *et al* doped Sr into indium oxide NWs by the electrospinning method to improve the electrical properties, thereby effectively controlling the carrier concentration in indium oxide NFs [30]. By simply controlling the concentration of Sr, In_2O_3 NFs with controllable electrical properties were obtained. Using 3.6 mol% Sr-In_2O_3 NFs as the channel material, an enhancement mode (E mode) FET was successfully constructed. By doping the Sr element, the off-state current of the device was not only reduced by at least 5 orders of magnitude, and the threshold voltage was adjusted to 0.3 V, but also its stability was improved. The carrier mobility of the device was 3.67 $cm^2\ V^{-1}\ s^{-1}$, the threshold voltage was between 0 and 5 V, and the current switching ratio was 10^7 to 10^8. In addition, when Al_2O_3 is used as a dielectric layer, the operating voltage is reduced to one-tenth of the original. This not only reduces costs and energy consumption, but also the stability of the equipment stays very high. Moreover, when a resistive load inverter is paired with the completed device, there is a good voltage gain. These results show that the preparation of NFs by the electrostatic spinning method has broad application prospectives in electronic products with low cost, low energy consumption, high efficiency and high performance (figure 6.9) [30].

6.4.3 Quaternary metal oxide NW transistor

Although amorphous IGZO materials have been widely used in various thin film transistor (TFT) applications, their device performance is still limited by inherent

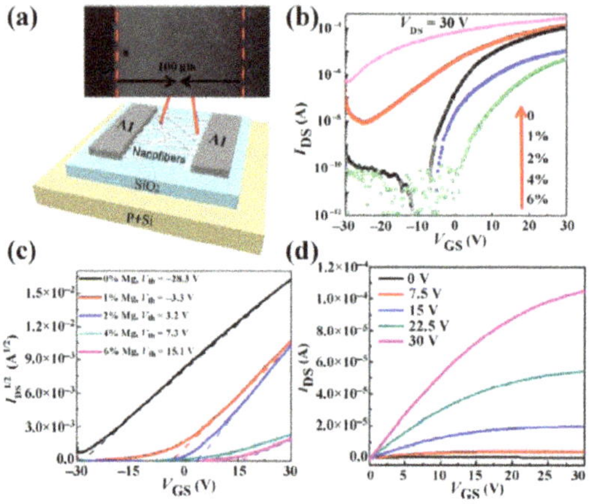

Figure 6.8. (a) (Top) SEM image and (bottom) schematic of a back-gated Mg-doped In_2O_3 NF FETs fabricated with Al electrodes and heavily doped silicon substrates covered with 150 nm thick thermally grown SiO_2. The SEM image illustrates the density of Mg-doped In_2O_3 NFs in the channel and when the density was 0.43 pillars/μm (spinning time of 30 s); the Mg-doped In_2O_3 NF FETs show the best performance. (b) Transfer curves for the backgated FETs with different Mg doping concentrations (0, 1, 2, 4, and 6 mol%). (c) $I_{DS}^{1/2}$ versus V_{GS} curves for the back-gated FET with different Mg doping concentrations (0, 1, 2, 4, and 6 mol%), demonstrating the D-mode FET being successfully transformed into an E-mode one. (d) Output curves of 2 mol% Mg-doped In_2O_3 NF FETs. Reproduced from [29] Tsinghua University Press.

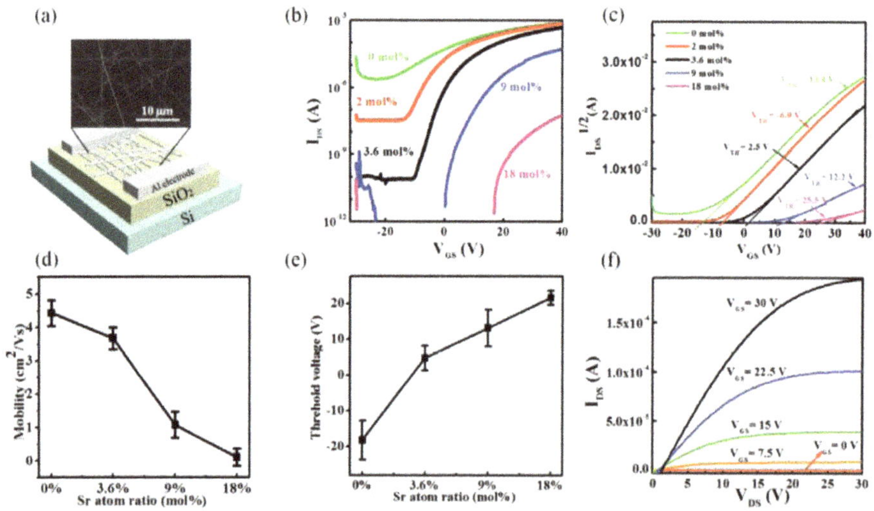

Figure 6.9. (a) Schematic diagram of NF-FET, inset is the SEM image of In_2O_3 NFs. (b) Transfer characteristic of undoped, 2, 3.6, 9, and 18 mol% Sr–In_2O_3 devices, respectively. (c) V_{TH} corresponding to the transfer curves in (b). (d) The corresponding average field effect mobility. (e) The corresponding average V_{TH} of all Sr–In_2O_3 devices. (f) Output characteristic of 3.6 mol% Sr–In_2O_3/SiO_2 device. [30] John Wiley & Sons. © 2019 WILEY-VCH Verlag GmbH & Co., KGaA, Weinheim.

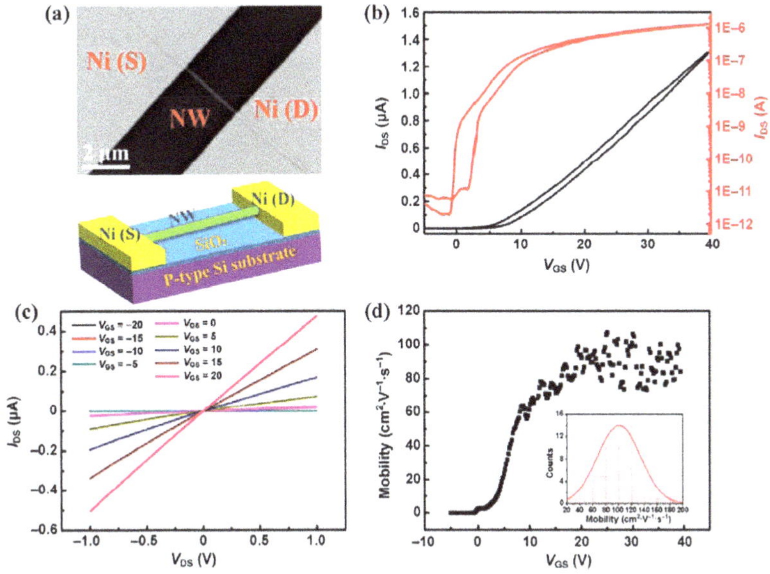

Figure 6.10. Electrical characterization of a typical FET based on a single $In_{1.8}Ga_{1.8}Zn_{2.4}O_7$ NW under vacuum. (a) SEM image and the schematic illustration of the FET device. (b) Transfer characteristics of the device ($V_{DS} = 1$ V). (c) Output characteristics of the device. (d) Field-effect electron mobility assessment for the same device presented in (b) and (c) with $V_{DS} = 1$ V. The inset shows the statistics of field-effect mobility based on 40 devices. Reproduced from [63] Tsinghua University Press.

material issues, specifically due to their amorphous nature [62]. Li *et al* successfully produced high crystal superlattice structure IGZO NWs with different Ga element concentrations by chemical vapor deposition (CVD) [63]. In this study, it was found that when the ratio of In:Ga:Zn:O is 1.8:1.8:2.4:7, the prepared NWs have the best performance when using FETs. The average electron mobility of the device is about 110 cm^2 V^{-1} s^{-1}, and the on/off current ratio is about 10^6. The introduction of the Ga element effectively adjusts the carrier concentration in the fiber. The high binding energy of Ga and oxygen will preferentially form gallium oxide. This reduces the oxygen vacancy in the NFs, thus reducing the carrier concentration and the carrier scattering, and improving the switching current ratio of the device. These results prove that the fabricated devices have good application prospects in the arena of next generation display technologies (figure 6.10) [63].

6.4.4 Other metal oxide NF transistors

The research on one-dimensional n-type metal oxide semiconductor NFs is becoming more and more extensive, but there are few studies on p-type nanostructures. The lack of effective manufacturing methods and relatively low field-effect performance have seriously hindered the further development of complementary logic circuits. In complementary logic circuits, both n-type and p-type semiconductors are necessary [64]. Therefore, there is an urgent need to develop p-type semiconductor NFs. Zhu *et al* prepared CuO NWs semiconductor by electrostatic

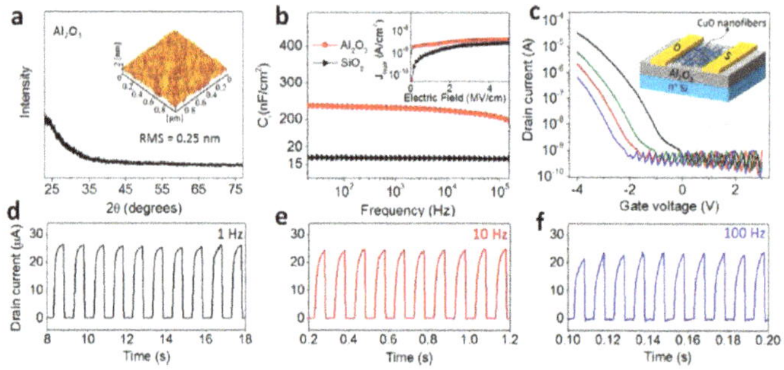

Figure 6.11. (a) The XRD pattern and AFM image of the solution-processed high-κ Al_2O_3 thin film. (b) The J_{leak} versus voltage and C_i versus frequency curves of Al_2O_3 and SiO_2 thin films. (c) The transfer curves of CuO nanofiber network FETs based on the Al_2O_3 dielectric with different V_{DS} values (i.e. 1, 2, 3, and 4 V). I_{DS} responses to a square wave signal applied to the gate voltage with the frequency of (d) 1 Hz, (e) 10 Hz, and (f) 100 Hz. Reprinted from [65], with the permission of AIP Publishing.

spinning and applied them in FET. They exhibited typical p-type properties [65, 66] as figure 6.11 shows.

When using SiO_2 as the gate dielectric, the carrier mobility is 10^{-2} cm^2 V^{-1} s^{-1}, and the current switching ratio is about 10^4. When Al_2O_3 is used instead of SiO_2 as the gate dielectric, the carrier mobility is increased to 3.5 cm^2 V^{-1} s^{-1}. The I_{on}/I_{off} is 3×10^5, and the lower operating voltage is 4 V. Devices based on CuO nanofibers have the advantages of light weight, low pressure, and high performance. They have great application potential in the next-generation chemical/biosensor and electronics (photoelectron) fields.

Matsubara *et al* successfully manufactured highly uniform and continuous NiO NWs by electrospinning [67]. Compared with other methods, electrospinning provides a convenient, scalable, and low-cost way to produce ultra-long NFs with ultra-long aspect ratios exceeding 10^5. Another advantage of electrospinning is that, because the sol–gel-based process can precisely control the doping concentration, it is possible to achieve enhanced conductivity by introducing an appropriate amount of Li^+ into NiO NWs. The assembled FET exhibits typical p-type conductivity, and the doping effect is greatly enhanced by Li doping (figure 6.12) [67].

6.5 Application of one-dimensional metal oxide semiconductor FET

6.5.1 Transparent and flexible electronic devices

The concept of flexible 'transparent electronics' will play a pivotal role in the fields of science, technology and commercial application because they can be used in fast consumer electronics, new energy and transportation [68]. For example, popular car head-up display systems, mobiles and computer display can use 'transparent electronics.' Optically transparent and mechanically flexible electronic circuits will be the center of the next-generation of display technology. Transistors based on polysilicon or amorphous silicon and widely used in current commercial displays, have the problem of low carrier mobility and non-transparent. This limits the further improvement of the refresh frequency and resolution of the display. The metal oxide

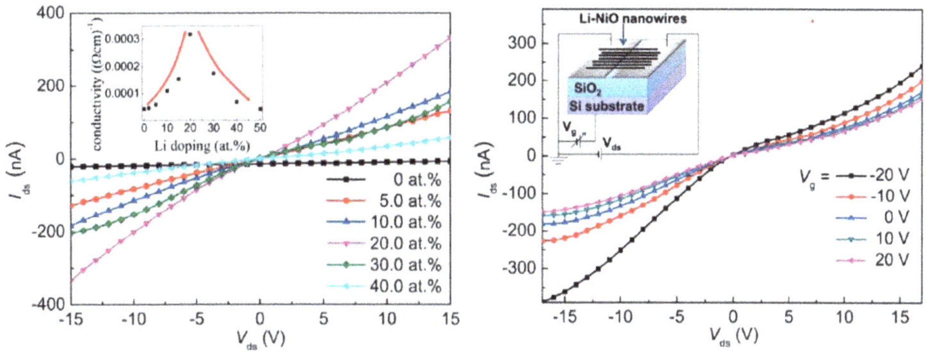

Figure 6.12. (a) Full range XRD of NiO NWs doped with different Li concentrations (0–50.0 at.%). (b) *I–V* curves of NiO nanowires with different Li concentrations. Inset: conductivity as a function of Li at.%. Reproduced from [67] with permission of The Royal Society of Chemistry.

nanomaterials such as ZnO, In_2O_3, and SnO_2 have proper forbidden band widths and high optical transmittance in the visible light range. Compared with the bulk and thin film materials, the 1D metal oxide nanomaterials will have huge application prospects in transparent flexible electronic circuits [15, 20, 68].

The development of optically transparent and mechanically flexible electronic circuits is a key step in the development of skin-like, wearable and display devices. A transparent transistor should have a transparent conductive electrode, dielectric layer, and metal oxide semiconductor layer. In 2007, Ju *et al* using graphic IZO as the gate, ALD deposited alumina dioxide as the dielectric layer, ZnO or In_2O_3 NWs as the semiconductor channel, ITO as the source/drain electrodes, prepared a fully transparent NWe transistor [69]. The optical properties of the device on the glass substrate are excellent, with optical transmittance of 83%. Using forced terephthalate (PET) as substrate, the optical transmittance of the device was 81%. In addition, the device had excellent mechanical flexibility and excellent electrical performance under bending conditions. Figure 6.13 shows a fully transparent and flexible In_2O_3 devices [69]. This research shows great potential of 1D metal oxide nanomaterials in the future development of transparent flexible devices.

6.5.2 Gas sensor devices

Sensing behavior is another important characteristic of metal oxides [56]. One-dimensional metal oxide NWs have a large specific surface area, abundant oxygen vacancies, and surface defects. The NFs can absorb gas from the surrounding environment and undergo a redox reaction, thereby changing the electrical conductivity of the NFs [11, 12, 15–17, 19, 56, 64, 70]. For n-type metal oxides, the reducing gas reacts with the adsorbed oxygen species and releases electrons into the metal oxide, resulting in increased conductivity, while the oxidizing gas uses a reverse process to capture electrons from the metal oxide, thereby reducing conductivity.

Carbon monoxide (CO) plays an important role in the chemical and metallurgical industry, but is toxic to humans [64]. Inhaling an excessive amount of high-concentration carbon monoxide in a short period of time results in carbon monoxide

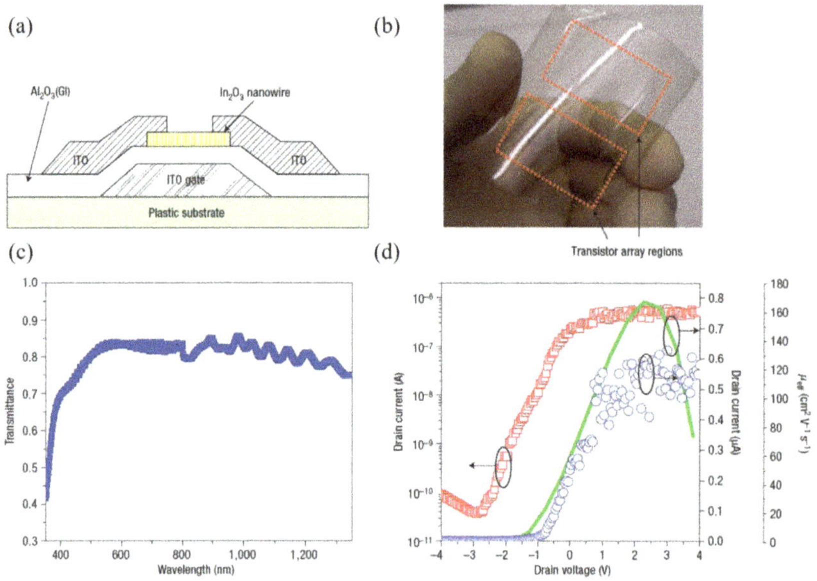

Figure 6.13. Fully transparent and flexible In_2O_3 NW transistors (NWTs). (a) Cross-sectional view of fully transparent and flexible NWT device structure consisting of a plastic substrate, patterned ITO gate electrode (120 nm), ALD-deposited Al_2O_3 gate insulator (50 nm), a single In_2O_3 nanowire (D/L ~20 nm/1.79 mm) for the active channel, and ITO for the source/drain electrodes (120 nm). (b) Image of In_2O_3 NWTs on the plastic substrate, showing the optical clarity and mechanical flexibility. (c) Optical transmission spectrum of region containing In_2O_3 NWTs on the plastic substrate (ITO(S/D)/In_2O_3 nanowires/Al_2O_3/ITO(G)/plastic substrates). (d) The I_{ds}–V_{gs} characteristic of In_2O_3 NWTs at V_d = 0.5 V. Blue, red and green data points correspond to linear-scale I_{ds}–V_{gs}, log-scale I_{ds}–V_{gs} and μ_{eff}, respectively. Arrows indicate appropriate axis. [69] (2007) Copyright. With permission of Springer.

poisoning and causes secondary oxygen depletion damage to the brain, heart, liver, kidneys, lungs, and other tissues. The detection of indoor carbon monoxide concentration is extremely important. Singh *et al* prepared Zn-doped In_2O_3 NWs by CVD, using a single NW as a semiconductor layer, and SiN_x as a dielectric layer. The transistors showed excellent electrical performance with on-state current of 8×10^{-6} A, a current switching ratio of about 10^6, and a carrier mobility of 139 $cm^2\ V^{-1}\ s^{-1}$. Using these transistors to detect CO gas at room temperature, gave results that showed that when compared with pure In_2O_3 NW transistors, Zn-doped In_2O_3 NW transistors have a more sensitive detection effect, and can selectively detect CO gas when in concentration with NO and NO_2 [71]. When exposed to CO gas at a concentration of 5 ppm, a significant threshold voltage shift of −3.5 V can be observed, as shown in figure 6.14 [71].

Ammonia is a colorless gas with irritating properties, mainly used in industrial production. It is very harmful to the human respiratory tract, eyes and skin. The detection of ammonia at room temperature has attracted widespread attention. Huang *et al* prepared polycrystalline $SrGe_4O_9$ nanotubes (NTs) through the electrospinning process [72]. The prepared NTs is a new type of gas sensing material, which can detect NH_3 at room temperature. The maximum sensing response of

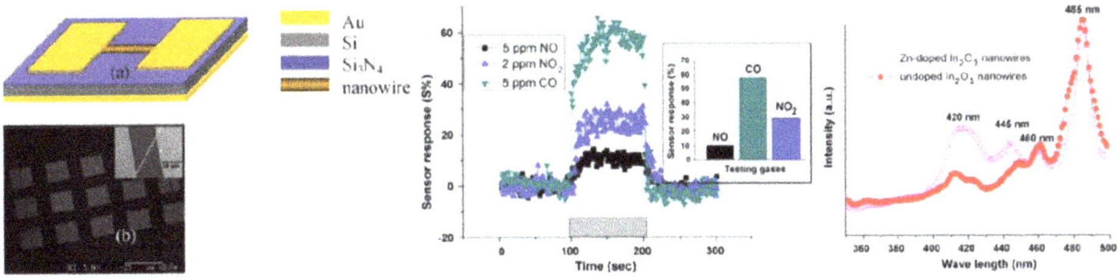

Figure 6.14. (a) Schematic diagram of single NWe FET, with Cr (10 nm)/Au (50 nm) serving as electrodes, SiN_x serving as gate electrode and highly doped (resistivity 0.005 Ω cm) n-type Si as back gate. (b) FESEM image of an array of devices and in the inset a single NW-FET device touching the two electrodes with a channel length of 18 μm. (c) Response of single Zn–In_2O_3 NW sensor to the different gases at a fixed $V_D = 3$ V and $V_G = -15$ V, sensor was operated at room temperature. (d) Photoluminescence spectrum measured from the as deposited Zn-doped and undoped In_2O_3 NWs with an excitation wavelength of 264 nm at room temperature. Reprinted from [71], Copyright (2010), with permission from Elsevier.

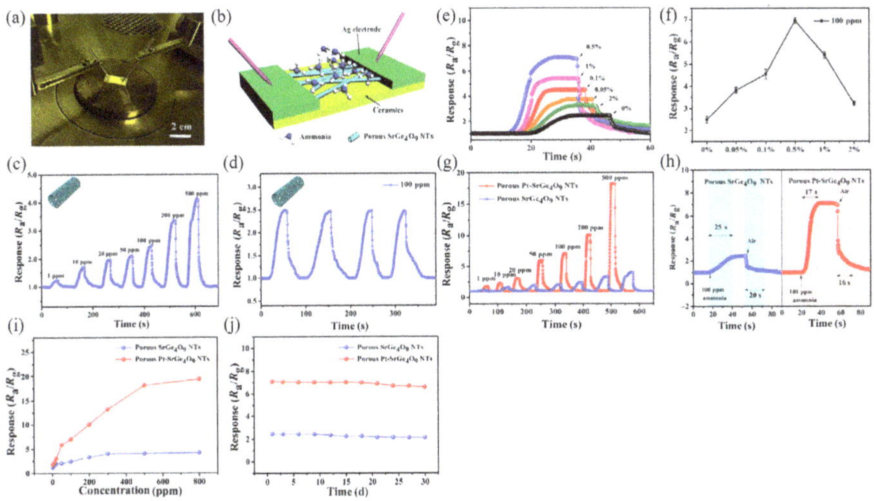

Figure 6.15. (a) Digital photograph, (b) schematic of the $SrGe_4O_9$ NT sensor. (c) Response curves of pure $SrGe_4O_9$ NT sensors to NH_3 gas with different concentrations. (d) Dynamic response of pure SrGe4O9 NT sensors to 100 ppm NH_3 at room temperature. (e) Response curves of the Pt-decorated $SrGe_4O_9$ NT sensors with various Pt concentrations. (f) Error curve of the Pt-decorated $SrGe_4O_9$ NT sensors with various Pt concentrations. (g) Sensitivity with respect to the ammonia concentration, in the range of 1–500 ppm. (h) Dynamic response–recovery curves to 100 ppm ammonia. (i) Gas response to ammonia with various concentrations. (j) Long-term response stability. Reproduced from [72] Tsinghua University Press.

$SrGe_4O_9$ NTs to 100 ppm NH_3 is 2.49. The maximum sensor response obtained is 7.08. Manufactured flexible gas sensors made from these NTs, show excellent flexibility in different bending states, mechanical stability, and sensing stability. Their performance is shown in the figure 6.15 [72]. These properties indicate that the prepared devices have a future in wearable electronic products.

6.5.3 Photodetector

A photodetector can be produced using the photoconductive effect of semiconductor materials. The photoconductivity effect refers to changes in the conductivity of irradiated material when irradiated. Photoconductive detectors are widely used in various fields of the military and national economy. When the irradiated photon energy $h\nu$ is equal to or greater than the forbidden band width E_g of the semiconductor, the photon can excite the electrons in the valence band allowing them to travel to the conduction band, thereby generating conductive electron–hole pairs. This is an intrinsic photoconductive effect. The long wave limit of intrinsic photoconductive materials is limited by the width of the forbidden band. This photoconductive effect is used in FET. When the semiconductor material in the channel is irradiated, the carrier concentration and mobility change, so that the current between the source and drain electrodes changes [18, 20, 73].

Because superlattice NWs have a unique energy band structure for effective photocarrier separation and collection, they have a great potential as active materials for high-performance optoelectronic devices. Li *et al* obtained InGaZnO NFs with superlattice structure and controllable stoichiometry through ambient pressure chemical vapor deposition [74]. Along the nanofiber axis, InGaO $(ZnO)^{4+}$ blocks and InO^{2-} layers are alternately arranged to form a periodic layered structure. When configured as a photodetector, it exhibits surprising performance. When the ratio of In:Ga:Zn:O is 1.8:1.8:2.4:7, a single device has a response of 1.95×10^5 A W^{-1} for ultraviolet light detection, and an external quantum efficiency as high as $9.28 \times 10^7\%$, The detection time is 0.93 s, and the decay time is 0.2 s. The reason for such excellent performance is due to the incorporation of Ga. The concentration of Ga has a great influence on the device. Ga reduces the oxygen vacancies in the fiber and reduces the carriers in the channel, which reduces the O^{2-} adsorption, thereby improving the photoelectric detection performance of the device. The optical transmittance of the material is high, so a completely transparent ultraviolet light detector can be produced (figure 6.16) [74].

One-dimensional nanostructured SnO_2 has been widely studied due to its large specific surface area, high optical permeability, and n-type semiconductor properties. Based on this, Liu *et al* used a low-cost electrospinning method to prepare SnO_2 NFs and NT arrays [18]. Using this as a channel material, an FET was prepared. By simply adjusting the content of Sn in the precursor solution, the electrical and optoelectronic properties of the fabricated device were easily adjusted. When 5 ml of precursor solution contains 0.2 g $SnCl_{2\cdot4}H_2O$, the prepared device shows the best performance, high saturation current and large current switching ratio. It also could detect deep ultraviolet light (wavelength 240–320 nm), including an ultra-high photocurrent of 307 μA, a high photosensitivity of 2003, respond ability of 214 A W^{-1} and detectivity of 2.19×10^{13} Jones. When fluorine-doped tin oxide (FTO) glass was used as the base material, it showed high optical transmittance and photosensitivity (figure 6.17) [18]. These results indicate that the one-dimensional SnO_2 prepared by electrospinning has application prospects in the next generation of functional electronic devices and optoelectronic devices.

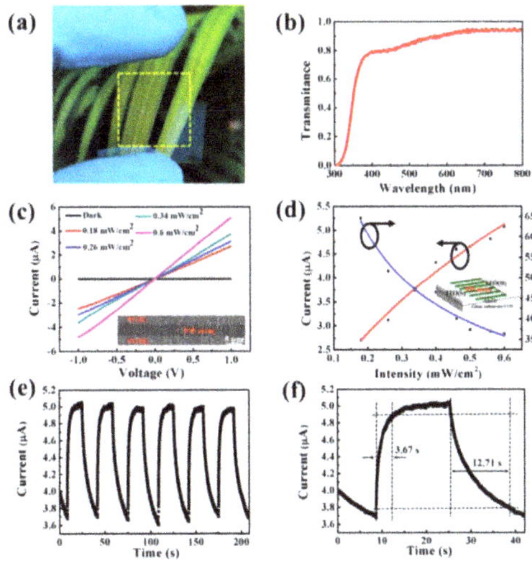

Figure 6.16. Transparent photodetector device based on $In_{1.8}Ga_{1.8}Zn_{2.4}O_7$ NW parallel arrays. (a) Optical image of the leaves covered by the as-fabricated device, while the device area is highlighted by the yellow dashed rectangle. (b) Transmittance measurement of the device. (c) *I–V* curves of the device under different power intensities of 261 nm light illumination and in the dark. The inset shows the top-view SEM image of the NW parallel arrayed device channel. (d) Dependence of the photocurrent and the responsivity on the light intensity. The inset gives the device schematic of the photodetector fabricated on glass substrates. (e) Photoresponse characteristics (current versus time) of the device. The light intensity is 0.6 mW cm^{-2}. (f) Corresponding high-resolution current versus time curve. The bias is 1 V and $V_G = -20$ V. Reprinted with permission from [74]. Copyright (2019) American Chemical Society.

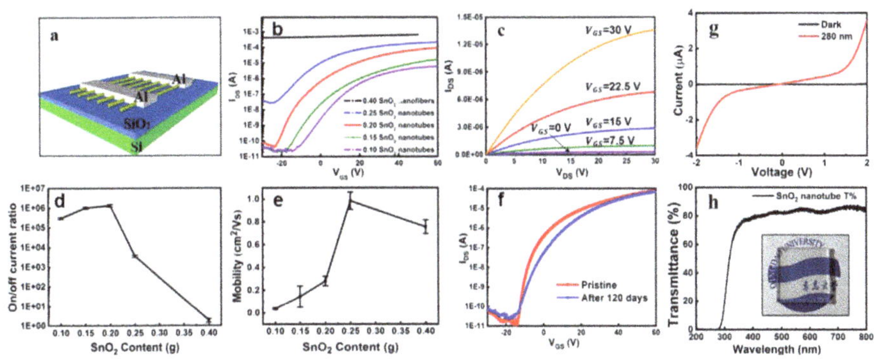

Figure 6.17. (a) FET structure of SnO_2 NTs array. (b) The output curve of the SnO_2 NTs FET. (c) Transfer curves of SnO_2 FETs with different Sn amount. (d) Average on/off current ratio of SnO_2 FETs of different Sn contents. (e) Average mobility of SnO_2 FETs with different Sn contents. (f) The stability of 0.2 SnO_2 NTs FET after 120 days. (g) The *I–V* curve of 0.2 SnO_2 nanotubes photodetector using FTO glass as substrate. (h) The optical transmittance curve of the FTO glass after being covered by 0.2 SnO_2 nanotubes. Reproduced from [18]. © IOP Publishing Ltd. All rights reserved.

6.5.4 Synapse-like devices

As a representative of the development of human science and technology, computers play an irreplaceable role in people's daily lives. With the rapid increase in the amount of information in human society, while the speed of computer calculations has increased, energy consumption has also increased rapidly. The human brain is essentially a very efficient and energy-saving information processing system. Adult brains have only about 20 W of power. This huge difference in energy consumption is mainly due to differences in the structure and information processing methods of computers and human brains. The human brain consists of a network of neurons formed by the cross-connection of approximately 10^{11} neurons. The connections between neurons are called synapses, and there are about 10^{15} synapses in the human brain. By changing the connection strength between neurons (called synaptic weights) to store and process information, the brain has a self-learning function to achieve the integration of storage and processing. In view of the characteristics of neural networks, in order to improve the information processing efficiency of computers, researchers have long begun to use traditional silicon-based CMOS processing devices to simulate neuron information processing methods. However, on the one hand, since the traditional silicon-based semiconductor devices are mostly volatile or binary nonvolatile memory devices, this is inconsistent with the simulated characteristics of neural synapses. On the other hand, with the development of CMOS technology for decades, the size of the device channel has entered more than a dozen nanometers, or even a few nanometers, not only facing extremely high manufacturing difficulties, but also getting closer to physical limits (such as thermal effects, quantum tunneling effect). Therefore, when building large-scale neural networks, CMOS devices face huge challenges in terms of high-density integration and energy consumption reduction. In order to overcome the difficulties faced by CMOS devices, it has become an important goal of scientific research and industry to find new materials and devices that meet the characteristics of neuromorphic computing [75–77].

Gou *et al* manufactured a synaptic transistor with a multi-face gate structure by using a single SnO_2 NW as a channel layer [78]. The in-plane-gate electrode is regarded as the presynaptic terminal and the SnO_2 nanowire channel with a Vds biased is used as the postsynaptic terminal. Due to the proton-related electric-double-layer (EDL) coupling of synaptic devices, it not only successfully mimics important synaptic functions including excitatory postsynaptic current (EPSC), paired-pulse facilitation (PPF), dynamic logic, adaptive and synaptic filtering. Moreover, the synaptic behavior can be adjusted in the dynamic range through multi-terminal adjustment of synaptic input. In addition, a SnO_2 NW synaptic transistor is irradiated with a light source as a light modulation terminal, and the observed neuromorphic function is dynamically modulated by optical density. Figure 6.18 shows a schematic diagram of a synaptic transistor and the electrical signals of the transistor when simulating the behavior of a synapse [78]. These excellent nanoscale synaptic transistors have broad application prospects in synaptic electronics.

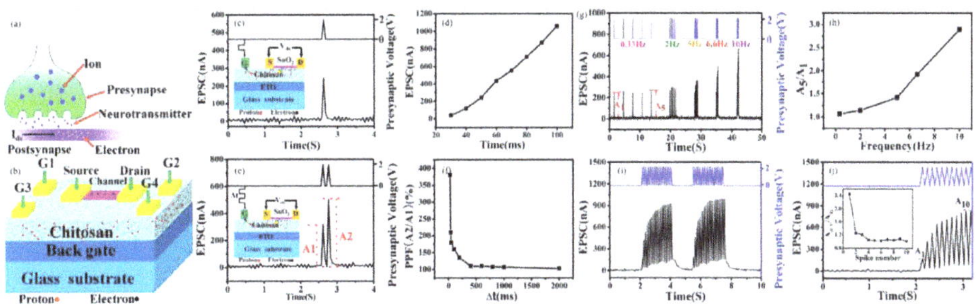

Figure 6.18. (a) Schematic image of a biological synapse simulated by synaptic transistors. (b) The schematic diagram of SnO_2 NW-based neuron transistor gated by the chitosan film on FTO glass substrate. (c) EPSC triggered by an in-plane presynaptic spike (2.0 V, 50 ms). (d) Spike duration-dependent EPSC, the amplitude of the spike is 2.0 V. (e) The EPSCs triggered by a pair of presynaptic spike with a time interval (Δt) = 100 ms. (f) PPF index (100% x A2/A1) plotted as a function of spike Δt. (g) EPSCs recorded in response to the stimulus train with different frequencies. (h) EPSC amplitude ratio (A_5/A_1) plotted as a function of presynaptic spike frequency. (i) A train of spikes (2.0 V, 50 ms, Δt = 50 ms) is applied on the in-plane-gate and EPSC is measured as a function of time. (j) The amplified sections of a 2.0–3.2 s range corresponding to adaptation training from 4 (i). Inset: the rate of A_{i+1}/A_i (A_{i+1} and A_i are the amplitudes of the previous and latter EPSC amplitude, respectively). Reproduced from [78] with permission of The Royal Society of Chemistry.

6.6 Summary and outlook

In summary, one-dimensional metal oxides have excellent optical, electrical and gas detection performance, and are the basis for building multi-functional and high-performance field-effect transistors. Through continuous adjustment and optimization of transistor performance, the application of transistors will be widely used in electronic display technology, sensing equipment, electronic memory, and logic circuits.

At present, there are still some problems in the preparation of one-dimensional metal oxide semiconductors. For example, the cost of preparing nanofibers using CVD molecular beam epitaxy, or other methods is relatively high, the equipment size is large, and it is difficult to prepare and grow on a large scale. The orientation of the fiber is inconsistent, and the fibers needs to be transferred to the substrate and then subjected to a photolithography process. The utilization rate of the raw material is also low. When using the electrospinning method to prepare nanofibers, high temperature annealing is required to remove the presence of the fiber polymer. If the required temperature is higher, normal transparent flexible substrates will deform or even decompose.

With the continuous development of science and technology, the focus of future research will be on transparent flexible electronic devices. The future research directions of FETs are as follows:

1. Discovering a method that can lower the annealing temperature, so that not only can transparent preparation be achieved, but also flexible electronic devices can be produced by large-scale printed electronic circuitry;
2. Designing a geometry of metal oxide nanofibers to realize storage or logic circuits, so the integration into electronic circuits increases;

3. Studying the growth mechanism of NWs, with the goal of controlling the surface morphology, and the concentration of defects, to obtain higher performance optical detectors, gas sensors and other electronic devices;
4. Develop p-type metal oxides with excellent performance. CMOS circuits based on n-type and p-type one-dimensional metal oxide semiconductors have broad prospects.

References

[1] Fiori G, Bonaccorso F, Iannaccone G, Palacios T, Neumaier D, Seabaugh A, Banerjee S and Colombo L 2014 Electronics based on two-dimensional materials *Nat. Nanotechnol.* **9** 768–79

[2] Li Y, Yang X-Y, Feng Y, Yuan Z-Y and Su B-L 2012 One-dimensional metal oxide nanotubes, nanowires, nanoribbons, and nanorods: synthesis, characterizations, properties and applications *Crit. Rev. Solid State Mater. Sci.* **37** 1–74

[3] Devan R, Patil R, Lin J H and Ma Y-R 2012 One-dimensional metal-oxide nanostructures: recent developments in synthesis, characterization, and applications *Adv. Funct. Mater.* **22** 3326–70

[4] Comini E, Baratto C, Faglia G, Ferroni M, Vomiero A and Sberveglieri G 2009 Quasi-one dimensional metal oxide semiconductors: preparation, characterization and application as chemical sensors *Prog. Mater Sci.* **54** 1–67

[5] Zhang H, Song L, Luo L, Liu L, Wang H and Wang F 2017 TiO_2/Sb_2S_3/P3HT based inorganic–organic hybrid heterojunction solar cells with enhanced photoelectric conversion performance *J. Electron. Mater.* **46** 4670–5

[6] Wang F, Zhang H, Liu L, Shin B and Shan F 2016 Synthesis, surface properties and optical characteristics of CuV_2O_6 nanofibers *J. Alloys Compd.* **672** 229–37

[7] Wang F, Zhang H, Liu L, Shin B and Shan F 2016 AgV_7O_{18}: a new silver vanadate semiconductor with photodegradation ability on dyes under visible-light irradiation *Mater. Lett.* **169** 82–5

[8] Wang F, Song L, Zhang H, Luo L, Wang D and Tang J 2017 One-dimensional metal-oxide nanostructures for solar photocatalytic water-splitting *J. Electron. Mater.* **46** 4716–24

[9] Yin H, Zhao Y L, Hua Q, Zhang J, Zhang Y, Xu X, Long Y, Tang J and Wang F 2019 Controlled synthesis of hollow α-Fe_2O_3 microspheres assembled with ionic liquid for enhanced visible-light photocatalytic activity *Front. Chem.* **7** 58

[10] Luo L, Zhang H, Song L, Liu L, Yi X, Tan M, Song J, Wang G and Wang F 2018 Phase transition induced synthesis of one dimensional $In_{1-x}Zn_xO_y$ heterogeneous nanofibers for superior lithium ion storage *Appl. Surf. Sci.* **470** 340–7

[11] Tong W, Wang Y, Bian Y, Wang A, Han N and Chen Y 2020 Sensitive cross-linked SnO_2:NiO networks for MEMS compatible ethanol gas sensors *Nanoscale Res. Lett.* **15** 35

[12] Wang Y, Liu C, Wang Z, Song Z, Zhou X, Han N and Chen Y 2019 Sputtered SnO_2:NiO thin films on self-assembled Au nanoparticle arrays for MEMS compatible NO_2 gas sensors *Sensors Actuators* B **278** 28–38

[13] Zhu X, Li Y, Zhang H, Song L, Zu H, Qin Y, Liu L, Li Y and Wang F 2020 High-performance field effect transistors based on large ratio metal (Al, Ga, Cr) doped In_2O_3 nanofibers *J. Alloys Compd.* **830** 154578

[14] Han N, Yang Z, Shen L, Lin H, Wang Y, Pun E Y B, Chen Y and Ho J C 2016 Design and fabrication of 1-D semiconductor nanomaterials for high-performance photovoltaics *Sci. Bull.* **61** 357–67

[15] Huang H, Liang B, Liu Z, Wang X, Chen D and Shen G 2012 Metal oxide nanowire transistors *J. Mater. Chem.* **22** 13428–45

[16] Li R, Chen S, Lou Z, Li L, Huang T, Song Y, Chen D and Shen G 2017 Fabrication of porous SnO_2 nanowires gas sensors with enhanced sensitivity *Sensors Actuators* B **252** 79–85

[17] Chen S, Lou Z, Chen D, Chen Z, Jiang K and Shen G 2016 Highly flexible strain sensor based on ZnO nanowires and P(VDF-TrFE) fibers for wearable electronic device *Sci. China Mater.* **59** 173–81

[18] Liu D, Li H, Song L, Zhu X, Qin Y, Zu H, He J, Yang Z and Wang F 2020 Modulating electrical and photoelectrical properties of one-step electrospun one-dimensional SnO_2 arrays *Nanotechnology* **31** 335202

[19] McAlpine M C, Ahmad H, Wang D and Heath J R 2007 Highly ordered nanowire arrays on plastic substrates for ultrasensitive flexible chemical sensors *Nat. Mater.* **6** 379–84

[20] Li Y, Qian F, Xiang J and Lieber C 2006 Nanowire electronic and optoelectronic devices *Mater. Today* **1** 18–27

[21] Seoane N, Loureiro A G and Kalna K 2020 Special issue: nanowire field-effect transistor (FET) *Materials* **13** 1845

[22] Jia C, Lin Z, Huang Y and Duan X 2019 Nanowire electronics: from nanoscale to macroscale *Chem. Rev.* **119** 9074–135

[23] Dasgupta N, Sun J, Liu C, Brittman S, Andrews S C, Lim J, Gao H, Yan R and Yang P 2014 Semiconductor nanowires-synthesis, characterization, and applications *Adv. Mater.* **26** 2137–84

[24] Jia X, Zhu X, Tian W, Ding Y, Tian X, Cheng B, Cheng L, Bai S and Qin Y 2020 Nanowire templated CVD synthesis and morphological control of MoS_2 nanotubes *J. Mater. Chem.* C **8** 4133–8

[25] Lee Y-H *et al* 2012 Synthesis of large-area MoS_2 atomic layers with chemical vapor deposition *Adv. Mater.* **24** 2320–5

[26] Wu J J and Liu S C 2002 Low-temperature growth of well-aligned ZnO nanorods by chemical vapor deposition *Adv. Mater.* **14** 215–8

[27] Li Y L, Kinloch I A and Windle A H 2004 Direct spinning of carbon nanotube fibers from chemical vapor deposition synthesis *Science* **304** 276–8

[28] Chhowalla M, Teo K B K, Ducati C, Rupesinghe N L, Amaratunga G A J, Ferrari A C, Roy D, Robertson J and Milne W I 2001 Growth process conditions of vertically aligned carbon nanotubes using plasma enhanced chemical vapor deposition *J. Appl. Phys.* **90** 5308–17

[29] Zhang H, Meng Y, Song L, Luo L, Qin Y, Han N, Yang Z, Liu L, Ho J C and Wang F 2018 High-performance enhancement-mode thin-film transistors based on Mg-doped In_2O_3 nanofiber networks *Nano Res.* **11** 1227–37

[30] Song L, Luo L, Li X, Liu D, Han N, Liu L, Qin Y, Ho J C and Wang F 2019 Modulating electrical performances of In_2O_3 nanofiber channel thin film transistors via Sr doping *Adv. Electron. Mater.* **5** 1800707

[31] Wang F *et al* 2018 ZnO nanofiber thin-film transistors with low-operating voltages *Adv. Electron. Mater.* **4** 1700336

[32] Yang X, Zu H, Luo L, Zhang H, Li J, Yi X, Liu H, Wang F and Song J 2020 Synergistic tungsten oxide/N, S co-doped carbon nanofibers interlayer as anchor of polysulfides for high-performance lithium-sulfur batteries *J. Alloys Compd.* **833** 154969

[33] Subbiah T, Bhat G S, Tock R W, Parameswaran S and Ramkumar S S 2005 Electrospinning of nanofibers *J. Appl. Polym. Sci.* **96** 557–69
[34] Frenot A and Chronakis I S 2003 Polymer nanofibers assembled by electrospinning *Curr. Opin. Colloid Interface Sci.* **8** 64–75
[35] Demir M M, Yilgor I, Yilgor E and Erman B 2002 Electrospinning of polyurethane fibers *Polymer* **43** 3303–9
[36] Bhardwaj N and Kundu S C 2010 Electrospinning: a fascinating fiber fabrication technique *Biotechnol. Adv.* **28** 325–47
[37] Matthews J A, Gary E W, Simpson D G and Bowlin G L 2002 Electrospinning of collagen nanofibers *Biomacromolecules* **3** 232–8
[38] Reneker D H and Yarin A L 2008 Electrospinning jets and polymer nanofibers *Polymer* **49** 2387–425
[39] Li D and Xia Y 2003 Fabrication of titania nanofibers by electrospinning *Nano Lett.* **3** 555–60
[40] Agarwal S, Joachim H W and Greiner A 2008 Use of electrospinning technique for biomedical applications *Polymer* **49** 5603–21
[41] Fridrikh S V, Jian H Y, Brenner M P and Rutledge G C 2003 Controlling the fiber diameter during electrospinning *Phys. Rev. Lett.* **90** 144502
[42] Fortunato E, Barquinha P and Martins R 2012 Oxide semiconductor thin-film transistors: a review of recent advances *Adv. Mater.* **24** 2945–86
[43] Hoffman R L, Norris B J and Wager J F 2003 ZnO-based transparent thin-film transistors *Appl. Phys. Lett.* **82** 733–5
[44] Carcia P F, McLean R S, Reilly M H and Nunes G 2003 Transparent ZnO thin-film transistor fabricated by rf magnetron sputtering *Appl. Phys. Lett.* **82** 1117–9
[45] Ionescu A M and Riel H 2011 Tunnel field-effect transistors as energy-efficient electronic switches *Nature* **479** 329–37
[46] Arnold M S, Avouris P, Pan Z W and Wang Z L 2003 Field-effect transistors based on single semiconducting oxide nanobelts *J. Phys. Chem.* B **107** 659–63
[47] Heo Y W, Norton D P, Tien L C, Kwon Y, Kang B S, Ren F, Pearton S J and LaRoche J R 2004 ZnO nanowire growth and devices *Mater. Sci. Eng.* R **47** 1–47
[48] Morisot F, Nguyen V H, Montemont C, Maindron T, Muñoz-Rojas D, Mouis M, Langlet M and Ternon C 2019 Al_2O_3, Al doped ZnO and SnO_2 encapsulation of randomly oriented ZnO nanowire networks for high performance and stable electrical devices *Nanotechnology* **30** 385202
[49] Ditshego N M J and Sultan S M 2019 3D simulation investigating ZnO NWFET characteristics *J. Nano Res.* **58** 40–8
[50] Ditshego N M J and Sultan S M 2019 Top-down fabrication process of ZnO NWFETs *J. Nano Res.* **57** 77–92
[51] Chang H, Lee D H, Kim H S, Park J and Lee B Y 2018 Facile fabrication of self-assembled ZnO nanowire network channels and its gate-controlled UV detection *Nanoscale Res. Lett.* **13** 413
[52] Chakraborty A, Pizzoferrato R, Agresti A, Matteis F D, Orsini A and Medaglia P G 2018 Wet-chemical synthesis of ZnO nanowires on low-temperature photo-activated ZnO-rGO composite thin film with enhanced photoconduction *J. Electron. Mater.* **47** 5863–9
[53] Zheng X, Sun Y, Yan X, Chen X, Bai Z, Lin P, Shen Y, Zhao Y and Zhang Y 2014 Tunable channel width of a UV-gate field effect transistor based on ZnO micro-nano wire *RSC Adv.* **4** 18378–81

[54] Willander M *et al* 2009 Zinc oxide nanorod based photonic devices: recent progress in growth, light emitting diodes and lasers *Nanotechnology* **20** 332001

[55] Zhang Z, Wang K, Zheng K, Deng S, Xu N and Chen J 2018 Electron bombardment induced photoconductivity and high gain in a flat panel photodetector based on a ZnS photoconductor and ZnO nanowire field emitters *ACS Photonics* **5** 4147–55

[56] Zhang D, Liu Z, Li C, Tang T, Liu X, Han S, Lei B and Zhou C 2004 Detection of NO_2 down to ppb levels using individual and multiple In_2O_3 nanowire devices *Nano Lett.* **4** 1919–24

[57] Xu Q, Liu X, Wan B, Yang Z, Li F, Lu J, Hu G, Pan C and Wang Z L 2018 In_2O_3 nanowire field-effect transistors with Sub-60 mV/dec subthreshold swing stemming from negative capacitance and their logic applications *ACS Nano* **12** 9608–16

[58] Park H, Yoon K R, Kim S K, Kim I-D, Jin J, Kim Y H and Bae B-S 2016 Highly conducting In_2O_3 nanowire network with passivating ZrO_2 thin film for solution-processed field effect transistors *Adv. Electron. Mater.* **2** 1600218

[59] Yan J, Chen Y, Wang X, Fu Y, Wang J, Sun J, Dai G, Tao S and Gao Y 2019 High-performance solar-blind SnO_2 nanowire photodetectors assembled using optical tweezers *Nanoscale* **11** 2162–9

[60] Liu H and Wan Q 2012 Low-voltage SnO_2 nanowire transistors gated by solution-processed chitosan-based proton conductors *Nanoscale* **4** 4481–4

[61] Wang X, Aroonyadet N, Zhang Y, Mecklenburg M, Fang X, Chen H, Goo E and Zhou C 2014 Aligned epitaxial SnO_2 nanowires on sapphire: growth and device applications *Nano Lett.* **14** 3014–22

[62] Sporea R A, Kham M N, Flewitt A J, Ravi S and Silva P 2019 Novel tunnel-contact-controlled IGZO thin-film transistors with high tolerance to geometrical variability *Adv. Mater.* **31** 1902551

[63] Li F, Yip S P, Dong R, Zhou Z, Lan C, Liang X, Li D, Meng Y, Kang X and Ho J C 2019 Crystalline InGaZnO quaternary nanowires with superlattice structure for high-performance thin-film transistors *Nano Res.* **12** 1796–803

[64] Hou L, Zhang C, Li L, Du C, Li X, Kang X-F and Chen W 2018 CO gas sensors based on p-type CuO nanotubes and CuO nanocubes: morphology and surface structure effects on the sensing performance *Talanta* **188** 41–9

[65] Zhu H, Liu A, Liu G and Shan F 2017 Electrospun p-type CuO nanofibers for low-voltage field-effect transistors *Appl. Phys. Lett.* **111** 143501

[66] Talapin D and Murray C 2005 PbSe nanocrystal solids for n- and p-channel thin film field-effect transistors *Science* **310** 86–9

[67] Matsubara K, Huang S, Iwamoto M and Pan W 2014 Enhanced conductivity and gating effect of p-type Li-doped NiO nanowires *Nanoscale* **6** 688–92

[68] Georgiou T *et al* 2013 Vertical field-effect transistor based on graphene–WS2 heterostructures for flexible and transparent electronics *Nat. Nanotechnol.* **8** 100–3

[69] Sanghyun J, Facchetti A, Xuan Y, Liu J, Ishikawa F, Ye P, Zhou C, Marks T J and Janes D B 2007 Fabrication of fully transparent nanowire transistors for transparent and flexible electronics *Nat. Nanotechnol.* **2** 378–84

[70] Kolmakov A and Moskovits M 2004 Chemical sensing and catalysis by one-dimensional metal-oxide nanostructures *Annu. Rev. Mater. Res.* **34** 151–80

[71] Singh N, Yan C and Lee P S 2010 Room temperature CO gas sensing using Zn-doped In_2O_3 single nanowire field effect transistors *Sensors Actuators* B **150** 19–24

[72] Huang T, Lou Z, Chen S, Li R, Jiang K, Chen D and Shen G 2017 Fabrication of rigid and flexible $SrGe_4O_9$ nanotube-based sensors for room-temperature ammonia detection *Nano Res.* **11** 431–9
[73] Yu X, Marks T and Facchetti A 2016 Metal oxides for optoelectronic applications *Nat. Mater.* **15** 383–96
[74] Li F *et al* 2019 High-performance transparent ultraviolet photodetectors based on InGaZnO superlattice nanowire arrays *ACS Nano* **13** 12042–51
[75] Chen Y, Qiu W, Wang X, Liu W, Wang J, Dai G, Yuan Y, Gao Y and Sun J 2019 Solar-blind SnO_2 nanowire photo-synapses for associative learning and coincidence detection *Nano Energy* **62** 393–400
[76] Chen H *et al* 2019 Time-tailoring van der Waals heterostructures for human memory system programming *Adv. Sci.* **6** 1901072
[77] Vu C-A and Chen W-Y 2019 Field-effect transistor biosensors for biomedical applications: recent advances and future prospects *Sensors* **19** 4214
[78] Gou G, Sun J, Qian C, He Y, Kong L-A, Fu Y, Dai G, Yang J and Gao Y 2016 Artificial synapses based on biopolymer electrolyte-coupled SnO_2 nanowire transistors *J. Mater. Chem.* C **4** 11110–7

Chapter 7

Semiconductor metal-oxide thin-film transistors for display

Zhihe Xia

Due to their process requirements being similar to the making of amorphous silicon thin-film transistors (TFTs) and their relatively higher field-effect mobility (μ_{FE}), metal-oxide (MO) TFTs are being deployed in the construction of active-matrix flat-panel displays (FPDs). In this chapter, the brief status and principle of active-matrix (AM) FPDs are introduced within comparing the MO TFTs with other silicon based competing technologies. In the ensuing section, the major issues of TFT architectures and channel material are addressed. Compared to the early state of developing MO TFTs, the major manufactories of MO TFTs in the display industry locate indium–gallium–zinc oxide (IGZO) as their core material. The latest practical IGZO TFTs reveal good characteristics in both AM liquid-crystal displays (LCD) and AM organic light-emitting diode (OLED) displays. Considering commonly n-type depletion mode and reliability issues of IGZO TFTs, specific circuits design for both pixels and scan drivers in AMLCD and AMOLED are introduced in the third section. Novel micro-LED displays emerging and new materials for wide-bandgap and p-type MO TFTs are anticipated to boost the growth of MO TFTs involving in the FPD industry.

7.1 Introduction

7.1.1 Current status of displays

Displays are becoming increasingly pervasive in our daily life. Essential as critical channels for man–machine interaction, electronic displays, such as computer monitors and televisions, are ubiquitous in modern information and entertainment systems. The total display market has remained over \$115 billion in the last five years. Modern FPDs have already replaced the venerable century-old cathode-ray-tube-based displays. FPDs, as the conduits to the information superhighway, are the more significant branch of electronic displays and encompass a growing number of

doi:10.1088/978-0-7503-2556-1ch7

electronic visual display technologies. Currently FPDs have over 90% of the display market.

FPDs appear all around us. Active-matrix liquid-crystal displays have dominated the TV and monitor market for years. The resolution of FPDs is being increased continuously and high-resolution FPDs have become must-haves not only in TV and personal computer (PC)/laptop applications, but also in the emerging markets for smart phones, tablets and e-book readers. Along with LCDs, the market for AM organic light-emitting diode displays is also growing very quickly and intense competition between AMLCDs and AMOLED displays has taken place in the smartphone market and is continuing into the TV market.

The basic difference between AMLCDs and AMOLED displays is how they realize a pixel. As we know, three original colors, red, green and blue, with different intensities, make up a tremendous number of colors. For the AMLCD, liquid-crystal itself cannot generate original colors, but it can be combined with color filters and polarizers to modulate the intensity of light. It can be considered as a 'passive' color management mode. For the AMOLED, various types of organic light-emitting diodes can generate the original color itself. It can be considered as an 'active' color management mode. The different color management results in different system constructions.

With early startup and mature production, AMLCDs are the leading technology in FPDs at present. It accounts for over 80% of the display market. The construction of an AMLCD is shown in figure 7.1(a) [2]. An AMLCD uses a liquid shutter, between two polarizers, to switch on and off the light, which is provided by the back-side light source. Typically, an AMLCD uses multiple LEDs as the light source. As the source is not uniform, a light guide system needs to be used to convert it to an area form to be used as a backlight unit. Light exiting from the light guide is transmitted through the prism sheet and diffuser, and is then polarized by the rear polarizer. Polarization of the light is changed by the field-induced orientation of the LC molecules so that the designated amount of light is excited when it passes the exit polarizer, converted to red, green or blue light by a color filter, and the exit light intensity is controlled by means of the exit polarizer and reaches the human eye.

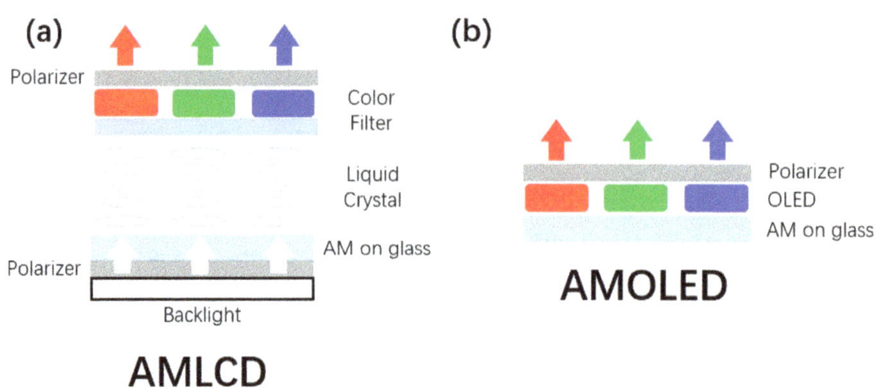

Figure 7.1. Examples of the display components used in (a) an AMLCD and (b) an AMOLED FPD [2].

Recently, AMOLEDs have become a rising challenger due to their outstanding properties. The working principle of OLEDs is totally different from that of LCDs. An LCD cell is a light valve that controls the amount of the backlight allowed to pass through the cell. In contrast, an OLED is a self-emitting device. The light in an AMOLED is derived from a self-emitting organic electroluminescent stack between the backplane and the encapsulation cover. Thus, the construction of an AMOLED (figure 7.1(b)) is much simpler compared to that of an AMLCD. In AMOLED displays, light emission takes place in an OLED device fabricated on the surface of a glass substrate. The reduced components and self-emitting property benefit the AMOLED with improved luminous efficiency, high brightness, excellent viewing angle and lower power consumption.

7.1.2 Active-matrix driving for FPDs

The current and foreseeable FPDs, whether based on LC or OLED, require an active-matrix backplane containing TFTs as the enabling active devices due to resolution limitation of conventional passive-matrix driving mode. An active-matrix display consists of an array of pixels (figure 7.2(a)); each containing at least an 'address' TFT controlling the optical state of its host pixel (figure 7.2(b)).

The address TFT in the pixel of an AMLCD (type I) plays the role of a 'switch'. The switch is controlled by two pulse signals generated by the scan and data drivers. The rows are scanned by a selected gate pulse (V_{sc}), and the frequency is determined by the fresh rate and resolution. The gate pulse of the selected scan line will turn on the addressing TFT of each pixel and at the same time the storage capacitor will be charged with the data voltage provided by the data driver, which will affect the light intensity of the pixel. The pixel voltage (V_{PX}) therefore increases, as shown in region I of figure 7.3. The pixels in other scan lines will not be affected. After one scan period, the addressing TFT will be turned off when the negative edge of the gate pulse is delivered, and the charge stored in the storage capacitors will be held. V_{PX} remains unchanged if the capacitor is large enough until the next frame time, and the status of the LC cell is therefore maintained, as shown in region II of figure 7.3. For the next frame time, V_{PX} is changed depending on the new V_{DT}, as shown in region III of figure 7.3.

Unlike the optical state of an LC pixel that can be voltage-switched by using a single switching transistor (figure 7.2(a)), the current-controlled emission of an OLED pixel (type II) requires at least two transistors and a storage capacitor (2T1C) to maintain a continuous diode current, as shown in figure 7.2(b). Sometimes three or more TFTs are used to compensate for the variation of the V_{th} and I_{on} to improve the display uniformity. The function of the address TFT in an AMOLED pixel is similar to that in an AMLCD pixel. It controls the signal data transfer. When the switching TFT is turned on, the signal data is transferred to the gate electrode of the driving TFT, and at the same time a corresponding driving current will be supplied to the OLED. The storage capacitor is charged simultaneously. The signal data will be maintained by the storage capacitor when the switching TFT is turned off.

A pixel is typically composed of three sub-pixels (figure 7.2(a)) of the red, green and blue primary colors. There is a delay T_{DL} for signals propagating along the scan-

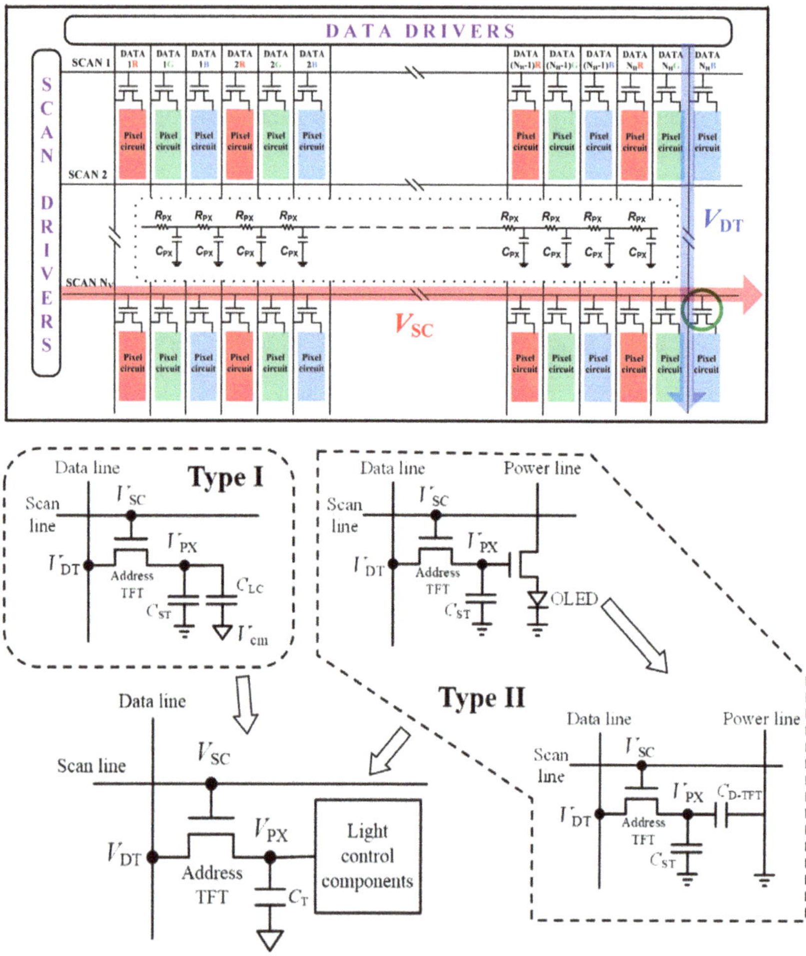

Figure 7.2. (a) A display panel shown schematically as a distributed resistor-capacitor network. (b) Electrical schematic of basic pixel level driving circuits of an AMLCD and an AMOLED. © (2020) IEEE. Reprinted, with permission, from [3].

and data-lines with distributed finite resistance (R_{PX}) and capacitive loading (C_{PX}) associated with each sub-pixel. The pixel located at the top-left corner of the display is in the immediate vicinity of the drivers and suffers from a negligible delay, whereas the 'farthest' one located at the lower-right corner (figure 7.2(a)) suffers from the longest delay. When an address TFT is switched on, C_T is charged. However, it takes a finite amount of charging time t_{CH} for the pixel voltage V_{PX} on C_T to approach V_{DT}. V_{PX} in turn determines the amount of light (hence the gray level) transmitted or emitted by the pixel. For proper driving, V_{DT} must not be changed until the address TFT is adequately switched off. Denoting T_{SC} and T_{DT} (figure 7.4) as the propagation delay respectively associated with the scan- and data-lines reaching the farthest pixel, and

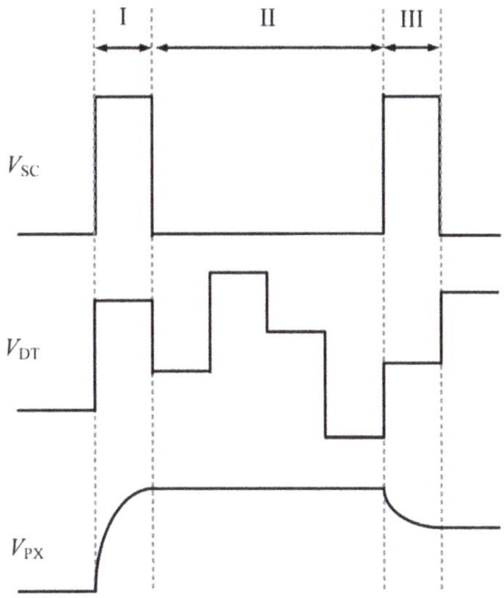

Figure 7.3. Basic time sequential response of V_{PX} to the V_{SC} and V_{DT}.

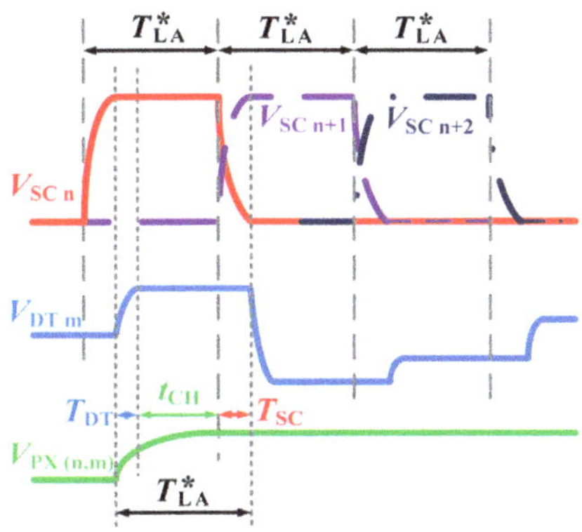

Figure 7.4. Timing diagrams illustrating the charging of a capacitor of a selected pixel in a matrix . © (2020) IEEE. Reprinted, with permission, from [3].

$$T_{LA} \equiv T_{SC} + T_{DT} + t_{CH} < T_{LA}^{*} \equiv \frac{1}{fN_V}, \tag{7.1}$$

where T_{LA}^{*} and T_{LA} are the respective maximum allowable and the actual line-address time, f is the frame rate and N_V is the number of scan-lines.

Signal propagation along an interconnect is determined largely by the product of its resistance (R) and capacitance (C), with a larger RC time-constant corresponding to a longer delay. The two main components making up C are the capacitance associated with the attached address-TFTs and that associated with the intersection with other interconnects. The TFT-related capacitance can be further decomposed to include an intrinsic gate-channel capacitance and a parasitic gate-S/D overlap capacitance, with respective values sensitive to the scale and architecture of the TFT.

In-pixel charging process would be approximately simplified with analysis of the amount of charge Q_{PX} stored on C_T and the charging current provided by the address TFT, one obtains

$$V_{PX} \approx V_{DT} - (V_{DT} - V_{PXo})e^{-\frac{t}{\tau_{CH}}}, \tag{7.2}$$

where $V_{PXo} \equiv V_{PX}|_{t=0}$ and the charging time-constant $\tau_{CH} \equiv [\mu_{FE}(V_{SC} - V_T - V_{DT})(\frac{W}{L})(\frac{C'_{GI}}{C_T})]^{-1}$. Parameters consisting of τ_{CH} are the field-effect mobility μ_{FE}, the channel width W and length L, the gate insulator capacitance per unit area C'_{GI} and the threshold voltage V_T.

7.1.3 TFT technologies in display

Different from the transistor in a traditional integrated circuit, fabricated in the bulk semiconductor, the TFT in an FPD is deposited on the glass substrate, requiring the deposition of a semiconductor film as the active channel layer. According to the material used as the active layer, it is divided into various TFT technologies. Silicon (Si) is the earliest material of choice in the FPD industry, in the form of amorphous Si (a-Si). Due to the presence of abundant electron traps in a-Si, the mobility of an a-Si TFT is normally below 1 $cm^2\ V^{-1}\ s^{-1}$. Although such limited mobility precluded a-Si from fabricating complex integrated circuits on glass, drastic improvement of the driving technology and the electrical performance of a-Si TFTs make it possible not only to switch pixels but to integrate the gate driver. a-Si has been capable of driving AMLCDs during the past two decades; however, the growth of the high-end commercial market with high resolution (>4K UHD), high frame rate (>240 Hz), and large size (>70 in.) requires a higher electrical performance than the a-Si TFTs can provide. It is closely beyond the capability of a-Si TFTs [4]. The paramount position of a-Si in the FPD backplane has been relentlessly challenged and partially replaced by an evolved silicon, low temperature poly-silicon (LTPS). In current FDP applications, Si-based TFT technologies are dominating the market with mature production and successful commercialization.

Pure a-Si is populated by a high density of localized trap states [5]. In order to effectively modulate its carrier concentration using doping [6] and electrical bias, an amount of hydrogen (H) is incorporated to passivate trap states, thus forming a-Si:H. Compatible with a large-area substrate up to a 10th generation mother glass (G10: ~$3 \times 3\ m^2$) [7], plasma enhanced chemical vapor deposition (PECVD) is the preferred film preparation technique with the deposition temperature is typically below 250 °C [8]. Compared to crystalline silicon, a larger ratio of on- to off-current ($I_{on}/I_{off} \sim 10^7$) is possible, because of the relatively larger bandgap of a-Si:H than that of crystalline silicon [9]. The field-effect mobility μ_{FE} for electrons is on the order of 0.5 – 1 $cm^2\ V^{-1}\ s^{-1}$, and a couple of orders of magnitude lower for holes [10]. This disparity in electron and hole μ_{FE} makes it unsuitable for implementing circuits with complementary n- and p-type TFTs, hence only the n-type variety is deployed in display applications. The low μ_{FE} also places a limit on how small a TFT can be scaled to satisfy a given current-drive requirement. Consequently, a-Si:H TFTs are more compatible with the implementation of displays containing relatively larger pixel sizes, rather than those demanding finer pixel dimensions.

LTPS can be prepared by either deposition directly in the polycrystalline form [11] or thermal crystallization from a-Si. The latter technique, including solid-phase crystallization [12], metal-induced crystallization [13] or excimer-laser crystallization (ELC) [14], is preferred because of the resulting larger average grain size, hence also better average film quality and device characteristics [15]. However, this is often accompanied by increased film roughness and non-uniformity in device characteristics, particularly compared to those obtainable in an a-Si:H thin film. Hitherto ELC is the technique that is industrially deployed but constrained by equipment availability to a maximum size of G8 ($\sim 2.5 \times 2.5$ m^2) mother glass. The μ_{FE} of electrons is on the order of $\sim 10^2$ cm^2 V^{-1} s^{-1} and that of holes is comparable but lower [16]. The higher values of μ_{FE} and their rough parity make it possible to implement a range of complementary peripheral circuits and to use a relatively smaller TFT to meet a given current-drive requirement. However, p-type TFTs are more commonly deployed within the pixel matrix of a flat-panel display because of their higher stability against hot-carrier stress [17]. Relative to those of an a-Si:H TFT, the I_{off} is higher and the I_{on}/I_{off} is generally one to two orders of magnitude smaller [18]. Multi-gate device architecture [19] is often employed to reduce I_{off}, thus resulting in a device footprint larger than that of the conventional single-gate variety. However, the positive attributes outweigh the relative drawbacks and make LTPS TFT the preferred device technology for implementing high-resolution mobile displays with fine pixel dimensions.

There is a large family of MO semiconductors [4], including single-cation binary oxides such as zinc oxide (ZnO) [20], indium oxide (In_2O_3) [21] and tin(IV) oxide (SnO_2) [22], etc; double-cation ternary oxides such as indium–zinc oxide (IZO) [23], zinc–tin oxide (ZTO) [24] and indium–gallium oxide (IGO) [25], etc; and triple-cation quaternary oxides such as IGZO [26–30] and indium–gallium–tin oxide (IGTO), etc. Steady progress in the development of MO device technologies has been made, starting from the investigation in the mid-1960s [31] into the use of binary oxides in the construction of TFTs to the realization of polycrystalline ZnO TFT at a temperature comparable to that used for the construction of a-Si:H TFTs while offering a much higher electron $\mu_{FE} \gg 10$ cm^2 V^{-1} s^{-1} [20]. Similar to a-Si:H, most MO semiconductors exhibit unipolar n-type conduction behavior.

The attributes cited for MO TFTs being particularly suitable for display applications are their low process temperature [32] compatible with common glass substrates; typically tens of cm^2 V^{-1} s^{-1} and between those of a-Si:H and LTPS, a μ_{FE} that fits the purpose; a large mobility gap and strongly localized valance band tail-states leading to a lower I_{off} [33] and a large optical bandgap resulting in higher transparency within the visible spectrum [33]. However, because of device instability issues when subjected to a combination of high electrical bias and short-wavelength illumination [34], the potential of realizing a higher light-transmission implied by the last attribute when using a MO semiconductor as a channel material is over-stated and currently unattained [35].

Three technologies are comprehensively compared in table 7.1 [36, 37]. First of all, oxide semiconductor materials are promising alternatives due to a relative high electron mobility (3–50 cm^2 V^{-1} s^{-1}) [26, 38, 39], which is large enough to fabricate a

Table 7.1. Comparison of MO and the incumbent silicon-based TFT technologies [36, 37].

Parameters	a-Si:H TFT	LTPS TFT	MO TFT
Mother glass	>G10	G8	>G10
Channel semiconductor	a-Si:H	ELC LTPS	IGZO
μ_{FE} ($cm^2\ V^{-1}\ s^{-1}$)	< 1	30–100	3–50
TFT uniformity	Good	Poor/better	Good
TFT polarity	n-Type	n, p-type	n-Type
Cost	Low	High	Low
Current-stress stability	Poor	Good	Good
Bias-stress stability	Adequate	Good	Adequate
Circuit integration	Limited	Yes	Yes
Process temperature (°C)	150–350	250–550	RT-450
Compatible pixel types	LC	LC, OLED	LC, OLED

high-speed transistor for driving the AMLCD mentioned above. In addition, oxide semiconductor materials have an amorphous phase in general, which shows uniform electrical properties even with a large-area substrate, which is strongly advantageous for display applications.

The second potential application for oxide semiconductor TFTs is the AMOLED. For mobile devices such as cellular phones, navigation, and notebook PCs, LTPS TFTs currently drive AMOLED because it exhibits a more sufficient and stable current supply into the OLED than the a-Si TFTs. However, the LTPS TFT has a problem with the spatial variation of electrical performances originating from the non-uniformity of grain sizes in the poly-Si film. Consequently, the LTPS TFTs are not suitable for AMOLED TVs. Even for a small size AMOLED (below 5 in.), a compensation circuit for pixels is required in order to obtain uniform image quality for the LTPS TFTs. As a backplane for the AMOLED TV, the stability of the oxide semiconductor TFT is superior to that of a-Si, and at the same time, the uniformity is much better than the LTPS TFTs. These aspects have motivated many demonstrations of AMOLEDs driven by MO TFTs [40].

7.1.4 Development trend of MO TFTs in display

Brief commercialization history is summarized in figure 7.5. Since the introduction of MO semiconductors such as crystalline and amorphous IGZO in 2003 and 2004, considerable work has been carried out on the study of amorphous oxide semiconductor devices. Due to their process compatibility with that of the conventional a-Si TFTs, the prototype of both AMLCD and AMOLED was announced within only three years. Since then, the major manufacturers have been working on the process issues and reliability testing for mass production of oxide-based FPDs. The first commercialization product of MO TFT based FPD was supplied by Sharp in 2012 for the retina display in Apple's new iPad. In the same year, Sharp also released its own-brand AQUOS smartphone, Mebius Pad and monitors. It is noteable that

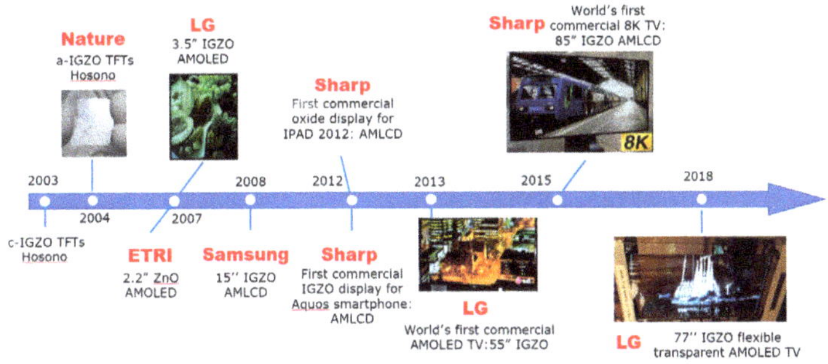

Figure 7.5. Brief history trend of MO TFTs applied for display applications.

the MO TFT based panel was successfully commercialized in small/medium size displays. To be further applied in large-size TV, one of the key issues is to develop a sputtering system for large-size uniform MO film deposition. As a large-size sputtering target is usually made from several pieces of ceramic bodies jointed by bonding, it would transfer the target joint pattern to the sputtered film and cause a uniformity issue. ULVAC solved the problem and supplied a monolithic IGZO target for G8.5 mother glass. In the same year, LG first commercialized flat/curved 55 in. OLED TV using a-IGZO TFT and white OLED. With resolution of the FPD continuously increasing, the advantages on mobility of MO TFTs over that of a-Si TFTs are obviously revealed. The world's first 8K TV was commercialized by Sharp using a-IGZO TFT panel in late 2015. With its mature top-gate self-aligned technology, LG is currently the only commercial manufacturer to product OLED TV. Recently, they revealed their remarkable 77 in. flexible and transparent AMOLED TV.

At present, the FPD development trend can be basically segmented by technology and display size. Currently it is in a reshaping process. The major factor that is driving the FPD market is the high adoption of advanced technology such as AMOLEDs over AMLCDs.

The ultimate display is to represent the real world to our eyes. This requires both immersion and vividness. Thus, for better immersion, one FPD development trend is 'larger than large'. With advantageous μ_{FE} over a-Si and better uniformity over LTPS, MO TFTs is the most promising leading technology for AMFPDs. Especially a-Si is meeting its μ_{FE} bottleneck applying to 8K display with high frame rate. The investment on MO manufacture is also speeding up. As summarized in table 7.2, the MO TFTs based panel production lines for TV products are entering the G10.5 with more big-brand participation.

For vividness, the trend is 'denser than dense'. The key quality factors for high-performance displays include high contrast, excellent color reproduction and flicker-free moving images. In the large size FPDs, we have already entered the 8K UHD era. This means over 33 million pixels in one frame. For the small size applications competition in the pixel density, pixel per inch (PPI), started as the retina display concept proposed by Apple with the iPhone series. Currently, the PPI of a high-end smartphone screen is over 500 PPI. This will be continually increased with the

Table 7.2. The global G10.5 lines for FPDs.

Company	Location	Technology	Generation
Sharp/Foxconn	Japan	IGZO/LCD	10.5
	Guangzhou, China	IGZO/LCD	10.5
	USA[a]	IGZO/LCD	10.5
LG	KOREA	IGZO/OLED	10.5
	Guangzhou, China	IGZO/OLED	8.5
CEC	Nanjing, China	IGZO/LCD	8.5
	Chengdu, China	IGZO/LCD	8.5
BOE[a]	Hefei, China	IGZO/OLED	8
CSOT[a]	Shenzhen, China	IGZO/OLED	11

[a] In plan.

requirement of virtual reality, which may need over 2000 PPI FPDs. A pixel density of over 2000 PPI means that the pitch of a single pixel is less than 12.5 μm. This requires TFTs with a channel length as short as 1.5 μm that still retains good on/off ratio performance. This is hardly achievable using LTPS technology. With further consideration of the aperture ratio limitation for AMLCD, the top emitting AMOLED is promising in virtual reality (VR) applications. Thus the MO TFT is a dominant candidate among other TFT technologies.

7.2 Issues of MO TFTs in display

7.2.1 TFT structures

Introduced in section 7.1, The TFT-related capacitance is sensitive to the scale and architecture of the TFT. In fabrication of a commercial FPD system, the configuration of TFTs is affected by the realistic requirement of pixel density, system aperture ratio and the amount of lithography process, etc. As shown in figure 7.6, there are major classes of configuration of MO TFTs: bottom-gate (BG) back-channel etch (BCE) structure and etch-stop (ES) structure, elevated-metal metal-oxide (EMMO) and top gate (TG) self-aligned structure with a technology with minimum lithography resolution of λ.

The BCE structure (figure 7.6(a)) is the same as that used in conventional a-Si TFTs, so it is easily adapted to process and design current TFT-LCD mass production. A typical BCE TFT fabrication process starts with the preparation and patterning (Mask 1) of the gate electrode. This is followed by the PECVD of a hydrogenated silicon nitride (a-SiN:H)/silicon oxide (SiO_x) gate dielectric. The MO active layer usually is sputtered as it is a suitable approach for large-size fabrication. A metal electrode layer is subsequently sputtered after the individual active islands for the TFTs are patterned (Mask 2). The stack of metal and contact layers on the islands are subsequently patterned (Mask 3), forming the S/D electrodes. While the metal electrode layer is often wet-etched. Any finite etch selectivity between the contact layer and the underlying MO layer would lead to the etching of a certain amount of the latter during the definition of the S/D electrodes including the contact

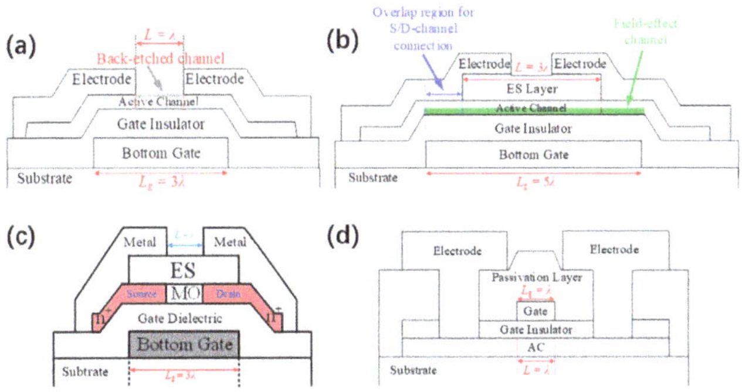

Figure 7.6. Schematic cross-sections of (a) a BCE, (b) an ES, (c) a conventional single-gate TG SA TFT architecture and (d) a double-gate TG SA TFT architecture typically employed to reduce the leakage current of an LTPS TFT. L_g denotes the length of the gate electrode in the direction of the channel current and L denotes the channel length.

layer. The portion of the active layer located away from the gate electrode is designated the 'back-channel' region, hence the structure resulting from this fabrication sequence is given the designation of 'back-channel etched' architecture (figure 7.6(a)). The etching of the back-channel invariably leads to the creation of localized defects. If unmitigated, these defects would lead to degradation of the TFT characteristics; including reduced μ_{FE} and device reliability, and increased I_{off} and threshold voltage V_{th} [42]. However, in spite of these obstacles, the BCE structure is one of the ultimate configurations for cost competitive TFTs.

The highest pixel density achievable in a display is constrained by the minimum-geometry TFT that can be realized for a given lithographic resolution limit λ. For a BCE TFT, the minimum channel length L_{min} is limited by the smallest lithographically definable separation between the S/D electrodes, hence $L_{min} = \lambda$. If the field-effect control of the channel is to be guaranteed, the minimum length of the gate electrode along the channel-length direction would be 3λ if an alignment accuracy of λ were assumed. This would lead to an overlap of λ between the gate and each of the S/D electrodes, contributing to the parasitic capacitance mentioned above.

The ES structure (figure 7.6(b)) was developed to overcome the mentioned drawbacks in the BCE structure by inserting an etch stopper layer (additional mask) between the MO and S/D electrode. The S/D electrode and passivation layer can be formed without changing the semiconductor properties, and the chip wet-etch process for the source/drain can be carried out based on a high selectivity between the electrodes and the ES layer. The stability and reliability of MO TFTs based on an ES structure are sufficiently fair.

For a minimum-geometry ES TFT, the separation between the S/D electrodes, the length of the ES (hence also the channel) layer and the length of the gate electrode are λ, 3λ and 5λ, respectively. The overlap between the gate and each of the S/D electrodes is 2λ. The improved characteristics of an ES TFT come with a number of drawbacks: the requirement of an extra mask-count implies a higher manufacturing cost; the larger gate and overlap parasitic capacitance leads to a longer propagation delay; and the larger device footprint puts a lower bound on the maximum achievable pixel density. It is for these reasons that the ES architecture is often an option of last resort, when adequate means of mitigating the etch-damage of the back-channel of a BCE TFT are not yet available.

A new TFT architecture (figure 7.6(c)) is presently described, in which the S/D metal electrodes are 'elevated' above an ES layer inserted into a BCE TFT (figure 7.6(a)) while maintaining the length of the gate electrode at 3λ. The edges of the channel are self-aligned to those of the S/D electrodes projected in the direction normal to that of the channel current, thus retaining $L_{\text{min}} = 1\lambda$ for a minimum-geometry TFT. The realization of the new architecture requires the provision of conducting S/D regions connecting the field-induced channel to the contacts with the S/D electrodes. Hitherto it has been difficult to realize this architecture for a silicon-based TFT without a significant increase in process complexity. However, it can be more readily realized for an MO TFT because of a unique annealing behavior of MO semiconductors [20]. The resulting structure is dubbed an EMMO TFT [43], offering the smaller gate-to-S/D overlap and device footprint of a BCE TFT (figure 7.6(a)), and the protection of the back-channel of an ES TFT (figure 7.6(b)).

In order to reduce the delay when driving a large-size display, a TG self-aligned TFT (figure 7.6(d)) with minimal parasitic capacitance has been deployed [40]. It has been mentioned that a TG TFT will exhibit relatively poorer performance and reliability compared to an ES TFT [44], plausibly due to the thermally-induced generation of defects [45] in the channel sealed under the gas-impermeable gate electrode. For AMLCD applications, however, because light from the backlight unit reaches the semiconductors without being blocked by the gate electrode, the stability of MO for light illumination should be strengthened. A comparison of the current structures in MO TFTs is summarized in table 7.3.

Besides, several novel structures were proposed for display application. Lee *et al* demonstrated that the performance of flexible a-IGZO TFT can be improved significantly by splitting the active layer and S/D electrodes (figure 7.7(a)). The on-current of the split TFTs is much higher than that of normal TFTs [46, 47]. The μ_{FE} are 22.5 and 74.3 $\text{cm}^2\ \text{V}^{-1}\ \text{s}^{-1}$ for normal and split TFTs, respectively. The split TFT

Table 7.3. Comparison of common TFT architectures.

Structures	BCE	ES	EMMO	TG SA
Channel damage	Yes	No	No	No
Capacitance	Medium (1λ overlap)	Large (2λ overlap)	Medium (1λ overlap)	Minimal (self-aligned)
TFT footprint ($L_{\text{min}} = 1\lambda$)	1λ	3λ	1λ	$1\lambda/2\lambda$ (single/double gates)
Masking steps (manufacturing cost)	3 (low)	4 (high)	4 (high)	4 (high)

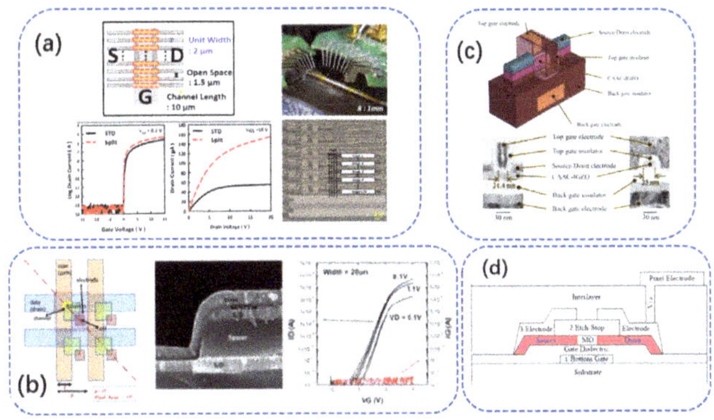

Figure 7.7. Reports on novel structures of MO TFTs. [46–48] John Wiley & Sons. © (2018, 2019) WILEY-VCH Verlag GmbH & Co., KGaA, Weinheim. © (2019, 2018) IEEE. Reprinted, with permission, from [49, 50].

exhibits the lower SS of 82 mV/dec, and low V_{th} of 0.2 V, indicating very low density of SiO_2/a-IGZO interface states. The PBTS can be also improved remarkably, and rolling stability can be also improved by the splitting. These excellent performance and stability are correlated to the improvement of top interface by the metal-F bond formation by NF_3 plasma for ES patterning and subsequent reduction in the interface state density. A 4-in. transparent AMOLED display was fabricated using IGZO TFT with split structures.

To achieve high pixel density, instead of scaling down channel length of planar oxide TFTs, oxide TFTs with vertical channel was proposed by Hwang *et al* [48]. Pixel array with pixel pitch of 1 μm can be designed with pattern of 0.5 μm line and space as shown in figure 7.7(b). Vertical channel oxide semiconductor was sputtered. To overcome the large aspect ratio of vertical channel region and reducing gate leakage currents, ALD deposition of gate insulator was preferred. Back-channel interface between active layer and spacer layer should be carefully engineered to improve the performance of vertical channel TFTs. The performance and cross-sectional SEM image of vertical channel oxide TFTs are shown in figure 7.7(b).

SEL also reported a c-axis-aligned crystalline (CAAC) IGZO TFT with a top-gate length of 21 nm [49] with a bottom-gate electrode (figure 7.7(c)). This work shows how small a CAAC-IGZO based TFT could be achieved. Meanwhile, prototyped devices have retained low off-state leakage current, high thermal stability. This also reveals the promising perspective of MO TFTs to be applied in high resolution displays.

4-Mask EMMO (4-EMMO in short) TFT technology combined benefits of the protection of the back-channel by the ES layer and the small footprint of a BCE TFT. However, the advantages are gained at the expense of an extra masking step employed for the definition of the ES layer, hence a higher manufacturing cost than that of a 3-mask BCE technology (table 7.3(a)). Li *et al* proposed 3-EMMO TFT technology [50]: instead of the active island being conventionally patterned using a separate mask, its postponed patterning is self-aligned to the combined patterns of the ES layer and the S/D electrodes (figure 7.7(d)). Since the need to accommodate lithographic misalignment between the active island and the ES layer or the S/D electrodes is eliminated, the footprint of a minimum 3-EMMO TFT can be made even smaller than that of an equivalent 4-EMMO TFT. 3-EMMO TFTs demonstrated good characteristics, stability and reliability.

7.2.2 MO materials

Brief introduced in the previous section, there is a large family of MO semiconductors as it could evolve from binary oxides (ZnO, In_2O_3, SnO_2) to double-cation ternary oxides such as IZO, ZTO and IGO, etc, and further to quaternary oxides, quintuple oxides and so on. With foundation of the remarkable IGZO system, many attempts have been made based on materials design formation of $X_1X_2X_3O$, X_1X_2YO and $X_1X_2X_3YO$, where X includes In, Zn, Ga, Sn and Y includes potential elements in the periodic table. The target is to discover a new system with better mobility and reliability. However, most of the attempts have faded out from the practical development research of FPD production.

The potential reasons could be that the IGZO system currently has the best balance between electrical performance and reliability [51]. In a-IGZO system, the In ions contribute to a large electron mobility and act as 'mobility enhancer.' Ga forms stronger chemical bond with oxygen and would suppress the formation of oxygen deficiency and the generation of conduction electrons and act as 'stabilizer' [51]. Zn is expected to stabilize amorphous structure. The replacement of Ga and Zn with other 'mobility enhancers' will degrade reliability which is specially emphasized in practical production. Meanwhile, the current FPD may not require so much on further enhancement of mobility of 10 $cm^2\ V^{-1}\ s^{-1}$. At present, commercial a-Si TFTs based AMLCD FPDs are at the level of 75 in., 8K, 60 Hz. Thus, for the same panel design, the MO TFTs of 10 $cm^2\ V^{-1}\ s^{-1}$ could support 3-generation-ahead FPDs compared to a-Si TFTs.

Currently, IGZO is the dominant and only commercial material applied in FPD production. By changing the composition of In:Ga:Zn, the mobility of IGZO film could range from 10 to 25 $cm^2\ V^{-1}\ s^{-1}$. Since the mass production of IGZO-TFT, Sharp claimed they are improving the mobility and stability of IGZO by generations (figure 7.9(a)). Meanwhile, they abandoned the ES structure to BCE structure due to the delay issues we discussed in the previous section. With good uniformity of V_{th} under 10% on G8 mother glass (figure 7.8(a)), the typical values of TFT characteristics are as follows: the mobility is 15 $cm^2\ V^{-1}\ s^{-1}$, V_{th} is 3.2 V, and SS is 0.4 V/dec. The off-current of TFT characteristics is less than 10 fA [52].

Further, the Sharp's cooperator SEL claim a new type of IGZO film called CAAC IGZO [53]. They claim that CAAC IGZO is closer to a single crystal than an amorphous material. It has *c*-axis (vertical direction) alignment like a single crystal, but its *a*–*b* plane (horizontal plane) has a mosaic pattern and no grain boundaries (figure 7.8(b)). The CAAC differs from a polycrystal, which generally has *a*-, *b*-, and *c*-axes oriented at random and with clear grain boundaries. CAAC IGZO does not have *a*- and *b*-axes alignment across the entire in-plain region resulting good uniformity in devices. The CAAC IGZO was reported to have smaller density of

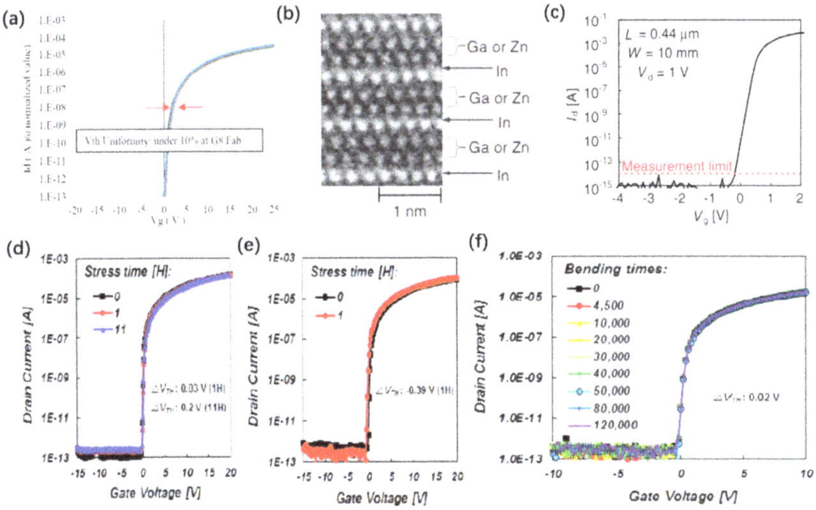

Figure 7.8. Commercial MO TFTs performance from Sharp, SEL and LG. (a) [52] John Wiley & Sons. © (2018) WILEY-VCH Verlag GmbH & Co., KGaA, Weinheim. (b), (c) Reproduced from [53]. © IOP Publishing Ltd. All rights reserved. (d)–(f) [54] John Wiley & Sons. © (2018) WILEY-VCH Verlag GmbH & Co., KGaA, Weinheim.

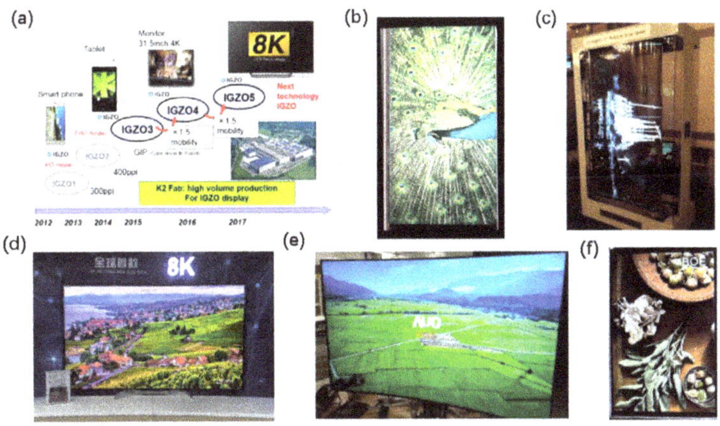

Figure 7.9. Latest products/prototype using MO TFTs from (a) Sharp [52], (b) SEL [55], (c) LG [56], (d) CSOT [57], (e) AUO [58] and (f) BOE [59]. (a) [52] John Wiley & Sons. © (2018) WILEY-VCH Verlag GmbH & Co., KGaA, Weinheim. (b) [55] John Wiley & Sons. © (2016) WILEY-VCH Verlag GmbH & Co., KGaA, Weinheim. (c) [56] John Wiley & Sons. © (2018) WILEY-VCH Verlag GmbH & Co., KGaA, Weinheim. (d) [57] John Wiley & Sons. © (2018) WILEY-VCH Verlag GmbH & Co., KGaA, Weinheim. (e) [58] John Wiley & Sons. © (2019) WILEY-VCH Verlag GmbH & Co., KGaA, Weinheim. (f) [59] John Wiley & Sons. © (2017) WILEY-VCH Verlag GmbH & Co., KGaA, Weinheim.

subgap states and higher quality density compared to normal IGZO film. It reveals an extremely low off-state leakage current of 1.35×10^{-22} A μm^{-1} (figure 7.8(c)). With this technique, SEL reveals prototypes of 4K 1058 PPI AMLCD and flexible 8K 664 PPI AMOLED (figure 7.9(b)).

Different from Sharp, LG adapt top-gate self-aligned structure for their IGZO TFTs to further reduce the parasitic capacitance and block the light illumination from bottom-emission OLED at the top side. They reported PI-based TFTs revealing good performance on average: V_{th} of 0.35 V, μ_{FE} of 10.4 $cm^2\ V^{-1}\ s^{-1}$ and SS of 0.23 V/dec [54]. The electrical properties of TFTs fabricated on PI are very similar to those of TFTs fabricated on glass. For the reliability test. Both positive bias temperature stability (PBTS) (figure 7.8(d)) and negative bias temperature illumination stability (NBTIS) (figure 7.8(e)) test were performed TFTs fabricated on polyimide substrate after laser lift-off process. Devices show slight negative V_{th} shift less than 0.5 V. Bending test (figure 7.8(f)) was performed with a radius of 30 mm to evaluate bending flexibility of TFTs. During 120 000 cycles of bending and unfolding processes, transfer characteristics of a TFT were measured at a flat state and its V_T was extracted in certain bending times. The initial V_{th} of TFT is 0.75 V and it varies until 0.77 V at the end of the bending test. With this mature technology, LG has announced world's first flexible and transparent AMOLED TV (figure 7.9(c)).

Currently, only Sharp and LG started mass production of MO based panels. However, the situation will soon change with BOE and CSOT announcing their production plans for AMOLED TV. With the mature prototype of IGZO and ITZO, TFT based FPDs show big manufacture (figure 7.9). The current oxide panel markets will be expected to be joined by more and more companies [52, 55–59].

7.3 Circuit design for MO TFTs in display

7.3.1 Pixel circuits

The prototype circuit of a pixel in an AM LCD is shown in figure 7.10(a). As the LC is driven by voltage, the requirement for the reliability of address TFT is not harsh as the voltage value is input by data driver and the address TFT is only to retained off to hold the pixel voltage. Small variation of V_{th} in the address TFT will not affect the light output of the pixel. The major concern of LCD pixel is the aperture ratio. As shown in figure 7.10, LC itself is not an emission device. When backlight goes through the AM panel, some light will be blocked by other components, such as TFT and signal lines (figure 7.10(b)). The aperture ratio affects the power consumption strongly in a transmissive LCD. For example, an LCD with an aperture ratio of 10% requires five times as much backlight power as an LCD with

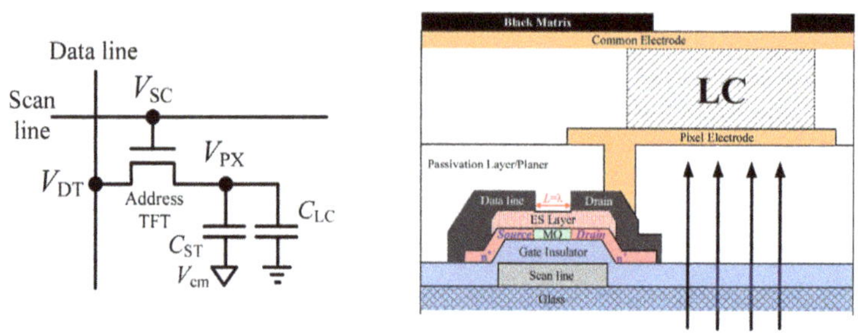

Figure 7.10. (a) Prototype pixel circuits of AMLCD and (b) cross-section of schematic of AMLCD pixel.

an aperture ratio of 50% to achieve the same luminance. Thus, the pixel circuit of LCD applied with MO TFTs will remain as simple as 1T1C.

For AMOLED (figure 7.11(a)), as the OLED is driven by current, the requirement for the reliability of the drive TFT is much more severe than that of the LCD. The current flowing in an OLED cell is determined with:

$$I_d = \frac{W}{2L}\mu_{FE}C'_{GI}(V_{PX} - V_{OLED} - V_T)^2$$

Small variation of V_T in the drive TFT will affect the brightness of the cell, as it is proportional to the driving current. As human eyes are very sensitive to the change of brightness, non-uniform brightness of AMOLED display was unacceptable. the compensation circuit must be addressed in the OLED pixel. The V_{th} shift of MO TFT under stress must be compensated to avoid deviations of OLED current.

Moreover, the consideration of normally-on characteristics depending on fabrication process is also crucial in the pixel design of AMOLED backplanes because it would limit the application of the MO TFTs for prior driving schemes with compensation of TFT V_{th} shift. Driving schemes which employed diode-connected compensation method operate by n-channel driving TFT with only positive V_{th} [60]. Therefore, they cannot function well for normally-on TFTs. To resolve the problem mentioned above, several pixel compensation circuits were proposed and are summarized in table 7.4. Mo *et al* proposed a five-TFT and two-capacitor pixel circuit which compensates for V_{th} shift of normally-on IGZO TFTs [61]. The scheme (figure 7.12(a)) fixes the gate node voltage and charges the voltage of source node of driving TFT during compensation period until driving TFT is turned off. V_{th} of the driving TFT thus is obtained regardless of negative value. Although having the ability to compensate for negative V_{th}, the driving scheme suffers from undesirable charging of gate-source voltage of driving TFT in transition between compensation stage and emission stage, leading to incomplete V_{th} compensation.

However, these circuits still suffer from TFT mobility variations. Kim *et al* present a novel IGZO TFT based voltage programmed pixel circuit for AMOLEDs (figure 7.12(b)), which can compensate the mobility variation as well as both positive and negative threshold voltage variations of driving TFT [62]. SEL also propose an external compensation method using external monitor lines (figure 7.12(c)). The

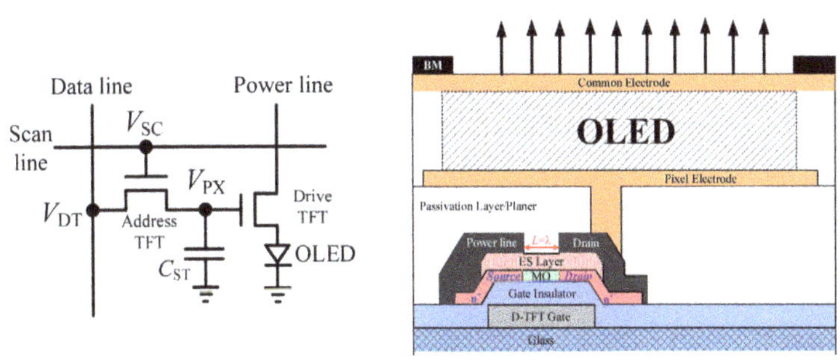

Figure 7.11. (a) Prototype pixel circuits of AMOLED and (b) cross-section of schematic of AMOLED pixel.

Table 7.4. Comparison between different circuit diagram of a AMOLED pixel.

In-pixel devices	Compensation function			Additional signal line
	V_T variation	μ_{FE} variation	OLED degradation	
5T2C [61]	Yes	No	No	3
6T1C [62]	Yes	Yes	No	2
3T1C [63]	Yes	Yes	No	1 (external compensation line)
5T2C [64]	Yes	No	Yes	3
4T1C [65]	Yes	No	Yes	1
4T2C [66]	Yes	Yes	No	2

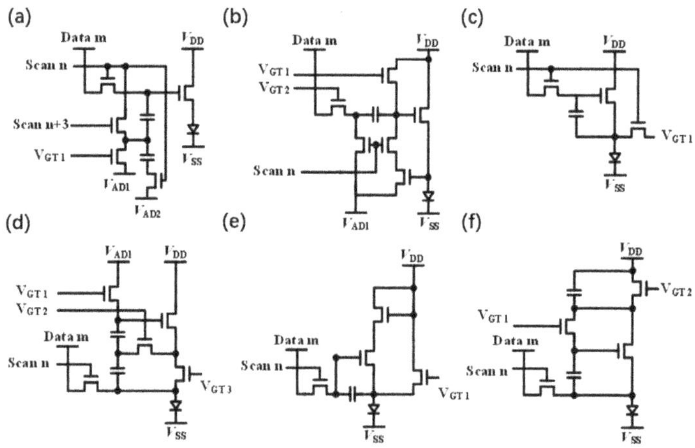

Figure 7.12. Various pixel circuit schematics for compensation in AMOLED using MO TFTs [55–60].

compensation for V_{th} and μ_{FE} could be calculated by a two-point measurement. With this design, the in-pixel devices could be as small as 3T1C [63]. It might be applied in high-resolution systems.

And furthermore, the luminance of OLED degrades gradually while the device is operated under long-term bias current, and the phenomenon causes severe image burn-in. Therefore, the maintenance of OLED luminance is also significant to extend the lifetime of AMOLED displays. Lin *et al* proposed a new pixel circuit (figure 7.12(d)) with designing source-follower structure to realize compensation for the V_{th} shift of IGZO TFTs and the drop of OLED luminance by separately sensing the threshold voltages of TFT and OLED [64]. Kim *et al* present a voltage programmed pixel circuit (figure 7.12(e)) compensating for V_{th} variation of TFTs and OLED degradation with a reduced number of signal lines [65], achieving a high aperture ratio and low power consumption to be applied for large size and high resolution AMOLED displays.

Lin *et al* proposed that the circuit (figure 7.12(f)) combines the parallel addressing scheme, threshold voltage shifts compensation, mobility, shifts compensation, and

OLED degradation compensation into one simple structure [66] and applied into a display prototype.

7.3.2 Driver circuits

Besides pixel panel, an AM FPD also needs scan drivers and data drivers for the whole panel to function. Compared to data signals currently with 8–10 bits, scan signals generally have only two states of high and low to turn on/off the pixel circuit. As a result, the required driving frequency of scan drivers is at ~hundreds of kilohertz level while that for data drivers will be over hundreds of megahertz. Considering mobility of ~10 $cm^2\ V^{-1}\ s^{-1}$, it is possible to integrate the scan drivers on the glass simultaneously within fabricating the pixel circuits. It is cost-effective and advantageous to achieve high resolution displays with narrow bezel.

A scan driver commonly comprises three parts: a shift register, a buffer, and a signal and power supply line. The key component is the shift register. Similar to a pixel compensation circuit, there exists the problem from n-type depletion mode oxide TFTs having negative threshold voltages resulting zero gate-source voltage cannot shut down the current path.

Several approaches have been established to cope with the depletion mode operation of oxide TFTs. Multiple low supplies are employed to ensure that the gate voltages of large size pull-up and pull-down TFTs are lower than the low level of an output signal. Lin *et al* proposes a low-power and stable gate driver with two clock sets and two low-voltage-level power sources (figure 7.13(a)) to completely turn off TFTs for low power consumption [67]. To remove the internal direct current paths, Kim *et al* propose a depletion-mode a-IGZO TFT shift register employing only one low-voltage-level power signal without an additional signal. They designed a shift register that employs the series-connected two-transistor structure for charging and discharging TFTs [68]. Park *et al* proposed a method that applies

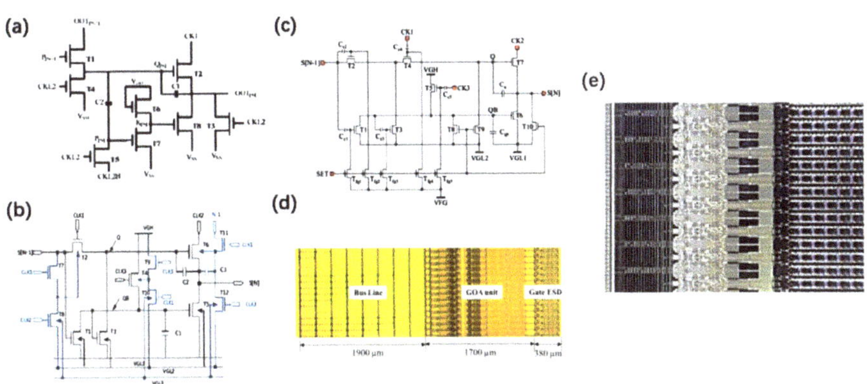

Figure 7.13. Various scan driver circuits on array schematics for using MO TFTs. (a) [67] John Wiley & Sons. © (2014) WILEY-VCH Verlag GmbH & Co., KGaA, Weinheim. (b) [69] John Wiley & Sons. © (2011) WILEY-VCH Verlag GmbH & Co., KGaA, Weinheim. (c) [70] John Wiley & Sons. ©(2011) WILEY-VCH Verlag GmbH & Co., KGaA, Weinheim. (d) ©(2019) IEEE. Reprinted, with permission, from [72]. (e) [73] John Wiley & Sons. ©(2018) WILEY-VCH Verlag GmbH & Co., KGaA, Weinheim.

dual gate oxide TFTs to shift some internal TFTs of shift registers (figure 7.13(b)) into enhancement mode on an additional back gate [69]. Kang *et al* also applied a floating gate method enabling a lower gate voltage (figure 7.13(c)) than any internal signals by means of capacitors connected to gates of some TFTs [70] to minimize the off-state current of the negative V_T TFT.

Recently, the IGZO based scan driver was applied to low-frame rate and high-resolution FPDs. Song *et al* present an oxide TFT gate driver circuit with the wide operating frequency range and controllable overlap interval for a variable frame rate AMOLED display [71]. The working range of gate driver circuit is from 1 to 240 Hz. The driver was applied to a 15 in. AMOLED panel. Qungang Ma proposed a design with a pulse gating scheme engaged into scan driver to avoid long term continuous bias stressing of the TFT. The reliability of the GOA circuit (figure 7.13(d)) is well proved and applied in an 8K AMLCD panel [72].

For AMOLED TV, the scan driver is more complicated for OLED displays as it needs sophisticated sensing and compensation. It not only drives gate lines sequentially, but also senses V_{th} and mobility fluctuations. LG recently reported its high-reliability integrated gate driver circuit (figure 7.13(e)) using IGZO TFTs with a lifetime longer than 60 000 h to drive 65 in. 4K AMOLED TV [73].

7.4 Summary and perspective

Since the first mass production of MO TFTs for displays, oxide semiconductor material has changed from a promising material to a practical material. Although there are still some intrinsic problems concerning the origin of instability of MO TFTs, the practical solution of processing and structure design could reduce and avoid the stress to ensure a sufficient lifetime for MO panels, such as adding an additional light-shield or improving the inner thermal reliability. The main advantages of IGZO TFTs such as scalability, uniformity, availability of low temperature process, and a wide range of material choice for gate insulators will make it the leading candidate in the 5G-8K era for both large-size high-resolution and flexible helmet-mounted display applications. With yield-climbing, the MO TFTs will become cost-effective and performance-dominant comparing to a-Si TFTs to further enlarge the share of TV markets.

There are also new emerging challenges and opportunities for MO TFTs in display. Firstly, the emerging mini/micro LED displays require more suitable driving TFT technologies. To improve HDR performance, high-end AMLCD FPD is changing LED-backlight to AM LED-backlight, that could be a customized application for MO TFTs. Meanwhile, MO TFTs might provide the minimum device size to supply the ultra-high PPI requirement of micro-LED, which could break through density limitation of evaporated OLED.

The goal of achieving CMOS by combining p-type and n-type oxide TFTs still needs to be accomplished. Although the monopolar channel has been proven enough for pixel TFTs in FPDs, CMOS is better used for peripheral and other logic circuits in scan/data drivers to reduce power consumption. The application of

LTPO has already verified its potential value. Recent progress of novel p-type material and improvement of SnO is anticipated.

Beyond displays, MO TFTs could also serve for other AM panels such as photodiode, pressure-, bio-, temperature- and chemical-sensors. Starting from display, MO TFTs will be more applicable in potential and promising flexible electronics.

References

[1] Tsujimura T 2012 *OLED Display Fundamentals and Applications* (New York: Wiley) p 106

[2] Park J-S 2015 *AOS TFTs for AMOLED TV Handbook of Visual Display Technology* (Berlin: Springer) pp 1–19

[3] Xia Z, Liu X, Shi R, Li J, Lu L, Kwok H-S and Wong M 2020 A timing model for the optimal design of a prototype active-matrix display *IEEE Trans. Electron Devices* **67** 3167–74

[4] Kamiya T, Nomura K and Hosono H 2010 Present status of amorphous In–Ga–Zn–O thin-film transistors *Sci. Technol. Adv. Mater.* **11117** 44305–23

[5] Shanks H, Fang C J, Ley L, Cardona M, Demond F J and Kalbitzer S 1980 Infrared spectrum and structure of hydrogenated amorphous silicon *Phys. Status Solidi* **100** 43–56

[6] Stutzmann M, Biegelsen D K and Street R A 1987 Detailed investigation of doping in hydrogenated amorphous silicon and germanium *Phys. Rev.* B **35** 5666–701

[7] Souk J, Morozumi S, Chen F and Bita I 2018 *Flat Panel Display Manufacturing* (New York: Wiley)

[8] Chen J, Cranton W and Fihn M 2012 *Handbook of Visual Display Technology* (Cham: Springer)

[9] Klazes R H, Van Den Broek M H, Bezemer J and Radelaarj S 1982 Determination of the optical bandgap of amorphous silicon *Philos. Mag.* B **45** 377–83

[10] Street R A 2005 *Hydrogenated Amorphous Silicon* (Cambridge: Cambridge University Press)

[11] Joubert P 1987 The effect of low pressure on the structure of LPCVD polycrystalline silicon films *J. Electrochem. Soc.* **134** 2541

[12] Bo X-Z, Yao N, Shieh S R, Duffy T S and Sturm J C 2002 Large-grain polycrystalline silicon films with low intragranular defect density by low-temperature solid-phase crystallization without underlying oxide *J. Appl. Phys.* **91** 2910–5

[13] Qi G J, Zhang S, Tang T T, Li J F, Sun X W and Zeng X T 2005 Experimental study of aluminum-induced crystallization of amorphous silicon thin films *Surf. Coat. Technol.* **198** 300–3

[14] Wu G M, Chen C N, Feng W S and Lu H C 2009 Improved AMOLED with aligned poly-Si thin-film transistors by laser annealing and chemical solution treatments *Physica* B **404** 4649–52

[15] Kim C W, Jung J G, Choi J B, Kim D H, Yi C, Kim H D, Choi Y H and Im J 2011 59.1: invited paper: LTPS backplane technologies for AMLCDs and AMOLEDs *SID Symp. Dig. Tech. Pap.* **42** 862–5

[16] Rath J 2003 Low temperature polycrystalline silicon: a review on deposition, physical properties and solar cell applications *Sol. Energy Mater. Sol. Cells* **76** 431–87

[17] Moon K C, Lee J H and Han M K 2005 The study of hot-carrier stress on poly-Si TFT employing C-V measurement *IEEE Trans. Electron Devices* **52** 512–7

[18] Matsuo T, Mori S, Ban A and Imaya A 2014 8.3: invited paper: advantages of IGZO oxide semiconductor *SID Symp. Dig. Tech. Pap.* **45** 83–6

[19] Tanaka T, Horie H, Ando S and Hijiya S 1991 Analysis of p/sup +/ poly Si double-gate thin-film SOI MOSFETs *Int. Electron Devices Meeting 1991* (IEEE) pp 683–6 (http://ieeexplore.ieee.org/document/235330/)

[20] Lu L and Wong M 2014 The resistivity of zinc oxide under different annealing configurations and its impact on the leakage characteristics of zinc oxide thin-film transistors *IEEE Trans. Electron Devices* **61** 1077–84

[21] Wang L, Yoon M-H, Lu G, Yang Y, Facchetti A and Marks T J 2006 High-performance transparent inorganic–organic hybrid thin-film n-type transistors *Nat. Mater.* **5** 893–900

[22] Presley R E, Munsee C L, Park C-H, Hong D, Wager J F and Keszler D A 2004 Tin oxide transparent thin-film transistors *J. Phys. D: Appl. Phys.* **37** 2810–3

[23] Chiang H Q, Wager J F, Hoffman R L, Jeong J and Keszler D A 2005 High mobility transparent thin-film transistors with amorphous zinc tin oxide channel layer *Appl. Phys. Lett.* **86** 013503

[24] Barquinha P, Pimentel A, Marques A, Pereira L, Martins R and Fortunato E 2006 Effect of UV and visible light radiation on the electrical performances of transparent TFTs based on amorphous indium zinc oxide *J. Noncryst. Solids* **352** 1756–60

[25] Presley R E, Hong D, Chiang H Q, Hung C M, Hoffman R L and Wager J F 2006 Transparent ring oscillator based on indium gallium oxide thin-film transistors *Solid State Electron.* **50** 500–3

[26] Nomura K, Ohta H, Takagi A, Kamiya T, Hirano M and Hosono H 2004 Room-temperature fabrication of transparent flexible thin-film transistors using amorphous oxide semiconductors *Nature* **432** 488–92

[27] Hanyu Y, Domen K, Nomura K, Hiramatsu H, Kumomi H, Hosono H and Kamiya T 2013 Hydrogen passivation of electron trap in amorphous In-Ga-Zn-O thin-film transistors *Appl. Phys. Lett.* **103** 202114

[28] Nomura K, Kamiya T, Hirano M and Hosono H 2009 Origins of threshold voltage shifts in room-temperature deposited and annealed a-In-Ga-Zn-O thin-film transistors *Appl. Phys. Lett.* **95** 2014–7

[29] Ide K, Kikuchi Y, Nomura K, Kimura M, Kamiya T and Hosono H 2011 Effects of excess oxygen on operation characteristics of amorphous In-Ga-Zn-O thin-film transistors *Appl. Phys. Lett.* **99** 093507

[30] Hosono H, Yasukawa M and Kawazoe H 1996 Novel oxide amorphous semiconductors: transparent conducting amorphous oxides *J. Noncryst. Solids* **203** 334–44

[31] Boesen G F and Jacobs J E 1968 ZnO field-effect transistor *Proc. IEEE* **56** 2094–5

[32] Liu J, Sonshine D A, Shervani S and Hurt R H 2010 Controlled release of biologically active silver from nanosilver surfaces *ACS Nano* **4** 6903–13

[33] Medvedeva J E, Buchholz D B and Chang R P H 2017 Recent advances in understanding the structure and properties of amorphous oxide semiconductors *Adv. Electron Mater.* **3** 1700082

[34] Ryu B, Noh H K, Choi E A and Chang K J 2010 O-vacancy as the origin of negative bias illumination stress instability in amorphous In-Ga-Zn-O thin film transistors *Appl. Phys. Lett.* **97** 2012–5

[35] Oh H, Yoon S M, Ryu M K, Hwang C S, Yang S and Park S H K 2010 Photon-accelerated negative bias instability involving subgap states creation in amorphous In-Ga-Zn-O thin film transistor *Appl. Phys. Lett.* **97** 1–4

[36] Park J S, Maeng W J, Kim H S and Park J S 2012 Review of recent developments in amorphous oxide semiconductor thin-film transistor devices *Thin Solid Films* **520** 1679–93

[37] Fortunato E, Barquinha P and Martins R 2012 Oxide semiconductor thin-film transistors: a review of recent advances *Adv. Mater.* **24** 2945–86

[38] Xia Z, Lu L, Li J, Feng Z, Deng S, Wang S, Kwok H S and Wong M 2017 Characteristics of elevated-metal metal-oxide thin-film transistors based on indium-tin-zinc oxide *IEEE Electron Device Lett.* **38** 894–7

[39] Song J H, Kim K S, Mo Y G, Choi R and Jeong J K 2014 Achieving high field-effect mobility exceeding 50 cm^2/Vs in In-Zn-Sn-O thin-film transistors *IEEE Electron Device Lett.* **35** 853–5

[40] Arai T and Sasaoka T 2011 49.1: invited paper: emergent oxide TFT technologies for next-generation AM-OLED displays *SID Symp. Dig. Tech. Pap.* **42** 710–3

[41] Ando M, Wakagi M and Minemura T 1998 Effects of back-channel etching on the performance of a-Si:H thin-film transistors *Japan J. Appl. Phys.* **37** 3904–9

[42] Lu L, Li J, Feng Z, Kwok H S and Wong M 2016 Elevated-metal-metal-oxide thin-film transistor: technology and characteristics *IEEE Electron Device Lett.* **37** 728–30

[43] Shin H-J *et al* 2015 7.1: invited paper: novel OLED display technologies for large-size UHD OLED TVs *SID Symp. Dig. Tech. Pap.* **46** 53–6

[44] Lu L, Li J and Wong M 2014 A comparative study on the effects of annealing on the characteristics of zinc oxide thin-film transistors with gate-stacks of different gas-permeability *IEEE Electron Device Lett.* **35** 841–3

[45] Jang J and Lee S 2018 24.1: invited paper: flexible oxide TFTs for bendable and flexible displays *SID Symp. Dig. Tech. Pap.* **49** 248–51

[46] Lee S, Chen Y, Kim H, Kim J and Jang J P 2019 14: distinguished poster: highly robust oxide TFT with bulk accumulation and source/drain/active layer splitting *SID Symp. Dig. Tech. Pap.* **50** 1263–6

[47] Hwang C-S *et al* 2018 46-2: invited paper: ultimate resolution active matrix display with oxide TFT backplanes for electronic holographic display *SID Symp. Dig. Tech. Pap.* **49** 610–2

[48] Kunitake H *et al* 2019 A *C*-axis-aligned crystalline In-Ga-Zn oxide FET with a gate length of 21 nm suitable for memory applications *IEEE J. Electron Devices Soc.* **7** 495–502

[49] Li J, Lu L, Xia Z, Kwok H S and Wong M 2018 Three-mask elevated-metal metal-oxide thin-film transistor with self-aligned definition of the active island *IEEE Electron Device Lett.* **39** 35–8

[50] Kamiya T and Hosono H 2015 Oxide TFTs *Handbook of Visual Display Technology* (Berlin: Springer) pp 1–28 (https://doi.org/10.1007/978-3-642-35947-7_52-2)

[51] Nomura K, Kamiya T, Ohta H, Hirano M and Hosono H 2008 Defect passivation and homogenization of amorphous oxide thin-film transistor by wet O_2 annealing *Appl. Phys. Lett.* **93** 10–3

[52] Hara Y, Kikuchi T, Kitagawa H, Morinaga J, Ohgami H, Imai H, Daitoh T and Matsuo T 2018 53-3: distinguished paper: IGZO-TFT technology for large-screen 8K display *SID Symp. Dig. Tech. Pap.* **49** 706–9

[53] Yamazaki S, Suzawa H, Inoue K, Kato K, Hirohashi T, Okazaki K and Kimizuka N 2014 Properties of crystalline In-Ga-Zn-oxide semiconductor and its transistor characteristics *Japan J. Appl. Phys.* **53** 04ED18

[54] Yoo W B *et al* 2018 54-2: flexible a-IGZO TFT for large sized OLED TV *SID Symp. Dig. Tech. Pap.* **49** 714–6

[55] Yoshitomi S, Niikura Y, Oe Y, Kubota D, Kusunoki K, Hirakata Y, Yamazaki S, Katayama M and Matsuo N 2016 37-2: 1058-ppi 4K LC display using a top-gate self-aligned CAAC-OS FET *SID Symp. Dig. Tech. Pap.* **47** 473–6

[56] Park C I *et al* 2018 54-1: distinguished paper: world 1st large size 77-inch transparent flexible OLED display *SID Symp. Dig. Tech. Pap.* **49** 710–3

[57] Zhao Y-C *et al* 2018 28-3: world's first 85-in. 120 Hz-driven 8K × 4K BCE IGZO GOA VA-LCD *SID Symp. Dig. Tech. Pap.* **49** 362–4

[58] Chang C *et al* 2019 62-3: a 120Hz 1G1D 8k4k LCD with oxide TFT *SID Symp. Dig. Tech. Pap.* **50** 882–4

[59] Lu X *et al* 2017 21-2: highly reliable amorphous indium-gallium-zinc-tin-oxide TFTs with back-channel-etch structure *SID Symp. Dig. Tech. Pap.* **48** 291–3

[60] Wu W-J, Zhou L, Yao R-H and Peng J-B 2011 A new voltage-programmed pixel circuit for enhancing the uniformity of AMOLED displays *IEEE Electron Device Lett.* **32** 931–3

[61] Mo Y G, Kim M, Kang C K, Jeong J H, Park Y S, Choi C G, Kim H D and Kim S S 2011 Amorphous-oxide TFT backplane for large-sized AMOLED TVs *J. Soc. Inf. Disp.* **19** 16

[62] Kim Y, Kanicki J and Lee H 2014 An a-InGaZnO TFT pixel circuit compensating threshold voltage and mobility variations in AMOLEDs *J. Disp. Technol.* **10** 402–6

[63] Kimizuka N and Yamazaki S 2016 *Physics and Technology of Crystalline Oxide Semiconductor CAAC-IGZO: Application to Displays* ed S Yamazaki and T Tsutsui (Chichester: Wiley)

[64] Lin C-L, Chen P-S, Lai P-C, Chu T-C, Wang M-X and Lee C-L 2016 Novel pixel circuit with compensation for normally-off/on a-IGZO TFTs and OLED luminance degradation *J. Disp. Technol.* **12** 1

[65] Kim D, Kim Y, Lee S, Kang M S, Kim D H and Lee H 2017 High resolution a-IGZO TFT pixel circuit for compensating threshold voltage shifts and OLED degradations *IEEE J. Electron Devices Soc.* **5** 372–7

[66] Lin C L, Lai P C, Lai P C, Chen P S and Wu W L 2016 Pixel circuit with parallel driving scheme for compensating luminance variation based on a-IGZO TFT for AMOLED displays *J. Disp. Technol.* **12** 1681–7

[67] Lin C-L, Du Y-W, Cheng M-H, Chen Y-C and Tu C-D 2014 P-26: a low-power gate driver using depletion-mode a-IGZO TFTs *SID Symp. Dig. Tech. Pap.* **45** 1039–42

[68] Kim B, Choi S C, Lee S Y, Kuk S H, Jang Y H and Kim C D *et al* 2011 A depletion-mode a-IGZO TFT shift register with a single low-voltage-level power signal *IEEE Electron Device Lett.* **32** 1092–4

[69] Park Y-S, Chung B, Kang C, Park S, Im K, Jeong J H, Kim B-H and Kim S S 2011 49.3: oxide TFT scan driver with dynamic threshold voltage control *SID Symp. Dig. Tech. Pap.* **42** 718–21

[70] Kang C-K, Park Y-S, Park S-I, Mo Y-G, Kim B-H and Kim S S 2011 4.2: integrated scan driver with oxide TFTs using floating gate method *SID Symp. Dig. Tech. Pap.* **42** 25–7

[71] Song E, Kang B, Han I, Oh K, Kim B and Nam H 2015 Depletion mode oxide TFT shift register for variable frame rate AMOLED displays *IEEE Electron Device Lett.* **36** 247–9

[72] Ma Q, Wang H, Zhou L, Fan J, Liao C, Guo X and Zhang S 2019 Robust gate driver on array based on amorphous IGZO thin-film transistor for large size high-resolution liquid crystal displays *IEEE J. Electron Devices Soc.* **7** 717–21
[73] Shin H-J *et al* 2018 28-2: a novel OLED display panel with high-reliability integrated gate driver circuit using IGZO TFTs for large-sized UHD TVs *SID Symp. Dig. Tech. Pap.* **49** 358–61

Chapter 8

Metal oxide transistor for tactile imaging

Shuhai Liu and Yong Qin

Designing and fabricating a meaningful tactile imaging system needs a wide, but integrated, knowledge of how tactile information is detected and transmitted at various stages of interaction via touch. This chapter mainly focuses on the state of the art of tactile sensing technologies based on the metal oxide transistor in recent years. Various new principle tactile sensing technologies are discussed along with their unique advantages. Particularly, metal oxide transistors, including piezotronic transistor together with piezo-phototronic light-emitting diode, and tribotronic transistor are introduced as the new tactile imaging technologies that currently attract much attention. The final part of this chapter briefly introduces a self-powered system as the new trend in tactile imaging.

8.1 Introduction

Designing and fabricating functional systems by nanodevice arrays is the essence of transforming basic scientific principles into applicable technology and productivity. With the rapid development of human artificial intelligence and the inevitably expanding markets, the past two decades have witnessed an urgent demand for a well-evolved manufacturing process of novel functional materials and innovative device structure design to achieve human–machine communication and robotics. Among various senses needed to perceive and react to the environment, the sense of touch is particularly important since it introduces awareness of the body and differentiating 'me' from 'not me'. Unlike other senses (e.g. vision, audio), it involves complex physical interaction, and plays an essential role in feeling contact properties such as material type, hardness, shape and many more [1]. With the continuous development of nanotechnology, much progress has been achieved in implementing strain or pressure sensors based on arrays of tactile pixels for mimicking the tactile sensing capabilities of human skin. Traditionally, conventional silicon complementary metal-oxide semiconductor (CMOS) based tactile imaging has excellent properties, such as high sensitivity, but the rigid and fragile nature hamper their

doi:10.1088/978-0-7503-2556-1ch8

applications in portable and implantable devices. In order to achieve flexible and sensitive tactile sensors with high-resolution, a variety of advanced functional materials and principles have been investigated and proposed for touch applications [2–6]. These tactile sensors can be mainly classified into piezoresistive sensor, piezoelectric sensor, capacitive sensor, piezotronic sensor and triboelectric sensor, etc.

Metal oxides are known to possess unique functionalities that are absent or inferior in other solid materials. Their nanostructures have emerged as an important class of materials with a rich collection of properties and great potential for device applications. These include high-mobility transistors, photonic devices, nanogenerators, non-volatile memories and strain sensors. Most recently, electronic components consisting of metal oxide transistors have drawn considerable attention as readout elements for detecting strain-induced property changes in the strain-sensitive media, and have inspired some very important applications in human–electronics interfacing, smart skin, and micro- and nanoelectromechanical systems [7].

Analyzing the present state of research in the area of tactile sensing, three kinds of metal oxide transistor are highly desirable for development of human–machine interface and precise detection of human activity. The first one is the transistor-based pressure sensor that mainly utilizes the dielectric capacitance of the transistor to build the relationship between the electrical properties and the external stimuli. The second one is the piezotronic transistor, a kind of device based on piezotronic effect, owing to the coupling between the piezoelectric and semiconductor property of metal oxide like ZnO. The final one is the tribotronic transistor based on triboelectric effect, which possesses both merits of high sensitivity and simplified fabrication. Unlike the conventional arrangement where transducer and electronics are separate entities, these three type devices are all ‘integral sensing unit’ compressing of transducer and electronic unit, which can be highly integrated on the same chip. These metal oxide transistors exhibit great advantages as fundamental components of pressure-sensitive units in electronics, and pave the way for a large-scale and integrated production of tactile imager, which will contribute to its application in many fields such as human–electronic interaction, electronic skin and smart MEMS.

In this chapter, the transistor-based tactile sensor with extra compressible gate electrode is firstly described in section 8.2. Then, as one that has been widely focused on in recent years, piezotronic and tribotronic transistors for tactile imaging will be, respectively, introduced in sections 8.3 and 8.4, including its principles and the latest research progress. These new principle devices are mainly based on piezotronic, piezo-phototronic and tribotronic effects, first proposed by Professor Zhong Lin Wang and gradually recognized in various fields during recent years, especially for adaptive electronics and optoelectronics [8, 9]. In addition, section 8.5 shows a newly proposed self-powered system with both merits of low power consumption and high detection sensitivity, which is the new trend in the field of tactile imaging. Finally, section 8.6 is dedicated to the conclusions.

8.2 Transistor-based tactile sensor with compressible gate electrode

Recently, significant breakthroughs have been achieved on a large-scale and flexible transistor-based tactile sensor based on different transduction methods that convert mechanical stimuli into electronic signals. Compared with the capacitor-based pressure sensors, transistor-based sensors offer signal amplification, higher sensitivity and lower power consumption [10, 11]. Recent efforts have led to notable developments for the transistor-based pressure sensor, in which the performance has been improved by integrating carbon nanotube, graphene, transition metal dichalcogenides or organic transistors with compressible pressure-sensitive rubber or micro-structured dielectrics gate electrode [11–15].

Miniaturization and integration are the development trends where tactile sensing will never change. It is crucial to optimize the structure of transistor-based tactile sensors to establish a direct interaction between the electrical properties of the transistor and the applied pressure; in other words, the device did not require an extra pressure-dependent electronic component. Only in this way can the scale of the pressure-sensitive component be effectively decreased. Considering this, Bao and co-workers fabricated a highly sensitive pressure-dependent transistor, as shown in figure 8.1(a), based on micro-structured compressible gate dielectrics instead of traditional rigid dielectric [16]. When subjected to a compressive stress, the compressible dielectric will deform and lead to a change in the dielectric capacitance of the transistor. As a result, the drain–source current I_{DS} of the transistor will

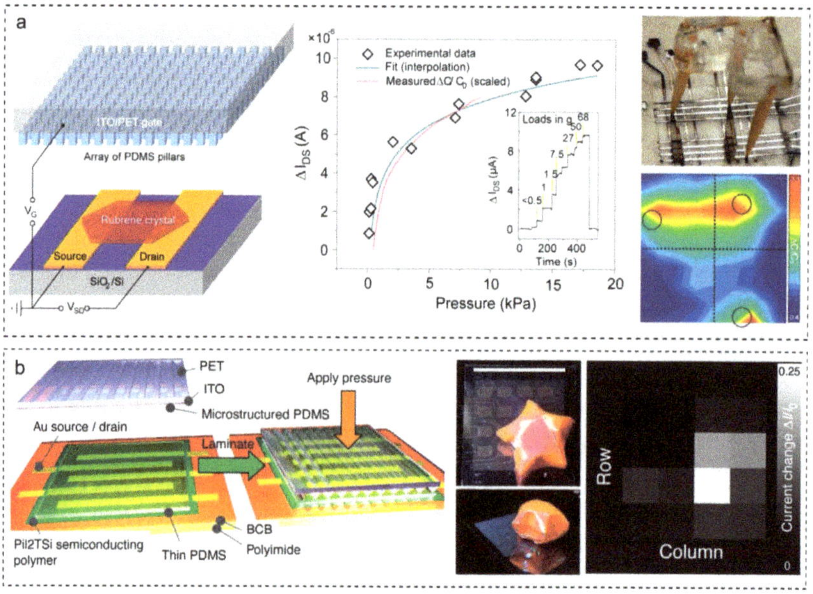

Figure 8.1. Transistor-based tactile sensor with compressible gate electrode. (a) Pressure-sensitive transistor-based sensor with micro-structured compressible gate dielectrics. [16] (2010) Copyright. With permission of Springer. (b) Large-scale flexible transistor-based tactile sensor array with the micro-structured elastic dielectric for tactile imaging. [10] (2013) Copyright. With permission of Springer.

change quickly when the device is under mechanical stimuli. Furthermore, based on this principle, a large-scale flexible pressure-sensitive sensor array in figure 8.1(b) is also successfully demonstrated for tactile imaging with both merits of high sensitivity (9.4 kPa^{-1}) and fast response (<10 ms) [10]. This structure optimization of applying micro-structured compressible gate dielectrics enhanced the performance of a transistor-based sensor, especially in monitoring weak pressure signals.

Following a similar design, a new class of transistor-based tactile sensor is proposed with a suspended gate or air gap gate. This air gap has obtained growing attention as an alternative dielectric layer in the transistor for tactile imaging. Figure 8.2(a) shows a transistor with suspended gate for ultrasensitive pressure detection [11]. This sensor has ultrahigh sensitivity of about 192 kPa^{-1} and low power consumption of less than 100 nW, which is able to monitor the press of a human finger. A demonstration of the sensor matrix of 8 × 8 pixels shows its capacity of spatial pressure mapping and motion monitoring. Shin and co-workers offered a special scheme to fabricate a pressure-sensitive field effect transistor (FET) as shown in figure 8.2(b) [14]. The air gap is formed by folding a substrate with two opposing plastic panels and a foldable elastic joint. The source/drain electrodes and channels were both patterned on one panel, while the gate electrode was placed on another panel. This scheme of operation not only reduced the hysteresis effect of the pressure sensors but also obtained a higher resolution with 50 × 50 pixels and wider tactile pressure sensing range (0.25–3000 kPa). These two tactile sensors exhibit high sensitivity and high resolution through the utilization of an air gap.

Most recently, Huang *et al* proposed new sensitive pressure sensors, as in figure 8.3, by integrating a conductive micro-structured air-gap gate with two-dimensional semiconductor transistors [15]. The air-gap gate can be used to create capacitor-based sensors (figure 8.3(a)) that have tunable sensitivity and

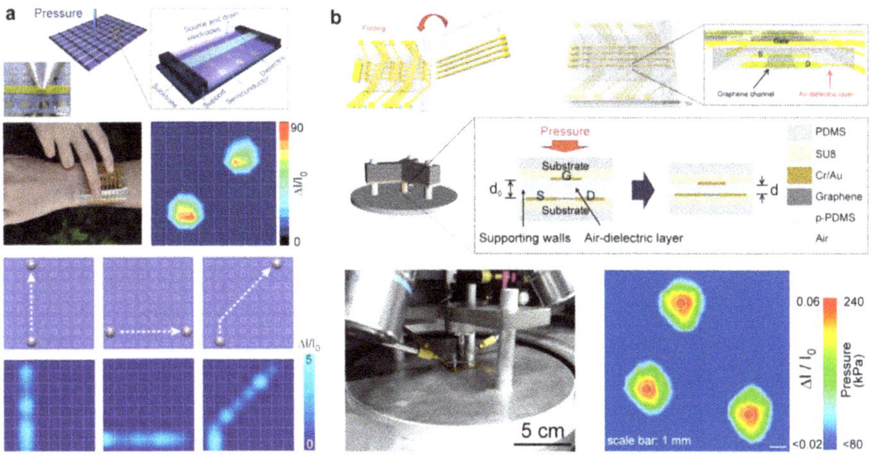

Figure 8.2. Transistor-based tactile sensor with suspended gate electrode. (a) Pressure-sensitive transistor-based sensor with suspended gates. [11] (2015) Copyright. With permission of Springer. (b) Pressure-sensitive field effect transistor array with air-gap gates for tactile imaging. [14] (2017) Copyright. With permission of Springer.

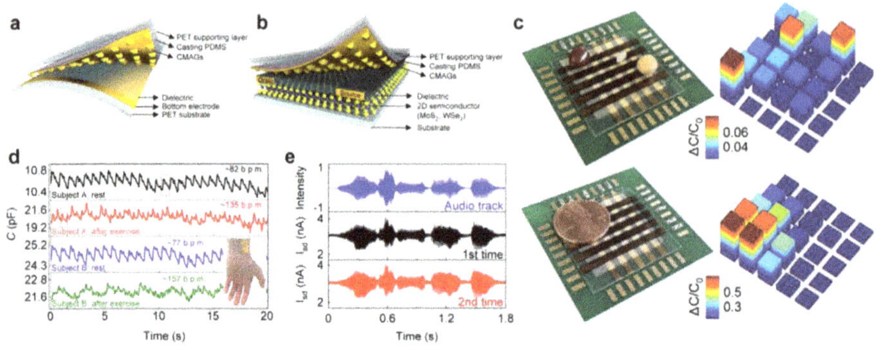

Figure 8.3. (a), (b) Schematic of the sensitive pressure sensors by integrating a conductive micro-structured air-gap gate with two-dimensional semiconductor transistors. Demonstration of the sensors for (c) static pressure mapping, (d) real-time pulse wave monitoring of the radial artery and (e) acoustic wave detection. [15] (2020) Copyright. With permission of Springer.

pressure-sensing range, exhibiting an average sensitivity of 44 kPa^{-1} in the 0–5 kPa regime and a peak sensitivity up to 770 kPa^{-1}. Furthermore, by employing the air-gap gate as a pressure-sensitive gate for two-dimensional semiconductor transistors (figure 8.3(b)), the pressure sensitivity of the device can be amplified to ~10^3–10^7 kPa^{-1} at an optimized pressure regime of ~1.5 kPa. The demonstration of the sensitive pressure sensors for static pressure mapping (figure 8.3(c)), real-time pulse wave monitoring (figure 8.3(d)) and acoustic wave detection (figure 8.3(e)) exhibits its high sensitivity, high stability together with its versatility.

In this section, we briefly introduced the latest developments of the transistor-based tactile sensor. It can be found that the optimization of the dielectric structure by introducing a compressible gate electrode plays a crucial role in the performance of transistor-based tactile sensor, and gradually becomes the new trend of the field effect transistor for imaging tactile. However, these tactile sensors still rely on extra gate electrodes and lack active sensing of external stimuli, which hinders the miniaturization and functionalization of tactile imaging to some extent. Next, we will introduce some new principles of tactile imaging technologies and their applications in the following two sections.

8.3 Piezotronic/piezo-phototronic tactile sensor

As we are gradually approaching the limits of materials and processing in electronic technology, a promising way above Moore's law to propose a seamless and adaptive interaction between functional electronics and their environment (for example, the human body) is crucial for advancing emerging electronic technologies, such as human–computer communication, robotics and artificial intelligence [8]. Since the electrical modification of carriers is the key physical process in state-of-the-art semiconductor technology, mechanical stimuli need to be converted to the electrical signals to be detected and then bring out a response. However, this kind of device not only requires complex integration of heterogeneous components, but also lacks a direct and active interaction between human/ambient and electronics [3]. To a certain extent, this limitation hinders the application of electronic technologies in electromechanical systems, particularly in tactile imaging. Piezotronic effect [17–19], as a new effect, was first proposed in 2006 by *Wang* to achieve the effective, adaptive and seamless interactions between electronic devices and the external mechanical stimuli, which utilizes the strain-induced piezoelectric polarization potential as the virtual gate to tune/control the electrical transport in electronic devices. New principle tactile sensors inspired by the piezotronic effect and the corresponding derived piezo-phototronic effect are gradually exhibiting huge potential in applications for tactile imaging.

8.3.1 Piezotronic effect

Based on the direct piezoelectric effect, piezoelectric semiconductor upon straining can generate polarization at its surface or the interfaces with other materials. Due to the depletion of carriers near the interface of a semiconductor, the produced polarization charges will not be completely electrically shielded by free charges, which could result in a redistribution of carriers and bending of the energy profile through Coulombic interactions [20]. This scheme of operation can utilize the external dynamic mechanical inputs to modulate the electronic transport across the interface. This coupling between semiconductor property and piezoelectricity in piezoelectric semiconductors is referred to as the piezotronic effect.

With the development in recent years, the concept of piezotronic effect has become more universal, and its scope of application has been expanded. Piezotronic effect now describes the law and phenomena that the strain-induced piezopotential can be taken as a 'gate' voltage to control the carrier transport properties of electronics by regulating the barrier height (BH) at the metal–semiconductor (M/S) contact, *p–n* junction, grain boundary and other interfaces. As shown in figure 8.4, compressive and tensile strains applied on the semiconductor with a *c*-axis pointing to the right side will, respectively, induce positive and negative piezopotential at the interfaces. As for the case of Schottky barrier formed by an *n*-type semiconductor and a metal with high work function, the incomplete shielded piezoelectric polarization charges at the semiconductor side can significantly change the Schottky barrier height (SBH) and other properties of the barrier. The positive piezopotential generated will attract the major carriers (for example, electrons for *n*-type ZnO) to the interface leading to a less depleted layer with decreased SBH (figure 8.4(a)); while the negative piezopotential will repel the major carriers away from the interface, resulting in an increased SBH (figure 8.4(b)). Therefore, the strain-induced polarization can effectively modulate the carrier transport across the metal–semiconductor Schottky contact [21].

As for the *p–n* junction, the electrostatic effect of the strain-induced piezoelectric polarization can be enhanced by the constructed depletion region, because of the diminished screening of free charges. For simplicity, only the band structure of homojunction with an *n*-type piezoelectric semiconductor are considered here (figures 8.4(c) and (d)). The strain-induced piezoelectric polarization will modulate the redistribution of charge carriers and hence influence the profile of the depletion

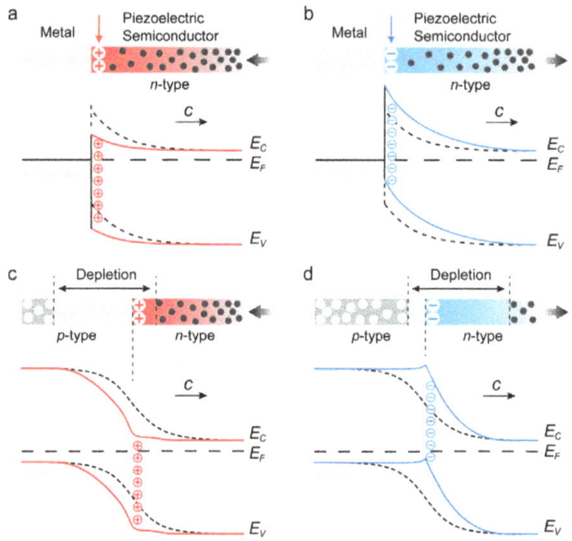

Figure 8.4. Piezotronic modification of the energy band diagrams of (a), (b) Schottky barrier and (c), (d) *p–n* junction. The dotted lines and solid lines (including red and blue lines) represent the energy profiles under strain and strain free, respectively.

layer and the local band diagram at the junction, effectively gating the electronic transport across the interface [22]. Upon straining of the *n*-type piezoelectric semiconductor, the depletion layer is shifted to the *p*-type semiconductor side by the positive piezoelectric polarization (figure 8.4(c)); conversely, the depletion region expands on the *n*-type semiconductor side by the negative piezoelectric charges (figure 8.4(d)). In addition to these two typical situations, the strain-induced polarization can also control the carrier transport across some other interfaces, such as double Schottky barrier and tunneling junction [23–25].

Therefore, owing to the induced piezopotential under external strain, the piezotronic effect can be taken advantage of by force/pressure/strain-gated devices, allowing the adaptive and seamless interactions between electronic devices and the external mechanical stimuli.

8.3.2 Piezotronic transistor as pressure sensor

As one of the most important applications of piezotronic effect, piezotronic transistor with a typical structure of metal-piezoelectric semiconductor-metal (for example, Pt–ZnO–Pt) demonstrates great potential in human–machine interfacing, particularly as a pressure sensor. Because the piezopotential generated in the piezoelectric semiconductor is used as the control 'gate' voltage, the piezotronic transistor is a two-terminal device without extra gate electrode, which simplifies the fabrication process of devices. According to the above introduction about piezotronic effect, it can be found that the electronic transport properties of the transistor based on piezotronic effect could be controlled by external stimuli. In other words,

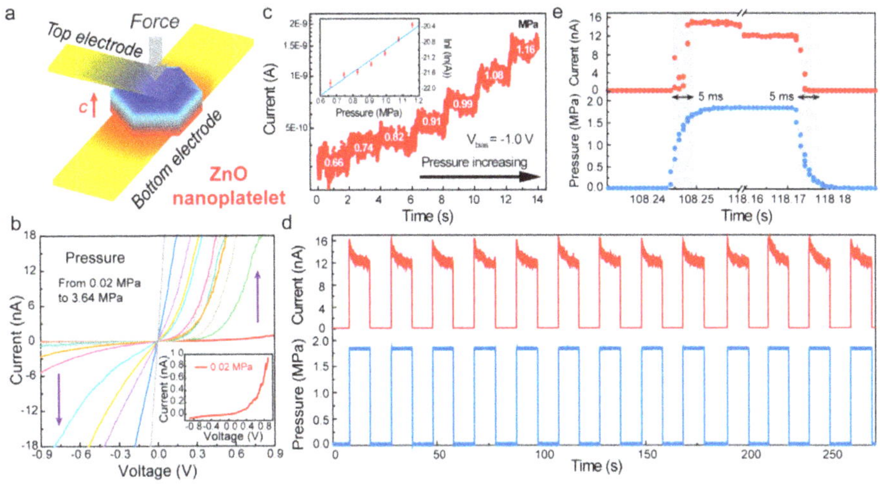

Figure 8.5. (a) Typical structure of piezotronic transistor based on ZnO nanoplatelet, and the corresponding (b) *I–V*, (c) *I–t* characteristics and (d), (e) dynamic response to external applied stimuli. [26] John Wiley & Sons. © (2017) WILEY-VCH Verlag GmbH & Co., KGaA, Weinheim.

by measuring the signal of changes in its electronic transport, we can in turn detect the applied force/pressure/strain on the piezotronic transistor. This makes the piezotronic transistor a kind of pressure-sensitive electronics.

Figure 8.5 exhibits the basic structure of piezotronic transistor, and the corresponding *I–V* characteristics and dynamic response to external applied stimuli [26]. It can be seen from figure 8.5(a) that the piezotronic transistor is usually in possession of a very simple two-terminal structure. The measured *I–V* and *I–t* characteristics exhibit critical behavior as a function of applied pressures, which manifests large modification of Schottky barrier profiles via piezotronic effect (figure 8.5(b)). Also, as shown in figure 8.5(c), the output signal of current shows a very good and reversible response with an on/off ratio of approximately 202 as the applied pressure varies between 0.02 and 1.84 MPa. Also, the current responds very quickly to the pressure change in time less than 5 ms, which is far less than the response time of human fingertips (~30–50 ms). This demonstrates the potential applications of the piezotronic transistor for tactile sensing with ultrafast responses.

Figure 8.6 exhibits the basic structures and corresponding performances of piezotronic transistors based on a metal oxide piezoelectric semiconductor with vertically aligned piezoelectric polarization *c*-axis [3, 26–31]. This kind of vertical piezotronic transistor with vertically aligned ZnO nanowire was previously proposed by Han *et al* in 2012 [27]. Due to the diverse methods and strong controllability of nanowire growth, early researches of piezotronic transistor mainly focused on various nanowires, especially ZnO nanowire, which is easy to bend when it is subjected to compressive stress (buckling effect). With deepening of researches, the device structure has been optimized to improve the performance. Here, a parameter pressure sensitivity is defined as $S = \Delta SBH/\Delta Pressure$ to characterize the performance of piezotronic transistors. In order to avoid the buckling effect of vertically

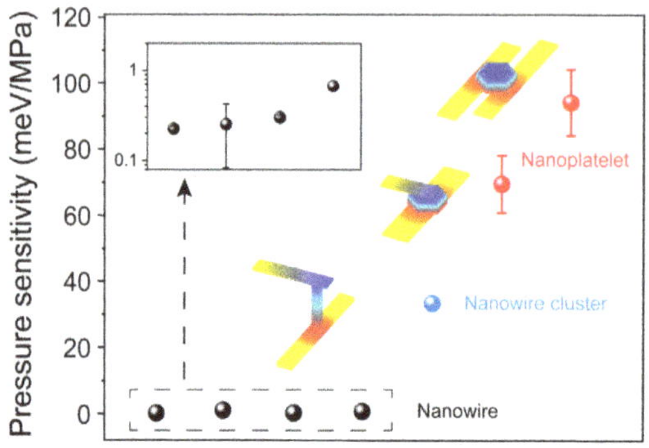

Figure 8.6. The performances of piezotronic transistors with different structures based on nanowire (black points), nanowire cluster (blue point) and nanoplatelet (red point).

aligned nanowire, ZnO nanoplatelet was adopted as the key piezoelectric semiconductor in a piezotronic transistor. This replacement increases the efficiency of piezoelectric polarization generation and thus enhances the device sensing ability of applied pressure. The piezotronic transistor based on ZnO nanoplatelet is achieved with a high pressure sensitivity of 60.97–78.23 meV MPa^{-1} [26], which is 90–116 times higher than that of the nanowire-based ones. Additionally, the piezoelectric semiconductor nanowire/nanoplatelet upon compressive stressing will produce negative and positive piezopotential, respectively, at the top and bottom Schottky contacts, resulting in opposite modulations between these two interfaces. The SBH of the top contact increases with the applied pressure, while the bottom one decreases. This will have a negative influence on the device performance. To solve this problem, a scheme of designing two electrodes on a single polar facet (bottom surface) of the nanoplatelets is proposed and further improves the transistor pressure sensitivity as high as 84.2–104.4 meV MPa^{-1} [31]. Some other strategies can also improve the performance of the piezotronic transistor, such as introducing alloying process [32] or creating pores [33] to suppress the screening of piezoelectric polarization.

In addition to the above-mentioned interface gating case, the piezopotential can also be directly used as the 'gate' voltage to tune/modulate the channel width of the field effect transistor, as illustrated in figure 8.7 [34]. The piezotronic transistor based on 2 nm ZnO nanosheet exhibits critical behavior with the applied pressures (figure 8.7(b)). Figure 8.7(c) gives the possible work mechanism. As the thickness of ZnO gradually decreases, the regulation of piezoelectric polarization on the semiconductor channel width gradually increases. This is also the case with the well-studied piezoelectric-oxide-semiconductor-field-effect-transistor for tactile imaging [35], which is most recently extended to the 2D device basing on the 2D ZnO nanosheet.

8.3.3 Piezotronic transistor for tactile imaging

The above introduced single piezotronic transistor composed of metal oxide piezoelectric semiconductor (for example, ZnO) can detect pressure/strain very

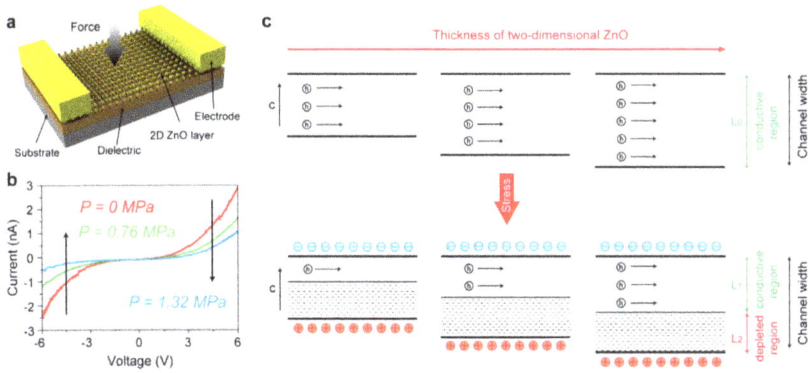

Figure 8.7. Channel width gating of 2D piezotronic transistor. (a) Schematic of 2D piezotronic transistor based on ZnO nanosheet. (b) The *I–V* characteristics as a function of applied pressure. (c) The regulation of piezoelectric polarization on the channel width, indicating an enhanced modulation as the thickness of ZnO gradually decreases. Reprinted from [34], Copyright (2019), with permission from Elsevier.

sensitively. Different from the traditional FETs with three terminals (source, drain and gate) based on voltage-controlled operation, which limits the spatial resolution when used as tactile sensors, the piezotronic transistor is a kind of two-terminal electronics with the strain-induced piezopotential replacing the gate voltage. This novel structure makes the piezotronic transistor a possession of great advantage in realizing tactile imaging with very high spatial resolution.

In 2013, Wu *et al* reported a kind of large-scale integrated circuit based on piezotronic transistor array by taking vertically aligned ZnO nanowire cluster as the active basic sensor unit for dynamic tactile imaging [3]. Owing to the non-centrosymmetric wurtzite structure, a piezopotential is produced along the ZnO nanowire when it is subjected to external pressure, which can be utilized to serve as the driving stimuli similar to the gate voltage in FETs and to tune the electrical transport properties at the interface between the nanowires and the electrodes (figure 8.8(a)). This makes it possible to decrease the number of electrodes, and thus significantly simplify the fabrication processes of a high-resolution tactile imaging device with piezotronic effect. In this work, they achieved 92 × 92 tactile sensors with a spatial resolution of 100 μm, a response time of 0.15 s, a sensitivity of around 2.1 μS kPa^{-1} and a sensing range of a few kPa to 30 kPa, which is well matched to the range of pressure that a human finger applies to sense texture and shape, 10–40 kPa. Each pixel is composed of an individual Au–ZnO–Au piezotronic transistor (figures 8.8(b) and (c)). Benefiting from a scanning multi-channel measurement system, such an 8464 piezotronic transistors tactile imaging device can map the strain or pressure distribution actively and directly by addressing each pixel. Therefore, strain/pressure distribution imaging can be achieved by measuring the output signal of current change for each pixel (figure 8.8(d)). Furthermore, as presented in figure 8.8(e), the devices can also be fabricated on flexible substrate and used to map the strain on a flexible device. The simplification of the matrix integration provides a significant route to the development of multifunctional and diversified large-scale smart sensors.

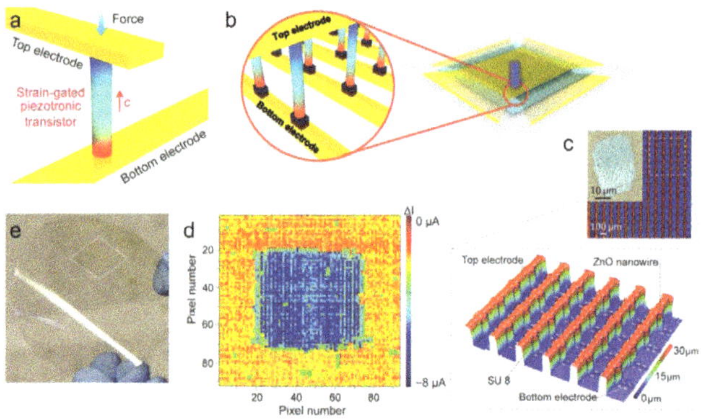

Figure 8.8. (a) A two-terminal piezotronic transistor based on vertically aligned ZnO nanowire. (b) Illustration of the design of a large-scale integrated circuit based on cross-bar piezotronic transistor array for tactile imaging. (c) Device structure of a piezotronics transistor array and a scanning electron microscopy image of an individual pixel in the transistor array. (d) Current response from a piezotronic transistor array for imaging the spatial distribution of applied stress. (e) Photograph of a piezotronic transistor array on a flexible substrate. [3] (2013) Copyright. With permission of Springer.

To shorten the response time of the tactile sensor, Pan *et al* proposed the first nanowire LED arrays for visible mechanosensory electronics (figure 8.9) with an enhanced response time of 90 ms based on the piezo-phototronic effect [36], which depends on the coupling among piezoelectricity, semiconductor charge transport and optical processes. Similar to the piezotronic effect that can directly convert mechanical input into electrical output signal, the new mechanism of the piezo-phototronic effect applied here can directly convert a mechanical stimulation into an optical signal. The *n*-ZnO nanowires grown on a *p*-GaN thin film form an array of *p–n* junction pixel light emitters. Piezoelectric polarization charges of nanowires upon compressive strain will be produced to form a local dip at the interface, which could trap holes and enhance the carrier injection/recombination rate. This will lead to a giant enhancement of the emitting light intensity (figure 8.9(b)). As a result, when a convex pattern 'PIEZO' was pressed on the pressure-sensitive LED arrays, a clear spatial pressure distribution could be observed by using a charge coupled device camera (figure 8.9(c)). It is worth noting that all of output signals of each pixel were read out in parallel, leading to the rapid mapping speed. This offers an effective scheme for tactile imaging based on optical communication technology. Additionally, a flexible ZnO NWs array LEDs with higher resolution was also fabricated by replacing organic the rigid *p*-GaN substrate by poly polystyrenesulfonic acid films. Similarly, the luminous intensity of each ZnO nanowire presents a linear relationship with the applied strain, which could be used to detect and map the trajectory profile spatially.

Most recently, toward high-sensitive and fast-response tactile imaging, a brand new piezotronic tunneling transistor with metal/insulator/piezoelectric semiconductor heterostructure (Pt/Al_2O_3/*p*-GaN) was developed by Qin and co-workers [25],

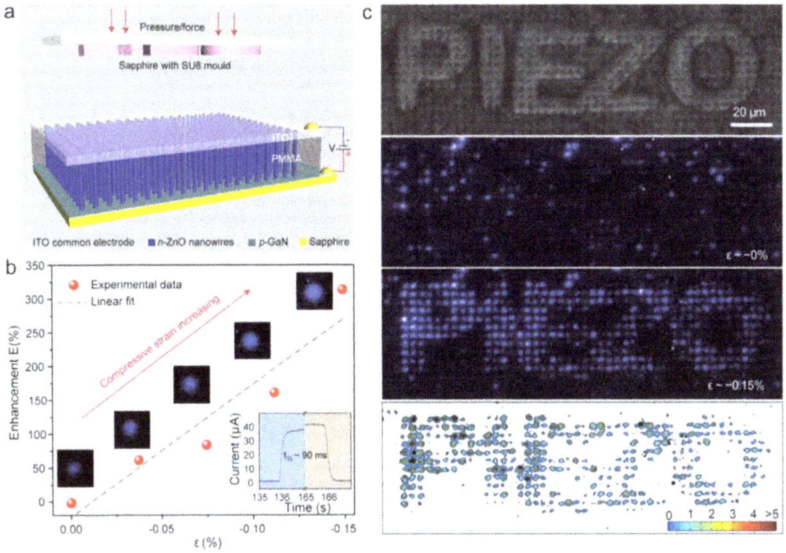

Figure 8.9. (a) Illustration of device design and experiment approaches for imaging pressure distribution using the piezo-phototronic effect. (b) Enhancement factor E of a NW-LED as a function of an applied compressive strain of up to $\varepsilon = -0.15\%$, together with the corresponding emitting light images. (c) High-resolution parallel tactile imaging of pressure/force distribution. [36] (2013) Copyright. With permission of Springer.

which can be gated by external mechanical stimuli via the piezotronic effect. In this transistor, both the height and the width of the tunneling barrier are first modulated by the external mechanical stimuli as a result of the piezotronic effect, and then, this continuous modification of Schottky barrier makes an obvious regulation of the dividing voltage on the Al_2O_3 insulator and hence the tunneling transport. This piezotronic cooperative regulation of the tunneling barrier height and width offers the possibility to improve the sensitivity of the device detection. Figure 8.10(a) shows the tactile imaging demonstration based on the piezotronic tunneling transistor tested using an atomic force microscopy (AFM) system in tapping scanning mode. Alternately small force (~18 nN) and large force (~370 nN) are applied to achieve quantum tunneling off state and on state at a fixed bias of 5.3 V. As can be seen from the current response mapping in figure 8.10(b), the contours of the different force regions are distinguished very well, which demonstrates the great potential of mechanically tunable tunneling junctions for tactile imaging. The on/off ratio has an average value of about 1.34×10^5, corresponding to the qualitatively estimated gauge factor of about 2.6×10^8. By changing the scanning speed and pixel size of the probe, the response time of tactile imaging can be evaluated to be about 4.38 ms (figure 8.10(c)), which is short enough to satisfy the common requirements for tactile sensing applications.

In conclusion, based on piezotronic effect, the piezotronic transistor with a two-terminal concise structure containing no extra gate electrode exhibits high performance in tactile imaging. The piezotronic tactile imaging can have high resolution,

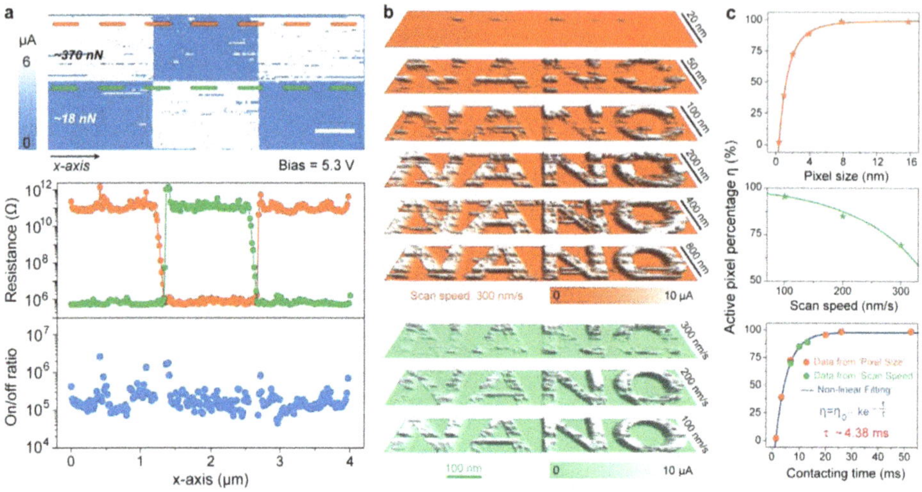

Figure 8.10. (a) Tactile imaging demonstration based on the piezotronic tunneling transistor shows an average on/off ratio of about 1.34×10^5. (b) 3D current response contour plot for imaging the spatial profile 'NANO' of applied force with various pixel sizes and scan speeds. (c) Active pixel percentage η versus the contacting time calculated from (b), indicating a response time of approximately 4.38 ms. [25] John Wiley & Sons. © (2019) WILEY-VCH Verlag GmbH & Co., KGaA, Weinheim.

fast response and high sensitivity, which gives it great potential for future electromechanical systems.

8.4 Tribotronic tactile sensor

In addition to the piezotronic transistor described above, another type of active metal oxide transistor without extra gate electrode based on tribotronics also shows great potential in tactile imaging. Tribotronics is an emerging field proposed by coupling triboelectricity and semiconductor, in which the triboelectric potential generated by contact-electrification could be used as a gate signal to tune/control the carrier transport in metal-oxide-semiconductor field effect transistor instead of the traditional gate voltage [37]. As the most basic application unit, the tribotronic transistor has been developed for human–machine interfacing with direct interactions between the external environment and electronics, which has demonstrated triboelectric controlling effects on electronics. By utilizing the pressure-induced contact electrification to produce triboelectric charge and corresponding tribopotential to switch the electric transport of each transistor as tactile pixels, tribotronic transistor array can be applied for tactile imaging and shows great potential in high-resolution mapping of strain/pressure.

8.4.1 Contact electrification

The contact-electrification effect is a universal phenomenon that commonly occurs in nature and in our lives at any time and any place for all materials, including lightning during thunder and the rubbing of a plastic rod and animal fur. It refers to two materials that are electrically charged after physical contact. However, contact-electrification is generally referred to as triboelectrification in conventional terms. In fact, triboelectrification is a convolution of contact-electrification and tribology, while contact-electrification is a physical effect that occurs only due to the contact of two materials without rubbing against each other, and tribology refers to mechanical rubbing between materials that always involves debris and friction. Actually, the triboelectric charges can be generated through physical contact without friction based on contact electrification.

The key parameters for contact-electrification, the surface charge density, the polarity, and the strength of the charges, are strongly dependent on the materials. The triboelectric series describes materials' tendency to generate triboelectric charges. Wang and co-workers quantified the triboelectric series of inorganic non-metallic materials (figure 8.11) and verified that the contact-electrification is an electronic quantum transition effect under ambient conditions [38]. From the matrix of materials' triboelectric charge densities and band structures, they also found that the triboelectric output is strongly related to the work functions of the materials. Within a broad range, materials at the opposite ends of the triboelectric series can be charged with giant triboelectric charge densities and produce a superior output performance. For example, quartz glass is a common electronegative material and makes it easy to generate negative triboelectric charges. In comparison, mica or metal (Au, Cu, Al, etc) is more likely to generate positive triboelectric charges.

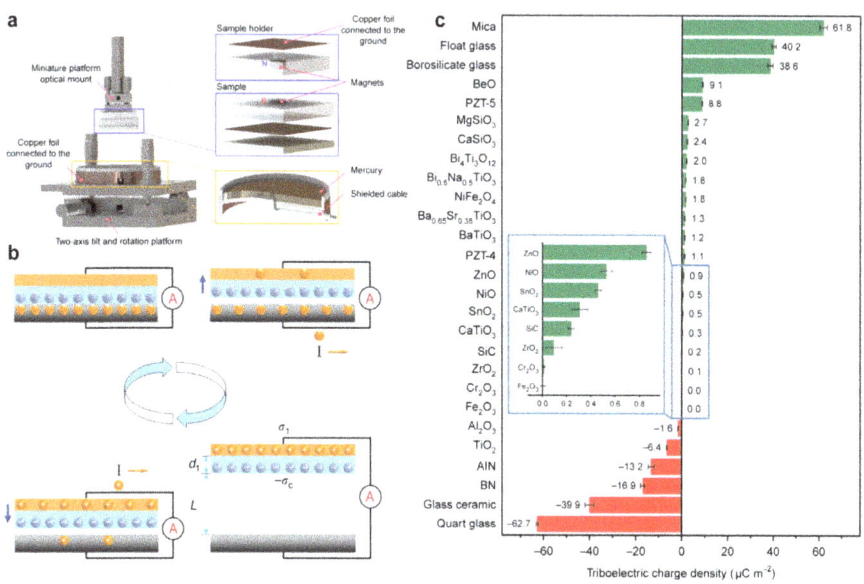

Figure 8.11. The triboelectric series of some inorganic non-metallic materials. (a) Experimental setup and (b) working mechanism of the measurement technique. (c) Quantified triboelectric series of some common inorganic non-metallic materials. The error bar indicates the range within a standard deviation. [38] (2020) Copyright. With permission of Springer.

The triboelectric nanogenerator [39] based on the contact electrification phenomenon has achieved substantial progress since its introduction in 2012. The physical model [40] of the basic mechanism proposed in 2017 is derived from Maxwell's displacement current, which is defined as follows

$$J_D = \frac{\partial D}{\partial t} = \varepsilon_0 \frac{\partial E}{\partial t} + \frac{\partial P}{\partial t}$$

herein, the second term $\partial P/\partial t$ represents the polarization-induced current, which is the origin of the nanogenerator. It can be effective in converting mechanical energy from ambient environment into electricity and has thus been extensively applied in various self-powered systems and triboelectric potential-modulated semiconductor devices, such as metal-oxide-semiconductor transistor.

8.4.2 Tribotronic transistor and work mechanism

Wang's group has developed the first tribotronic transistor [6] based on contact electrification through the combination of triboelectric nanogenerator and a back-gate silicon-on-insulator (SOI) metal-oxide-semiconductor FET (MOSFET), in which an external friction material contacts the gate electrode for electrification and a tribo-potential can be applied between the gate and source electrodes. When a positive tribo-potential is applied, the channel width of the tribotronic transistor is reduced and the drain–source current is thus decreased. Similarly, the channel width of the tribotronic transistor can be widened and the drain–source current is thus enhanced.

Moreover, the metal-insulator-semiconductor structure in MOSFET can be simplified for a mobile triboelectric layer and a dielectric insulator layer in tribotronic transistor, which can be directly used for contact electrification. The typical tribotronic transistor without the gate electrode is exhibited in figure 8.12 [41]. In this simple structure, a mobile triboelectric layer (material 'A') is directly contacted and separated with the dielectric insulator layer (material 'B', SiO_2 for example) for electrification by the external force to produce a tribo-potential as the gate voltage. Figure 8.12(a) shows the detailed working principle of the device. Here we take the mobile layer with stronger ability to attract electrons than the insulator layer as an example. When they contact with each other by an external force, the mobile layer could have negative charges and the insulator layer has positive charges. At this time, there is no effect on the channel carrier concentration for the electrostatic equilibrium. When the mobile layer slowly separates from the insulator surface by an external force, an internal electric field is formed, so that holes are repelled from the interface of the insulator and the semiconductor layers, generating a depletion zone in the conductive channel. As a result, the depletion zone will be enlarged, and the drain–source current I_{DS} will be reduced until a maximum separation distance of the mobile layer is reached. However, when the mobile layer starts to come back to the original station, the hole carriers accumulate at the interface of the semiconductor and the insulator layers, which depresses the internal charge polarization and decreases the depletion zone. Therefore, the holes

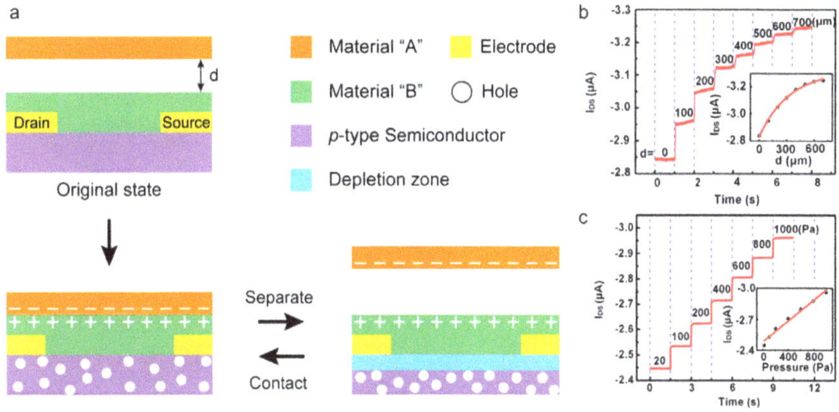

Figure 8.12. (a) Working principle diagram of tribotronic transistor. (b) I_{DS}–t output characteristics of a typical tribotronic transistor with different separate distance d. (c) I_{DS}–P output characteristics of a typical tribotronic transistor with different input pressures. Reproduced with permission from [41], Copyright (2017), American Chemical Society.

concentration and the I_{DS} in the interface of semiconductor and insulator are restored. A typical relationship between I_{DS} and the separation distance d and the applied external pressure P of a tribotronic transistor is shown in figures 8.12(b) and (c), in which the current improved with decreased separation distance by external mechanical stimuli. Moreover, when a metal, Cu for example, is selected as the mobile layer to be rubbed with the dielectric insulator layer, a tribotronic transistor without gate electrode in the enhancement mode can be demonstrated, which also exhibits an excellent modulation ability.

Different from the traditional transistor with three-terminal structure controlled by external applied voltage, the tribotronic transistor is a two-terminal semiconductor device similar to the piezotronic transistor, in which there are only source and drain. It can be controlled by inner electrostatic potential generated by contact/friction/force/sliding. These various mechanical driving methods determine that the tribotronic transistor can be used for active sensing with a large range, which has a wider application prospect than the traditional transistor. Therefore, as a new principle pressure-sensitive device, tribotronic transistor demonstrates the active mechanosensation of electronics for external environment.

8.4.3 Tribotronic transistor for tactile imaging

The tribotronic transistor has built a direct interaction mechanism between the external environment and electronics, providing a new method for sensors, human–machine interaction devices, and active control devices. As each tribotronic transistor acts as a pressure sensor, a large number of transistors assemble to form a transistor array can be applied to detect and sense a force matrix with a certain distribution, that is, to achieve tactile imaging.

Figure 8.13 shows a 10 × 10 tribotronic transistor array fabricated for demonstration of active tactile sensing [42]. Each transistor unit (pixel, 0.5 × 0.5 mm^2) in

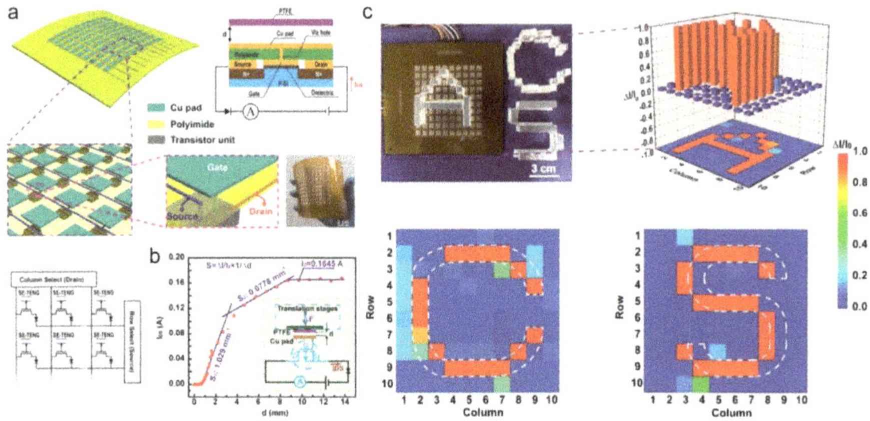

Figure 8.13. Tribotronic transistor array fabricated for demonstration of active tactile imaging. (a) Schematic structure of a 10 × 10 tribotronic transistor array. (b) I_{DS}–t output characteristics of an individual tribotronic transistor with different separate distance d. (c) Tribotronic transistor array as a fully integrated system for tactile sensing and map reconstruction. Reproduced with permission from [42], Copyright (2016), American Chemical Society.

the array is made up of an n-type transistor and a commercial diode, and the source electrode of the n-type transistor is electric connected with the anode of the diode. This structure makes sure that the I_{DS} could flow across the transistor unidirectionally. The I_{DS} of each tribotronic transistor can be individually regulated, thereby inducing a local bias as the conventional 'gate' signal (V_{GS}). Initially, a mobile layer (PTFE film) acts as the triboelectric layer to fully contact with the Cu pad of each tribotronic transistor by an external force. Since the PTFE is possessed of stronger electron affinity, positive triboelectric charges are induced on the surface of Cu pad, while equal opposite charges are transformed to the PTFE film. At this time, the positive and negative charges coinciding with each other will have no influence on the transistor. A narrow conductive channel across the source and drain electrode is corresponding to an 'OFF' state with a small I_{DS}. Then, when the PTFE film separates from the Cu pad, a positive 'gate' bias will be individually created at the Cu pad and the gate electrode due to the induced potential drop by the electrostatic induction. This enables electrons to flow freely from the source to the drain electrodes. As the distance between the mobile PTFE layer and Cu pad increases gradually, there are three different states of I_{DS} depending on the distance, in which there is an original quick increase within 3 mm, and then a slow growth between 3 and 9 mm, and finally a saturation status beyond 9 mm (figure 8.13(b)). Considering that I_{DS} changes versus mechanical stimuli, the tribotronic transistor array can be used to achieve force matrix sensing and dynamic motion monitoring. Figure 8.13(c) confirms its tactile imaging capability, in which the contour of objects could be accurately identified and further imaged by the 2D current mapping. Different from the traditional sensor arrays that are complex and costly to manufacture, the tribotronic transistor array with the inexpensive and practical polymer films for contact electrification, which has potential application in the field of flexible/wearable electronics and active tactile imaging systems.

In addition to the typical structure described above, there are some new advances in tribotronic transistor for tactile imaging. Figure 8.14 shows a mechanosensation-active matrix based tribotronic transistor array fabricated by directly placing the ion gel dielectric layer on graphene channel [43]. The ion gel in each pixel both acts as the dielectric insulator layer and the triboelectric layer; while graphene plays the roles of transistor channel and source–drain electrodes. A tribo-potential induced by the contact electrification between external mobile triboelectric layer and ion gel could be used to modulate the transistor performance. When a typical material, Cu film selected as an example, fully contacts with the ion gel, positive charges are left on the surface of Cu film and equal opposite charges are induced on the ion gel surface for the stronger electron-capture ability of the ion gel, which makes no effect on the transistor channel for the electrostatic equilibrium. When the positively charged Cu film separates from the ion gel, the cations are attracted to compensate for the accumulated negatives charges on the ion gel surface, while the anions migrate to the ion-gel/graphene interface, which is equal to applying a negative 'gate' bias on the transistor. The drain–source current I_{DS} is improved with the increased distance between the external mobile triboelectric layer and ion gel. In addition, the mobile triboelectric layer can also be Al, skin, PET, PTFE, FEP and so on (figure 8.14(b)). Based on this mechanism, the tribotronic transistor can be utilized as a force sensor and used to recognize the material category of the mobile layer, as shown in figure 8.14(c). By assembling 4×4 tribotronic transistors together, tactile imaging can be achieved, which is demonstrated by placing the 'L' shaped polyacrylate on the sensing pixels. The accurate tactile imaging profile confirms that

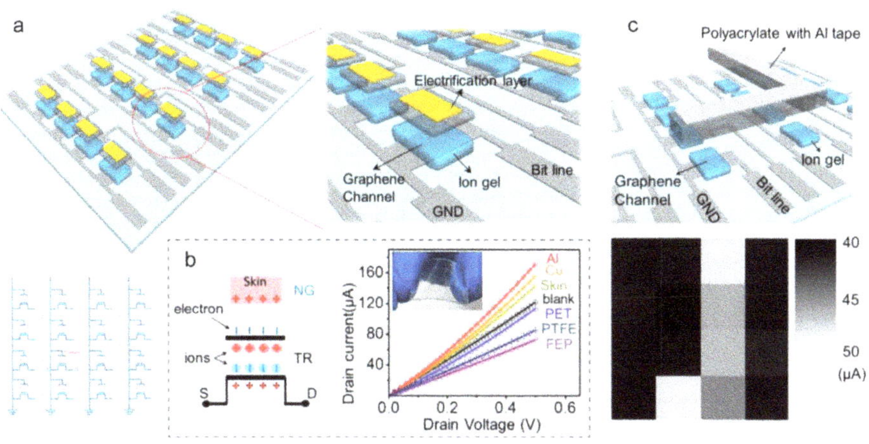

Figure 8.14. New design of mechanosensation-active matrix for tactile imaging based on direct-contact tribotronic transistor array composing of an ion gel dielectric layer. (a) Schematic illustration of the mechanosensation-active matrix and the corresponding circuit diagram. (b) Output characteristics of an individual tribotronic transistor contact with different friction materials (separation distance is 1 mm). (c) 2D imaging of the sensing output currents for the matrix. Reproduced with permission from [43], Copyright (2018), American Chemical Society.

the tribotronic transistor matrix is valuable and instructive for the application of a skin-inspired electronic system.

In previous reports, the interactive interfaces between external environment and electronics in current tribotronic tactile sensing are all in the macro scale, which has limited its integration and further applications. Based on this consideration, most recently, the nanoscale tribotronic transistor [44] is demonstrated by contact-mode atomic force microscopy and scanning Kevin probe microscopy. The modulation effect still exists when the size scales down to the micro or nano range (figure 8.15). Although this work only validated the feasibility of the miniaturization of tribotronic transistor in principle, it offers great prospects in nanoscale transistor, micro/nano-electronic circuit and nanoelectromechanical system (NEMS), which will push the tactile imaging devices based on tribotronic transistor toward the miniaturization and functionalization.

In conclusion, tribotronic transistors, based on the contact electrification effect, have shown great potential applications in tactile sensing. Due to its simple structure and active working mode, together with a wide selection of materials, the tribotronic transistor is a new direction for the development of metal oxide transistors for tactile imaging in future years.

8.5 New trends in tactile imaging: self-powered system

There is the common concern of energy consumption of a device for researchers. In recent years, a self-powered system [45] based on the piezoelectric/triboelectric nanogenerators has been proposed, which provides novel methods and ideas for the development of tactile sensor arrays based on metal oxide transistor matrix. Figure 8.11(b) in the previous section shows a typical triboelectric nanogenerator that can both act as a generator or a self-powered sensor. The following figures 8.16

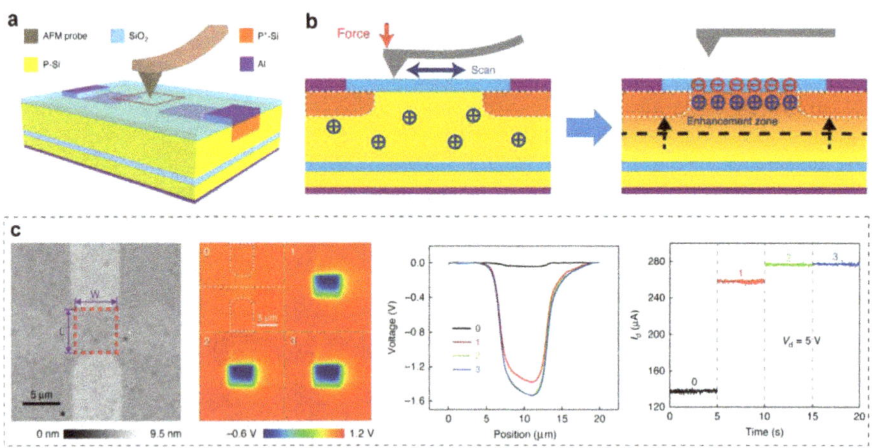

Figure 8.15. Nanoscale tribotronic transistor. (a) Schematic illustration of the nanoscale tribotronic transistor gated by nanoscale triboelectrification. (b) Schematic working principle of the nanoscale tribotronic transistor. (c) The effect of contact cycles (0, 1, 2, 3 cycles) on the characteristics of the nanoscale tribotronic transistor. [44] (2020) Copyright. With permission of Springer.

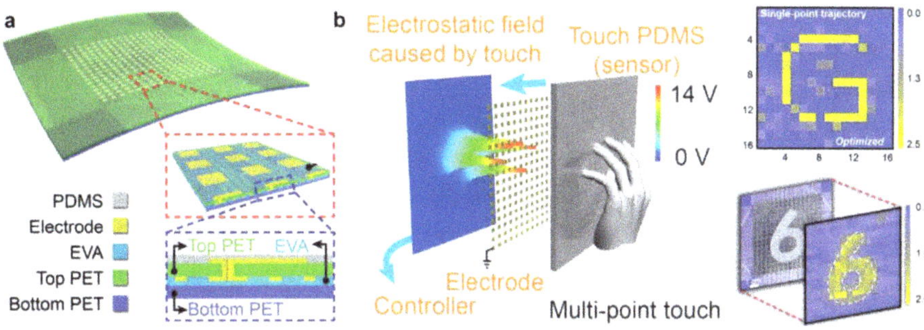

Figure 8.16. Large-scale, pressure-sensitive self-powered sensor matrix with 16 × 16 pixels for the dynamic tactile imaging. (a) Schematic of a self-powered matrix with 16 × 16 sensors. Insets: partially enlarged tilted view and cross-sectional view of the structural design of the top and bottom, respectively. (b) Demonstration of the self-powered tactile imaging. [46] John Wiley & Sons. © (2016) WILEY-VCH Verlag GmbH & Co., KGaA, Weinheim.

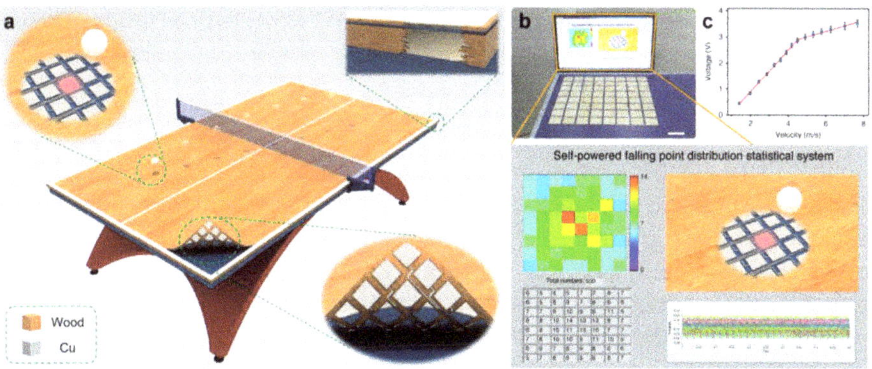

Figure 8.17. Flexible and durable high-performance triboelectric nanogenerator for self-powered sensing. (a) Schematic illustration of the flexible self-powered sensing and smart ping-pong table. (b) Dynamic demonstration of the self-powered falling point distribution statistical system. Scale bar: 6 cm. (c) The summarized relationship and linear fitting between output voltages and velocities. Error bars represent standard deviations. [47] (2019) Copyright. With permission of Springer.

and 8.17 give two samples [46, 47]. The working principle is to use nanogenerators to produce electrical signals under mechanical stimuli, and then in turn utilize the measured electrical signals to detect the driving mechanical stimulation. This scheme of operation does not require additional battery power supply, so it is called a self-powered system.

Figure 8.16(a) exhibits a large-scale, pressure-sensitive self-powered sensor matrix with 16 × 16 pixels for the dynamic detection of single-point and multipoint touching [46]. Also, a high-resolution cross-type sensor matrix with 32 × 20 pixels was fabricated by using cross-locating technology for more rapid tactile imaging. The simplification of the device structure significantly reduces the addressing lines from $m \times n$ to $m + n$. Compared with the traditional large-scale tactile sensors, the self-powered sensor matrix exhibited more advantages, such as low-power consumption, easy preparation, low cost, and high sensitivity.

Figure 8.17 shows a kind of a flexible and durable high-performance wood-based triboelectric nanogenerator for self-powered sensing in sports equipment and athletic big data analytics [47]. An 8×8 array was developed and set on the surface of the table for detecting the falling point of ping-pong balls. As a ball impacts on the table surface, the pixel sensor at the contact position will generate an output signal of voltage. Through the multi-channel data acquisition method, real-time voltage signals of each sensor pixel could be detected simultaneously. After signal detection, real-time statistic results can be displayed in the program. Figure 8.17(b) shows the self-powered falling point distribution statistical system. Also, this self-powered tactile imaging matrix can detect the velocity of ping-pong balls (figure 8.17(c)).

As can be seen from the above two examples, the self-powered system has great potential in flexible applications for tactile imaging, because it not only does not require external power supply, but also has a wide range of material options. This new scheme combined with metal oxide transistor is expected to simulate new type strain/pressure sensors with both merits of low power consumption and high performance.

8.6 Summary

In this chapter, we have exhibited several examples to demonstrate the applications of metal oxide transistor in the field related to tactile imaging. The demonstrated tactile sensors include the transistor-based sensor with extra compressible gate electrode, the pressure-dependent piezotronic and tribotronic transistors, and finally the self-powered system. Among them, the transistor-based sensor with extra compressible gate electrode exhibits high performance, especially in the sensitivity and the applications of weak signal monitoring, and gradually becomes the trend of the tactile sensor based on traditional field effect transistor. Different from the above one that relies on extra gate electrodes, the pressure-dependent piezotronic and tribotronic transistors are both electronics that can actively sense the external mechanical stimuli without the need for extra gate control signals, in which the piezoelectric potential or triboelectric potential could be regarded as a gate voltage. This simplification of structure and advantage of working mode offers the possibility of large-scale integration and high-sensitivity detection. In addition, owing to the advantage of low power consumption, the self-powered system based on tribo-/piezoelectric nanogenerators shows great advantages and is gradually becoming the new trend in tactile imaging. In summary, new structural optimizations and the emergence of new principles provide new ideas and developments for metal oxide transistor-based tactile imaging, making it one step further towards a wide range of applications in a variety of industrial fields.

References

[1] Dahiya R S and Valle M 2012 *Robotic Tactile Sensing: Technologies and System* (Berlin: Springer)

[2] Kim D-H *et al* 2011 Epidermal electronics *Science* **333** 838–43

[3] Wu W, Wen X and Wang Z L 2013 Taxel-addressable matrix of vertical-nanowire piezotronic transistors for active and adaptive tactile imaging *Science* **340** 952–7

[4] Wang S *et al* 2018 Skin electronics from scalable fabrication of an intrinsically stretchable transistor array *Nature* **555** 83–8

[5] Yu X *et al* 2019 Skin-integrated wireless haptic interfaces for virtual and augmented reality *Nature* **575** 473–9

[6] Zhang C, Tang W, Zhang L, Han C and Wang Z L 2014 Contact electrification field-effect transistor *ACS Nano* **8** 8702–9

[7] Ramanathan S 2010 *Thin Film Metal-Oxides* (New York: Springer)

[8] Wu W and Wang Z L 2016 Piezotronics and piezo-phototronics for adaptive electronics and optoelectronics *Nat. Rev. Mater.* **1** 1–17

[9] Pan C, Zhai J and Wang Z L 2019 Piezotronics and piezo-phototronics of third generation semiconductor nanowires *Chem. Rev.* **119** 9303–59

[10] Schwartz G, Tee B C-K, Mei J, Appleton A L, Kim D H, Wang H and Bao Z 2013 Flexible polymer transistors with high pressure sensitivity for application in electronic skin and health monitoring *Nat. Commun.* **4** 1–8

[11] Zang Y, Zhang F, Huang D, Gao X, Di C-a and Zhu D 2015 Flexible suspended gate organic thin-film transistors for ultra-sensitive pressure detection *Nat. Commun.* **6** 1–9

[12] Sekitani T *et al* 2009 Organic nonvolatile memory transistors for flexible sensor arrays *Science* **326** 1516–9

[13] Takei K, Takahashi T, Ho J C, Ko H, Gillies A G, Leu P W, Fearing R S and Javey A 2010 Nanowire active-matrix circuitry for low-voltage macroscale artificial skin *Nat. Mater.* **9** 821–6

[14] Shin S-H *et al* 2017 Integrated arrays of air-dielectric graphene transistors as transparent active-matrix pressure sensors for wide pressure ranges *Nat. Commun.* **8** 14950

[15] Huang Y-C *et al* 2020 Sensitive pressure sensors based on conductive microstructured air-gap gates and two-dimensional semiconductor transistors *Nat. Electron.* **3** 59–69

[16] Mannsfeld S C B, Tee B C K, Stoltenberg R M, Chen C V H H, Barman S, Muir B V O, Sokolov A N, Reese C and Bao Z 2010 Highly sensitive flexible pressure sensors with microstructured rubber dielectric layers *Nat. Mater.* **9** 859–64

[17] Gao Y and Wang Z L 2007 Electrostatic potential in a bent piezoelectric nanowire. The fundamental theory of nanogenerator and nanopiezotronics *Nano Lett.* **7** 2499–505

[18] Wang Z L 2007 Nanopiezotronics *Adv. Mater.* **19** 889–92

[19] Wang Z L 2012 *Piezotronics and Piezo-Phototronics* (Berlin: Springer)

[20] Wang Z L 2010 Piezopotential gated nanowire devices: piezotronics and piezo-phototronics *Nano Today* **5** 540–52

[21] Zhang Y, Liu Y and Wang Z L 2011 Fundamental theory of piezotronics *Adv. Mater.* **23** 3004–13

[22] Liu Y, Niu S, Yang Q, Klein B D B, Zhou Y S and Wang Z L 2014 Theoretical study of piezo-phototronic nano-LEDs *Adv. Mater.* **26** 7209–16

[23] Raidl N, Supancic P, Danzer R and Hofstaetter M 2015 Piezotronically modified double schottky barriers in ZnO varistors *Adv. Mater.* **27** 2031–5

[24] Keil P, Trapp M, Novak N, Frömling T, Kleebe H-J and Rödel J 2018 Piezotronic tuning of potential barriers in ZnO bicrystals *Adv. Mater.* **30** 1705573

[25] Liu S, Wang L, Feng X, Liu J, Qin Y and Wang Z L 2019 Piezotronic tunneling junction gated by mechanical stimuli *Adv. Mater.* **31** 1905436
[26] Liu S, Wang L, Feng X, Wang Z, Xu Q, Bai S, Qin Y and Wang Z L 2017 Ultrasensitive 2D ZnO piezotronic transistor array for high resolution tactile imaging *Adv. Mater.* **29** 1606346
[27] Han W, Zhou Y, Zhang Y, Chen C-Y, Lin L, Wang X, Wang S and Wang Z L 2012 Strain-gated piezotronic transistors based on vertical zinc oxide nanowires *ACS Nano* **6** 3760–6
[28] Zhou Y S *et al* 2012 Vertically aligned CdSe nanowire arrays for energy harvesting and piezotronic devices *ACS Nano* **6** 6478–82
[29] Han X, Du W, Yu R, Pan C and Wang Z L 2015 Piezo-phototronic enhanced UV sensing based on a nanowire photodetector array *Adv. Mater.* **27** 7963–9
[30] Zhao Z, Pu X, Han C, Du C, Li L, Jiang C, Hu W and Wang Z L 2015 Piezotronic effect in polarity-controlled GaN nanowires *ACS Nano* **9** 8578–83
[31] Liu S, Wang L, Wang Z, Cai Y, Feng X, Qin Y and Wang Z L 2018 Double-channel piezotronic transistors for highly sensitive pressure sensing *ACS Nano* **12** 1732–8
[32] Wang C H, Lai K Y, Li Y C, Chen Y C and Liu C P 2015 Ultrasensitive thin-film-based $Al_xGa_{1-x}N$ piezotronic strain sensors via alloying-enhanced piezoelectric potential *Adv. Mater.* **27** 6289–95
[33] Su Y-L, Gupta K, Hsiao Y-L, Wang R-C and Liu C-P 2019 Gigantic enhancement of electricity generation in piezoelectric semiconductors by creating pores as a universal approach *Energy Environ. Sci.* **12** 410–7
[34] Wang L *et al* 2019 2D piezotronics in atomically thin zinc oxide sheets: interfacing gating and channel width gating *Nano Energy* **60** 724–33
[35] Dahiya R S, Lorenzelli L, Metta G and Valle M 2010 POSFET devices based tactile sensing arrays *Proc. 2010 IEEE Int. Symp. Circuits and Systems* (Piscataway, NJ: IEEE)
[36] Pan C, Dong L, Zhu G, Niu S, Yu R, Yang Q, Liu Y and Wang Z L 2013 High-resolution electroluminescent imaging of pressure distribution using a piezoelectric nanowire LED array *Nat. Photonics* **7** 752–8
[37] Zhang C, Li J, Han C B, Zhang L M, Chen X Y, Wang L D, Dong G F and Wang Z L 2015 Organic tribotronic transistor for contact-electrification-gated light-emitting diode *Adv. Funct. Mater.* **25** 5625–32
[38] Zou H *et al* 2020 Quantifying and understanding the triboelectric series of inorganic non-metallic materials *Nat. Commun.* **11** 1–7
[39] Fan F-R, Lin L, Zhu G, Wu W, Zhang R and Wang Z L 2012 Transparent triboelectric nanogenerators and self-powered pressure sensors based on micropatterned plastic films *Nano Lett.* **12** 3109–14
[40] Wang Z L 2017 On Maxwell's displacement current for energy and sensors: the origin of nanogenerators *Mater. Today* **20** 74–82
[41] Zhao J *et al* 2017 Flexible organic tribotronic transistor for pressure and magnetic sensing *ACS Nano* **11** 11566–73
[42] Yang Z W, Pang Y, Zhang L, Lu C, Chen J, Zhou T, Zhang C and Wang Z L 2016 Tribotronic transistor array as an active tactile sensing system *ACS Nano* **10** 10912–20
[43] Meng Y, Zhao J, Yang X, Zhao C, Qin S, Cho J H, Zhang C, Sun Q and Wang Z L 2018 Mechanosensation-active matrix based on direct-contact tribotronic planar graphene transistor array *ACS Nano* **12** 9381–9
[44] Bu T, Xu L, Yang Z, Yang X, Liu G, Cao Y, Zhang C and Wang Z L 2020 Nanoscale triboelectrification gated transistor *Nat. Commun.* **11** 1–8

[45] Wang Z L 2018 Nanogenerators, self-powered systems, blue energy, piezotronics and piezo-phototronics – a recall on the original thoughts for coining these fields *Nano Energy* **54** 477–83
[46] Wang X, Zhang H, Dong L, Han X, Du W, Zhai J, Pan C and Wang Z L 2016 Self-powered high-resolution and pressure-sensitive triboelectric sensor matrix for real-time tactile mapping *Adv. Mater.* **28** 2896–903
[47] Luo J *et al* 2019 Flexible and durable wood-based triboelectric nanogenerators for self-powered sensing in athletic big data analytics *Nat. Commun.* **10** 1–9

IOP Publishing

Semiconducting Metal Oxide Thin-Film Transistors

Ye Zhou

Chapter 9

Phototransistors based on metal oxide semiconductors

Lin-Bao Luo and Can Fu

As a representative example of photodetectors, the phototransistor with typical photogating effect, has attracted intensive research interest due to relatively superior device performance over its counterparts. In the past decade, a huge number of phototransistors with various device geometries and exceptionally high photoconductive gain or external quantum efficiency have been reported. The majority of these devices are usually assembled by integrating a number of metal oxide semiconductors with other organic or inorganic semiconductor materials in the form of either thin film or nanostructures. This chapter presents a comprehensive review of metal oxide semiconductor based phototransistors for either UV light or visible light in the past decade. Different kinds of phototransistors, which are fabricated on varied semiconductors including binary metal oxides (e.g. ZnO, SnO, TiO_2), ternary metal oxides (e.g. InZnO, ZnON), and quaternary metal oxides (e.g. ZnInSnO, InGaZnO). The operation mechanism of these metal oxide based devices are discussed systematically. What is more, some emerging techniques which have been widely adopted to optimize device performance are also discussed. Lastly, the future challenges and perspectives in this field are also provided.

9.1 Introduction

A phototransistor is typically device that has a much larger area of the base and collector junction (figures 9.1(a) and (b)). Basically, it is composed of three-layer semiconductors layers, which resembles a conventional transistor except for the lack of a base terminal. Unlike traditional photodiodes, most of phototransistors usually display relatively higher responsivity and high gain, yet with persistent photoconductivity effect (PPC), which means the response speed is very slow. What is more, compared to avalanche photodiode, it avoids noise-related problems and high voltage [1–7]. Very recently, a new type of phototransistor geometry has been developed

doi:10.1088/978-0-7503-2556-1ch9

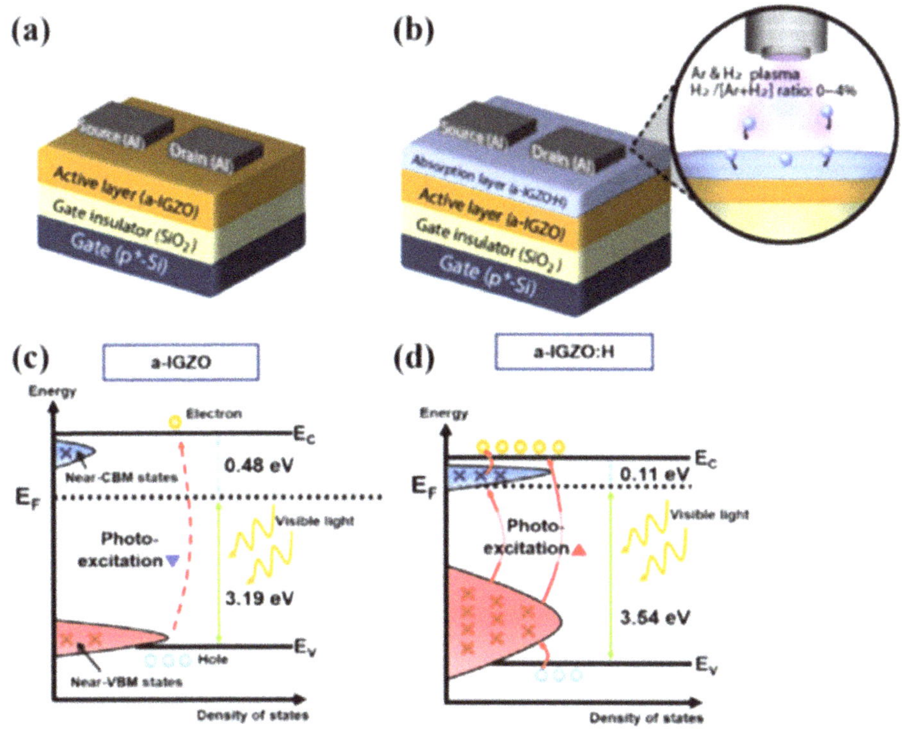

Figure 9.1. Schematic illustration of a-IGZO-based phototransistors (a) without HAL and (b) with HAL. The working mechanism of this type of devices (c) a-IGZO for the active layer and (d) a-IGZO: *H* for the absorption layer under different light illuminations . Reprinted with permission from [13], Copyright (2018) American Chemical Society.

consisting of double layers of a heterojunction, where one layer serves as light absorbing media and another layer as charge carrier transport channel. When such a simplified optoelectronic device is shined on by light illumination, electron–hole pairs are produced in the light absorbing layer, and subsequently, one type of carrier drifts toward the transport channel, leaving another type of carrier trapped in the optical absorbing layer due to the alignment of energy band levels between two materials. The photocarriers are then collected by source–drain electrodes at the opposite ends of the channel, giving rise to the formation of photocurrent.

Apparently, the operation mechanisms of phototransistors are based on the quantum photoelectric effect. The long detecting wavelength limit (λ), which is usually named as cutoff wavelength, is directly related to energy transition (ΔE) in the device operation with the following equation: $\lambda = \frac{hc}{\Delta E} = \frac{1240}{\Delta E\ (\text{eV})}(\text{nm})$, where h is the Planck's constant and c is the speed of light [2], ΔE is the transition of energy levels in relation to the bandgap of the semiconductors. In most cases, ΔE is equal to the barrier height for Schottky photodiodes or transition energy between an impurity (defect) level and the band edge. In the past decade, with the rapid development in materials synthesis and device fabrication, various functional semiconductors, in particular metal oxide semiconductors, have been employed to assemble a number of phototransistors with exceptional device performance. For example, many thin-film-transistors (TFTs) phototransistors based on amorphous oxide semiconductors (AOS) have been used to detect ultraviolet (UV) or blue light because AOS usually has a wide bandgap of about 3.0 eV [13].

Formation of sub-gap states with doped absorption layer can provide various photoexcitation possibilities under light with relatively low energy, which is very important to broaden the detection wavelength [8]. Deliberate generation of defects within the channel layer and stacking various materials with additional light absorbing layers are two methods that can extend the sensitivity of the metal oxide phototransistors to longer wavelength [9–14]. Kang and co-workers have demonstrated visible light phototransistor composed of amorphous IGZO absorption layer doped by hydrogen [13]. As revealed in figures 9.1(c) and (d), the hydrogen-doped absorption layer (HAL) resulted in formation of sub-gap states between the conduction band minimum (CBM) and the valence band maximum (VBM). Under illumination, these defect levels may provide various excitation paths for photon transitions. Due to the existence of these sub-gap states, more photogenerated carriers are formed, leading to the detection of visible light ranging from 400 to 700 nm, with a high responsivity of 1932.6 A W^{-1} (635 nm light illumination), which was much higher than the device with oxygen vacancy induced sub-state mechanism.

Furthermore, Chung *et al* have recently proposed a solution-processed method to grow oxide absorption layer (SAL) that can introduce oxygen vacancy induced sub-bandgap states (figures 9.2(a) and (b)) [15]. Figure 9.2(c) shows band structure of the IGZO and SAL. It is seen that the energy band levels of the IGZO layer obtained by either sputtering or solution process methods are nearly the same. The difference of CBM levels between the IGZO and SAL was only 0.18 eV, suggesting that the phototransistor has good carrier transport characteristics. The working mechanism of

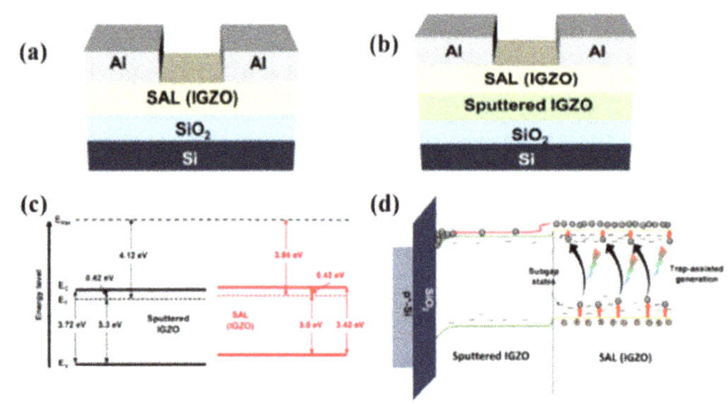

Figure 9.2. Schematic illustration of a-IGZO-based phototransistors. (a) With the solution-processed oxide absorption layer (SAL) for the channel layer and (b) sputtered IGZO and SAL for the channel layer. (c) Band structure of sputtered IGZO and SAL. (d) The working mechanism of sputtered IGZO TET with SAL for the absorption layer under the appropriate light illumination. Reprinted with permission from [15], Copyright (2019) American Chemical Society.

this device is illustrated in figure 9.2(d). Various defect states that are close to the CBM and VBM levels in the sub-gap were formed, as a result of organic residues and oxygen related defects in SAL. When shone by appropriate illumination, electrons in the valence band were easily trapped in these sub-gap states. Compared with electrons in valence band, electrons in sub-gap states were excited to the conduction band by a photon with lower energy, thus expanding the wavelength detection range. Although IGZO had a wide bandgap (3.7 eV), these sub-gap states contributed to the photoelectron production in SALs under visible light of long wavelength. Furthermore, the optimized a-IGZO phototransistor with HAL inserted in the channel layer exhibited a responsivity of 206 A W^{-1} and switched on and off at 0.05 Hz for 7200 s under monochromatic red light (635 nm) illumination.

Recently, Kim *et al* have developed a phototransistor based on a-GIZO with mesh-structured TiO_2 (m-TiO_2) [8]. By inserting the m-TiO_2 between the channel and the gate, the concentration of V_o was increased (figure 9.3(a)). As a consequence, sub-gap states whose absorption energy was lower than the actual bandgap of the channel material was increased, and finally the wide wavelength visible light (405–635 nm) was successfully detected. As shown in figures 9.3(b) and (c), the V_o was formed in highly defective interface (HDL)-GIZO-based phototransistor channel and more sub-bandgap states were formed within the IGZO bandgap. As a result, the IGZO phototransistor with m-TiO_2 displayed superior optoelectronic characteristics such as a high responsivity of 178.66 A W^{-1} and sensitivity of 1.04×10^5 under red light illumination.

Thanks to the combined benefits of multifunctional materials, hybrid heterostructures phototransistors have been recently receiving surging research interest. Compared with devices solely composed of homojunction, phototransistors composed of hybrid heterojunction could take advantage of the synergistic benefits of

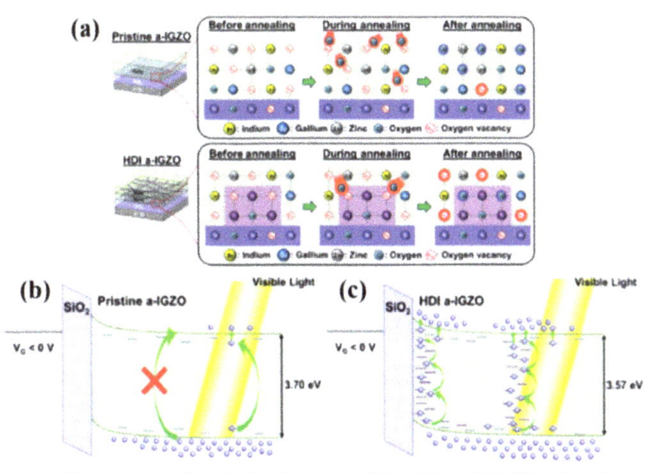

Figure 9.3. (a) The interface of pristine a-GIZO and SiO_2 in the top and the HDI a-GIZO in the image below. (b) The working mechanism of the pristine a-IGZO TET and (c) the HDI a-GIZO TFT under the visible light illumination. Reproduced from [8] with permission of The Royal Society of Chemistry.

both materials [16, 17], and therefore they usually exhibited high responsivity, high specific detectivity, high photoconductive gain, and even low energy consumption [18, 19]. By the difference in device geometry, the majority of the hybrid heterostructures based phototransistors could be classified into the following five categories: (i) graphene (Gr)-metal oxide semiconductor phototransistor; (ii) quantum dot (QD)-metal oxide semiconductor phototransistor; (iii) perovskite-metal oxide semiconductor phototransistor; (iv) conducting metal oxides–oxide semiconductor; (v) organic bulk heterojunction. For instance, quantum-dots oxide (QDO)–metal oxide hybrid phototransistors usually had broad waveband sensitivity from deep-UV to visible light, high responsivity from ~10^3 to 10^8 A W^{-1} and quick response speed of about 10 ms [20–27]. For example, Liu *et al* observed efficient charge transfer between IGZO and CdSe QD (figure 9.4(a)), which was facilitated by oxygen molecule absorption and the ligand (tri-*n*-octylphosphine oxide) that was modified with n-type CdSe QDs [26]. Moreover, Schottky barrier was formed between QDs and IGZO through ligand connection, which directly determined the size of the barrier. As depicted in figure 9.4(c), when shone by visible illumination, the photoexcited electrons were transferred to IGZO layer through ligand. The device exhibited responsivity with a maximum value of about 10^4 A W^{-1} at 450 nm. However, when illuminated as in figure 9.4(b), the device showed an obvious PPC effect.

Young and co-workers have reported a hybrid phototransistor by capping a layer of $CH_3NH_3PbI_3$ onto an IGZO thin film (figure 9.4(d)) [28]. The as-assembled

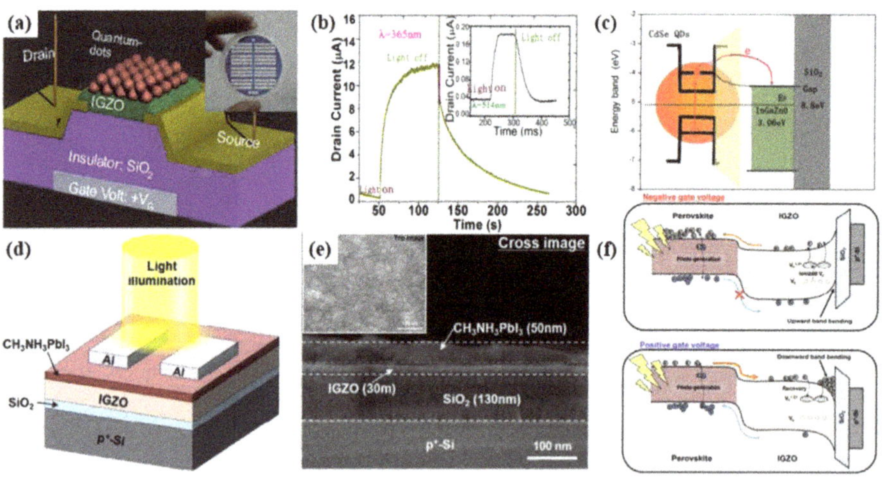

Figure 9.4. (a) Schematic of the QDs/IGZO hybrid phototransistor and the inset shows the photography of photo-induced thin film transistor. (b) Under 365 nm light, the fast quenching of the QDs and the PPC effect, the photocurrent response toward the illumination on/off test of 514 nm light (upper right). (c) The energy band and charge transfer schematic of the QDs/IGZO hybrid phototransistor. Reprinted from [26], with the permission of AIP Publishing. (d) Schematic of the IGZO hybrid phototransistor. (e) Cross-sectional SEM image of the heterostructure and SEM image of $CH_3NH_3PbI_3$ films (inset). (f) Energy band diagram with negative and positive V_{GS} under light illumination. Reprinted with permission from [28], Copyright (2018) American Chemical Society.

device was sensitive to a wide range of illumination from UV to visible with decent performance parameters in terms of a high responsivity of 61 A W^{-1} and specific detectivity of 9.42×10^{10} Jones. In addition, the hybrid heterostructure did not show any PPC effect that was usually observed in metal oxide phototransistors. As the IGZO layer and $CH_3NH_3PbI_3$ film exhibited a large difference in the valence band maximum energy level shown (figure 9.4(f)), such a $CH_3NH_3PbI_3$-IGZO phototransistor was characterized by efficient prevention of photoexcited holes transported between them. As a consequence, under a negative V_{GS}, most of photoexcited electrons were readily transferred in a perovskite layer because of upward band bending, which results in a high off-current under illumination.

9.2 Performance parameters

Hitherto, a large number of phototransistors have been realized by utilizing different metal oxide semiconductors. Even though their working principle varied from one device to another, their response capability should be assessed in order to compare the device performance in a quantitative manner. Table 9.1 lists some of the key performance parameters that are often used to evaluate the performance of phototransistors.

9.3 Binary metal oxide semiconductors

9.3.1 ZnO

Adapted from Liang *et al* with permission from The Royal Society of Chemistry.
ZnO is one of the most pivotal II–VI compound semiconductor materials and has gained substantial research interest due to its excellent properties in terms of a wide direct bandgap of ~3.37 eV, high exciton binding energy at room temperature and so on [29, 30]. Due to these excellent properties and its easy processing, a variety of ZnO-based photonic and optoelectronic devices have been fabricated for various UV light related applications [31]. In recent years, ZnO-based hybrid heterojunction has been widely studied for phototransistors application. Typically, three types of ZnO-based hybrid heterojunction have been investigated in these phototransistors, such as ZnO hybrid heterojunction (Gr/ZnO QD, Gr/ZnO NW and so on), ternary ZnO-based hybrid heterojunction (a-ZTO/c.c.-CdS/a-ZTO, IZO/PbS QD and so on), quaternary ZnO-based hybrid heterojunction (Ta_2O_3/ZITO, MoS_2/IGZO, Gr QD/IGZO and so on). This section will be mainly focused on the fabrication of ZnO hybrid heterojunction for UV phototransistor application. For convenience, our discussion starts from heterojunction preparation methodology, followed by device performance characterization and operation mechanism analysis.

9.3.1.1 Fabrication of ZnO based heterojunction phototransistor
ZnO has been usually combined with Gr or organic semiconductor materials to form hybrid heterojunction phototransistors. The most convenient way to obtain a Gr–ZnO heterojunction is to simply transfer a large-area Gr sheet onto the surface of ZnO via an aqueous transfer process [32]. For example, chemical vapor deposition (CVD)-derived Gr was spin-coated with 5 wt% polymethylmethacrylate (PMMA) in

Table 9.1. Performance parameters of phototransistors.

	Quantity	Symbol	Unit	Definition
Electrical properties	Subthreshold swing	SS	V/dec	Subthreshold swing is defined as the gate voltage required to change the drain current by one order of magnitude and extracted from the following equation: $SS = \frac{dV_G}{d(\log(I_D))}$
	Field-effect mobility	μ_{FET}	$cm^2\ V^{-1}\ s^{-1}$	The field-effect mobility is extracted from the following equation: $\mu_{FET} = \left(\frac{2}{C_{OX}}\right) \cdot \left(\frac{L}{W}\right) \cdot (G_{max})^2$; where C_{ox}, L, W are the capacitance of the gate insulator, the channel length, the channel width, respectively. G_{max} is the maximum value of the derivative of the square root of the drain current with respect to the square root of the gate voltage, $G_{max} = \max\left(\frac{\partial\sqrt{I_D}}{\partial\sqrt{V_G}}\right)$
	Number of defects	N_t	$cm^{-2}\ eV^{-1}$	N_t is extracted from the following equation: $N_t = \frac{C_{OX}}{q}\left(\frac{q \cdot SS \cdot \log \wp(e)}{kT} - 1\right)$; where q is the electron charge, k is the Boltzmann constant, T is absolute temperature.
Optoelectronic properties	Photoconductive gain	G	Unitless	The ratio between the number of electrons collected by the electrode and the number of incident photons.
	Dark current	I_{dark}	A	Current flowing in the photodetector in the absence of light illumination.
	Response speed	Γ_r, Γ_d	s	Rise time (Γ_r) and decay time (Γ_d) are defined as the time which is needed for the photoresponse to increase from 10% to 90% or drop from 90% to 10% of its peak value.
	Responsivity	R	$A\ W^{-1}$	The ratio between the photocurrent and the incident optical power on the phototransistor. PR is extracted from the following equation: $R = \frac{J_{photo}}{P_{photo}}$; where J_{photo} is the current density during light illumination and P_{photo} is the incident laser power density.
	Sensitivity	PS	Unitless	PR is extracted from the following equation: $PS = \frac{I_{photo}}{I_{dark}}$; $I_{photo} = I_{illuminated} - I_{dark}$.
	Noise-equivalent power	NEP	$WHz^{-1/2}$	The minimum impinging optical power required to achieve a signal-to-noise ratio of unity in a 1 Hz bandwidth.
	Detectivity	D^*	Jones	A parameter for comparing the sensitivity between photodetectors with different devices. It is defined as: $D^* = \frac{(A\Delta f)^{1/2}}{NEP}$ or $D^* = \frac{PR}{(2qJ_{dark})^{1/2}}$; where A is the device area, Δf is the band width, q is the absolute value of the charge of an electron (1.6×10^{-19} C).

chlorobenzene, and then the underlying Cu foils were removed using a Marbles's solution. Afterwards, the ZnO was soaked in deionized water, and slowly lifted to mount the Gr onto ZnO. In fact, such an aqueous transfer method has been also utilized to obtain other van der Waals Gr-semiconductor (e.g. Ga_2O_3, Si, GaAs and so on) heterojunctions [33–35]. On the other hand, a number of synthetic techniques such as CVD, electrochemical deposition method, hydrothermal method, metal-organic vapor-phase epitaxy (MOVPE), pulsed laser deposition (PLD), etc have also been employed to grow ZnO thin film or nanostructures directly on Gr. Liu's group reported the synthesis of ZnO nanorod arrays with high optical qualities on few-layer Gr sheets by using zinc nitrate hexa-hydrate and hexamethylenetetramine as precursors under hydrothermal conditions [36]. In fact, in similar mild conditions, ZnO nanoparticles and nanowire arrays could also be combined with Gr oxide to form ZnO–Gr oxide nano-composites, which found important applications in photodetectors from the visible to near-infrared range [37]. Electrochemical deposition with apparent advantages including simplicity, low cost, and rapid growth at a low temperature, has also been a popular method for the synthesis of ZnO nanostructures-Gr heterojunction [38]. For example, dense and vertical ZnO nanotubes (NTs) have been successfully grown on Gr using a seedless electrochemical deposition process [39]. Compared with other low-temperature methods, ZnO obtained from high-temperature route was often characterized by high-quality crystallinity with a relatively low density of defects and good transport properties. For this reason, high-temperature methods were frequently used to fabricate high-quality Gr–ZnO heterojunctions for optoelectronic device applications [40, 41]. As an example, CVD has proved to be a promising route to grow 1D ZnO nanostructure arrays (e.g. nanorods [42], and nanowires [43, 44]) onto Gr layers. The growth temperature was kept in the range from 450 to 900 °C. MOVPE represents another popular high-temperature method that has been widely adopted to grow ZnO on a Gr surface [45].

Apart from Gr, organic semiconductors such as small molecules and organic polymers have been used for assembly of phototransistors. All-solution processed fabrication with typical advantage of simplicity and low cost has also become a popular method for the synthesis of organic thin film on ZnO thin film [46]. For example, Pattanasattayavong *et al* reported a dye-sensitized (D 102 dye) ZnO phototransistor through an all-solution process [47]. Even though ZnO in many different nanoscale forms could be easily grown [48], it however could bring about a native point defect (e.g. surface defects, Zn interstitials, and oxygen vacancy) during the growth process [49, 50]. Various studies have shown that these defects could lead to not only n-type conductivity of ZnO but also exhibited very strong blue or green emissions, which is useful for solid-state white lighting devices [51].

9.3.1.2 Phototransistors

Gr–ZnO UV light phototransistors were geometrically characterized by two metal electrodes on the Gr layer, which suggested a completely different operation mechanism than that of photodiode. Under UV light illumination, the Gr could act as a carrier transport channel, while the ZnO was used as the photo absorbing

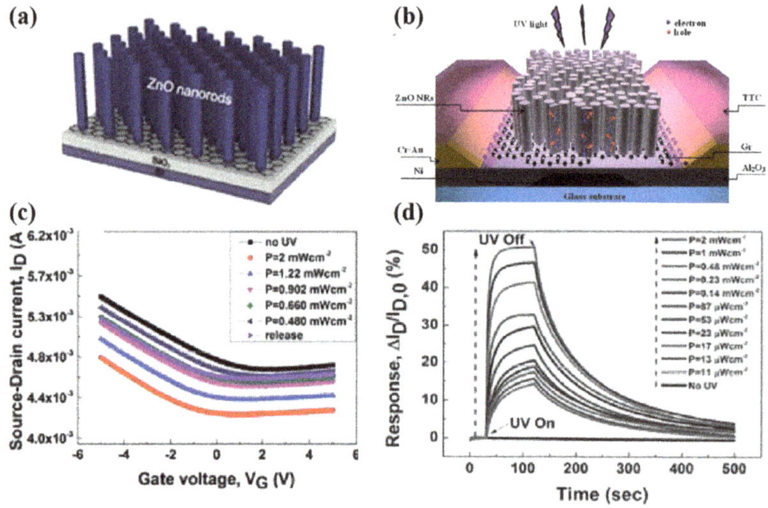

Figure 9.5. (a) Growth of ZnO nanostructures on Gr through a hydrothermal process . Reproduced from [36]. (b) Schematic of a Gr/ZnO UV phototransistor. (c) Transfer characteristics of the Gr/ZnO UV phototransistor under UV light with different intensities; (d) phototransistor of this device at a bias voltage of 1V. Reprinted with permission from [55], Copyright (2015) American Chemical Society.

material [52, 53]. Although a built-in electric field was also formed at the Gr–ZnO interface, it was nevertheless not used to separate the photo-generated electron–hole pairs, as often in Gr–ZnO UV photodiodes. Dang *et al* reported a hybrid channel UV phototransistor composed of Gr–ZnO NW arrays (figures 9.5(a) and (b)) [36, 54]. Device analysis revealed that the phototransistor had a high sensitivity and selectivity to near UV light (figure 9.5(c)). Additionally, the device could work at a low voltage and show good reproducibility, high responsivity and photoconductive gain (figure 9.5(d)). Flexible and sensitive Gr–ZnO NW hybrid UV phototransistors were also successfully assembled on polyethylene terephthalate (PET) substrates [55]. From the negative shift of the Dirac point in the transfer characteristics, the responsivity was calculated to be 3×10^8 V W^{-1}, which could be ascribed to the transfer of electrons from ZnO to Gr with a transferred electron concentration of 6×10^{12} electrons per mW cm^{-2}. Remarkably, the flexible Gr–ZnO hybrid phototransistor had excellent device stability, even after 10 000 cycles of bending at a strain of 0.5%.

Similarly, highly dense ZnO NW arrays could be grown onto Gr structures including three-dimensional Gr foam and rGO to form sensitive UV light phototransistors [56, 57]. For instance, by using a resistive thermal evaporation technique, Boruah *et al* fabricated a ZnO NWs–Gr foam hybrid structure that showed an excellent UV photoresponse, with a relatively slow response and recovery times of 9.5 and 38 s, respectively, and an EQE as high as 2490.8%. The sensitivity of the device could be understood as follows: under UV illumination, energy of electron–hole pairs were greater than or equal to the bandgap of ZnO, which causes the oxygen molecules to desorb, thus the free holes are captured by oxygen ions on the surface. A decrease in the depletion layer led to an increase in free charge carrier concentrations, which increased the conductivity of the device.

Combination of ZnO with QDs/nanoparticles and Gr could lead to phototransistor with high-efficiency exciton dissociation through atomistic design and engineering of the heterojunction [58, 59]. This geometry has exhibited distinctive advantages over conventional devices because of not only the unique QDs that can yield unique optoelectronic properties such as higher light absorption, charge mobility, and spectra tenability, but also the exciton dissociation and charge transfer at the heterojunction interface with favorable interfacial electronic structures for the generation of carriers. Son *et al* successfully fabricated a highly sensitive colloidal Gr–ZnO QD UV phototransistor using a wet spin-coating method. The UV device with ZnO QDs uniformly distributed between the voids of the surface circumferences of the Gr layer displayed a high on/off ratio of 1.1×10^4, and a rise time of 2 s [60]. By systematically analyzing the optoelectronic characteristics of a device with a similar structure [61], Guo *et al* have observed that the sensitive photoresponse to UV light was largely due to the charge transfer: in the dark, oxygen molecules adsorbed on the surface of the QDs and captured free electrons from the n-type ZnO QDs, resulting in a decrease in the conductance of the ZnO QDs. Under UV illumination, photo-generated holes will migrate to the surface and discharge the adsorbed oxygen ions through surface electron–hole recombination [O_2 (ad) + $h^+ \rightarrow$ O_2 (g)], the unpaired electrons thus enhanced the conductance [62]. It should be pointed out that the mobility of Gr transistors on SiO_2 were orders of magnitude lower than that of the free-standing Gr, which adversely influenced the performance of the above Gr–ZnO QDs phototransistors [63].

A feasible method that inserted an organic self-assembled monolayer (SAM) between SiO_2 and Gr as a spacer has been reported, in an effort to improve the carrier mobility of Gr. As an example, Shao *et al* have demonstrated an ultrasensitive UV phototransistor featuring an organic SAM sandwiched between an inorganic ZnO QD decorated Gr channel and SiO_2 (figures 9.6(a)–(d)) [21]. Since the electron transit-time in the channel was considerably decreased and the room-temperature mobility of the CVD derived Gr was increased by one order of magnitude, the recirculation of the electron can increase the responsivity and photoconductive gain of the transistor to ~10^8 A W^{-1} and ~3×10^9 (figure 9.6(e)), respectively, which were even comparable to some commercial devices.

To further improve the reproducibility as well as the sensitivity of the phototransistors, Gong *et al* have developed a simple precursor printing technology, via which a clean van der Waals interface was formed between the ZnO QDs and the Gr layer, enabling high-efficiency exciton dissociation and charge transfer across the Gr–ZnO QD interfaces [64]. Owning to the strong light–mater interaction, quantum confinement of the QDs, and high charge mobility of the Gr, the device exhibited ultrahigh responsivity of up to 9.9×10^8 A W^{-1} and a high detectivity of 7.5×10^{14} Jones. By using a similar precursor printing technology, a wafer-size Gr–ZnO nanoparticle UV phototransistor was also fabricated [65]. The device achieved a responsivity of 1000 A W^{-1}, and a photoconductive gain of 1.8×10^4. These values were much smaller than that of the printable Gr–ZnO QD device mentioned above, but were at least

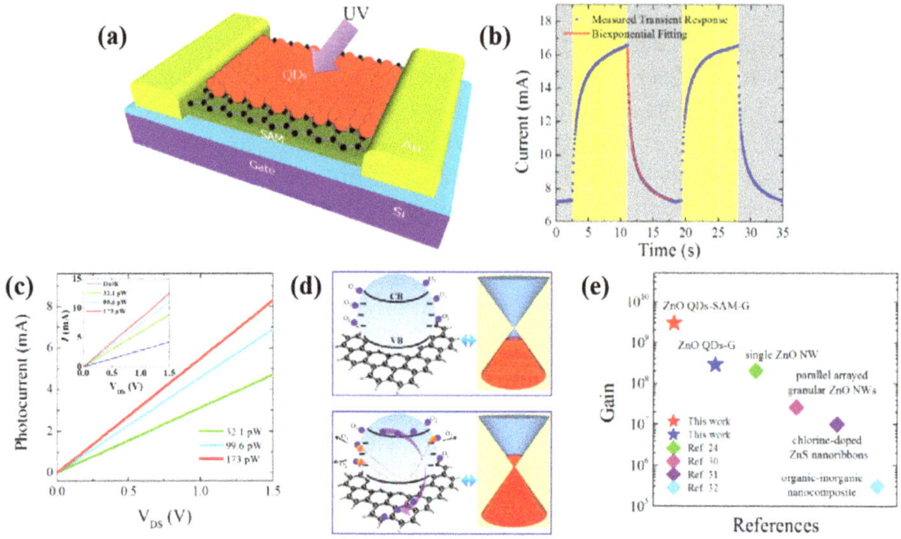

Figure 9.6. (a) Schematic illustration of QD/ZnO hybrid phototransistor; (b) transient response of the UV phototransistor under laser at 335 nm; (c) photocurrent of the device under different illumination; (d) schematic illustration of the operation mechanism of the ultrasensitive Gr/ZnO QDs UV phototransistor; (e) comparison of the photoconductive gain of the present device and previously reported devices. Reprinted with permission from [21], Copyright (2015) American Chemical Society.

one order of magnitude higher than that of previously reported UV photodetectors. What is more, Choi *et al* have proposed a simple and effective method to achieve the SnO_2/ZnO NPs bilayer UV phototransistor [66]. By virtue of high carrier mobility of SnO_2 (6.05 $cm^2\ V^{-1}\ s^{-1}$) and the lower band of SnO_2 than that ZnO NPs layer, the device exhibited high on/off current ratio of about 10^8, low subthreshold swing of 0.34 V/dec. as well as a high detectivity of 7.79×10^{13} Jones.

9.3.2 Other binary metal oxide semiconductors

Besides ZnO, other binary metal oxide semiconductors including TiO_2 [67–71], SnO [72], In_2O_3 [12, 73], have been combined with hybrid heterostructures for optoelectronic applications as well. TiO_2 possesses a distinctive electronic structure with a large bandgap (anatase: 3.2 eV, rutile: 3.0 eV), a comparably high lifetime of photoexcited electrons, making it possible for the application in photodetectors [68]. Unlike TiO_2, SnO had a direct optical bandgap (~2.8 eV) as well as a small indirect bandgap (~0.7 eV) resulting in the potential application in visible light detection. Recently, TiO_2 based hybrid heterostructures, such as Gr/TiO_2, perovskite/TiO_2 nanotubes (NT), and $MAPbI_3$/SnO have attracted considerable attention in optoelectronic applications, including solar cells and photodetectors [74, 75]. In this part, only discussion of TiO_2 or SnO based hybrid heterostructures for phototransistor application is entertained.

9.3.2.1 Fabrication of TiO_2 or SnO based hybrid heterojunction

The preparation of SnO hybrid heterojunction phototransistor was quite similar to that of ZnO base devices. Guan *et al* reported on the synthesis of SnO thin film via a current reactive magnetron sputtering at room temperature [72]. To achieve a SnO hybrid heterojunction, they used sequential vapor deposition and DMF vapor annealing, through which $MAPbI_3$ perovskite layer was deposited onto the SnO layer. Moreover, An inkjet printable ionic solution technique has also been proposed to fabricate Gr titanium oxide photoconductor, which was divided into three steps [68]. (i) Preparation of titanium dioxide sol. (ii) preparation of composite solution for Gr and titanium dioxide sol. (iii) drop-casting or inkjet printing on substrates and then two-stage sintering for Gr/TiO_2 hybrid film.

9.3.2.2 Phototransistors

Like Gr/ZnO hybrid structures, Gr/TiO_2 heterostructures as building blocks have been used to fabricate UV light phototransistors. Manga *et al* have developed a solution-processed route to assemble a rGO/TiO_2 hybrid phototransistor that was printable and spin-castable [68]. The as-prepared hybrids devices exhibited distinct photoresponse in UV–visible wavelength region with peak sensitivity at ~320 nm, but virtually no photoresponse was observed in pure rGO or TiO_2. Under UV illumination, photo-generated electrons in TiO_2 were injected into rGO due to appropriate band alignment, whereas photo-generated holes were driven to the surface of TiO_2 by band bending as a result of the presence of oxygen absorption. The specific detectivity could reach 2.3×10^{12} and 9.4×10^{11} Jones at UV and visible regions, respectively. What was more, the devices exhibited a fast response speed with $\sim$100 ms rise and fall times. The great advantages held by this technique such as easy chemical manipulation, aqueous-based solution-processability and printability for large-scale screen printing, signified its considerable innovation potential for future printable optoelectronic devices. To improve the visible photoresponse and extend the response region to IR wavelength, a Gr/PbSe QDs/TiO_2 hybrid photodetector was fabricated via a solution-based process [69]. The devices exhibited an I_{light}/I_{dark} ratio of 10^4, which outperforms devices made of Gr/PbSe (10^2), Gr/TiO_2 (10^2) composites and PbSe QDs (10). The responsivity was attributed to efficient electron injection from the photoexcited PbSe QDs to TiO_2 or Gr. As a result of the strong optical absorption across UV–visible to IR region, and the efficient carrier separation and facilitated transport property, the multicomponent composite-based devices have demonstrated specific detectivity of 3×10^{13} and 5.7×10^{12} Jones in visible and IR regions, respectively. Afterwards, Zheng and co-workers studied the optoelectronic properties of chemically exfoliated Gr/TiO_2 NPs heterostructures and found a reversible photoresponse in the visible spectrum region [70]. In the device, negatively charged oxygen anions were formed for the reaction between active electrons and oxygen molecules absorption at TiO_2 surface. The resultant reduce in scattering sites and release of holes into the Gr channel caused a low conductance of the n-type Gr in dark. In contrast, under light illumination, the desorption of oxygen molecules released the electrons back to the Gr channel and therefore formed the photocurrent.

A variety of binary metal oxides in the form of either thin films or nanostructures have been employed for assembling perovskite-based hybrid phototransistors [12, 72–75]. These devices typically show responsivity in the range of tens of $m A W^{-1}$ to tens of $A W^{-1}$, and response speed ranging from milliseconds to

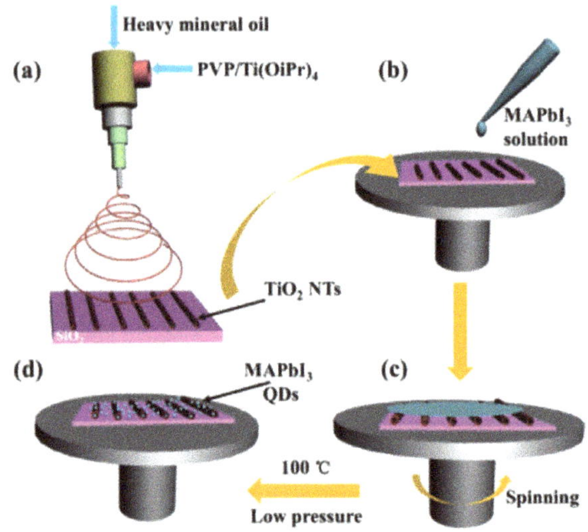

Figure 9.7. (a) The fabrication of TiO_2 NTs using co-electrospinning; (b)–(d) the fabrication of $MAPbI_3$ QDs/NTs using spin-coating. [71] John Wiley & Sons. © (2017) WILEY-VCH Verlag GmbH & Co., KGaA, Weinheim.

several seconds. As an example, Zheng *et al* incorporated $MAPbI_3$ QDs onto the surface of aligned TiO_2 NTs via a coaxial electrospinning technique [71]. As shown in figure 9.7(b), the as-prepared hybrid phototransistors showed obvious photoresponse to a wide range of illumination from UV to visible regimes, and the responsivity in visible region was improved by nearly three orders of magnitude compared with a pure TiO_2 NTs-based detector. According to the band alignment shown in figure 9.7(c), a type-II heterostructure was actually formed at the QD/NT interface, allowing electrons to move toward the perovskite side while holes moved toward the TiO_2 side upon light illumination. Thus, the spatial separation of photogenerated carriers dramatically suppressed charge carrier recombination and improved photoresponse performance. Specifically, the device attained responsivity of 1.3 and 0.2 A W^{-1}, specific detectivity of 2.5×10^{12} and 3.8×10^{11} Jones under 350 and 700 nm illumination, respectively. What is more, such a heterostructure device demonstrated high transparency over 85% in the visible region and excellent mechanical flexibility with no obvious degradation in performance even after 200 cycles of repeated bending at 90°. Furthermore, Guan *et al* have successfully developed a phototransistor by combining SnO thin film with $MAPbI_3$ over-layer (figure 9.8(a)) [72]. With strong photogating effect, the as-assembled device exhibited abnormal photoelectric behavior compared with a pure SnO based device. Their working mechanism could be divided into the following three cases: (i) under an appropriate illumination, perovskite over-layer produced hole–electron pairs. Applying a positive bias on the ITO electrode could reduce the interfacial energy barrier and lead to the transition of electrons to the p-type SnO over-layer. The electrons recombined with the holes in the p-type SnO, resulting in decrease in the off-state current for this device. (ii) When no bias

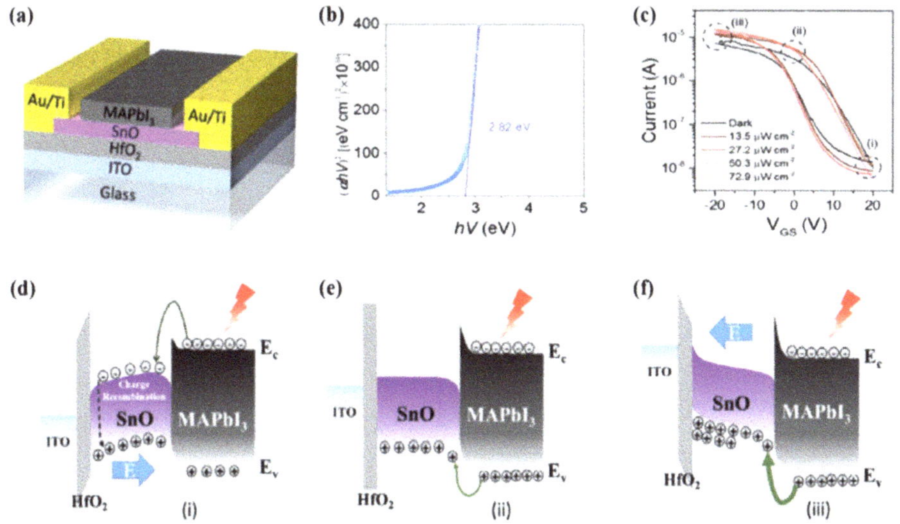

Figure 9.8. (a) Device structure of the $MAPbI_3$/SnO phototransistor; (b) the direct bandgap of SnO thin film; (c) transfer curve of the $MAPbI_3$/SnO phototransistor corresponding to three specific points; (d)–(f) schematic illustration of the band alignment and charge transfer at three specific points. [72] John Wiley & Sons. © (2019) WILEY-VCH Verlag GmbH & Co., KGaA, Weinheim.

was applied, almost no electrons can flow into the SnO over-layer due to the high energy barrier. Under this circumstance, holes would transfer from perovskite layer to SnO over-layer, leading to a high on-state current. (iii) Under a negative bias voltage, on-state current was higher than before (figure 9.8(d)). Compared with pure SnO phototransistor, the field-effect mobility increased from 3.55 to 5.53 cm^2 V^{-1} s^{-1} and the on/off ratio increased from 519 to 2.7×10^3 under dark.

9.4 Ternary metal oxide semiconductors

Zinc oxide (ZnO), as a typical example of amorphous binary metal oxide has been usually applied in the optoelectronic field due to its relatively higher field-effect mobility than that of hydrogenated amorphous silicon (a-Si:H) [78]. To achieve other functional applications, many ZnO-based metal oxide semiconductors, such as InZnO (IZO), ZnSnO, MgZnO, GaZnO (GZO), and InGaZnO (IGZO) have been synthesized [79–86]. In general, these multi-element films are often prepared by co-sputtering method, through which the bandgap of these films can be easily tailored by adjusting the content of each component [87]. This section mainly introduces the ternary metal oxide phototransistors obtained by doping zinc oxide, such as InZnO, ZnON [88], and ZnSnO, and then the ternary metal oxide related phototransistors will be introduced in detail in the next section.

9.4.1 Fabrication of ternary metal oxide semiconductors hybrid heterojunction

Rim *et al* have reported the assembly of a bulk heterojunction (BHJ)/ZnON hybrid phototransistor, in which the ZnON film as channel was deposited on $SiO_2/p^{++}Si$ by

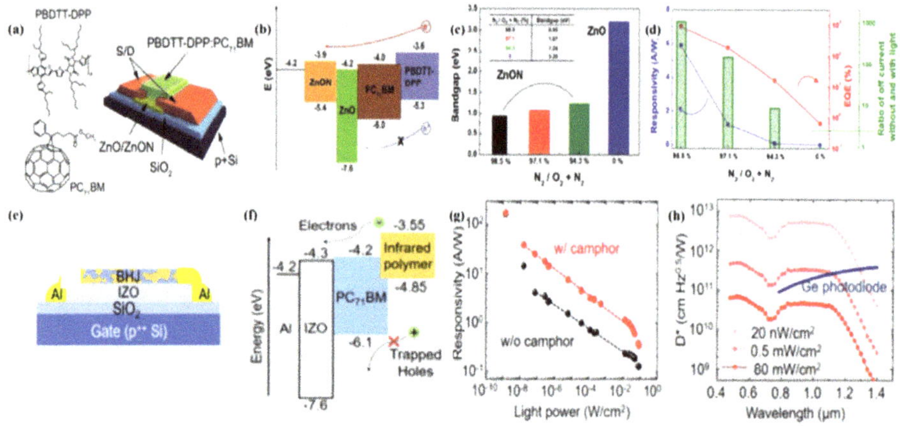

Figure 9.9. (a) Chemical structures of PBDTT-DPP, $PC_{71}BM$ and phototransistors with ZnON as the electron transporting layer; (b) energy bandgap illustrating electron transfer from the BHJ to ZnON, whereas holes are trapped in BHJ. Reprinted with permission from [88], Copyright (2016) American Chemical Society. (c) Bandgap of ZnON and ZnO films for different nitrogen gas flow ratios; (d) responsivity (R), external quantum efficiency (EQE), and I_{on}/I_{off} at various nitrogen ratios of ZnON and ZnO at different V_{GS}. Reprinted with permission from [87], Copyright (2018) American Chemical Society. (e) Schematic of the bulk heterojunction (BHJ) polymers-IZO hybrid structure and (f) energy bandgap alignment of a spontaneous IZO buffer layer formed at $PC_{71}BM$ interface; (g) responsivity and (h) detectivity under different light power. Reprinted with permission from [95], Copyright (2019) American Chemical Society.

direct current reactive sputtering [88]. Subsequently, the source/drain electrodes as well as indium tin oxide were deposited by electron beam, and then BHJ layer was spin-coated on ZnON TFT. Moreover, Lee and co-workers have further studied the relationship between N content and energy bandgap of ZnON film (figure 9.9(c)). Another way to achieve binary metal oxide hybrid heterojunction structure was the all-solution method. For instance, Feng's group have successfully synthesized an amorphous ZnSnO by solution fabrication process [86]. The Zn precursor solution and Sn precursor solution were mixed at the molar ratio of Zn:Sn = 4:7 and aged for 24 h, followed by spin-coating on the clean substrate. Finally, ZnSnO thin film was obtained by annealing in air ambient at 300 °C for 30 min Compared with ZnSnO single layer, the preparation of the binary metal oxide hybrid structure is more complex. Cho and co-workers have reported a-ZTO/c.c.-CdS/a-ZTO hybrid full-color phototransistors by solution processed method (figure 9.10(b)) [89]. By using the same method, Sharma *et al* have also prepared the IZO/PbS QD hybrid heterojunction phototransistor with Li_5AlO_4 as the dielectric layer [90].

9.4.2 Phototransistors

In order to extend the detection region of metal oxide based phototransistor, some groups tried to find a way to acquire narrow bandgap materials, which contained new orbital state formed by the deactivation of oxygen vacancy by locating the deep level V_o region above the oxygen 2p orbital by non-oxide anion [91, 92]. In 2016, Rim and co-workers achieved a phototransistor with BHJ/ZnON hybrid

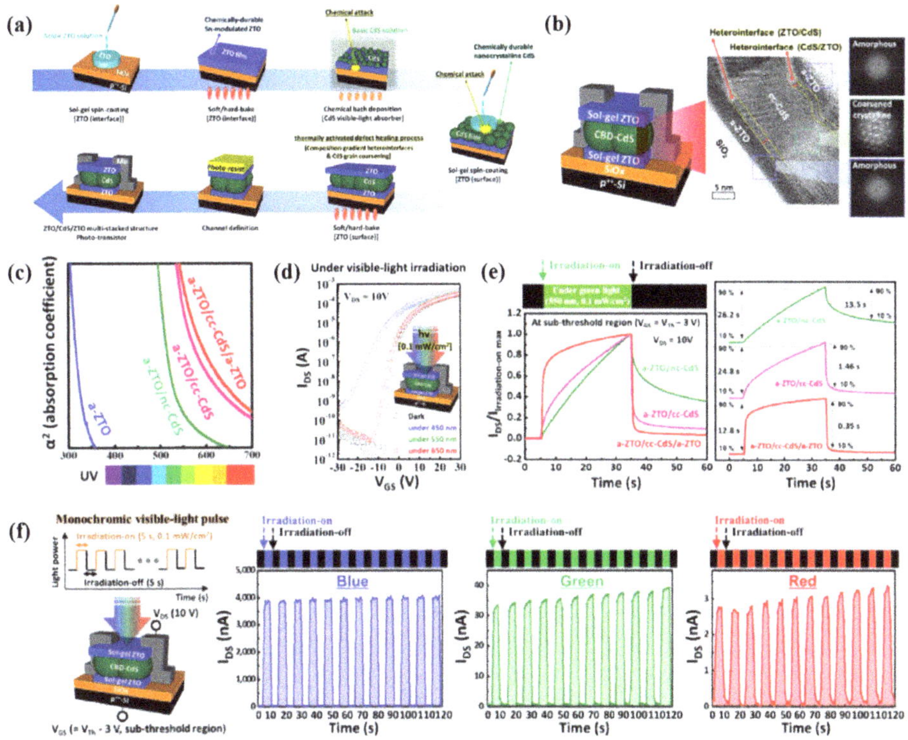

Figure 9.10. (a) All-solution process flow for chemically stable multistacking and thermally activated defect healing; (b) TEM image and diffraction patterns of cross-section; (c) UV–vis absorption spectra of different ZTO-based hybrid structures; (d) static photoresponse of a-ZTO/c.c.-CdS/a-ZTO phototransistor; (e) comparison of photosensing performance of various phototransistors at V_{GS} = −8 V and V_{TH} = −3 V: (f) reproducible real-time full-color detection with clear photocurrent generation and fast phototransient speeds under dynamic vis-light pulses. [89] John Wiley & Sons. © (2018) WILEY-VCH Verlag GmbH & Co., KGaA, Weinheim.

hetero-interface that could detect the optical wavelength range from the UV to near-infrared (NIR) (figures 9.9(a) and (b)) [88]. The bandgap of ZnON was estimated to be 1.5 eV, slightly larger than previously reported values [93, 94]. Lee *et al* have tried to shed light on the exact mechanism by studying the bandgap of ZnON film as a function of various N/O ratios (figure 9.9(c)) [87]. It was revealed that the ZnON thin film was actually composed of Zn_3N_2 and ZnO and with the increase of N/O ratio, bandgap of ZnO thin film would decrease which was reasonable because of decrease in chemical bonding ratio between ZnO (3.2 eV) and Zn_3N_2 (0.9 eV). In figure 9.9(d), among the three different types of phototransistors, the device with 98.5% nitrogen ratios of ZnON and ZnO exhibited the best performance with a responsivity of 5.90 A W^{-1}, EQE of 885% and I_{off}/I_{on} of 1.04×10^3. Compared with ZnO/ZnON phototransistor, BHJ/ZnO/ZnON device display ultrasensitive photonsensing property in the range from 650 to 850 nm. The corresponding responsivity, the special detectivity of and linear dynamic range were estimated to be 1.7×10^3 A W^{-1} 10^{12} Jones 1222.3 dB, respectively. The superior device performance was related to the following two factors: the effective exciton dissociation and the increased diffusion length of BHJ-PBDTT-DPP: PC71BM were beneficial for the generation of electrons and holes (figure 9.9(b)). On the other hand, ZnO, ZnON and BHJ were obviously matched, which helped to transfer electrons and to hinder holes through ZnO of spontaneous passivation, while ZnON narrow bandgap could not block holes. Similarly, Kim *et al* have also reported an all solution-processed BHJ/IZO phototransistor with detection range from 500 to 1400 nm [95]. The detectivity and a dynamic range of BHJ/IZO photo-TRs was as high as 5×10^{12} Jones and 127 dB, respectively.

In addition, Sharma *et al* have proposed a new PbS QD phototransistor using a wide bandgap Li_5AlO_4 as the gate medium of TFT, in which the loosely bonded Li^+ could move up and down and form a vertically polarized film under external bias [90]. The IZO/PbS QD heterojunction device showed a responsivity of 4.5×10^{-4} A W^{-1}, and a response time of 2.2 s. Apart from this, other phototransistors with quantum dots as absorption layers have also been developed. Cho *et al* have employed an all-solution process to achieve another visible light phototransistor with multilayer-stacked structure, which CdS QD layer located between two layers of ZTO thin film [89]. In comparison with ZTO/CdS hybrid heterojunction, the thermally healed ZTO/CdS films (a-ZTO/c.c.-CdS/a-ZTO and a-ZTO/c.c.-CdS) composition-gradient heterointerfaces (figure 9.10(b)), was helpful for ultrasensitive, fast dynamic, and full-color photodetection [89]. As shown in figure 9.10(c), the thermally healed ZTO/CdS films have optical band edge in the red light region (λ = 640 nm). This red-light sensitive heterojunction promoted detection covering the entire spectral rang of visible. Furthermore, it was found that the device exhibited remarkable sensitivity of 8.5×10^6 and 7.3×10^4 under blue and green illumination (figure 9.10(d)). In contrast to a-ZTO/nc-CdS and a-ZTO/c.c.-CdS phototransistors, the thermally healed ZTO/CdS films with multi-stacked structure exhibited slow photodecay speed (Γ_{decay} = 0.35 s) and smaller PPC effect (figure 9.10(e)). Conventional pure oxide or hybrid phototransistors inevitably required an artificial compulsory recovery to overcome inherent PPC effect. However, in figure 9.10(f), a-ZTO/c.c.-CdS/a-ZTO phototransistors showed repeatable real-time full-color detection, and they have exhibited fast photodecay speed and clear photocurrent generation under a dynamic visible-light pulse.

9.5 Quaternary oxide semiconductors

9.5.1 Fabrication of quaternary oxide hybrid heterojunction

Similar to a-Si based TFTs, a device composed of amorphous quaternary metal oxide semiconductors (AOS) which included zinc indium tin oxide (a-ZITO) and indium gallium zinc oxide (a-IGZO) often exhibited large-area uniformity. Moreover, it had a low off-current (I_{off}) in comparison with low-temperature poly-Si-based TFTs [96]. However, owing to its relatively large bandgap (about 3.0 eV), the above AOS had high transparency in visible light ranges from 700 to 400 nm. Therefore, a-GIZO and a-ZITO were usually used as channel layer of ultraviolet or blue light photodetectors [97, 98]. In order to expand the scope of light detection, several research groups tried to use hybrid heterojunction device structures stacked with additional light absorption layers [99–102]. Nanostructured semiconductor materials are a promising candidate for light absorption due to their unique optoelectronic properties [103–105]. In addition, two-dimensional materials, perovskite and polymeric capping layer could also absorb a wide range of light because of their narrow bandgap [106]. Typically, there are two methods to obtain hybrid heterojunction. Firstly, the method of direct coating by solution on the surface of quaternary metal oxide was the most convenient way to achieve a quaternary metal oxide hybrid heterojunction. Secondly, the fabrication of Gr/quaternary metal oxide hybrid heterojunction structure could be realized by simply transferring Gr layer on IGZO film which was deposited via an RF sputtering system using an InGaZnO target. To induce semiconductor characteristics, the above sample was annealed at 300 °C in air for 1 h [8].

9.5.2 Phototransistors

In previous studies, quaternary metal oxide semiconductors (such as a-IGZO and ZITO) hybrid heterojunction phototransistors were mainly comprised of conducting metal oxides as the charge carrier transport channel, and wide bandgap semiconductors as absorbing layer, as illustrated in figure 9.11(a) [4, 5]. The thickness of a-IGZO or a-ZITO was around 50 nm, while the employed wide bandgap semiconductors were usually Ta_2O_5 or Ga_2O_3 thin films with ten to hundreds of nanometers in thickness. The wide bandgap semiconductor also served as the gate dielectric in these phototransistors. The power consumption of these devices could

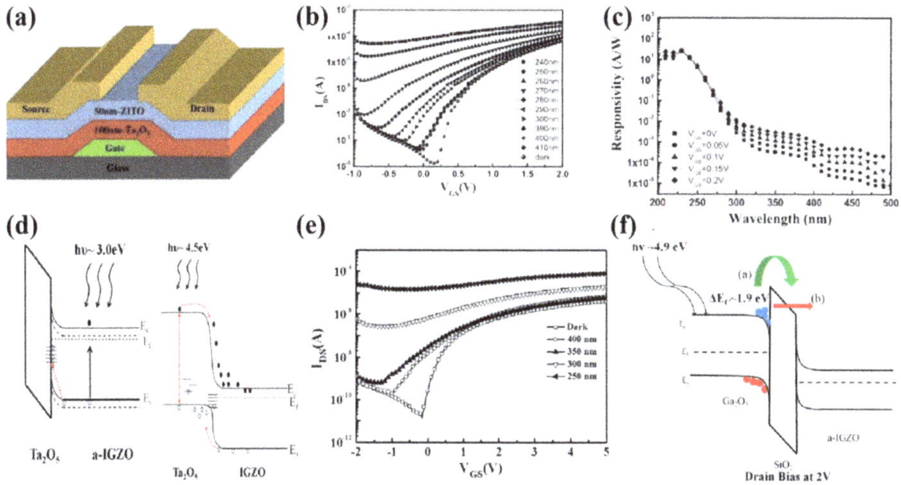

Figure 9.11. (a) Schematic diagram of a heterojunction phototransistor with a metal oxide as the charge transport channel, and a UWBG semiconductor as the DUV light absorbing media; (b) transfer characteristics of a Ta_2O_5/a-ZITO heterojunction phototransistor under different illuminations; (c) spectral response of the device at varied gate biases; (d) energy band diagram of the Ta_2O_5/a-ZITO phototransistor under illuminations. © (2012) IEEE. Reprinted, with permission, from [4]. (e) Transfer characteristics of a Ga_2O_3/SiO_2/a-IGZO heterojunction phototransistor under illuminations with different wavelengths; (f) energy band diagram under light illumination. Reprinted from [5], with the permission of AIP Publishing.

be reduced when ultrathin high-*k* materials were introduced between the ternary metal oxide layer and gate by physical vapor deposition (PVD) [83]. Some high-*k* materials, such as Al_2O_3, Ta_2O_5, and HfO_2, have proved to be helpful for improving the subthreshold swing (SS) and reducing the threshold voltage (V_T) [8]. Remarkably, after the interfacial passivation, the devices exhibited high current on-off ratio in the range of 10^4–10^6, with low dark currents on the order of 10^{-10}–10^{-9} A without illumination. In fact, the dark current could be further reduced to as low as 10^{-11} A by increasing oxygen partial pressure during the preparation of the Ta_2O_5 or Ga_2O_3. As figures 9.11(b) and (c) show, the detectors displayed high sensitivity with DUV/visible rejection ratio (R_{250}/R_{420}, R_{250}/R_{390}) reaching 10^4–10^6. In addition, the responsivity was as high as 1.38–6.4 A W^{-1}, suggesting a photoconductive gain in the device. The working process of this type of device could be understood as follows (figure 9.11(d)). Under near UV illumination with energy of 3.0 eV, a small number of electron–hole pairs were produced in the a-IGZO layer. However, the number of photo electrons was also low yield because of the low absorption coefficient of a-ZGIO. When shone by DUV illumination with photon energy above 4.5 eV, which was higher than the wide bandgap semiconductor (Ta_2O_5 or Ga_2O_3), electron–hole pairs were produced on both the wide bandgap semiconductor and a-IGZO/a-ZITO. According to the work principle of semiconductor energy band theory and heterogeneous medium structure, electrons in the wide bandgap semiconductor would drift toward the a-GIZO/a-ZITO, while holes diffused to the valence band of a-GIZO/a-ZITO or were trapped in the wide bandgap semiconductor. As a consequence, the photogenerated carriers were collected by the source–drain electrodes, producing the photocurrents.

To further improve the sensitivity of such devices, some research teams have proposed a-IGZO ultraviolet phototransistors with double-stack dielectric. For instance, Chang *et al* have proposed a strategy to introduce ultrathin SiO_2 interlayer between Ga_2O_3 gate dielectric and a-IGZO thin film to reduce the leakage current and suppress the carriers injected into the channel. The dark current could be further reduced to as low as 10^{-11} A by employing a SiO_2 thin-film interlayer (figure 9.11(e)) [5]. The TFT with the Ga_2O_3/SiO_2 bilayer gate insulator exhibited typical transistor behavior with an I_{on}/I_{off} of 1.08×10^5, SS of 0.49 V/dec, and deep-UV/visible of 1.0×10^4. To illustrate how the electrons transmit, the energy band sketch of these structures at 2 V drain bias was given (figure 9.11(f)) [5]. Under the UV illumination with the above 4.9 eV of photon energy, electron–hole pairs were produced on Ga_2O_3 thin-film and accumulated at the edge of Ga_2O_3 under the action of electric field. The barrier height between Ga_2O_3 and SiO_2 was very high, and the tunneling probability of carriers in the direct tunneling mechanism was very low. However, under a drain bias 2 V, the SiO_2 layer forms a triangular barrier height. Furthermore, a large number of photogenerated carriers gathered on the interface of Ga_2O_3–SiO_2 and then entered the channel through trapezoidal barrier. This Fowler–Nordheim tunneling effect increased the tunneling probability of carriers.

In the past several years, organic–inorganic halide perovskites have drawn significant attention and emerged as one of the most exploited candidate materials

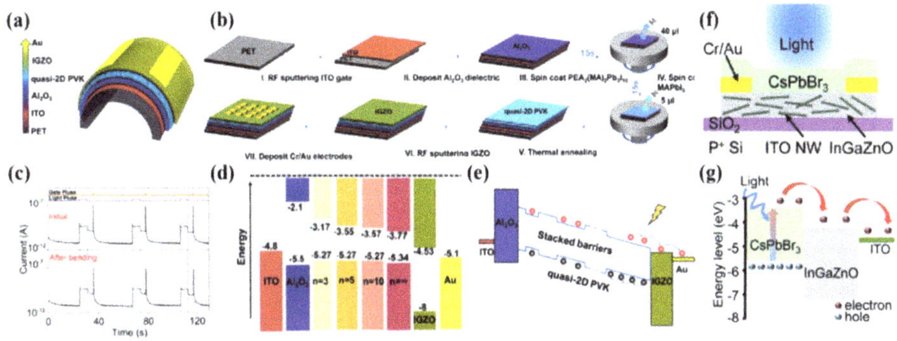

Figure 9.12. (a) Schematic illustration of 2D PVK/IGZO phototransistor; (b) the fabrication process of 2D PVK/IGZO phototransistor; (c) current–time curve of the device after 300 bending cycles; the schematics of the band alignment (d) and (e) charge transfer. [107] John Wiley & Sons. © (2019) WILEY-VCH Verlag GmbH & Co., KGaA, Weinheim. (f) Schematic illustration of $CsPbBr_3$/ITO NW/IGZO phototransistor; (g) energy band alignment and charge transfer of $CsPbBr_3$/ITO NW/IGZO phototransistor. [108] John Wiley & Sons. © (2020) WILEY-VCH Verlag GmbH & Co., KGaA, Weinheim.

for cost-effective and high-performance photovoltaics and optoelectronics. Their great promise in optoelectronics originates from the appealing electrical and optical properties, including long carrier lifetime (~270 ns) and carrier diffusion length (up to ~175 μm in single-crystals), high carrier mobility (~2320 $cm^2\ V^{-1}\ s^{-1}$), low exciton binding energy (~2 meV) as well as wide optical absorption across the UV–visible to NIR spectrum. Therefore, in the past decades, IGZO-based phototransistors with perovskites (PVKs) as absorbance layer have drawn increasing attention. Liao's group reported an ultrasensitive and highly stable photodetector with 2D-PVKs/a-IGZO hybrid heterostructure on flexible PET substrate (figure 9.12(a)) [107]. Figure 9.12(b) shows the detailed fabrication process. Briefly, 50 nm indium tin oxide (ITO) bottom gate electrode was deposited on PET flexible substrate by magnetron sputtering. Then, 100 nm thick Al_2O_3 dielectric layer was grown by atomic layer deposition (ALD), followed by spin coating of quasi-2D-PVKs. In this way, metal halides layers tended to extend parallel to the substrate, forming organic–inorganic alternating layers. As a result, this 2D PVKs film had a multiple quantum wells structure, in which the insulating organic layers played the barrier and the inorganic layers acted as well (figure 9.12(d)). In this case, as n value was adjusted, the band structure of 2D-quasi-PVKs thin film would be changed accordingly. As a result, photoexcited electrons can effectively transfer from 2D-quasi-PVKs thin film to IGZO layer (figure 9.12(e)). At the same time, the valence band maximum of all 2D-quasi-PVKs was lower than that of IGZO film, which facilitated the holes moving to the insulating layer. Further device performance analysis revealed that the quasi-2D-PVKs/IGZO heterojunction had an ultrahigh responsivity (3.1×10^5 A W^{-1}), outstanding specific detectivity (5.1×10^{16} J), wide-band optical response (457–1064 nm) and unique optoelectronic storage. These results signify the great potential of the quasi-2D-PVKs/IGZO phototransistor in the next generation of flexible optoelectronic devices applications. Later on, the same research group exhibited the structure in which the indium tin oxide nanowires (ITO

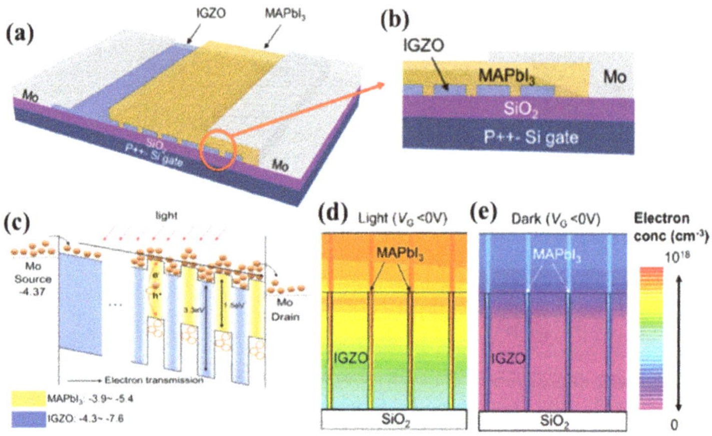

Figure 9.13. (a) Schematic illustration of $MAPbI_3$/IGZO phototransistor with nanostructured TFT; (b) cross-sectional schematic of $MAPbI_3$ and IGZO interface; (c) band alignment and charge transfer; (d), (e) TCAD simulation of $MAPbI_3$/IGZO phototransistor. Reprinted with permission from [109], Copyright (2019) American Chemical Society.

NWs) as electron pumps integrated into InGaZnO channel of the $CsPbBr_3$ thin film (figure 9.12(f)) [108]. As shown in figure 9.12(g), when the type II band alignment was illuminated by UV light, the photo-generated carriers at the interface of $CsPbBr_3$ layer and IGZO thin film could be separated effectively. Meanwhile, appropriate band alignment could enhance the mobility of collected carriers, leading to an exceptionally high responsivity of 4.9×10^6 A W^{-1}.

By integrating nanostructured $MAPbI_3$ with IGZO, some groups have achieved several hybrid phototransistors with responsivity of 10^5 A W^{-1} and high detectivity values up to 10^{16} Jones (figures 9.13(a) and (b)) [109]. Such a relatively high on-off ratio was associated with low dark current due to the low conductivity of $MAPbI_3$. Besides, the energy barrier in the heterostructure led to a very low off-current (~10^{-9} A) between the collector and the emitter, which was also the reason for the higher on-off ratio. As shown in figures 9.13(c)–(e), when illuminated, photo-generated carriers accumulated in the nanogaps with $MAPbI_3$, which induced high concentration of electron at the IGZO layer. By contrast, the electron accumulation in the IGZO layer is negligible in dark. In addition, the device exhibited broadband photoresponse from 457 to 1064 nm. Compared with thin film, perovskite nanostructures with higher crystallinity usually possessed much lower recombination rate of photo-generated carriers, and these structures were beneficial for photodetection application.

Apart from the perovskite/IGZO heterojunction, sensitive phototransistors could be achieved by choosing Gr (Gr)/IGZO hybrid heterojunction as building block. Typically, the Gr/IGZO based devices exhibited positive or negative photoresponses depending on the type of both Gr doping and photocarriers, and the responsivity achieved in these devices were tens of A W^{-1}. For example, a high I_{light}/I_{dark} ratio of up to 10^7 and long-lasting PPC with enduring lifetime up to years have been observed in Gr/amorphous IGZO hybrid [110]. The presence of Gr sheets could facilitate the dissociation of photoexcited electron–hole pairs generated within IGZO and the migration of the electrons from IGZO to Gr, thus reducing the carrier recombination. The fast transport of electrons through the Gr channel due to its high carrier mobility was helpful to achieve a large photocurrent. Additionally, holes had a great probability to be trapped and localized at deep level and/or defect trap centers, which prevented the recombination of holes with free electrons, causing an ultra-long recovery time of photocurrent. Gr QD, as a derivative of Gr layer, has been incorporated into IGZO-based phototransistors, which had a maximum responsivity of 897 A W^{-1} and a high I_{light}/I_{dark} ratio of up to 10^6. Under illumination, the photo-generated electrons in the QDs can easily transfer to the underlying IGZO and transport through the channel, with photo-generated holes trapped inside the QDs. These trapped holes could act as effective gate, giving rise to the negative shift of the threshold voltage of the IGZO phototransistor [111].

Although the a-ZIGO-based phototransistor with various absorption layer (perovskite, QD, organic) can achieve high gain and fast switching speed, its application in photoelectric sensor was limited due to the large leakage current. To reduce the leakage currents and improve the sensitivity and wide dynamic range of the device, Hwang *et al* proposed a phototransistor with p–n hetero-junction composed of functionalized sc-SWCNT and a-IGZO based on a flexible substrate (figure 9.14(a)) [112]. There were three advantages of the hybrid heterostructure: (1) good exciton dissociation of photo-generated electrons and the effective charge transfer; (2) relatively low recombination of functionalized sc-SWCNT region; (3) ease with the photo-induced charge modulation of a-IGZO region. Performance analysis revealed that the sc-SWCNT/a-IGZO device was capable of detecting broad ranges of spectrum (400–780 nm) with high responsivity

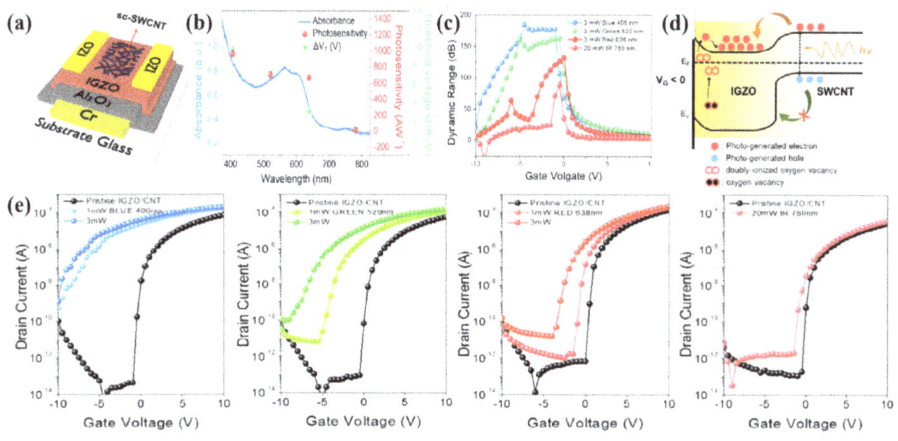

Figure 9.14. (a) Schematic illustration of sc-SWCNT/IGZO phototransistor with Al_2O_3 layer; (b) the absorbance, photosensitivity and threshold voltage shifts of sc-SWCNT/IGZO phototransistor; (c) dynamic range of sc-SWCNT/IGZO phototransistor under red, green, blue and IR illumination; (d) the schematics of the band alignment and charge transfer; (e) the sc-SWCNT/IGZO phototransistors under different light illumination at 10 V drain voltage. Reproduced from [112] with permission of The Royal Society of Chemistry.

and threshold voltage shift (figure 9.14(b)). Specifically, the heterogeneous phototransistors exhibited 100–180 dB of dynamic range in the operating V_{DS} and V_{GS} (figure 9.14(d)).

9.6 Conclusion and challenges

Metal oxide semiconductor based hybrid heterstructures have proved to be a promising application in assembling various high-performance phototransistors that could overcome intrinsic limitations of metal oxide semiconductors. In this chapter, we have presented a comprehensive overview of representative achievements in a number of phototransistors composed of ZnO, ternary metal oxide semiconductor and quaternary metal oxide based hybrid heterostructures. Particular attention has been given to some critical issues that pertain to the device design, device performance and physics, and the processing techniques for performance optimization. Through optimization of various approaches, these devices can demonstrate not only enhanced optoelectronic performance, but also some novel functionalities, which paves the way for them towards many practical applications.

The ZnO-based hybrid heterostructure phototransistors can be easily classified into the following three categories, including Gr/ZnO hybrid heterostructure, organic semiconductor/ZnO, and nanostructure/ZnO. Hybrid phototransistors composed of Gr and ZnO usually work on the photogating effect, where the ZnO is used as the light absorbing media and Gr function as the conducting channel for carrier transport and circulation. These devices can exhibit remarkable photoresponse characteristics with ultrahigh responsivities exceeding 10^9 A W^{-1}, unfortunately with slow response speed. In order to meet the requirement of some special application that needs broadband or spectral selective detection, the device geometries of these phototransistors have been further optimized. In general, the performance of some sensing materials and new materials, including organic and nanostructure materials, are employed to widen the application of the devices. For example, as the heterogeneity of nanostructured materials (NWs, QDs, 2D, semiconductors, etc) can greatly affect the uniformity of the device performance, more precise control of not only the morphology, hierarchical, crystallinity, and orientation assembly, but also physical and chemical properties, especially carrier transport characteristics, is highly demanding. So far, the phototransistors in reported literature can typically attain broadband or spectral selective photodetectors in terms of large photosensitivity (I_{light}/I_{dark} ratio), low dark current, respectable responsivity, high rejection ratio, sharp response cutoff, and relatively rapid response speed, as summarized in table 9.2.

Generally, multi-metal element oxides, such as ternary metal oxide film and quaternary metal oxide film, are obtained by doping zinc oxide (ZnO) with one or two cations. In addition to the similarity of optical properties with zinc oxide (ZnO), the bandgap of this amorphous thin film can be controlled varying the content of each component. However, it should be noted that oxide defects are very common in these oxide semiconductors. In some cases, the replacement of gallium (Ga) cations with some of the oxygen vacancies can weaken the PPC effect and increase the frame rate of the imaging device array. So far, a large number of perovskite nanostructure/multi-

Table 9.2. Summary of performance parameters of binary metal oxide-based phototransistors.

Device structure	I_{light}/I_{dark} ratio	R (A W^{-1})	D^* (Jones)	Response speed (s)	Ref.
Gr/ZnO QD	—	1.1×10^8	5.1×10^{13}	—	[21]
Gr/ZnO QD	1.7	9.9×10^8	7.5×10^{14}	5	[64]
Gr/ZnO QD	340	1000	—	—	[65]
Gr/ZnO NW	1.30	3×10^5	—	3.2	[54]
Gr/ZnO NW	1.25	2.5×10^6	—	10	[55]
Gr foam/ZnO NW	2	8	—	9.5	[56]
rGO-ZnONW	5	—	—	0.1	[61]
ZnO NW/PbS QD	—	0.15	9.9×10^8	—	[20]
ZnO/Au NP	6×10^8	1.51×10^5	2.05×10^{15}	—	[76]
$Al_2O_3/CdS/Al_2O_3/ZnO/Al_2O_3$	—	52.9	—	—	[14]
ZnOEP/ZnO	10^3	8.7×10^{-2}	—	—	[77]
Dye-sensitized (D 102 dye)/ZnO	10^6	10^4	—	—	[47]
Perovskite/SnO	2.7×10^3	1.83×10^3	2.11×10^{13}	—	[72]
Gr/PbSe QDs/TiO_2	10^4	—	3×10^{13}	50/750 ($\times 10^{-6}$)	[69]
$MAPbI_3$ QDs/TiO_2 NTs	—	0.2	3.8×10^{11}	7/4	[71]

element oxide hybrid phototransistors with impressive photoresponse performance have been realized. Multi-element oxide phototransistors based on $CsPbBr_3$ perovskite thin films or nanostructures exhibit responsivity values ranging from several to 10^6 A W^{-1}. Moreover, 3D $MAPbI_3$ perovskite thin films/multi-element oxide hybrid heterostructure phototransistors exhibit responsivity maximum values of several tens A W^{-1}. On the other hand, the phototransistors with 2D $(PEA)_2(MA)_{n-1}Pb_nI_{3n+1}$ perovskite as light absorbing layer have achieved peak responsivity as high as 10^5 A W^{-1} and detectivity as high as 10^{16} Jones at 457 nm illumination. However, devices with ultrahigh responsivities usually display very slow response speed. Very recently, flexible phototransistors based on perovskite/functional oxide material heterostructures have also been reported, which normally show excellent flexibility and remarkable mechanical stability and bending durability, suggesting great promise for flexible, stretchable, and bendable optoelectronic applications. To further enhance their performance parameters, in particular the responsivity and gain, metal oxide materials have been integrated with other functional materials including Gr, rGO, 2D TMDs, nanowires, colloidal QDs, organic to form hybrid heterostructure phototransistors. The important performance parameters of some representative hybrid phototransistors based on IGZO-based hybrid heterojunctions are summarized in table 9.3.

In spite of the above achievements, there are still many challenges that need to be tackled prior to the commercialization of oxide semiconductor–based heterostructure phototransistors. For example, perovskite/metal oxide semiconductors hybrid heterojunction phototransistors show low PPC compared with pure metal oxide

Table 9.3. Summary of performance parameters of IGZO-based phototransistors.

Device structure	Wavelength (nm)	I_{light}/I_{dark} ratio	R (A W^{-1})	D^* (Jones)	Rise/fall time (ms)	SS (V/dec.)	μ_{FE} (cm^2 V^{-1} s^{-1})	V_{th} (V)	Year	Ref.
Ta_2O_5/IGZO	250	10^6	6.4	—	—	0.32	—	0.6	2012	[5]
Ga_2O_3/SiO_2/IGZO	250	1.1×10^4	3.2	—	—	0.46	—	0.5	2013	[6]
Al_2O_3/IGZO/Al_2O_3	240	10^7	—	—	—	—	20.07	0.72	2016	[2]
Hydrogen-doped a-IGZO	400–700	3.85×10^6 at 532 nm	1932.6 at 532 nm	6.93×10^{11} at 532 nm	—	—	16.15	2.29	2018	[14]
SAL(IGZO)/IGZO	532	8.34×10^6	206	4.23×10^{11}	—	0.37	7.08	1.78	2019	[15]
MoS_2/IGZO	520	10^3	1.7	—	—	—	6.8–13.5	—	2015	[11]
Gr QD/IGZO	340	10^6	897	—	—	0.15	12.8	−0.1	2015	[25]
CdSe-QDs/IGZO	450	2.38×10^5	7×10^4	—	/30	0.39	31.7	−2.34	2014	[26]
(CdSe) QDs/IGZO	635	—	1.35×10^4	—	—	—	—	—	2015	[24]
CdSe-QDs/IGZO	638	—	8.3×10^3	4.2×10^{17}	270/90	0.226	22.1	—	2019	[31]
m-TiO_2/IGZO	635	1.04×10^5	178.66	1.25×10^{10}	/5	0.54	6.22	1.91	2020	[8]
Gr/IGZO film	350	2.0×10^7	22.8	—	—				2015	[111]
$MAPbI_3$ film/IGZO	385	10^6	0.025	10^{10}	40/100	1.33	12.9	−9.67	2017	[113]
$MAPbI_3$ film/IGZO	400–635	3.48×10^6 at 532 nm	61 at 532 nm	9.42×10^4 at 532 nm	680/600	—	—	—	2018	[28]
quasi-2D PVK film/IGZO	457–1064	3.3×10^5 at 457 nm	> 10^5 at 457 nm	5.1×10^{16} at 457 nm	—	—	18.6	—	2019	[107]
$CsPdBr_3$ NW/IGZO	470	1.2×10^4	3.794	—	2/96	—	—	—	2018	[114]
GNR/$CsPdBr_3$ QD/IGZO	512	$>10^3$	2.3×10^5	7.5×10^{14}	3.5/4.6	0.9	13	3	2018	[115]
$CsPdBr_3$/IGZO-TIO NW	457	3.4×10^4	4.9×10^6	7.6×10^{13}	450/550	—	—	—	2020	[108]
sc-SWCNT/IGZO	400–780	~10^8	9.6×10^2	4×10^{14}	—	—	~5	0.3	2017	[112]
PBDTT-DPP:PC_{61} BM/IGZO	400–800	6.6×10^6 at 532 nm	1.7×10^{12} at 532 nm	3.9×10^{12} at 532 nm	—	~0.42	~7.06	~1.51	2015	[116]

phototransistors. It is necessary to develop chemically stable perovskite materials, which allows one to fundamentally address this issue. According to previous reports, there are several useful approaches to improve long-term stability of perovskite materials: (i) to replace organic cations with inorganic ones such as cesium to produce all inorganic perovskites; (ii) to mix MA, FA, and cesium cations and iodine, bromine, and chlorine anions to produce perovskites with more thermally and structurally stable compositions; (iii) to use stoichiometry-tuned reduced-dimensionality (quasi-2D) perovskites as the light harvesting materials. Furthermore, metal oxide hybrid heterojunction could be manufactured by using all-solution-processes on flexible substrate to extend their application in flexible and wearable optoelectronics. To realize practical applications, besides the performance improvement, long-term stability and durability, environment-friendly and cost-effective processing, as well as large-scale production and integration are some of the most important issues that need to be tackled in future research.

References

[1] Sze S M and Ng K K 2007 *Physics of Semiconductor Devices* (New York: Wiley)

[2] Knobelspies S, Daus A, Cantarella G, Petti L, Münzenrieder N, Tröster G and Salvatore G A 2016 Flexible a-IGZO phototransistor for instantaneous and cumulative UV-exposure monitoring for skin health *Adv. Electron. Mater.* **2** 1600273

[3] Im H, Hong S, Lee Y, Lee H and Kim S 2019 A colorimetric multifunctional sensing method for structural-durability-health monitoring systems *Adv. Mater.* **31** 1807552

[4] Chiu C J, Shih S S, Weng W Y, Chang S J, Hung Z D and Tsai T Y 2012 Deep UV Ta_2O_5/zinc-indium-tin-oxide thin film photo-transistor *IEEE Photonics Technol. Lett.* **24** 1018–20

[5] Chang T H, Chiu C J, Weng W Y, Chang S J, Tsai T Y and Huang Z D 2012 High responsivity of amorphous indium gallium zinc oxide phototransistor with Ta_2O_5 gate dielectric *Appl. Phys. Lett.* **101** 261112

[6] Chang T H, Chiu C J, Chang S J, Tsai T Y, Yang T H, Huang Z D and Weng W Y 2013 Amorphous InGaZnO ultraviolet phototransistors with double-stack Ga_2O_3/SiO_2 dielectric *Appl. Phys. Lett.* **102** 221104

[7] Chang S J, Chang T H, Weng W Y, Chiu C J and Chang S P 2014 Amorphous InGaZnO ultraviolet phototransistors with a thin Ga_2O_3 layer *IEEE J. Sel. Top. Quantum Electron.* **20** 125–9

[8] Kim D, Kim Y G, Kang B H, Lee J H, Chung J and Kim H J 2020 Fabrication of indium gallium zinc oxide phototransistors via oxide-mesh insertion for visible light detection *J. Mater. Chem.* C **8** 165–72

[9] Wang C J, You H C, Ou J H, Chu Y Y and Ko F H 2020 Ultraviolet photodetecting and plasmon-to-electric conversion of controlled inkjet-printing thin-film transistors *Nanomaterials* **10** 458

[10] Williams K J, Nelson C A, Yan X, Li L S and Zhu X 2013 Hot electron injection from graphene quantum dots to TiO_2 *ACS Nano* **7** 1388–94

[11] Yan L, Du X, Liu C, Zhang S and Zhou H 2019 Narrow bandgap Pb–Sn perovskites/InGaZnO hybrid phototransistors for near-infrared detection *Phys. Status Solidi* a **216** 1900417

[12] Mottram A D, Lin Y H, Pattanasattayavong P, Zhao K, Amassian A and Anthopoulos T D 2016 Quasi two-dimensional dye-sensitized In_2O_3 phototransistors for ultrahigh responsivity and photosensitivity photodetector applications *ACS Appl. Mater. Interfaces* **8** 4894–902

[13] Kang B H, Kim W G, Chung J, Lee J H and Kim H J 2018 Simple hydrogen plasma doping process of amorphous indium gallium zinc oxide-based phototransistors for visible light detection *ACS Appl. Mater. Interfaces* **10** 7223–30

[14] Ahn C H, Kang W J, Kim Y K, Yun M G and Cho H K 2016 Highly repeatable and recoverable phototransistors based on multifunctional channels of photoactive CdS, fast charge transporting ZnO, and chemically durable Al_2O_3 layers *ACS Appl. Mater. Interfaces* **8** 15518–23

[15] Chung J, Tak Y J, Kim W G, Kang B H and Kim H J 2019 Artificially fabricated subgap states for visible-light absorption in indium–gallium–zinc oxide phototransistor with solution-processed oxide absorption layer *ACS Appl. Mater. Interfaces* **11** 38964–72

[16] Cheng C C, Zhan J Y, Liao Y M, Lin T Y, Hsieh Y P and Chen Y F 2016 Self-powered and broadband photodetectors based on graphene/ZnO/silicon triple junctions *Appl. Phys. Lett.* **109** 053501

[17] Yang H Y, Son D I, Kim T W, Lee J M and Park W I 2010 Enhancement of the photocurrent in ultraviolet photodetectors fabricated utilizing hybrid polymer-ZnO quantum dot nanocomposites due to an embedded graphene layer *Org. Electron.* **11** 1313–7

[18] Lee H, An N, Jeong S, Kang S, Kwon S, Lee J and Lee S 2017 Strong dependence of photocurrent on illumination-light colors for ZnO/graphene Schottky diode *Curr. Appl. Phys.* **17** 552–6

[19] Shao D, Sun X, Xie M, Sun H, Lu F, George S M and Sawyer S 2013 ZnO quantum dots-graphene composite for efficient ultraviolet sensing *Mater. Lett.* **112** 165–8

[20] Zheng Z, Gan L, Zhang J, Zhuge F and Zhai T 2017 An enhanced UV–Vis–NIR an d flexible photodetector based on electrospun ZnO nanowire array/PbS quantum dots film heterostructure *Adv. Sci.* **4** 1600316

[21] Shao D, Gao J, Chow P, Sun H, Xin G, Sharma P and Sawyer S 2015 Organic–inorganic heterointerfaces for ultrasensitive detection of ultraviolet light *Nano Lett.* **15** 3787–92

[22] Shin S W, Lee K H, Park J S and Kang S J 2015 Highly transparent, visible-light photodetector based on oxide semiconductors and quantum dots *ACS Appl. Mater. Interfaces* **7** 19666–71

[23] Wang S S, Rong R, Jin L Z, Yang S S, Li Y X, Zhang H and Huang W 2018 Variable segment roles: modulation of the packing modes, nanocrystal morphologies and optical emissions *Nanoscale* **10** 13310–4

[24] Zhu H, Liu A, Shan F, Yang W, Zhang W, Li D and Liu J 2016 One-step synthesis of graphene quantum dots from defective CVD graphene and their application in IGZO UV thin film phototransistor *Carbon* **100** 201–7

[25] Bonaccorso F, Sun Z, Hasan T and Ferrari A C 2010 Graphene photonics and optoelectronics *Nat. Photon.* **4** 611

[26] Liu X, Yang X, Liu M, Tao Z, Dai Q, Wei L and Nathan A 2014 Photo-modulated thin film transistor based on dynamic charge transfer within quantum-dots-InGaZnO interface *Appl. Phys. Lett.* **104** 113501

[27] Kim J, Kwon S M, Kang Y K, Kim Y H, Lee M J, Han K and Park S K 2019 A skin-like two-dimensionally pixelized full-color quantum dot photodetector *Sci. Adv.* **5** eaax8801

[28] Tak Y J, Kim D J, Kim W G, Lee J H, Kim S J, Kim J H and Kim H J 2018 Boosting visible light absorption of metal-oxide-based phototransistors via heterogeneous In–Ga–Zn–O and $CH_3NH_3PbI_3$ films *ACS Appl. Mater. Interfaces* **10** 12854–61

[29] Baskoutas S and Bester G 2011 Transition in the optical emission polarization of ZnO nanorods *J. Phys. Chem.* C **115** 15862–7

[30] Comini E, Faglia G, Sberveglieri G, Pan Z and Wang Z L 2002 Stable and highly sensitive gas sensors based on semiconducting oxide nanobelts *Appl. Phys. Lett.* **81** 1869–71

[31] Wang Z L 2004 Nanostructures of zinc oxide *Mater. Today* **7** 26–33

[32] Nie B, Hu J G, Luo L B, Xie C, Zeng L H, Lv P and Yu Y Q 2013 Monolayer graphene film on ZnO nanorod array for high-performance schottky junction ultraviolet photodetectors *Small* **9** 2872–9

[33] Kong W Y, Wu G A, Wang K Y, Zhang T F, Zou Y F, Wang D D and Luo L B 2016 Graphene-β-Ga_2O_3 heterojunction for highly sensitive deep UV photodetector application *Adv. Mater.* **28** 10725–31

[34] Wang M Z, Liang F X, Nie B, Zeng L H, Zheng L X, Lv P and Luo L B 2013 TiO_2 nanotube array/monolayer graphene film Schottky junction ultraviolet light photodetectors *Part. Part. Syst. Charact.* **30** 630–6

[35] Luo L B, Xie C, Wang X H, Yu Y Q, Wu C Y, Hu H and Jie J S 2014 Surface plasmon resonance enhanced highly efficient planar silicon solar cell *Nano Energy* **9** 112–20

[36] Kim Y J, Yoon A, Kim M, Yi G C and Liu C 2011 Hydrothermally grown ZnO nanostructures on few-layer graphene sheets *Nanotechnology* **22** 245603

[37] Liu H, Sun Q, Xing J, Zheng Z, Zhang Z, Lü Z and Zhao K 2015 Fast and enhanced broadband photoresponse of a ZnO nanowire array/reduced graphene oxide film hybrid photodetector from the visible to the near-infrared range *ACS Appl. Mater. Interfaces* **7** 6645–51

[38] Aziz N, Nishiyama T, Rusli N, Mahmood M, Yasui K and Hashim A 2014 Seedless growth of zinc oxide flower-shaped structures on multilayer graphene by electrochemical deposition *Nanoscale Res. Lett.* **9** 337

[39] Xu C, Kim B S, Lee J H, Kim M, Hwang S W, Choi B L and Whang D 2012 Seed-free electrochemical growth of ZnO nanotube arrays on single-layer graphene *Mater. Lett.* **72** 25–8

[40] Ali A, Ali K, Yang Y J, Jo J and Choi K H 2014 Rapid fabrication of graphene/ZnO composite thin film *Japan J. Appl. Phys.* **53** 05HA01

[41] Zeng Y, Zhao Y and Jiang Y 2014 Investigate the interface structure and growth mechanism of high quality ZnO films grown on multilayer graphene layers *Appl. Surf. Sci.* **301** 391–5

[42] Lin J, Penchev M, Wang G, Paul R K, Zhong J, Jing X and Ozkan C S 2010 Heterogeneous graphene nanostructures: ZnO nanostructures grown on large-area graphene layers *Small* **6** 2448–52

[43] Cheng S H, Yeh Y C, Lu M L, Chen C W and Chen Y F 2012 Enhancement of laser action in ZnO nanorods assisted by surface plasmon resonance of reduced graphene oxide nanoflakes *Opt. Express* **20** A799–805

[44] Luo L B, Yang X B, Liang F X, Jie J S, Li Q, Zhu Z F and Wang L 2012 Transparent and flexible selenium nanobelt-based visible light photodetector *CrystEngComm* **14** 1942–7

[45] Noh Y Y, Kim D Y and Yase K 2005 Highly sensitive thin-film organic phototransistors: effect of wavelength of light source on device performance *J. Appl. Phys.* **98** 074505

[46] Nam S, Seo J, Park S, Lee S, Jeong J, Lee H and Kim Y 2013 Hybrid phototransistors based on bulk heterojunction films of poly (3-hexylthiophene) and zinc oxide nanoparticle *ACS Appl. Mater. Interfaces* **5** 1385–92

[47] Pattanasattayavong P, Rossbauer S, Thomas S, Labram J G, Snaith H J and Anthopoulos T D 2012 Solution-processed dye-sensitized ZnO phototransistors with extremely high photoresponsivity *J. Appl. Phys.* **112** 074507

[48] Janotti A and Van de Walle C G 2008 Sources of unintentional conductivity in InN *Appl. Phys. Lett.* **92** 032104

[49] Zeng H, Cai W and Liu P *et al* 2008 ZnO-based hollow nanoparticles by selective etching: elimination and reconstruction of metal-semiconductor interface, improvement of blue emission and photocatalysis *ACS Nano* **2** 1661

[50] Schmidt-Mende L and MacManus-Driscoll J L 2007 ZnO–nanostructures, defects, and devices *Mater. Today* **10** 40–8

[51] McCluskey M D and Jokela S J 2009 Defects in ZnO *J. Appl. Phys.* **106** 10

[52] Konstantatos G, Badioli M, Gaudreau L, Osmond J, Bernechea M, De Arquer F P G and Koppens F H 2012 Hybrid graphene–quantum dot phototransistors with ultrahigh gain *Nat. Nanotechnol.* **7** 363–8

[53] Kumar M, Noh Y, Polat K, Okyay A K and Lee D 2015 Metal–semiconductor–metal UV photodetector based on Ga doped ZnO/graphene interface *Solid State Commun.* **224** 37–40

[54] Dang V Q, Trung T Q, Kim D I, Duy L T, Hwang B U, Lee D W and Lee N E 2015 Ultrahigh responsivity in graphene–ZnO nanorod hybrid UV photodetector *Small* **11** 3054–65

[55] Dang V Q, Trung T Q, Duy L T, Kim B Y, Siddiqui S, Lee W and Lee N E 2015 High-performance flexible ultraviolet (UV) phototransistor using hybrid channel of vertical ZnO nanorods and graphene *ACS Appl. Mater. Interfaces* **7** 11032–40

[56] Boruah B D, Mukherjee A, Sridhar S and Misra A 2015 Highly dense ZnO nanowires grown on graphene foam for ultraviolet photodetection *ACS Appl. Mater. Interfaces* **7** 10606–11

[57] Biroju R K, Tilak N, Rajender G, Dhara S and Giri P K 2015 Catalyst free growth of ZnO nanowires on graphene and graphene oxide and its enhanced photoluminescence and photoresponse *Nanotechnology* **26** 145601

[58] Yang H, Li J, Yu D and Li L 2016 Seed/catalyst free growth and self-powered photo-response of vertically aligned ZnO nanorods on reduced graphene oxide nanosheets *Cryst. Growth Des.* **16** 4831–8

[59] Zhan Z, Zheng L, Pan Y, Sun G and Li L 2012 Self-powered, visible-light photodetector based on thermally reduced graphene oxide–ZnO (rGO–ZnO) hybrid nanostructure *J. Mater. Chem.* **22** 2589–95

[60] Ick Son D, Yeon Yang H, Whan Kim T and Il Park W 2013 Photoresponse mechanisms of ultraviolet photodetectors based on colloidal ZnO quantum dot-graphene nanocomposites *Appl. Phys. Lett.* **102** 021105

[61] Son D I, Yang H Y, Kim T W and Park W I 2015 Transparent and flexible ultraviolet photodetectors based on colloidal ZnO quantum dot/graphene nanocomposites formed on poly (ethylene terephthalate) substrates *Composites* B **69** 154–8

[62] Guo W, Xu S, Wu Z, Wang N, Loy M M T and Du S 2013 Oxygen-assisted charge transfer between ZnO quantum dots and graphene *Small* **9** 3031–6

[63] Chen J H, Jang C, Adam S, Fuhrer M S, Williams E D and Ishigami M 2008 Charged-impurity scattering in graphene *Nat. Phys.* **4** 377–81

[64] Gong M, Liu Q, Cook B, Kattel B, Wang T, Chan W L and Wu J Z 2017 All-printable ZnO quantum dots/graphene van der Waals heterostructures for ultrasensitive detection of ultraviolet light *ACS Nano* **11** 4114–23

[65] Liu Q, Gong M, Cook B, Ewing D, Casper M, Stramel A and Wu J 2017 Transfer-free and printable graphene/ZnO-nanoparticle nanohybrid photodetectors with high performance *J. Mater. Chem.* C **5** 6427–32

[66] Choi H, Seo S, Lee J H, Hong S H, Song J, Kim S and Lee S 2018 Solution-processed ZnO/ SnO_2 bilayer ultraviolet phototransistor with high responsivity and fast photoresponse *J. Mater. Chem.* C **6** 6014–22

[67] Liu H Y and Huang R C 2018 A study of anatase TiO_2-based thin film phototransistors by non-vacuum thin film deposition method *IEEE Trans. Electron Dev.* **65** 2517–24

[68] Manga K K, Wang S, Jaiswal M, Bao Q and Loh K P 2010 High-gain graphene-titanium oxide photoconductor made from inkjet printable ionic solution *Adv. Mater.* **22** 5265–70

[69] Manga K K, Wang J, Lin M, Zhang J, Nesladek M, Nalla V and Loh K P 2012 High-performance broadband photodetector using solution-processible PbSe–TiO_2–graphene hybrids *Adv. Mater.* **24** 1697–702

[70] Zheng K, Meng F, Jiang L, Yan Q, Hng H H and Chen X 2013 Visible photoresponse of single-layer graphene decorated with TiO_2 nanoparticles *Small* **9** 2076–80

[71] Zheng Z, Zhuge F, Wang Y, Zhang J, Gan L, Zhou X and Zhai T 2017 Decorating perovskite quantum dots in TiO_2 nanotubes array for broadband response photodetector *Adv. Funct. Mater.* **27** 1703115

[72] Guan X, Wang Z, Hota M K, Alshareef H N and Wu T 2019 P-type SnO thin film phototransistor with perovskite-mediated photogating *Adv. Electron. Mater.* **5** 1800538

[73] Han G, Cao S, Yang Q, Yang W, Guo T and Chen H 2018 High-performance all-solution-processed flexible photodetector arrays based on ultrashort channel amorphous oxide semiconductor transistors *ACS Appl. Mater. Interfaces* **10** 40631–40

[74] Ge M, Cao C, Huang J, Li S, Chen Z, Zhang K Q and Lai Y 2016 A review of one-dimensional TiO_2 nanostructured materials for environmental and energy applications *J. Mater. Chem.* A **4** 6772–801

[75] Roy P, Berger S and Schmuki P 2011 TiO_2 nanotubes: synthesis and applications *Angew. Chem. Int. Ed.* **50** 2904–39

[76] Jin Z, Gao L, Zhou Q and Wang J 2014 High-performance flexible ultraviolet photoconductors based on solution-processed ultrathin ZnO/Au nanoparticle composite films *Sci. Rep.* **4** 1–8

[77] Wang F X, Lin J, Liu Y Q, Wu H D and Pan G B 2014 Facile fabrication and enhanced photoresponse of hybrid heterojunction films of znoep nanowire network/zno nanoparticles *Org. Electron.* **15** 844–9

[78] Ogo Y, Nomura K, Yanagi H, Kamiya T, Hirano M and Hosono H 2008 Amorphous Sn–Ga–Zn–O channel thin-film transistors *Phys. Status Solidi* a **205** 1920–4

[79] Song J I, Park J S, Kim H, Heo Y W, Lee J H, Kim J J and Choi B D 2007 Transparent amorphous indium zinc oxide thin-film transistors fabricated at room temperature *Appl. Phys. Lett.* **90** 022106

[80] Han D, Chen Z, Cong Y, Yu W, Zhang X and Wang Y 2016 High-performance flexible tin-zinc-oxide thin-film transistors fabricated on plastic substrates *IEEE Trans. Electron Dev.* **63** 3360–3

[81] Nomura K, Ohta H, Takagi A, Kamiya T, Hirano M and Hosono H 2004 Room-temperature fabrication of transparent flexible thin-film transistors using amorphous oxide semiconductors *Nature* **432** 488–92
[82] Jeon S, Song I, Lee S, Ryu B, Ahn S E, Lee E, Kim Y, Nathan A, Robertson J and Chung U I 2014 Origin of high photoconductive gain in fully transparent heterojunction nanocrystalline oxide image sensors and interconnects *Adv. Mater.* **26** 7102–9
[83] Li J Y, Chang S P, Hsu M H and Chang S J 2017 Photo-electrical properties of MgZnO thin-film transistors with high-k dielectrics *IEEE Photon. Techn. Lett.* **30** 59–62
[84] Yang C P, Chang S P, Chang S J, Chen S X, Hsu M H, Tung W J and Weng W Y 2018 Bandgap engineered ultraviolet photodetectors with gallium-zinc-oxide via co-sputtering method *ECS J. Solid State Sci. Technol.* **7** Q3083–8
[85] Hosono H 2006 Ionic amorphous oxide semiconductors: material design, carrier transport, and device application *J. Non-Cryst. Solids* **352** 851–8
[86] Feng L, Yu G, Li X, Zhang J, Ye Z and Lu J 2016 Solution processed amorphous ZnSnO thin-film phototransistors *IEEE Trans. Electron Dev.* **64** 206–10
[87] Lee H M, Jeong H J, Ok K C, Rim Y S and Park J S 2018 Near-infrared photoresponsivity of ZnON thin-film transistor with energy band-tunable semiconductor *ACS Appl. Mater. Interfaces* **10** 30541–7
[88] Rim Y S, Ok K C, Yang Y M, Chen H, Bae S H, Wang C and Yang Y 2016 Boosting responsivity of organic–metal oxynitride hybrid heterointerface phototransistor *ACS Appl. Mater. Interfaces* **8** 14665–70
[89] Cho S W, Kim Y B, Jung S H, Baek S K, Kim J S, Lee M and Kim Y H 2018 All-solution-processed metal oxide/chalcogenide hybrid full-color phototransistors with multistacked functional layers and composition-gradient heterointerface *Adv. Opt. Mater.* **6** 1800196
[90] Sharma A, Chourasia N K, Sugathan A, Kumar Y, Jit S, Liu S W and Pal B N 2018 Solution processed Li_5AlO_4 dielectric for low voltage transistor fabrication and its application in metal oxide/quantum dot heterojunction phototransistors *J. Mater. Chem.* C **6** 790–8
[91] Seung Rim Y, Jeong W, Du Ahn B and Jae Kim H 2013 Defect reduction in photon-accelerated negative bias instability of InGaZnO thin-film transistors by high-pressure water vapor annealing *Appl. Phys. Lett.* **102** 143503
[92] Jeon K, Kim C, Song I, Park J, Kim S, Kim S and Kim D H 2008 Modeling of amorphous InGaZnO thin-film transistors based on the density of states extracted from the optical response of capacitance-voltage characteristics *Appl. Phys. Lett.* **93** 182102
[93] Lee E, Benayad A, Shin T, Lee H, Ko D S, Kim T S and Park G S 2014 Nanocrystalline ZnON; high mobility and low band gap semiconductor material for high performance switch transistor and image sensor application *Sci. Rep.* **4** 4948
[94] Ok K C, Jeong H J, Kim H S and Park J S 2014 Highly stable ZnON thin-film transistors with high field-effect mobility exceeding 50 cm^2/Vs *IEEE Electron Dev. Lett.* **36** 38–40
[95] Kim H, Wu Z, Eedugurala N, Azoulay J D and Ng T N 2019 Solution-processed phototransistors combining organic absorber and charge transporting oxide for visible to infrared light detection *ACS Appl. Mater. Interfaces* **11** 36880–5
[96] Shin K Y, Tak Y J, Kim W G, Hong S and Kim H J 2017 Improvement of electrical characteristics and stability of amorphous indium gallium zinc oxide thin film transistors using nitrocellulose passivation layer *ACS Appl. Mater. Interfaces* **9** 13278–85

[97] Lee H, Chang K S, Tak Y J, Jung T S, Park J W, Kim W G and Kim H J 2017 Low-temperature activation under 150 °C for amorphous IGZO TFTs using voltage bias *J. Inf. Disp.* **18** 131–5

[98] Lim Y, Hwang N, Lee J, Lee S and Yi M 2019 Improved electrical performance of SiO_2-doped indium zinc oxide thin-film transistor *J. Nanosci. Nanotechnol.* **19** 1470–3

[99] Yoon J, Bae G Y, Yoo S, Yoo J I, You N H, Hong W K and Ko H C 2020 Deep-ultraviolet sensing characteristics of transparent and flexible IGZO thin film transistors *J. Alloys Compd.* **817** 152788

[100] Tao Z, Liu X, Lei W and Chen J 2018 High sensitive solar blind phototransistor based on ZnO nanorods/IGZO heterostructure annealed by laser *Mater. Lett.* **228** 451–5

[101] Wang J, Chen Y, Kong L A, Fu Y, Gao Y and Sun J 2018 Deep-ultraviolet-triggered neuromorphic functions in In-Zn-O phototransistors *Appl. Phys. Lett.* **113** 151101

[102] Yang J, Kwak H, Lee Y, Kang Y S, Cho M H, Cho J H and Kim H 2016 MoS_2–InGaZnO heterojunction phototransistors with broad spectral responsivity *ACS Appl. Mater. Interfaces* **8** 8576–82

[103] Barbagiovanni E G, Strano V, Franzò G and Mirabella S 2016 The role of Zn vacancies in UV sensing with ZnO nanorods *Appl. Phys. Lett.* **109** 143104

[104] Wu C, He Z, Lu J, Dai J and Xu C 2016 Energy band modification for UV photoresponse improvement in a ZnO microrod-quantum dot structure *RSC Adv.* **6** 687–91

[105] Ma M, Kang Z, Liao Q, Zhang Q, Gao F, Zhao X and Zhang Y 2018 Development, applications, and future directions of triboelectric nanogenerators *Nano Res.* **11** 2951–69

[106] Yoo H, Kim W G, Kang B H, Kim H T, Park J W, Choi D H and Kim H J 2020 High photosensitive indium–gallium–zinc oxide thin-film phototransistor with a selenium capping layer for visible-light detection *ACS Appl. Mater. Interfaces* **12** 10673–80

[107] Wei S, Wang F, Zou X, Wang L, Liu C, Liu X and Liao L 2019 Flexible quasi-2D perovskite/IGZO phototransistors for ultrasensitive and broadband photodetection *Adv. Mater.* 1907527

[108] Hou Y, Wang L, Zou X, Wan D, Liu C, Li G and Liao L 2020 Substantially improving device performance of all-inorganic perovskite-based phototransistors via indium tin oxide nanowire incorporation *Small* **16** 1905609

[109] Huang K *et al* 2019 Nanostructured high-performance thin-film transistors and phototransistors fabricated by a high-yield and versatile near-field nanolithography strategy *ACS Nano* **13** 6618–30

[110] Dai M K, Liou Y R, Lian J T, Lin T Y and Chen Y F 2015 Multifunctionality of giant and long-lasting persistent photoconductivity: semiconductor–conductor transition in graphene nanosheets and amorphous InGaZnO hybrids *ACS Photon.* **2** 1057–64

[111] Pei Z, Lai H C, Wang J Y, Chiang W H and Chen C H 2014 High-responsivity and high-sensitivity graphene dots/a-IGZO thin-film phototransistor *IEEE Electron Device Lett.* **36** 44–6

[112] Hwang I, Kim J, Lee M, Lee M W, Kim H J, Kwon H I and Park S K 2017 Wide-spectral/dynamic-range skin-compatible phototransistors enabled by floated heterojunction structures with surface functionalized SWCNTs and amorphous oxide semiconductors *Nanoscale* **9** 16711–21

[113] Du S, Li G, Cao X, Wang Y, Lu H, Zhang S and Zhou H 2017 Oxide semiconductor phototransistor with organolead trihalide perovskite light absorber *Adv. Electron. Mater.* **3** 1600325

[114] Sun M, Fang Q, Zhang Z, Xie D, Sun Y, Xu J and Zhang Y 2018 All-inorganic perovskite nanowires–ingazno heterojunction for high-performance ultraviolet–visible photodetectors *ACS Appl. Mater. Interfaces* **10** 7231–8

[115] Liu X, Kuang W, Ni H, Tao Z, Huang Q, Chen J, Liu Q, Chang J and Lei W 2018 A highly sensitive and fast graphene nanoribbon/$CsPbBr_3$ quantum dot phototransistor with enhanced vertical metal oxide heterostructures *Nanoscale* **10** 10182–9

[116] Rim Y S, Yang Y, Bae S H, Chen H, Li C, Goorsky M S and Yang Y 2015 Ultrahigh and broad spectral photodetectivity of an organic–inorganic hybrid phototransistor for flexible electronics *Adv. Mater.* **27** 6885–91

IOP Publishing

Semiconducting Metal Oxide Thin-Film Transistors

Ye Zhou

Chapter 10

Semiconductor metal oxide thin film transistor for non-volatile memory

Zhi Ye and Ning Zhang

In the scaling down of the Si MOSFET according to Moore's law, the physical limitation of the feature size will appear after a few years. In order to keep semiconductor industrial and commercial development in the more-than-Moore's period, it is significant to improve the density of the integration circuits. Vertical architecture would be one of the feasible methods. The electronic memory also obeys the law. Faster and higher density memory is always our pursuit. In this chapter, we propose non-volatile memory based on a non-silicon metal oxide thin film transistor. Owing to the unique properties of the metal oxide thin film transistor (TFT), such as low process temperature, wide bandgap, low off current and so on, the TFT based memory could be a promising candidate in terms of 3D or lager-area glass-substrate memory. Various TFT based memories have been investigated, and the mechanism of the hysteresis of these devices has been discussed in detail. The floating gate TFT based memory with different charge-trap layers attracts most researchers' attention. So far, ZnO is still considered as the better charge trapping material in metal oxide TFT due to its good compatibility and charge capture ability. Although nanocrystal charge-trap memory, ferroelectric memory and resistive memory based on metal oxide TFTs exhibit memory characteristics, there is a long way to achieve their practical application. The design and realization of the TFT based memories' array and corresponding circuits on panel or 3D non-volatile memory will be the developing trend in the future.

10.1 Overview

10.1.1 Introduction

Memory is a semiconductor device that can store digital information and is widely used in many digital products such as computers and mobile phones. As a memory element, it stores information through '1' and '0' in the binary code. According to

doi:10.1088/978-0-7503-2556-1ch10

whether the information can continue to be saved after power off, there are two kinds of memory: non-volatile memory (NVM) and volatile memory (VM). Figure 10.1 shows the basic classification of the semiconductor memories. Among them, typical representatives of the semiconductor VM include static random memory (SRAM) and dynamic random memory (DRAM), and flash memory in NVM, etc.

Nowadays, non-volatile memory has occupied a very important position in daily life due to its high storage density, fast programming and erasing speed, low power consumption, and low manufacturing cost. Among them, flash memory is one of the most extensively used NVMs, and has been closely related to our lives.

The concept of non-volatile floating gate memory was first proposed by Sze and Kahng in 1967 [1]. Then in 1984, Masuoka *et al* designed the fundamental structure of the flash memory cell based on the non-volatile semiconductor memory with floating gate structure, which is a massively used flash memory technology till now [2].

Figure 10.2 shows a schematic diagram of the structure of a traditional floating gate flash memory. The basic structure is divided into three layers from the substrate upwards: the tunneling layer, the charge-trap layer (CTL) and the blocking layer. Compared to the traditional field effect transistor, a floating gate is added in the flash memory for charge storing. For a normal transistor, the source/drain (S/D) current (I_D) is controlled by the gate voltage, and the threshold voltage (V_{th}) is fixed. If a floating gate is inserted between the tunneling and the blocking layer of the gate insulator of this transistor, the threshold voltage will be not fixed. When a larger voltage is applied to the gate electrode, the electrons transporting in channel may obtain energy and go through the tunneling layer into the floating gate layer and be stored therein. Due to the barrier of the tunneling layer and the blocking layer, these

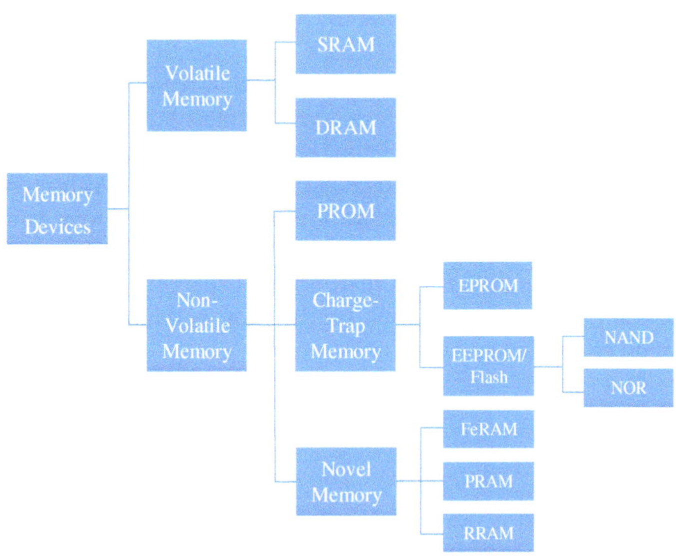

Figure 10.1. Classification of the semiconductor memories.

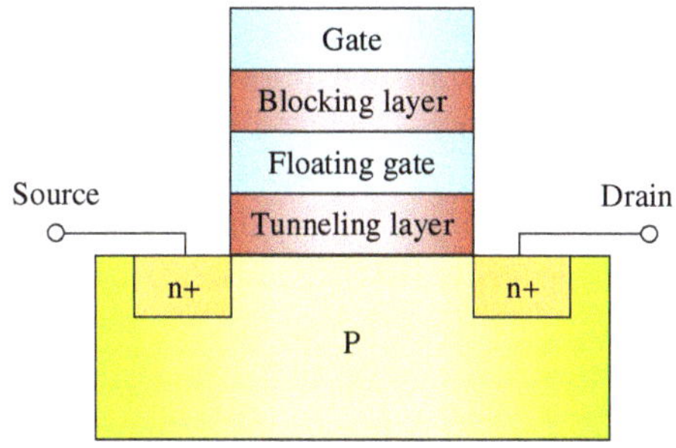

Figure 10.2. Cross section diagram of the floating gate memory.

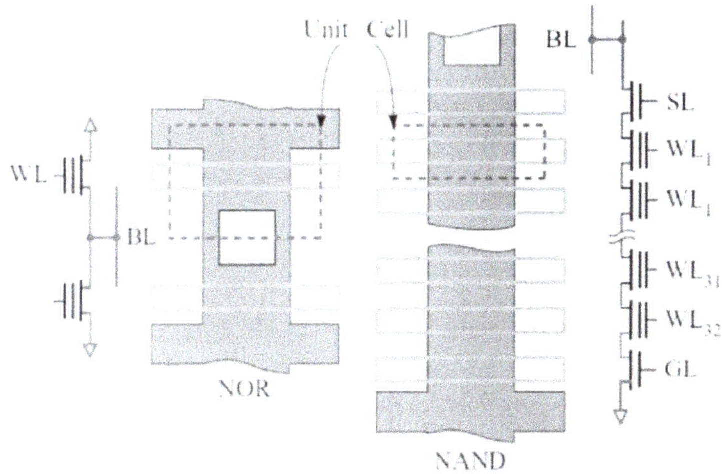

Figure 10.3. The layout and circuit schematic of NOR and NAND memory cells.

electrons will be trapped in the floating gate without being lost after removing the gate voltage. These electrons generate a negative electric field to channel, and the electric field will affect the control of the gate voltage on the I_D, thereby the threshold voltage shifting positive. In contrast, when a larger negative gate voltage is applied, the electrons will be pushed into channel from the floating layer. The threshold voltage shifts negative. By changing the threshold voltage of the device, we can achieve 0 and 1 signal conversion to complete the storage of information.

10.1.2 Classification of floating gate memory

There are different types of floating gate memory, which are mainly distinguished according to their data erasing and writing mechanisms. Among them, NOR memory and NAND memory are the two most common types [3]. As shown in figure 10.3, in the NOR memory, a single memory cell is directly connected to the address line (word line) and data line (bit line) of the memory array. In the NAND memory, several memory cells form a whole block, and are connected to the address lines and the data line through the selective transistor. Therefore, compared with

NOR memory, NAND memory has a higher integration, but in terms of reading speed, NOR memory is dominant. In comparison, the read data bandwidth of the NAND memory is only 18–25 Mb s^{-1}, while the read data bandwidth of the NOR memory can be more than 100 Mb s^{-1}.

The NOR memory uses hot electron injection and Fowler–Nordheim (F–N) tunneling to perform the program/erase (P/E) operation [4]. The erasing mechanism adopted by the NOR memory causes the operation speed of the device to be very slow. According to the characteristics of the fast reading speed and slow writing speed, the NOR memory is used in a programmable read-only memory that often needs to read data but rarely modifies data, such as the program codes in mobile phones, computers and other electronic devices. Because the NAND memory transistors are connected in series with the address lines and the data line, there is smaller drain current, resulting in a relatively slow reading speed. Although the data writing speed is still slow for the single transistor in NAND memory, the NAND memory can realize the parallel operation of the data writing, and hence achieve a good write bandwidth. In addition, the erasing time of the entire data block of the NAND memory is quite slow to 2 ms [5, 6]. These characteristics indicate that the NAND memory is suitable for large-capacity data storage related fields, such as music, video and large file.

10.1.3 The main technical indicators of floating gate memory

There are many parameters to measure the performance of charge trapped memory. The main parameters are: memory window, program/erase speed, endurance and data retention characteristics.

(1) Memory window

In a charge trap memory, the charge storage characteristics of the device can be expressed by using a threshold voltage window or a capacitance–voltage (*C–V*) curve window [7]. The difference of the threshold voltage of the device in the two states of writing and erasing is the memory window. The *C–V* window is equivalent to the threshold voltage window in measuring the charge storage capacity of the device. The only difference between the two is the measurement method. During the measurement of the *I–V* curve, the difference of the voltage between the program and erase states of the device is the memory window. The larger the memory window, the stronger the charge trap capacity of the device. The memory window is determined by the density of charges stored in the CTL, and the thickness of the tunneling layer and barrier layer.

(2) Program/erase (P/E) speed

The programming/erasing speeds of the charge trapping memory (CTM) reflect the data transfer speed of the device. In general, the P/E speed of the CTM is related to the device structure and operating voltage of the device.

In the device structure design, the thickness of the tunneling layer is the most important factor to affect the P/E speed. The thinner the thickness of the tunneling layer, the faster the P/E speed of the device. Because the width of the barrier that the electron needs to cross is reduced, the tunneling process becomes easy to occur. However, the thinner tunneling layer will lead to the deterioration of the device data retention. Therefore, in the choice of the thickness of the tunneling layer, there is also a balance between P/E speed and retention capability. In addition, the material of the tunneling layer will also affect the P/E speed of the device. For high-k materials with the same thickness, the voltage drop on the barrier layer is small, and the P/E speed will increase accordingly.

In terms of operating voltage, the P/E speed of the device becomes faster as the operating voltage increases. The larger the operating voltage, the stronger the driving force of the electric field received by the electrons. So the shorter the tunneling time is required, leading to the faster P/E speed. However, as the operating voltage increases, the power consumption of the device increases accordingly. Therefore, P/E speed and device power consumption need to be comprehensively considered in adjusting the operating voltage.

(3) Endurance

Endurance refers to whether the memory window of the flash remains unchanged after multiple P/E operations. After many P/E operating cycles, some leakage paths will be generated in the tunneling layer, resulting in the easy loss of electrons in the CTL. In this case, the memory window is reduced gradually and eventually the device fails to the store the data. The endurance of the device is greatly affected by the quality of the tunneling layer. Therefore, the tunneling layer with few defects and good interface quality can effectively ensure the good endurance.

(4) Retention

The data retention characteristic refers to the time that the charges stored in the charge trapping memory can be retained without applying a voltage. In order to meet the practical demand, it is generally required that the data retention characteristics must achieve 10 years or more.

10.2 Silicon-based TFT floating gate memory

10.2.1 SONOS structure

In the early days, from the first generation α-Si TFT to p-Si TFT, flash memory was mainly based on silicon technology. The typical representative among them is the Si/oxide/nitride/oxide/Si (SONOS) memory. As shown in figure 10.4, the SONOS type memory is developed from the metal–nitride–oxide–silicon (MNOS) structure [8–10] and its basic structure is poly-Si/oxide/nitride/oxide/Si [11].

At present, SONOS memory usually uses Si_3N_4 material as the charge trapping layer because it can be well compatible with traditional processes. The top silicon oxide in the SONOS structure can be used to block the charges injecting from gate electrodes to silicon nitride. It can be programmed in a low voltage, because the thinner charge-trap layer could be realized in the SONOS structure. Meanwhile, the

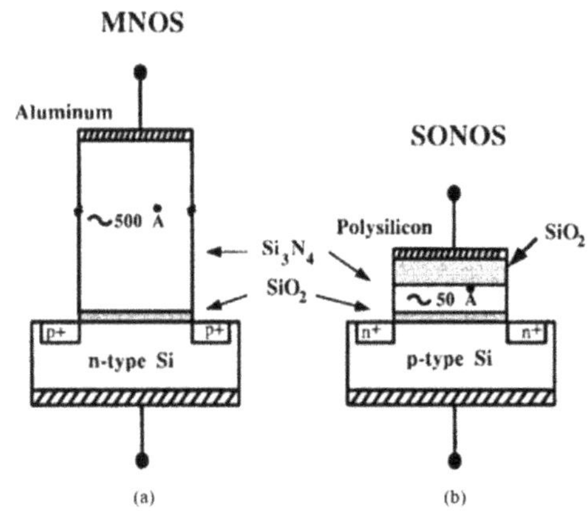

Figure 10.4. The conventional device structure: (a) MNOS, (b) SONOS. © (1997) IEEE. Reprinted, with permission, from [11].

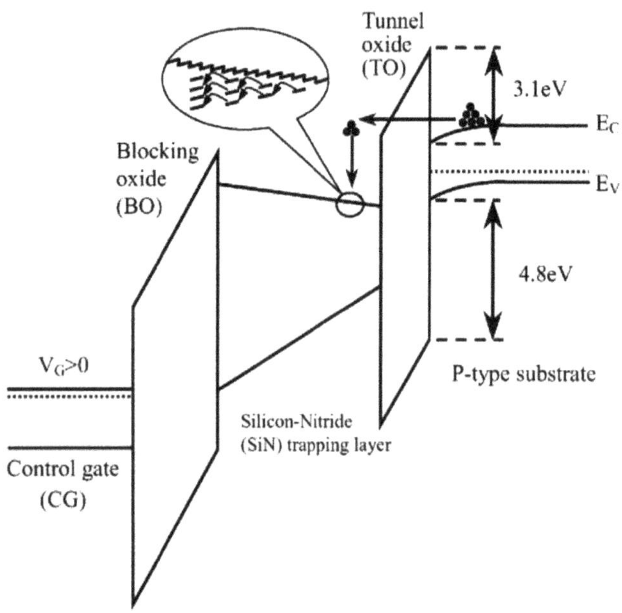

Figure 10.5. The grading energy bands of SiN_x inserted as charge trapping layer. © (2004) IEEE. Reprinted, with permission, from [12].

memory retention and fatigue properties have been improved due to the protection of the two oxide layers.

Thus, SONOS flash memory has been extensively researched. The researchers have reported better performance by optimizing materials and structures. Chen *et al* proposed a technology to adjust the bandgap of the nitride by changing the Si and N concentration [12]. A grading band energy structure is realized in the nitride charge-trap layer from Si-rich to N-rich, as shown in figure 10.5. This structure device has lower trapping efficiency but wide memory window at the range of the low speed programming, as shown in figure 10.6.

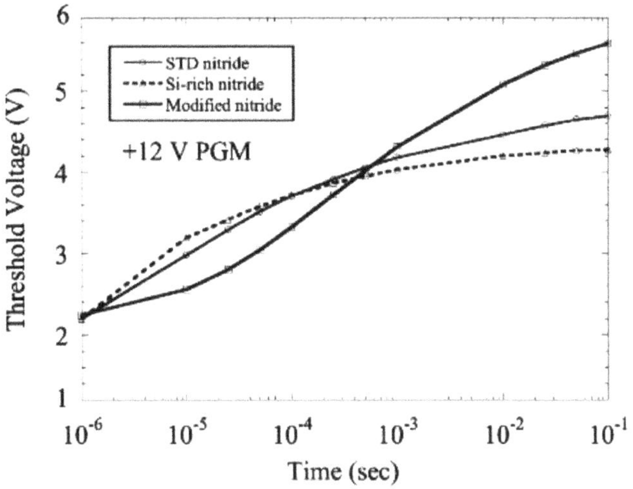

Figure 10.6. Threshold voltage shift after +12 V program for SONOS device. © (2004) IEEE. Reprinted, with permission, from [12].

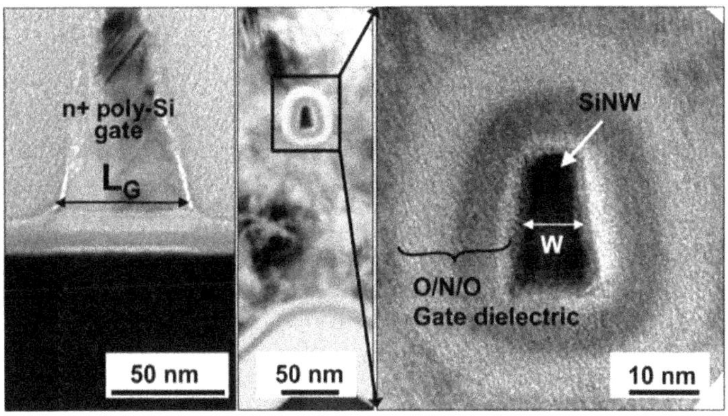

Figure 10.7. The TEM photos of the ONO junction-less transistors. © (2011) IEEE. Reprinted, with permission, from [15].

With the continuous development of technology, the device size becomes smaller and smaller. When the device size is reduced to 40 nm, Moore's law is no longer applicable because this size is close to the limit size that the floating gate type memory can reach [13]. Therefore, if you want to develop a new type of flash memory, it must make a breakthrough in the device structure [14].

Choi *et al* use junction-less transistors to reduce the scale-limitation in floating gate memory. It demonstrates the junction-less transistor with W/L ratio of 10 nm/50 nm [15]. The junction-less transistor consists of the N or P heavy homogeneously doped Si nanowires (SiNW) with a gate electrode. Compared to traditional transistors, it is easily fabricated and free from S/D junctions. These inherent advantages make the proposed idea feasible for the application of a flash based on the junction-less transistor [16, 17]. Figure 10.7 shows the horizontal and vertical transmission electron microscopy (TEM) images of the junction-less transistor.

The transfer curves exhibit well P/E performance level with an on/off ratio of $>10^5$ and good gate controllability. It also has 1.9 V threshold voltage at V_D values to 0.05 V and 1 V for the junction-less device with the ONO structure. The SONOS memory cell based on junction-less transistor with fast P/E behaviors and long retention tine has been demonstrated. The memory based on junction-less transistor has the potential to be the ideal choice for the future 3D NAND Flash under the 20 nm node.

The Si_3N_4 material is most commonly served as a charge-trap layer material in SONOS memory, but the defect density of Si_3N_4 is not very high. Although new processes and structures can be used to optimize flash memory performance, with the development of storage technology, the nature of the material itself has made it unable to meet the application requirements. In order to enhance the data storage density and the device data retention performance, the use of high dielectric materials as the charge storage layer is a viable solution, such as HfO_2, Al_2O_3, and ZrO_2 [18–20].

10.2.2 SOHOS structure

Here, high permittivity materials are used as charge-trap layers instead of silicon nitride, which is so called the Si/oxide/high-k/oxide/Si (SOHOS) structure [21, 22]. Compared with the traditional SONOS devices, this structure has excellent charge storage capacity, faster programming speed and less over erasure problem at low voltage. However, the charge retention capacity of SOHOs is worse than that of SONOS. The combination of the Al_2O_3 and HfO_2 could obtain better memory performance, such as P/E speed and data retention. The reason is that the alumina charge-trap layer behaves with good retention characteristics but slow P/E speed. Different trapping materials (nitride, Al_2O_3, HfO_2 and HfAlO) are fabricated and compared as shown in figure 10.8.

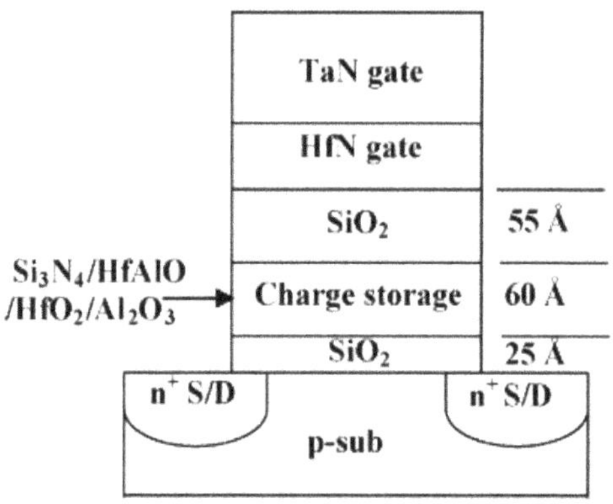

Figure 10.8. Cross-section of the SOHOS and SONOS transistor structures. © (2006) IEEE. Reprinted, with permission, from [18].

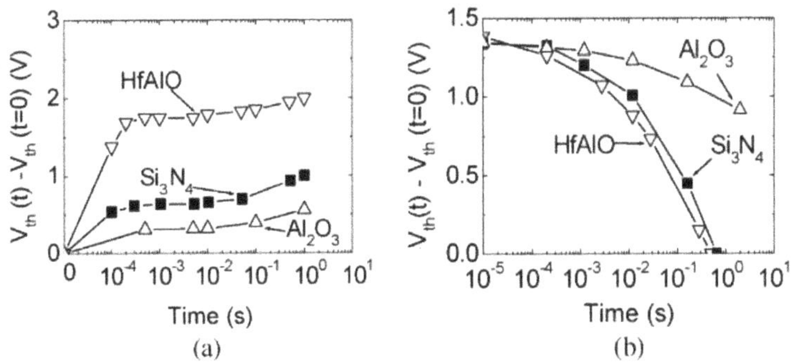

Figure 10.9. (a) Programming, and (b) erasing properties of the floating gate devices with nitride, HfAlO and alumina charge-trap materials. © (2006) IEEE. Reprinted, with permission, from [18].

Compared with traditional SONOS flash memory, SOHOS flash memory shows faster P/E speed (see figure 10.9), with great date retention and P/E cycling endurance properties.

10.3 Metal oxide TFT based charge-trap memory

10.3.1 Mechanism of the electrical hysteresis

As people continue improving the requirements of the charge-trap memory and more and more in-depth understanding of polysilicon materials, it shows that polysilicon is not an ideal channel material for TFT memory. Due to many grain boundaries in the polysilicon channels, the trapped electrons will be induced into inverse charges, which would cause the open voltage shift. In addition, the TFT memory based on the polysilicon channel will show severe threshold voltage fluctuation after being subjected to the bias stress of the gate, thereby affecting the reliability of the device. The forbidden band width of polysilicon is in the range of 1–2 eV and the mobility is less than 1 $cm^2\ V^{-1}\ s^{-1}$.

In this way, oxide semiconductors have gradually attracted people's attention. The oxide semiconductors mainly include zinc oxide (ZnO) and amorphous gallium–indium–zinc oxide (IGZO), and their bandgap are generally 3–4 eV, and the carrier mobility can reach 10–40 $cm^2\ V^{-1}\ s^{-1}$ [23–25]. Therefore, the new oxide semiconductor materials have larger forbidden band widths and higher carrier mobility than traditional semiconductors. This allows devices that use oxide semiconductors as channels to exhibit better electrical stability and higher breakdown voltage.

In particular, ZnO can not only be used as a semiconductor channel of memory, but also as a charge-trap layer to capture the electron. In some articles, the mechanism of the electrical hysteresis of the metal oxide TFT is revealed [26].

Figures 10.10(a) and (b) show the cross-section schematics and micrograph of passivated zinc oxide TFTs. The S/D of the self-align structure TFTs was formed by argon, boron, or hydrogen doped ZnO thin films. The ZnO channel layer was treated with the appropriate number of hydrogen or fluorine to passivate the inherent defects [27–29]. Figure 10.10(c) shows the transfer curves of the passivated

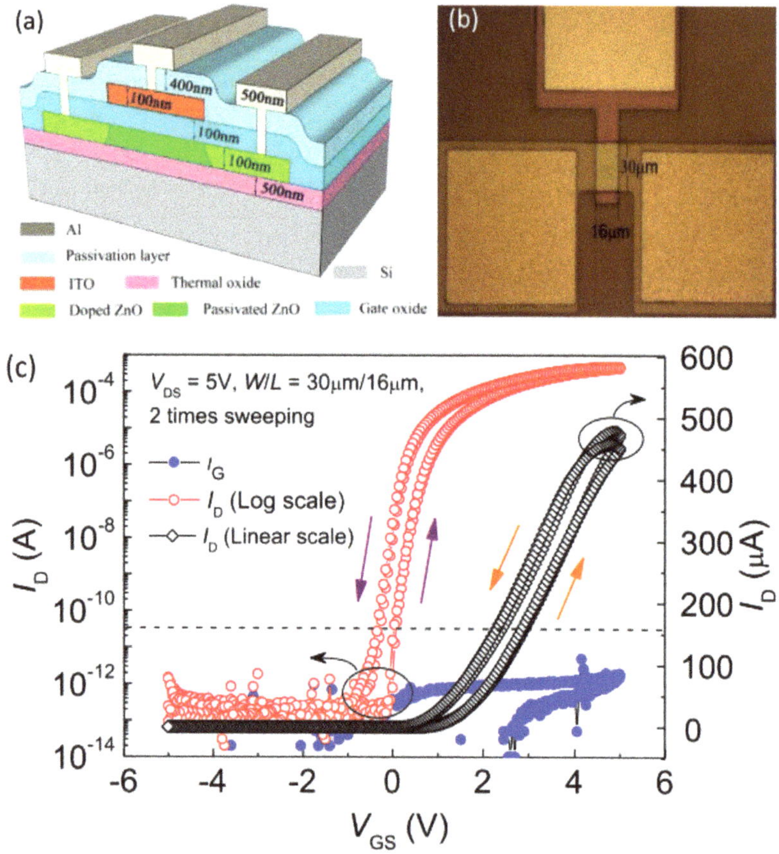

Figure 10.10. (a) Schematic cross-section, (b) micrograph and (c) transfer curves of the self-align zinc oxide TFTs. © (2017) IEEE. Reprinted, with permission, from [26].

ZnO TFTs tested in dark with sweeping gate voltage from −5 V to +5 V. It exhibits positive V_{th}, low gate leakage current, high field effect mobility and high ON/OFF drain current ratio. The counter-clockwise hysteresis was also observed in the I–V characteristics. Since this case always appeared metal oxide TFTs, the mechanism and origin of the hysteresis should be determined and analyzed.

Firstly, two kinds of hysteresis directions of the $I_D V_{GS}$ are always found in field effect transistors: clockwise and counter-clockwise. For the clockwise hysteresis, charge trapping in gate oxide, interface or channel is the main reason. As seen in figure 10.11(a), V_{th0} denotes the origin V_{th} when V_{GS} scans forward from −5 to +5 V. With the increase of V_{GS}, many carriers are captured by the defects close to the interface of channel and dielectric. When V_{GS} scans backward from +5 to −5 V, the V_{th} shifts positive (V_{th1}), leading to clockwise hysteresis. The fixed electrons provide a negative electrical field that a part of V_{GS} must be used to overcome. For the counter-clockwise hysteresis, mobile charge and electrical dipole are the probable factors. Figures 10.11(b) and (c) show the energy diagram of these two cases. When

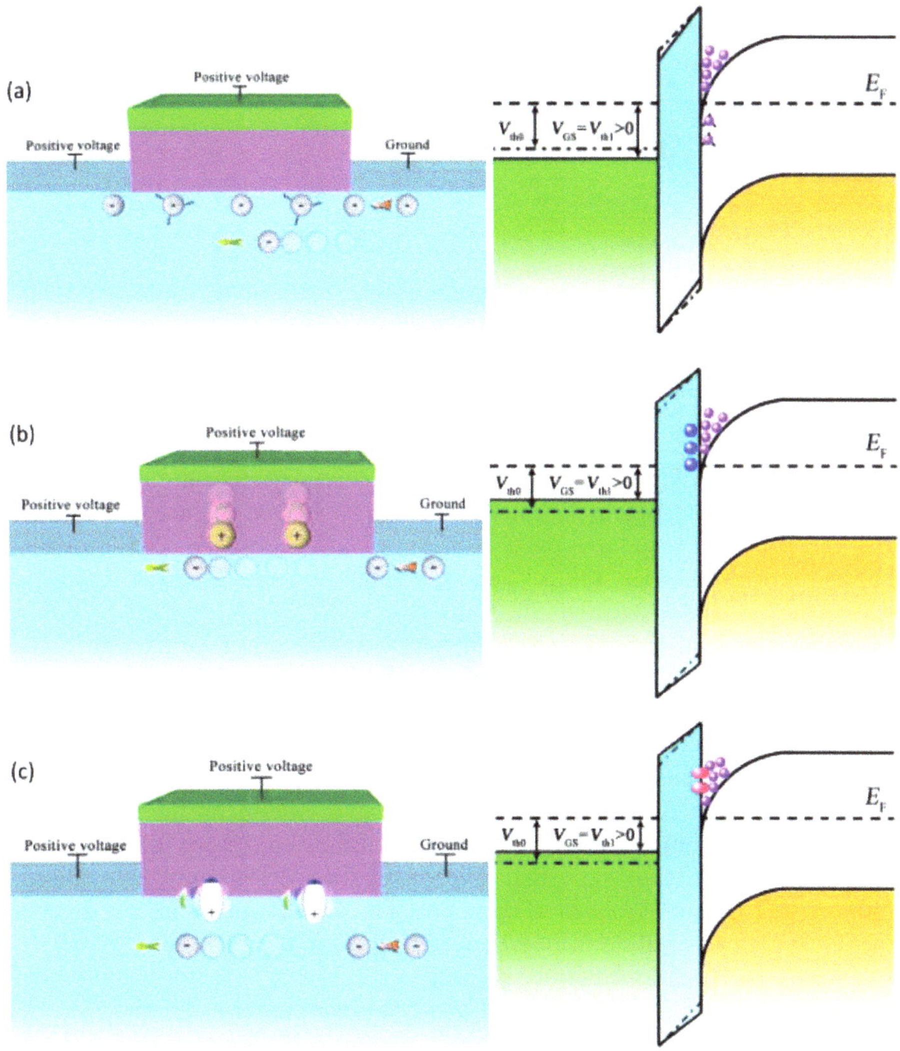

Figure 10.11. Schematic model and energy band drawing of the hysteresis caused by (a) charge trap, (b) mobile charge, and (c) electrical dipole. © (2017) IEEE. Reprinted, with permission, from [26].

V_{GS} sweeps from −5 to +5 V, mobile charge in dielectric will be pushed far away from the gate electrode. Meanwhile, electrical dipole turns over complying with the gate electrical field. When V_{GS} sweeps from +5 to −5 V, the V_{th1} is smaller than the V_{th0}, since the positive electrical field formed by these charge could give extra field, reducing flat band energy. The counter-clockwise hysteresis shown in figure 10.10(b) is consistent with the assumption that there is a dielectric moving charge or electric dipole.

Secondly, the cause of the hysteresis phenomenon of ZnO thin film transistors is studied experimentally. Figure 10.12(a) shows the effect of $I_D V_{GS}$ of the naked

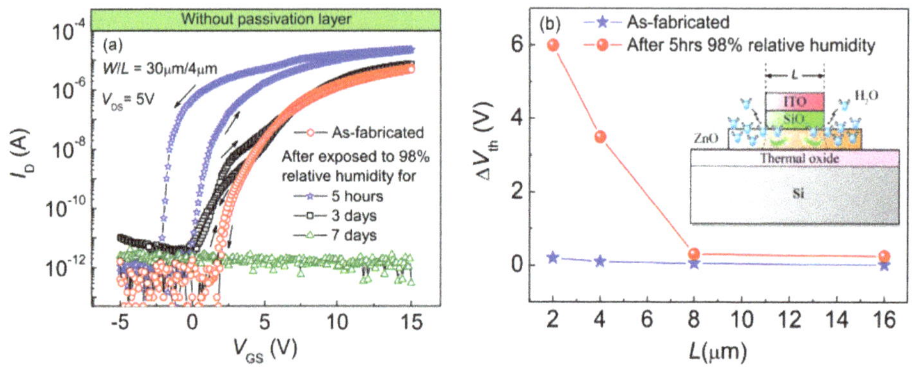

Figure 10.12. (a) The $I_D V_{GS}$ of the device under moist environment for different times. (b) Effect of ΔV_{th} of the device exposed to 98%RH on the channel length. © (2017) IEEE. Reprinted, with permission, from [26].

device under moist environment of 98%RH at RT on exposure duration. The W/L ratio of the testing transistor is 30 μm/4 μm. The ΔV_{th} is used as the evaluation of the shift of the V_{th}.

For as-fabricated device, the ΔV_{th} is very small and the direction is clockwise, indicating little charge traps in the transistor. After being exposed to a humid environment for 5 h, the ΔV_{th} and the direction becomes larger and counter-clockwise, respectively. Figure 10.12(b) shows the dependence of V_{th} on L which V_{th} will increase sharply when L decreases below 8 μm. The case can directly prove that H_2O diffuses from sidewall to center of the channel dominating the mechanism. The device with shorter channel is easily diffused by water, which has larger ΔV_{th}. The results show that the counter-clockwise hysteresis is caused by the water induced dipoles. It is noted from figure 10.12(a) that the subthreshold slope (SS) of the device decreases after 5-h steam exposing. The passivation effect may be realized by decomposing H_2O into H ions or –OH bonds. The defects in channel are passivated by these impurities, resulting in negative shift of IV and improvement of SS.

The mechanism mentioned above provides a theoretical basis for the non-volatile memory based on the TFT. For the charge-trapping memory, the direction of the hysteresis is clockwise, since the electrons are trapped in the interface between the channel and the gate oxide. In [30], the scientists demonstrate a flash fabricated in a single ALD process. The ZnO thin film is fabricated in ALD as both the channel and the CTL. Figure 10.13(a) depicts a cross-sectional view of the structure of a full ALD memory cell. Figure 10.13(b) shows a TEM image of a memory device based on ZnO-TFT.

Figure 10.14(a) shows the energy band diagram of the fabricated structure. Direct tunneling and F–N tunneling models are used in the whole programming and erasing cycle, which makes electrons tunnel from the channel to the CTL. The I_D–V_{GS} characteristics of the programming and erasing states are exhibited in figure 10.14(b), which shows a memory window of 2.12 V. The V_{th} migration obtained by TCAD is in good agreement with V_{th} migration in experiment.

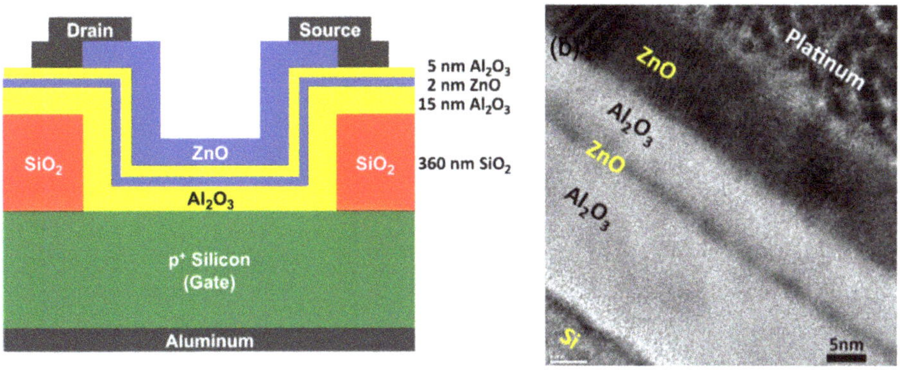

Figure 10.13. Cross sectional (left) schematic and (right) TEM image of the ALD ZnO TFT based memory. © (2012) IEEE. Reprinted, with permission, from [30].

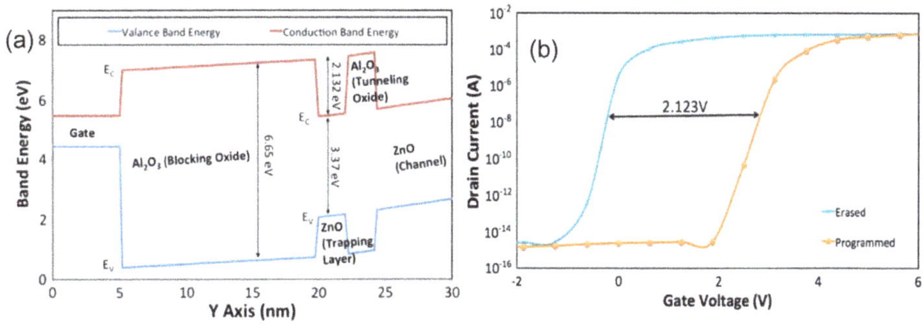

Figure 10.14. (a) Simulated energy bands of the gate-oxide-channel structure. (b) Calculated I_D–V_{GS} after P/E operation. © (2012) IEEE. Reprinted, with permission, from [30].

10.3.2 Metal oxide TFT based floating gate memory

In recent years, with the information explosion and the advent of big data era, the demands of NVM devices increased rapidly. Among them, the charge-trapping memory based on metal oxide thin film transistor stands out from the novel memory devices due to its excellent electrical characteristics especially for large area electronics [6–8]. For the charge-trapping memory, the charge-trapping layer plays an important role in capturing charges from the channel due to the F–N tunneling effect. Therefore, the selection of charge the materials of CTL is crucial to enhance the memory characteristic [31]. Here, the floating gate memory with metal oxide TFTs is discussed according to the different charge-trap materials as follows.

• *ZnO charge trapping layer* [32–35]

As mentioned above, Oruc *et al* demonstrates a memory device with ZnO films both as channel and CTL in a single ALD process. Firstly, a 15 nm thickness Al_2O_3 blocking layer is deposited followed by a 2 nm thickness ZnO CTL, a 5 nm thickness Al_2O_3 tunneling layer, and an 11 nm thickness ZnO channel. To check the memory

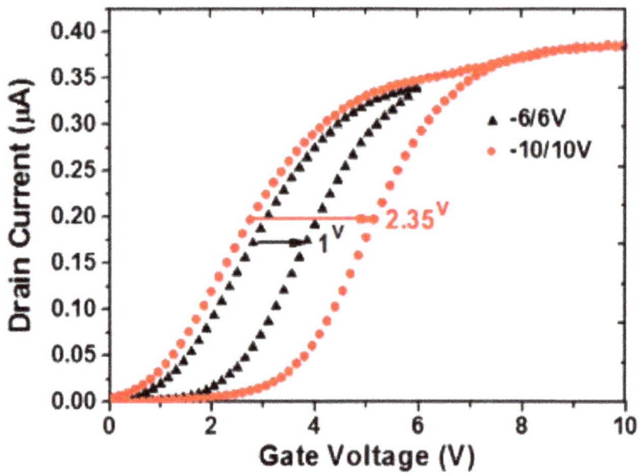

Figure 10.15. The transfer curves tested under different V_{GS} sweeping ranges. © (2012) IEEE. Reprinted, with permission, from [30].

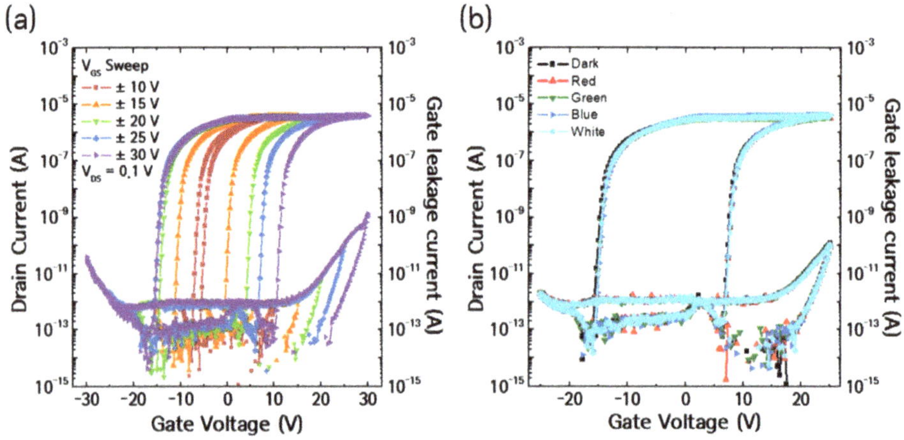

Figure 10.16. The hysteresis of the transfer curves under (a) different V_{GS} sweeping ranges and (b) different light wavelengths. © (2015) IEEE. Reprinted, with permission, from [36].

characteristic of floating gate thin film transistor, the hysteresis transfer curve was measured. Figure 10.15 shows the typical hysteresis phenomenon of ± 6 and ± 10 V voltage scanning, thus verifying the storage effect. It is noted that there is a 2.35 V memory window which is swept from −10 to 10 V. The control sample without ZnO CTL shows a hysteresis of <0.6 V due to unexpected charge trapping in the interface between the channel and the gate oxide.

The transparent charge-trapping memory TFTs with a top-gate structure consisted of IGZO channel and ZnO CTL was proposed [36]. A tri-layered stack structure was consisted of 5 nm Al_2O_3 as blocking layer 30 nm ZnO as CTL, and 3 nm Al_2O_3 layers as tunneling layer, which were consecutively deposited by ALD. As shown in figures 10.16 and 10.17, the device has outstanding memory

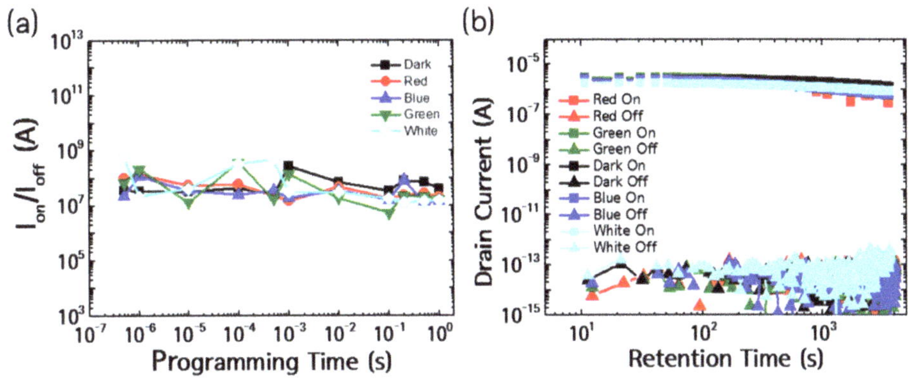

Figure 10.17. (a) The P/E speed and (b) the retention characteristics of the memory under different light illumination conditions. © (2015) IEEE. Reprinted, with permission, from [36].

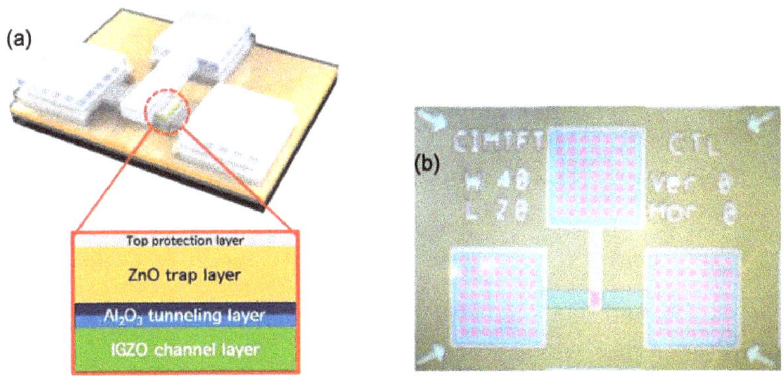

Figure 10.18. (a) Cross sectional schematic diagram, (b) microscope photo of CTMs. © (2016) IEEE. Reprinted, with permission, from [37].

performances including a large memory window of 25.8 V, high programming speed, and long retention lifetime. In addition, compared to other works, this device has strong light stability which only has a small threshold voltage drift of 0.4 V under a blue wavelength after 10 min.

A flexible CTM based on TFT using all oxide layers was fabricated on PEN substrate [37]. 10-nm Al_2O_3, 30-nm ZnO, and 3-nm Al_2O_3 layers were successively deposited by ALD as the tunneling, charge-trapping, and protection layers, respectively. All process environments were controlled below 180 °C. And the hybrid barrier layer is introduced to improve the surface roughness. All stacks of memory devices are oxide films, including IGZO channel, ZnO CTL and Al_2O_3 tunneling layer and ITO transparent electrode. Among them, multi-layer tunneling layer is designed to improve the stability of the device, as shown in figure 10.18. The experimental results are shown in figure 10.19. The memory has excellent memory characteristics including large memory window of 25 V, fast programming speed and long retention time. It shows that this work has great potential in the field of transparent electronics and flexible electronics.

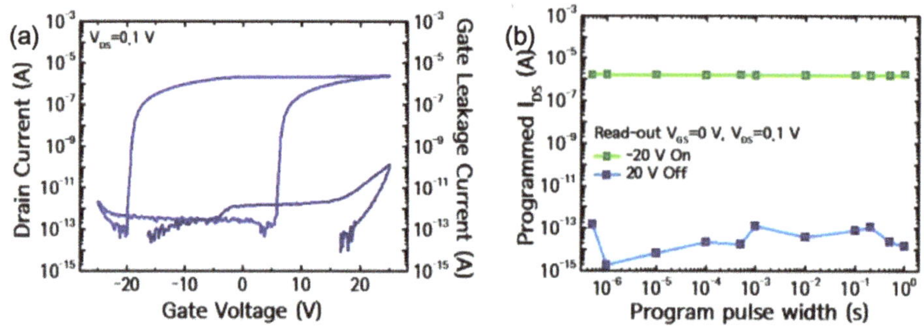

Figure 10.19. (a) I_D–V_{GS} memory window and (b) P/E speed of flexible memory. © (2016) IEEE. Reprinted, with permission, from [37].

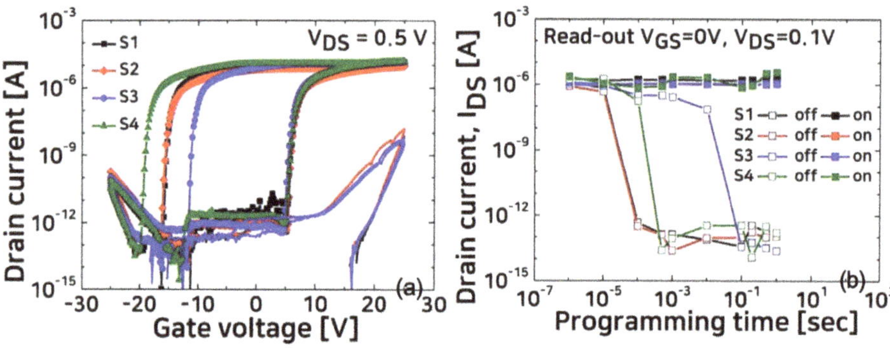

Figure 10.20. (a) Comparison of I_D–V_{GS} curves and (b) P/E speeds of different samples using hafnium oxide/zinc oxide stack charge-trap layer. © (2019) IEEE. Reprinted, with permission, from [38].

A charge trap memory based on TFT with IGZO active channel is fabricated by using HfO_2/ZnO stacked CTL. High-k materials may be reliable candidates for the CTL of oxide CTM TFT because of their many advantages such as high dielectric constant and high charge-trapping efficiency due to their inherent defects. In addition, it has been reported that the programming speed and memory window are improved by introducing a novel high-k charge trap layer. There have been many works introducing the high-k materials as charge-trapping layers such as Al_2O_3, Ta_2O_5, HfO_2, and so on. To study the effect of the thickness of HfO_2 layers inserted between ZnO layers as charge-trapping layer of stacked structure on device characteristics, HfO_2 thin films with thickness of 2 nm were embedded once (S1), twice (S2) and four times (S3), and 4 nm thickness HfO_2 (S4) layer was introduced twice between ZnO layers [38]. As shown in figure 10.20, TFTs using a stacked charge-trapping layer with a 4 nm thickness HfO_2 layer exhibit good storage performances, including a large memory window of 25 V and a fast programming/erasing speed because of a sufficient thickness to provide a large number of holes to capture the electrons. However, a relatively narrow memory window of 16 V and a slower programming speed of 100 ms are obtained for a memory device using a charge-trapping layer with four stacked 2 nm HfO_2 layers. The 2-nm-thick HfO_2 layer is only used as a dielectric film for TFT because the thickness of the film is so thin that the capture of electron charge is inactivated. This structure shows that the characteristics of memory device with different charge trapping layer structures is caused by the design change of the CTL structure when the interface quality inside the gate stack is prepared.

- *IGZO charge trapping layer*

The NVM with ZnO thin film as a charge-trap layer shows exciting characteristics for the flash memory applications. However, ZnO materials also have some defects, such as poor thermal stability, which limits the reliability of devices at high temperature. In addition, ZnO is usually prepared by ALD method, this process is relative complicated and easier to be polluted by water and oxygen in the air, which will affect the quality of the films. Alternatively, IGZO thin films can be a

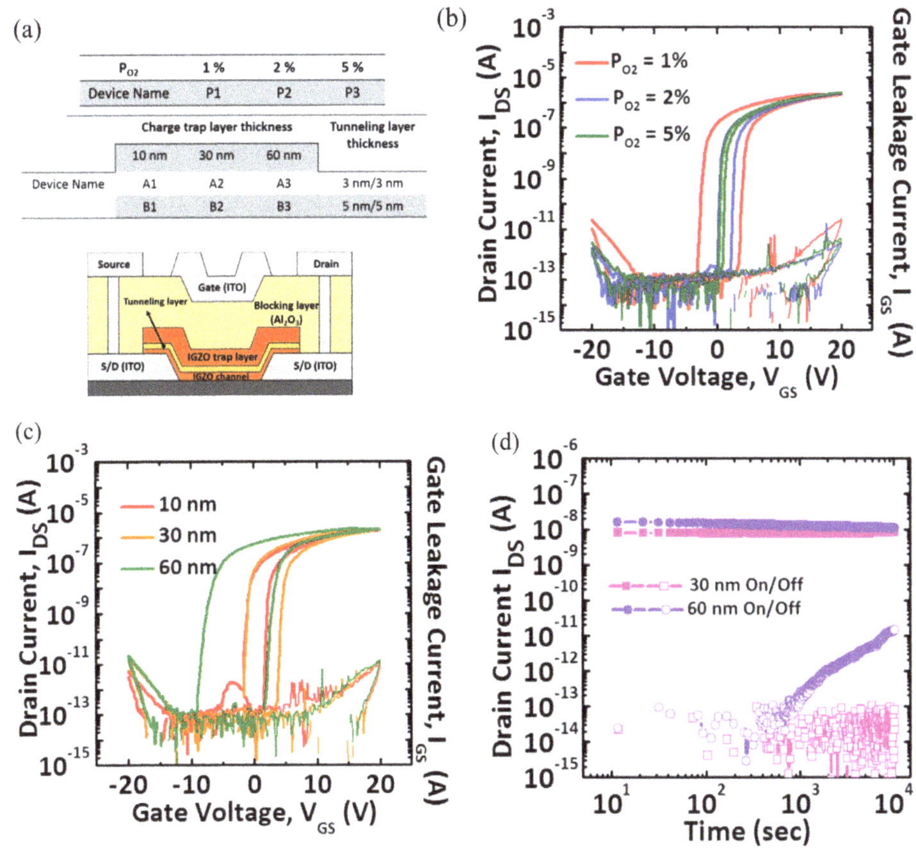

Figure 10.21. (a) Schematic cross sectional diagram of the device with IGZO active layer and charge trapping layer. The transfer curves of the samples made with (b) different oxygen pressure and (c) with various charge trapping layer thicknesses of 10, 30, and 60 nm. (d) Data retention of the IGZO TFT memory with 30 and 60 nm thick charge trapping layers. © (2016) IEEE. Reprinted, with permission, from [31].

suitable material as the charge trapping layer of the NVM based on TFT, which can be grown by sputtering. Yun *et al* fabricated the charge-trap memory TFTs with IGZO thin films as channel and CTLs, as shown in figure 10.21 [31]. In order to get the best process parameters, our experiment is divided into the following three plans: (1) selecting appropriate P_{O2} conditions to deposit IGZO charge trapping layer; (2) optimal combination of double-layer Al_2O_3 tunneling oxide; (3) film thickness of charge trap layer. Firstly, the memory window exhibited a trend of increasing from 1% to 5% with P_{O2}, and the TFT with the charge-trapping layer deposited at 1% P_{O2} showed the fastest programming speed and high switching current ratio. The result shows that the IGZO film deposited with 1% P_{O2} was most suitable for the CTL. The effect of the thickness of multi-layer tunneling oxide was studied from device characteristic and process margin. Although decreasing the thickness of the tunneling layer can enhance the programming speed and reduce the working voltage, the channel layer will be seriously damaged in the etching process which makes the memory unable to work. Therefore, it was set as 5 nm/5 nm to meet the requirements of memory characteristic. The memory based on the IGZO-TFT with 30-nm-thick IGZO charge-trapping layer showed the best performance including large memory window and the high I_{on}/I_{off} ratio. The thickness of the charge-trapping layer had an important effect on the number of trap sites in IGZO CTL.

The results showed that the optimum conditions of charge-trap memory TFT should be determined as a 1% P_{O2} and 30-nm-thick charge-trapping layer thickness. Under ±20 V programming events, the memory windows are maintained to be 6.2 V. Even with 1-μs programming pulse, the programmed $I_{ON/OFF}$ could be set to 2.85×10^5. In addition, the initial programmed current on/off (about 3.65×10^5) did not decrease significantly, even after 10^4 s, which has a great retention characteristic. The memory has a fast programming speed of up to 1 μs and excellent data retention, which proves that IGZO thin film is suitable as a CTL for the memory based on the TFT, which has the potential to replace the SONOS type memory.

Novel multi-level cell TFT memory with an a-IGZO CTL and channel were reported in [39]. The initial device was defined as the original state, which can be switched to the programming state (PS) after a 12 V gate voltage for 10 ms, and applied −15 V voltage for 10 ms to the erasing state. When a positive bias voltage is applied to the gate, electrons pass through the channel layer into the charge storage layer due to F–N tunneling, thus the device exhibits good storage characteristics. As shown in figure 10.22, after 100 cycles, the TFT memory has a memory window of 2.4 V and also has an erasure window of 2.66 V.

- *Sm_2O_3 and $SmTiO_3$ charge trapping layers*

Recently, due to the promising thermal stability and great electrical performances, RE oxide films are considered to be ideal materials for NVM devices. Among RE oxide materials, Sm_2O_3 thin film has high dielectric constant and wide energy bandgap of, which makes it one of the most suitable candidates as a CTL for charge-trapping memory. However, due to the high hygroscopicity, the formation of hydroxide film was found on the surface of RE oxide. In order to avoid this problem, the high stability of RE oxide film can be achieved by adding Ti atoms to

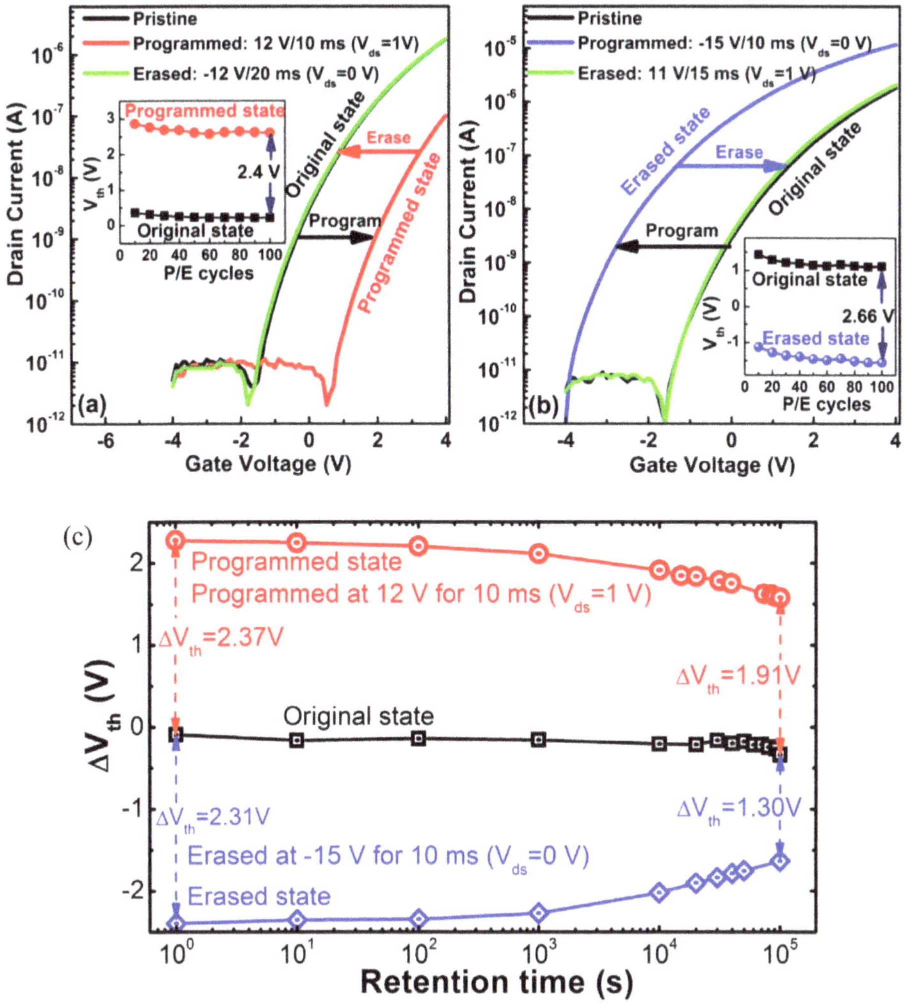

Figure 10.22. (a) Programming and erasing properties of the device from original state to (a) programmed state and (b) erased state. (c) Charge retention behaviors of the multi-level memory cells. © (2015) IEEE. Reprinted, with permission, from [39].

the RE film. It is found that Ti doped RE oxide films have higher dielectric constant and lower leakage current than undoped RE oxide films. Her *et al* also compared the Sm_2O_3 and $SmTiO_3$ thin films as a CTL in MOHOI non-volatile memory [40]. The flash memory based on the IGZO-TFT with a $SmTiO_3$ CTL showed a larger memory window and a long retention time which only loses ~15% charges after 10^5 s, compared with a Sm_2O_3 CTL. The experimental results show that the charge-trapping memory with $SmTiO_3$ thin film based on the TFT has a good memory performance which can have potential to become an alternative material in the future.

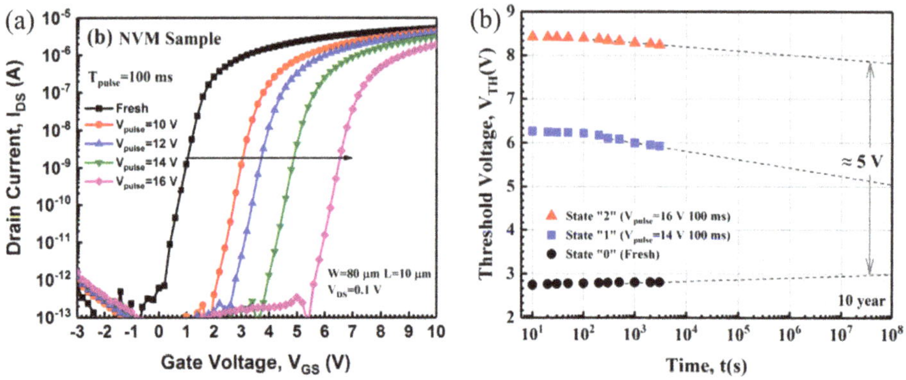

Figure 10.23. (a) The transfer characteristics of the TFTs under various writing voltage. (b) The endurance behaviors of the TFTs with TiAlO as charge trapping layer. © (2018) IEEE. Reprinted, with permission, from [41].

TiAlO charge trapping layer

Recently, in order to improve the programming and retention characteristics of devices, high-k materials have been proposed as CTL of the memory. Among them, TiO_2 has a high dielectric constant, which is beneficial for reducing the operation voltage of the memory, but its trap states density is low which is unsuitable to get a large memory window. Therefore, TiAlO began to enter the field of vision. It is a kind of material with a high trap density and a relative high dielectric constant, which has the advantages of low preparation conditions so that can be used in the field of flexible electronic devices. Zhang proposed and fabricated a memory device based on IGZO-TFT using the high-k TiAlO film as the charge trapping layer [41]. As shown in figure 10.23, the device exhibited excellent programming performances with a memory window of 5.74 V under a 16 V operation voltage. And after 10 years, it also maintains a memory window of about 5 V which has only ~15% charge loss. This work suggests that the high-k TiAlO memory based on the IGZO-TFT has great potential in the field of flexible electronic devices in the future.

• *Zn-doped Al_2O_3 (ZAO) charge trapping layer*

Although memory based on a-IGZO-TFT exhibits excellent programming and charge retention performances, it is difficult to erase only by electrical erasure. This is because a-IGZO is an intrinsic n-type semiconductor, which can only provide electrons for charge transport in the channel. In order to improve the erasing efficiency, a large number of holes must be generated in a-IGZO channel. Several methods have been proposed to enhance the erasing efficiency of the memory, such as light irradiation and light-assisted electrical erasure. A-IGZO-TFT memory is programmed by using stack of Al_2O_3/ZAO/Al_2O_3, and then it is erased by ML irradiation of various wavelengths with various negative gate bias and positive wavelength gate pulses at the same time [35]. At fixed gate bias, irradiation at 500 nm wavelength will greatly increase the erasure window (ΔV_{th}):

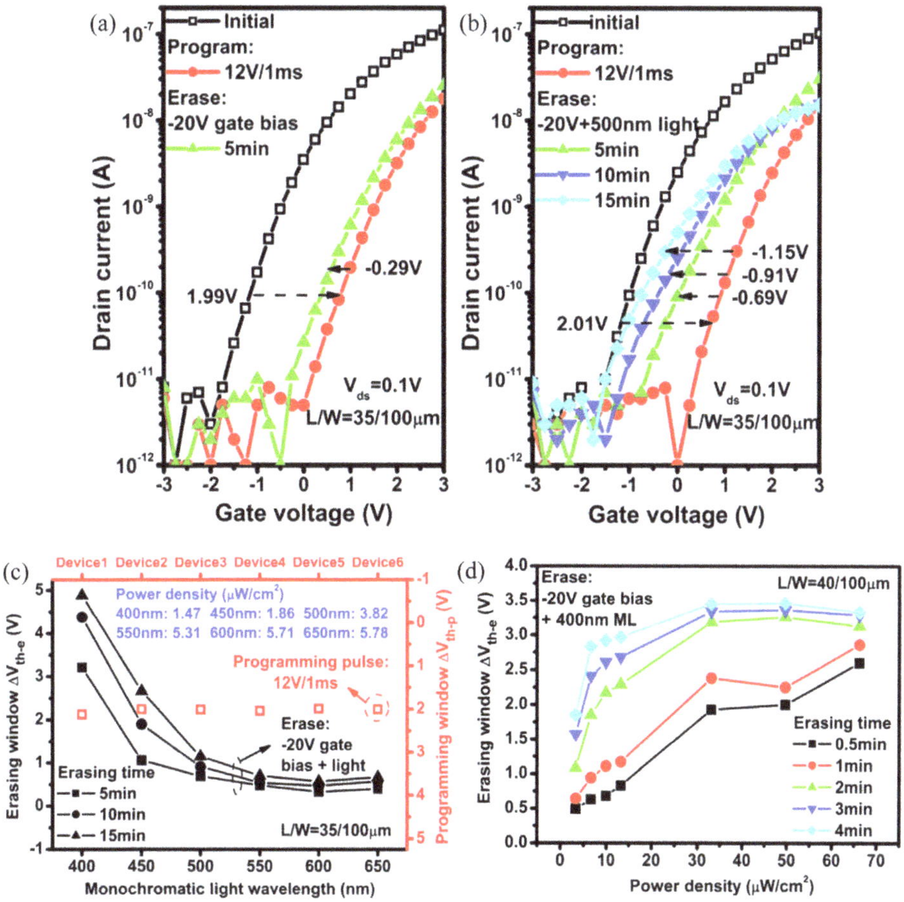

Figure 10.24. The transfer curves of the written device erased under (a) negative gate voltage and (b) different assisted light wavelength. (c) Dependence of ΔV_{th} on light wavelength and times. (d) Effect of V_{th} on power density of light illumination. © (2014) IEEE. Reprinted, with permission, from [35].

at a −20 V gate bias and a 5 min erasure time, illumination with ml at 400 nm will increase ΔV_{th} from 0.29 to 3.21 V, as shown in figure 10.24. ΔV_{th} increases with the increase of gate bias, erasure time and ml power density, especially at short milliliter wavelengths.

10.3.3 Nanocrystal (quantum dot) charge-trap memory

Driven by the continuous shrinking of the CMOS device size and the continuous optimization of the metal electrode/high-k process, memory based on high-k material has received widespread attention. Nanocrystal/quantum dots memory has recently become the focus of research. As shown in figure 10.25, compared to traditional floating gate flash memory, it change the original floating gate to a nanocrystal discrete structure to store charge, which has the following advantages [42, 43]:

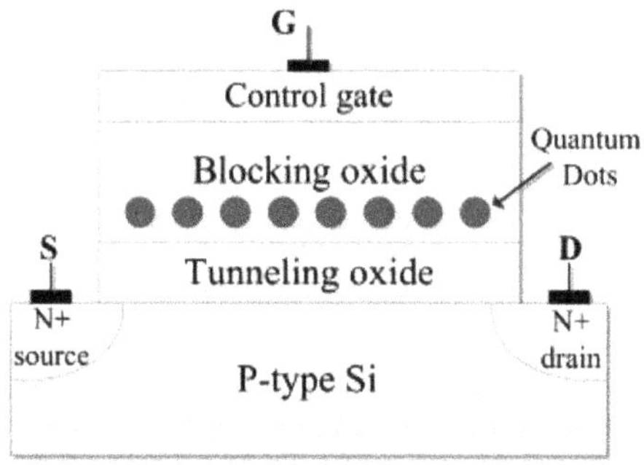

Figure 10.25. The structure of nanocrystal memory.

1. Better data retention. Discrete charge storage inhibits the free movement of charge in the storage layer, reduces charge loss due to tunnel layer quality problems, and improves device retention characteristics [44].
2. Lower device operating voltage and power consumption. The charge retention performance of the nanocrystal/quantum dot floating gate memory allows the device to use a thinner tunneling layer, thereby reducing the voltage required for the device to operate, and at the same time reducing power consumption because the device only operates a small portion of the charge in the nanocrystal [45].
3. Faster programming/erasing speed. The reduction in the thickness of the tunneling layer makes the programming/erasing speed faster. This is because when the same programming/erasing voltage is applied, the current of the device will increase, increasing the reading and writing speed of the device [46].

In [47], the researchers have successfully fabricated and proposed a p–n-diode (PND) structure of SONOS TFT memory with embedded Si-NCs which had excellent retention performance. This structure avoids the shortcomings of traditional SONOS memory devices, including large DIBL, acute punch through effect and severe SCE. In addition, when Si-NCs is embedded in the nitride layer, the deep-trapped energy levels can be created in Si-NCs and the interface between the nitride and Si-NCs. As a result, data retention is enhanced and the storage window is improved. Figure 10.26 shows the cross section of a memory device with Si-NCs based on the TFT. In order to sandwich Si-NCs in Si_3N_4 layer, Si_3N_4 (4-nm-thick)/Si-NCS/Si_3N_4 (4-nm-thick) trap layers were deposited by LPCVD.

Figure 10.27(a) shows the program programming characteristics at different operating voltages. The programming conditions were set as $V_G = 11$, 13, 15 V, $V_N = 0$ V and $V_P = -5$ V. When voltage is applied to the gate, electrons can tunnel from the P+ valence band to depletion region by BTBT, as shown in figure 10.27(b). Under the reverse junction bias voltage ($V_P = -5$ V), the horizontal electric field will accelerate the tunneling electron. Also in figure 10.27(b), hot electrons can be injected into the charge-trapping layer and captured by a positive gate voltage

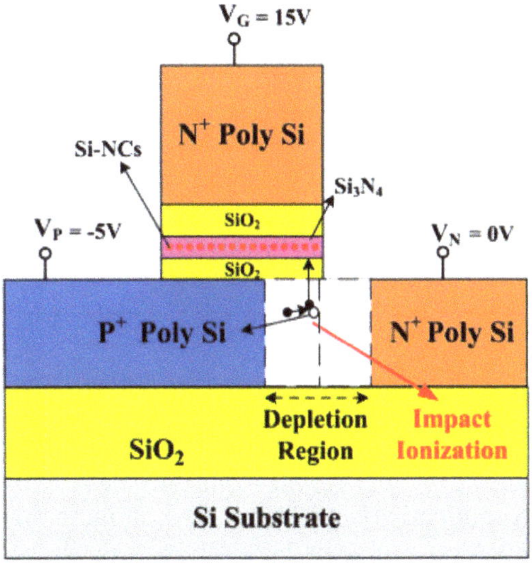

Figure 10.26. Schematic cross section view of the p–n-diode TFT NVW with silicon nanocrystals. © (2010) IEEE. Reprinted, with permission, from [47].

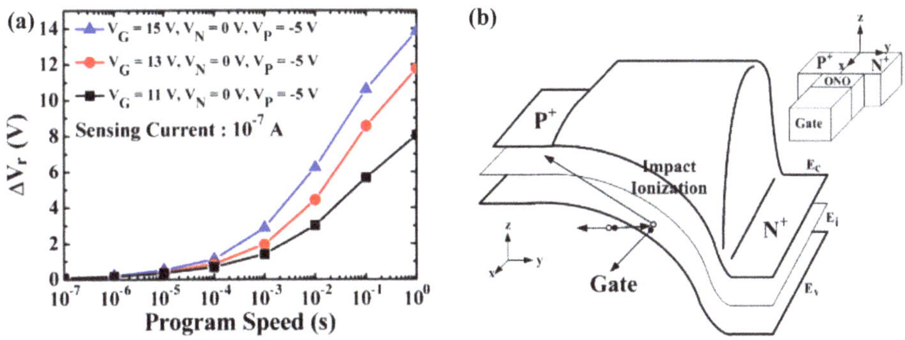

Figure 10.27. (a) Dependence of ΔV_T on program speed and (b) 3-D energy structure of the p–n-diode TFT NVWs. © (2010) IEEE. Reprinted, with permission, from [47].

bias. However, hot holes can also be injected into the charge-trapping layer and captured by a negative gate voltage bias. Due to the increase of vertical electric field, larger gate voltage bias tends to lead to faster programming and erasing speed. After programming, when the detection current is 10^{-7} A, it is found that ΔV_T has a 12 V memory window. This memory window has greatly exceeded the traditional SONOS flash without nanocrystals [48].

Obviously, polycrystalline silicon nanocrystals also have their own limitations. First of all, in the industrial world, high-k dielectric materials have gradually replaced silicon dioxide as the gate dielectric material under the new technology node. It means that the main source of technological advantages for choosing silicon nanocrystals has lost its support. The interface problem between high-k dielectrics and silicon materials has always been one of the core problems that continue to

Table 10.1. Some parameters of main nanocrystal metal candidate materials.

Metal material	Work function	Annealing temperature	Easy to form metal conductive filaments
Au	5.1 eV	650 °C	Hard
Ag	4.3 eV	550 °C	Hard
Pt	5.65 eV	950 °C	Hard
W	4.5 eV	1100 °C	Easy
Ni	5.15 eV	500 °C	Easy
Co	5 eV	750 °C	Easy

shrink the thickness of gate dielectrics. Polycrystalline silicon as a nanocrystal will produce many interface states at the interface with the high-k dielectric of the tunneling layer, thereby making charge transport easier. In addition, although polysilicon has a high conductivity, it is still a semiconductor in essence after all, so it will cause serious depletion problems [49]. When polysilicon is used as a silicon nanocrystal material for charge storage cells, similar problems will also occur. Strong depletion will make it difficult to store charge in the nanocrystals. Moreover, the reduction in size of silicon nanocrystals will make the single-electron Coulomb blocking effect and quantum confinement effect very obvious, making the size reduction unsustainable.

Therefore, it is imperative to use other conductive materials to replace silicon nanocrystals. Metal materials have become the natural choice. The advantages of metals include the high density of states near the Fermi level and the relatively small impact of metal nanocrystals on the Coulomb blocking effect and quantum confinement effects compared to semiconductor materials. Similarly, at this point, metal nanocrystals that also provide more size reductions are possible in the future.

As candidates for nanocrystal materials, many different types of metal materials have received attention and research, including Au, Ag, Pt, W, Ni, Co, etc [45, 50]. The following table 10.1 lists the main performance parameters of these metals, including work function, annealing temperature required to form nanocrystals, whether it is easy to form metal conductive filaments, etc.

It can be seen from the table that some materials such as Ag, Pt and Au are relatively stable due to their chemical properties, and do not easily react with oxygen ions to form oxides or conductive filaments. Choosing such a material as a nanocrystal is beneficial to the storage performance of the nanocrystal and the leakage current control of the oxide layer. In addition, the formation of Pt and W nanocrystals requires high temperature annealing, which poses a great challenge to the process compatibility of nano devices. Moreover, high-k dielectrics tend to crystallize at high temperatures, which will significantly affect the reliability of high-k dielectrics. For example, the temperature at which HfO_2 begins to crystallize is generally around 700 °C. For Ag and W, their work function is low and the potential well is relatively shallow, which is not conducive to P/E process and charge retention in the storage state.

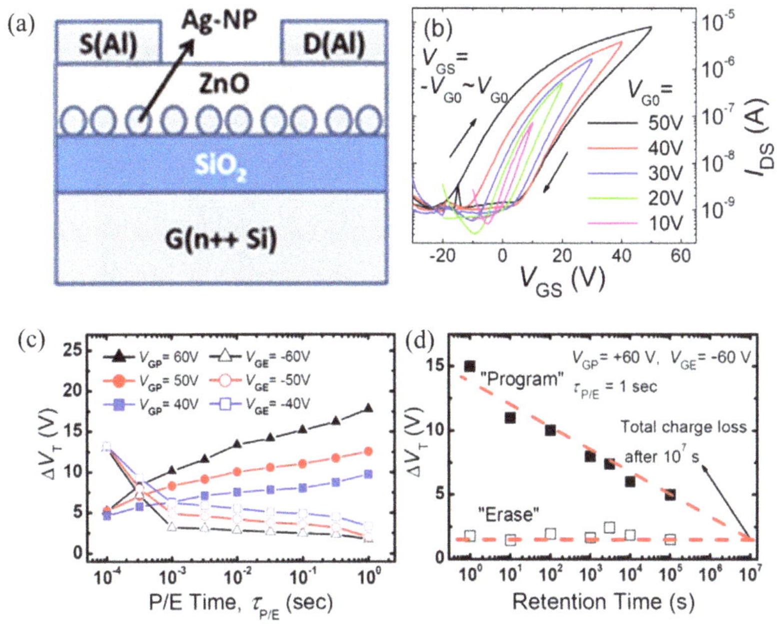

Figure 10.28. (a) Schematic cross section, (b) sub-hysteresis, (c) P/E speed behaviors and (d) data retention of the zinc oxide/Ag nanoparticles TFT-NVMs. © (2008) IEEE. Reprinted, with permission, from [51].

In the year of 2008, it used Ag nanoparticles as charge trapping nodes and embedded them into the interface between active layer and insulating layer of ZnO-TFT to prepare non-volatile memory [51]. As shown in figure 10.28, the transfer characteristics of the device show a large clockwise hysteresis, which is proportional to the scan range of the gate bias. The measurement results exhibit that the device has a fast programming/erasing speed as short as 10^{-4} s. After 10^5 s, the retention of the original memory window is measured to be 27%, which is expected to last to 10^7 s, which indicates that the memory performance is acceptable.

Recently, due to the high density near the Fermi level, strong charge limitation and a wide range of work functions, Ni nanocrystals seem to be more attractive. The a-IGZO-TFT memory device with Ni nanocrystals is fabricated by ALD in this work [52]. And the effect of tunneling layer thickness (i.e. Al_2O_3, 8 and 12 nm) on the memory characteristics is studied. In figure 10.29, after 5 s programming at 18 V, the 8 nm tunneling Ni-QD device has 4.8 V memory window, which reflects the excellent storage performance of the device. In order to explore and improve the device erasing efficiency, we use different erasure methods, including electrical erasure, monochromatic light erasure and monochromatic light assisted electrical erasure. The experimental results show that the highest erasing efficiency can be obtained by monochromatic light assisted electrical erasure. Under the condition of –20 V, the maximum erasing window can reach 9.11 V after 100 s monochromatic light illumination. The device also has excellent retention life whose memory window can reach 4.7 V at room temperature after 10 years. Though increasing

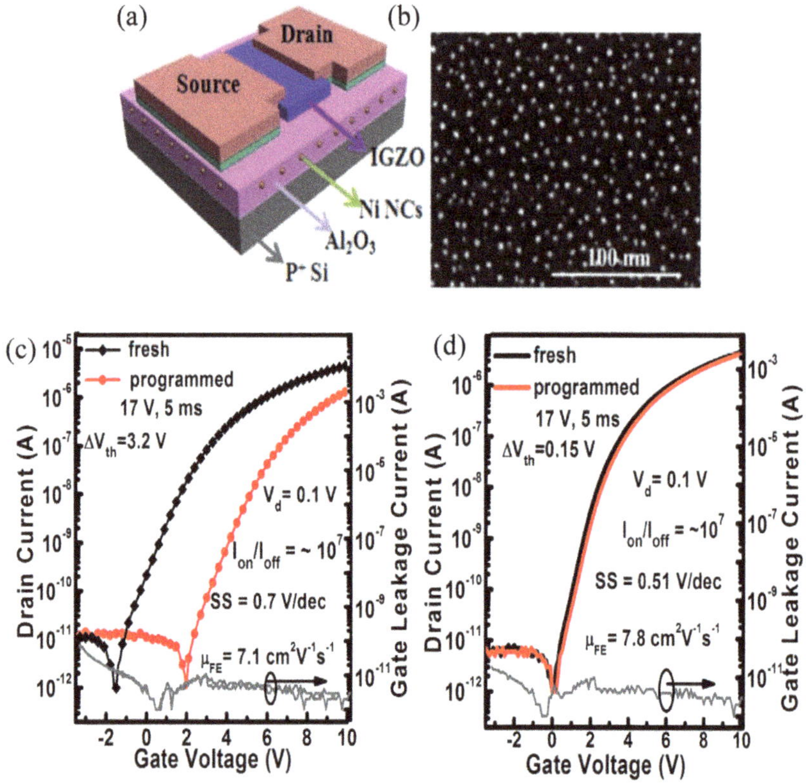

Figure 10.29. (a) Schematic cross section of the TFT-memory. (b) SEM image of Ni-NCs on Al_2O_3 film. Transfer characteristics of (c) the device with an 8 nm thick tunneling layer and (d) the control device. © (2017) IEEE. Reprinted, with permission, from [52].

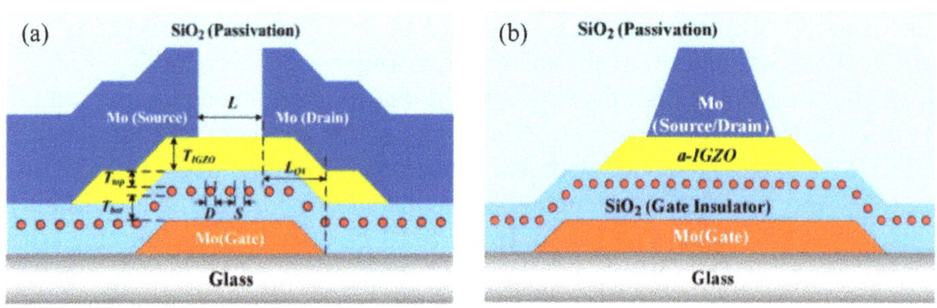

Figure 10.30. (a) X-direction cross-sectional view and (b) Y-direction cross-sectional view of the transparent memory device. © (2011) IEEE. Reprinted, with permission, from [53].

the thickness of the tunneling layer can enhance the data retention, it also damages the programming and erasing efficiencies slightly.

In other work [53], a-IGZO TFT based non-volatile transparent flash devices were fabricated with Au nanocrystal charge-trapping layer.

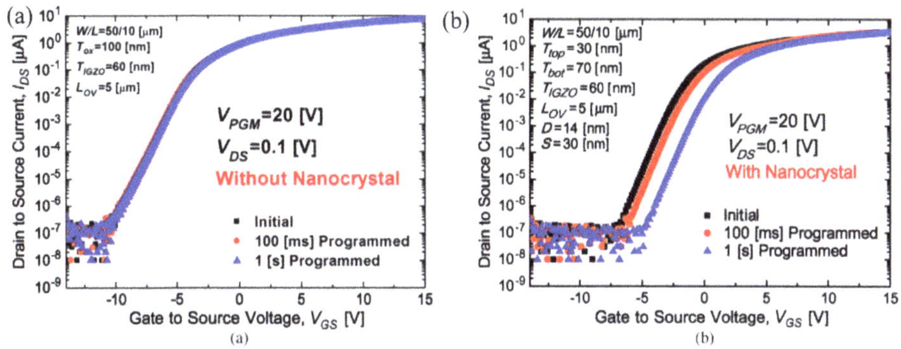

Figure 10.31. I_D–V_{GS} curves of the transparent TFT NVMs (a) without and (b) with the Au NC charge trapping layer. © (2011) IEEE. Reprinted, with permission, from [53].

Figure 10.30 shows the 3D structure of a-IGZO based transparent memory device with inverted interleaved Au nanocrystals. A flash memory with an IGZO channel and Au nanocrystal charge-trapping layer is fabricated on glass substrate.

In order to verify the performance of Au nanocrystalline memory devices, we apply the same programming operation to devices with and without Au nanocrystalline charge-trapping layer at the interface between the tunneling layer and the insulator. The experimental results are exhibited in the memory performance of the nanocrystalline device in figure 10.31. The program is executed under the condition that the gate voltage is applied to the 20 V and the programming time is 1 s.

As shown in figure 10.31, compared with the device without Au nanocrystalline charge-trapping layer, the TFT device with Au nanocrystal storage layer has a clear memory window under the same programming conditions. It is obvious that the Au nanocrystal layer works as a charge-trapping layer for the transparent flash memory based on a-IGZO TFTs.

10.4 Other novel memories

At present, the floating gate flash memory (charge-trapping type memory) is still the mainstream in the industry. However, a series of problems such as high power consumption, small storage density and device size reaching physical limits have seriously hindered its development. In the last two decades, people have proposed a variety of new types of non-volatile memory, including magnetic random access memory (MRAM) [54, 55], ferroelectric random access memory (FeRAM) [56, 57], phase-change random access memory (PRAM) [58] and resistive random access memory (RRAM) [59], etc.

10.4.1 FeRAM

FeRAM is a kind of non-volatile memory that will not lose information when power is off. It has the advantages of high speed, low power consumption and radiation resistance. It has shown great development potential and good application prospects in the fields of IC cards, mobile phones, embedded microprocessors, aerospace and military applications. It has received extensive attention from the scientific and technological circles and industries of various countries.

Ferroelectric is a kind of residual polarization material, which can be switched between two directions by electric field. The combination of non-volatile polarization and field driven switch makes it a natural non-volatile memory which solves the problem of slow programming speed of other common non-volatile memories [60]. Ferroelectric materials have attracted much attention because of their broad application prospects. After years of research, the current mainstream ferroelectric materials mainly include $PbZr_xTi_{1-x}O_3$ (PZT) and $Sr_{1-y}Bi_{2+x}Ta_2O_9$ (SBT).

PZT is currently the most researched and most widely used. The advantage is that it can be prepared by sputtering and metal organic vapor deposition (MOCVD) at a lower temperature, and has the advantages of larger residual polarization, cheap raw materials and lower crystallization temperature. The disadvantage is that there is the problem of fatigue degradation, and the lead content can also cause pollution to the environment. And the biggest advantage of SBT is that there is no fatigue degradation problem, it also will not cause environmental pollution. However, its high process temperature increases the difficulty of process integration, and the degree of residual polarization is small.

The above two materials are relatively mature but have their own defects. Recently, hafnium oxide and its doping have gradually entered people's field of vision as novel ferroelectric materials [57].

FeRAM can be divided into FeRAM based on ferroelectric capacitors and ferroelectric field effect transistors (FeFET). FeRAM memory cells are composed of ferroelectric capacitors and field effect transistors. The structure of the storage cell of the ferroelectric memory is similar to that of the DRAM. As shown in figure 10.32(a), the storage capacitor of the DRAM is replaced by a ferroelectric capacitor. The upper electrode is connected to the bit line (BL) and the lower electrode is connected to the plate line (PL). The polarization inversion of the ferroelectric film is used to write and read data. The basic unit structure of

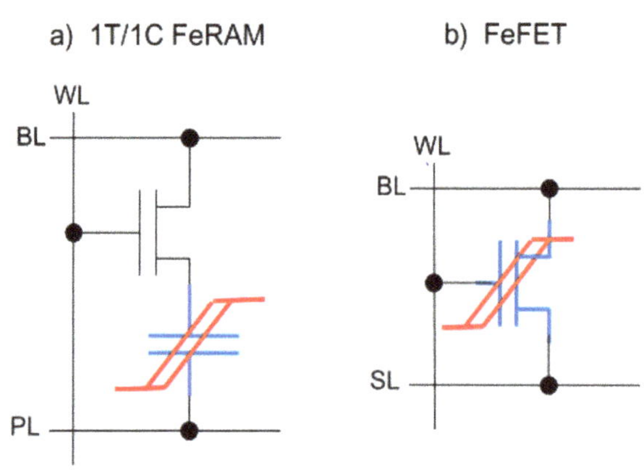

Figure 10.32. Two types of ferroelectric memory: (a) one transistor/one capacitor FeRAM, (b) FeFET. © (2019) IEEE. Reprinted, with permission, from [57].

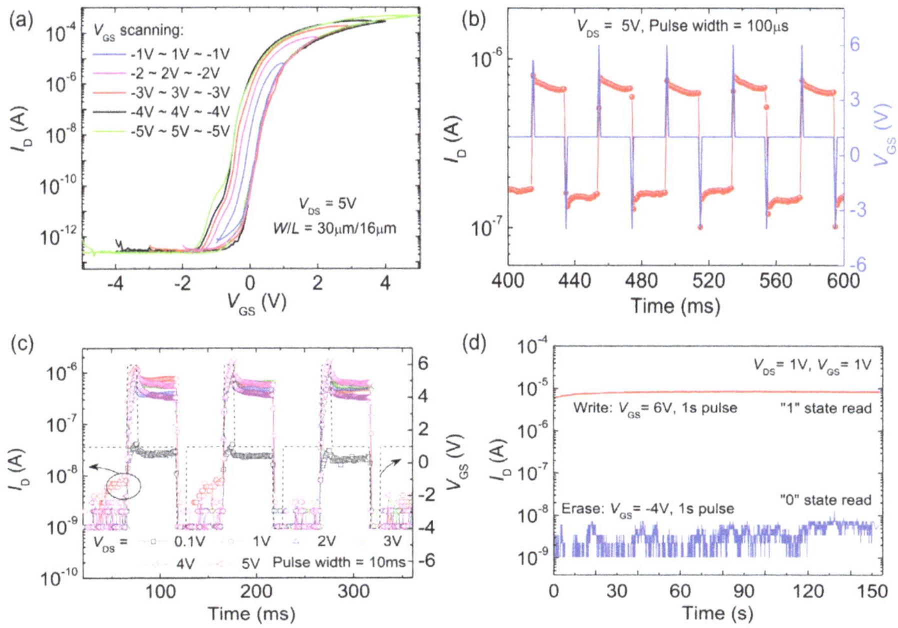

Figure 10.33. (a) Effect of the sub-hysteresis on the V_{GS} sweeping range. (b) I_D waveform obtained using pulse width of V_{GS} of 100 μs and (c) different V_{DS}. (d) Data retention properties of the ZnO TFT based NWM. © (2017) IEEE. Reprinted, with permission, from [26].

ferroelectric memory mainly includes 2T–2C and 1T–1C structures (figure 10.32). Compared with the 2T–2C structure, the 1T–1C structure area is reduced by half, which is the future development direction of high-density ferroelectric memory. And in figure 10.32(b), compared with the traditional MOSFET, FeFET only uses ferroelectric thin film material instead of SiO_2 as the gate dielectric. Essentially, FeFET uses the polarization reversal of the ferroelectric crystal to modulate the conductivity of the conduction channel to achieve data storage.

However, there are always two unsolved problems in the whole development process of FeFET: the depolarization field in the ferroelectric thin film material and the excessive gate leakage current, which seriously affects the retention performance of the memory.

In our previous works, the ZnO TFTs with water dipole in the interface between the channel and the gate oxide exhibit counter-clockwise hysteresis, which is similar to the mechanism of the ferroelectric memory. Hence, the memory characteristics of the devices have been measured according to the FeRAM. Figure 10.33(a) reveals the effect of the sub-hysteresis on different scanning bias ranges. Although the V_{th} decreases with the scan voltage narrows, the memory window is still obvious when the V_{GS} scanning range is ±1 V. The writing and reading operation of the ZnO memory element using V_{GS} pulse width of 100 μs has been revealed in figure 10.33(b). Figures 10.33(c) and (d) have checked the operation V_{DS} low to 0.1 V and the retention the electric memory element. The states are still unchanged after P/E operation and long-time duration, exhibiting good data retention behaviors.

10.4.2 RRAM

Resistive random access memory (RRAM) is a NVM device which can record and store data information according to changes in resistance. The developing of the RRAM began from the late 1990s with various complex metal-oxides, such as the perovskite oxides of $SrTiO_3$ [61], $SrZrO_3$ [62], and then the binary metal-oxides, such as NiO [63] and TiO_2 [64]. And the device structure is simply an oxide material sandwiched between two metal electrodes, called the MIM structure. At present, plenty of binary metal oxides have been found to exhibit resistive switching behavior. Most of them are transition metal oxides, and some are lanthanide series metal oxides. In the past few years, some of these materials have attracted great attention from the industry and have been widely studied such as HfO_2, TiO_2 and NiO.

Currently, HfO_x high-k material often has been used as active layer or gate oxide for high performance TFTs. At the same time, HfO_x, which is rich in defects, is also an excellent RRAM material. The Pt/HfO_x/TiN structure is always used to realize the bistable switching characteristics. Among them, TiN acts as an oxygen storage layer. The schematic diagram of a RRAM cell is shown in figure 10.34(a), which consists of an HfO_x film and a transistor (1T–1R). Figure 10.34(b) demonstrates the TEM image of the stacked layers composed of TiN/Ti/HfO_x/TiN. And the I–V curve for the resistive-switching of the RRAM (including 1R and 1T1R) is shown in figure 10.35(a). The obtained bipolar RRAM has high-speed operation (<10 ns), long lifetime, reliable switching endurance (>10^6 cycles, figure 10.35(b)), and high yield (~100%). So far, HfO_x is one of the most promising RRAM materials developed.

Yang *et al* reported an IGZO-TFT that can be operated either as transistor or RRAM device [65]. Before the molding process, we observe the transfer characteristics of the current–voltage curve, and measure the resistance switching (RS) characteristics after the molding process. These RS characteristics show two

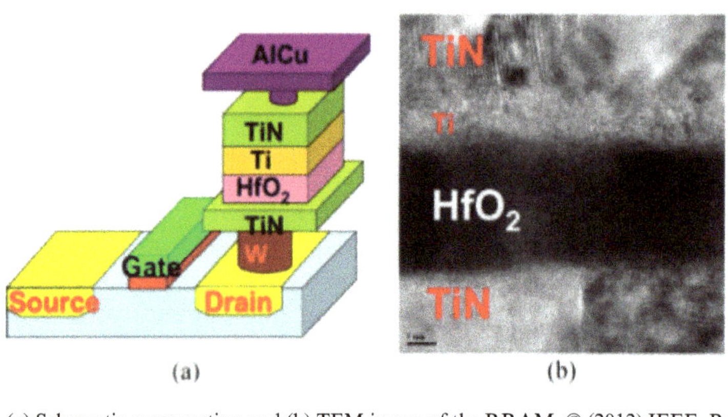

Figure 10.34. (a) Schematic cross section and (b) TEM image of the RRAM. © (2012) IEEE. Reprinted, with permission, from [59].

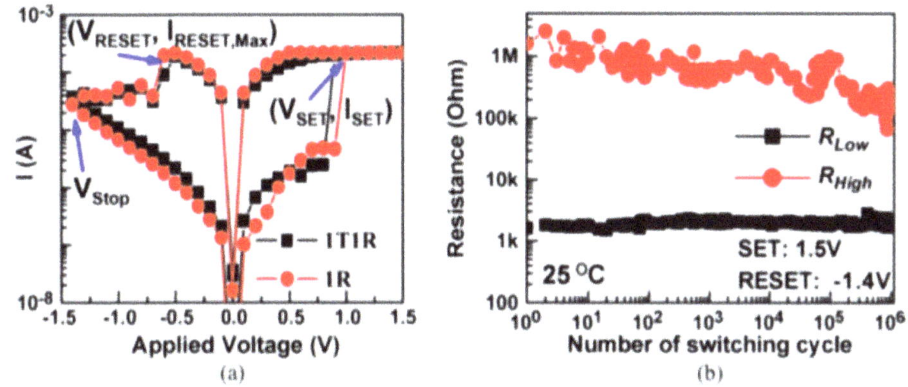

Figure 10.35. (a) RRAM characteristic and (b) switching endurance of the TiN/TiO_x/HfO_x/TiN devices. © (2012) IEEE. Reprinted, with permission, from [59].

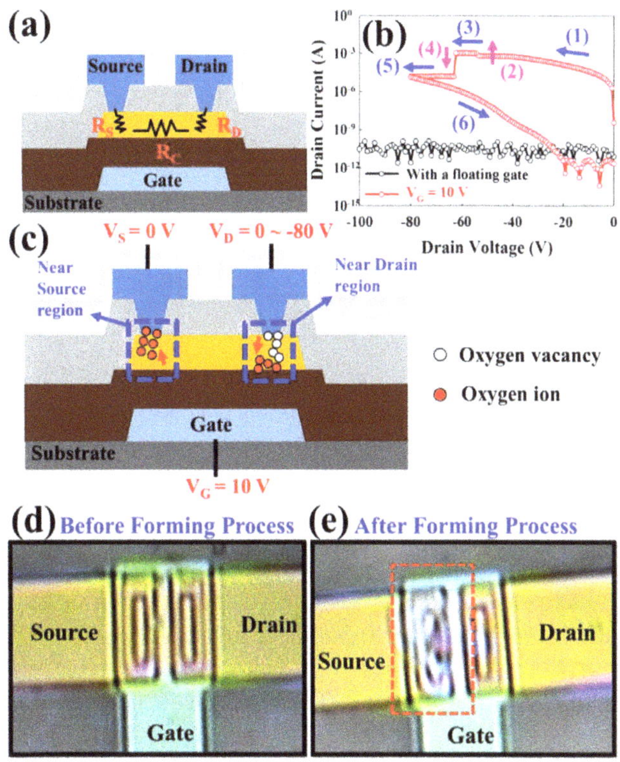

Figure 10.36. (a) Schematic cross-section and equivalent circuit of the RRAM sample. (b) IV curves of the operation process. (c) Mechanism of the conductive path. Top view of the device (d) before and (e) after forming process. © (2014) IEEE. Reprinted, with permission, from [65].

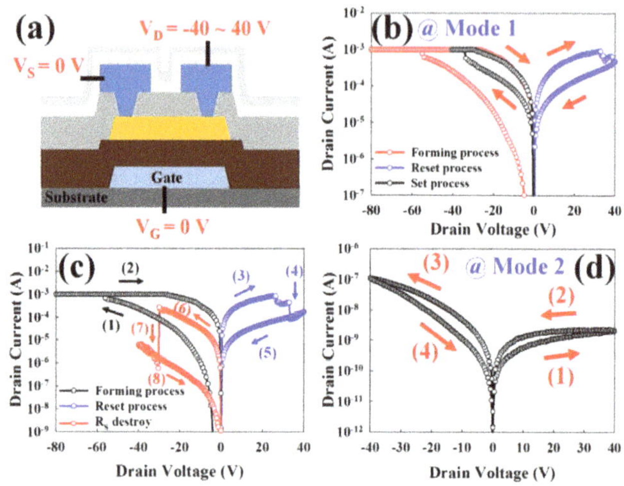

Figure 10.37. (a) Cross-section of the device with testing condition. (b) Current–voltage characteristics of (b) mode-1, (c) mode-1 to mode-2, and (d) mode-2. © (2014) IEEE. Reprinted, with permission, from [65].

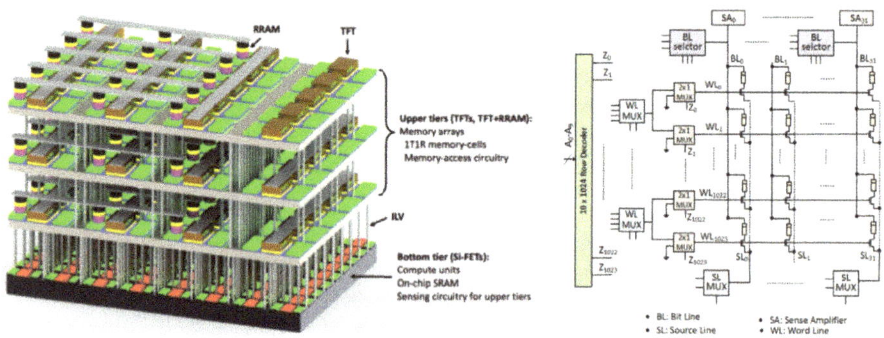

Figure 10.38. The introduced monolithic (left) 3D architecture and (right) circuit schematic. © (2020) IEEE. Reprinted, with permission, from [66].

behaviors which are determined by different mechanisms. As shown in figure 10.36, the switching behavior of mode 1 is determined by conductive paths, which are mainly composed of oxygen vacancies. And the mode-2 is determined by the thickness of rich-oxygen film. The two RS characteristics are due to the migration of oxygen ions near the S/D regions. As seen in figure 10.36(b), RS behavior shows bipolar switching characteristics after 1 mA compliant current process. When the drain voltage scans from 0 to +40 V, the I_D decreases. While the drain voltage changes from 0 to −40 V, the current increases. As shown in figure 10.37(c), the current decreases rapidly in step-7 after several switching cycles. And then, the RS characteristics change to mode-2 in figure 10.37(d). In addition, since the current can be controlled by gate bias to reduce the energy consumption, this provides an idea to reduce the power consumption of devices when they are used with a-IGZO TFT.

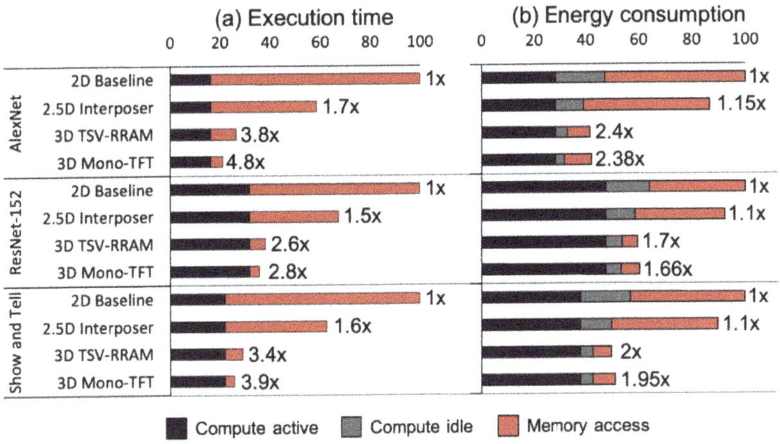

Figure 10.39. (a) Execution times and (b) power consumption of each workload/structure pair. These data are normalized to refer to the traditional 2-D circuits. © (2020) IEEE. Reprinted, with permission, from [66].

Nowadays, with the rapid development of the artificial intelligence industry, the data-centric workload is increasing, and the inefficiency of memory access in the computing system has been exposed. Monolithic chip 3D integration is a feasible method to overcome this shortcoming. It improves the memory density through the close vertical connection structure, so as to realize large-scale access between memory and computing cells. TFTs and RRAM can achieve monolithic 3D integration because of their low temperature preparation conditions. Felfel *et al* explore TFT based on ZnO and RRAM based on HfO_2 to build 1T1R memory system in the upper layer [66]. As shown in figure 10.38, the memory system based on TFT is stacked on top of Si-FET, which includes computing unit and SRAM. The simulations for the data-centric workloads show that TFT can still achieve significant energy-efficiency advantages and acceleration compared with baseline and TSV based 3D computing systems, as shown in figure 10.39. With the further development of TFT, higher density of RRAM based on TFT can be obtained. Therefore, TFT provides a promising alternative to single-chip 3D systems, which will demand further research at the device, circuit and architecture levels.

10.5 Conclusion and future development

In the chapter, we discussed various non-volatile memories based on the metal oxide TFTs. The floating gate memory is still mainstream in our current lives. Many metal oxide TFTs based floating gate memories have been fabricated with different CTLs. Among these materials, the ZnO CTL has good process compatibility and charge capture ability, among which the metal oxide memory TFTs exhibit the best performance of symmetrical memory window (25 V) and programming speed (500 μs). The other charge-trap type memory, nanocrystal non-volatile memory has some serious problems for practical application, such as the small memory window, poor retention and slow writing/erasing speed. In the novel memory field, metal oxide TFTs combined with RRAM have been proposed and fabricated. The

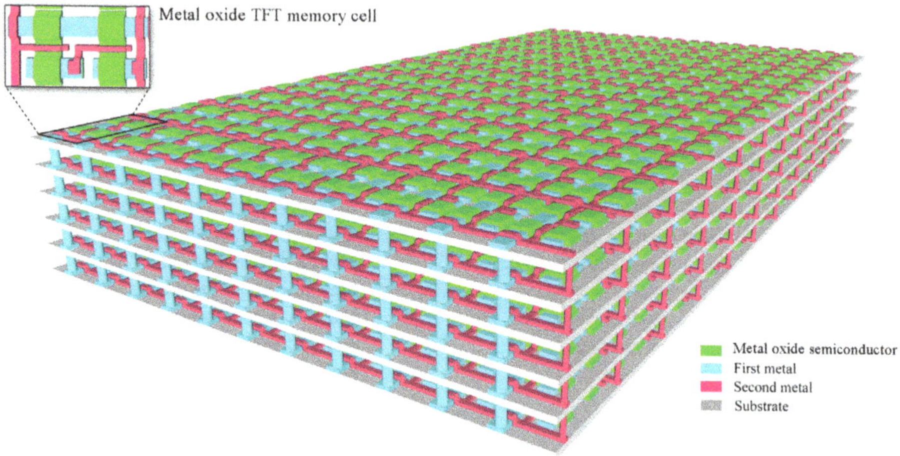

Figure 10.40. Schematic mode of the future 3-D NVM based on metal oxide TFT electronic memory element. © (2017) IEEE. Reprinted, with permission, from [26].

FeRAM with its ultra-fast P/E speed will be used as a new logic unit, showing the high density integration circuits of storage and calculation. Although the FeRAM based on metal oxide TFT has been reported rarely, there are many special properties and chances for the application of TFT based FeRAM. To date, most works just focus on the characteristics of single memory TFT devices. The whole performance of the TFT memory cell and array with corresponding circuits should be measured and checked systematically.

Due to the special advantages of the metal oxide TFT memory, such as low temperature process, transparency, and larger mobility, it is suitable for application on multi-layer memory. In theory, to increase the density of the memory, the oxide memory can be built on the Si CMOS circuit layer by layer. Figure 10.40 shows the schematic diagram of a three-dimensional non-volatile memory composed of several layers. Each layer is manufactured by a similar process and connected by metal wires and holes. The metal oxide TFT memory cell could be floating gate memory, RRAM, FeRAM and any information storage devices. In the future, the multi-layer non-volatile memory would have potential to achieve ultra-large scale storage density for the boundless demands of consumer electronic products.

References

[1] Kahng D and Sze S M 1967 A floating gate and its application to memory devices *Bell Syst. Tech.* J **46** 1288–95

[2] Masuoka F, Asano M, Iwahashi H, Komuro T and Tanaka S 1985 A 256K flash EEPROM using triple polysilicon technology *1985 IEEE Int. Solid-State Circuits Conf. Digest of Technical Papers (13–15 February 1985)* vol 28 pp 168–9

[3] Bez R, Camerlenghi E, Modelli A and Visconti A 2003 Introduction to flash memory *Proc. IEEE* **91** 489–502

[4] Lai S K 2008 Flash memories: successes and challenges *IBM J. Res. Dev.* **52** 529–35

[5] Irom F and Nguyen D N 2007 Single event effect characterization of high density commercial NAND and NOR nonvolatile flash memories *IEEE Trans. Nucl. Sci.* **54** 2547–53

[6] Pon H 2006 Technology scaling impact on NOR and NAND flash memories and their applications *2006 8th Int. Conf. Solid-State and Integrated Circuit Technology Proc. (23–26 October 2006)* pp 697–700

[7] Kapetanakis E *et al* 2000 Charge storage and interface states effects in Si-nanocrystal memory obtained using low-energy Si+ implantation and annealing *Appl. Phys. Lett.* **77** 3450–2

[8] Keshavan B V and Lin H C 1968 MONOS memory element *1968 Int. Electron Devices Meeting (23–25 October 1968)* pp 140–2

[9] Bostock D 1976 1024-bit fully decoded MNOS non-volatile memory *ESSCIRC 76: 2nd European Solid State Circuits Conf. (21–24 September 1976)* pp 42–3

[10] Takahashi Y and Ohnishi K 1993 Estimation of insulation layer conductance in MNOS structure *IEEE Trans. Electron Devices* **40** 2006–10

[11] White M H, Yang Y, Ansha P and French M L 1997 A low voltage SONOS nonvolatile semiconductor memory technology *IEEE Trans. Compon. Packag. Manuf. Technol.* A **20** 190–5

[12] Chen T S, Wu K H, Chung H and Kao C H 2004 Performance improvement of SONOS memory by bandgap engineering of charge-trapping layer *IEEE Electron Device Lett.* **25** 205–7

[13] Kinam K and Jungdal C 2006 Future outlook of NAND flash technology for 40 nm node and beyond *2006 21st IEEE Non-Volatile Semiconductor Memory Workshop (12–16 February 2006)* pp 9–11

[14] Kinam K 2005 Technology for sub-50 nm DRAM and NAND flash manufacturing *IEEE Int. Electron Devices Meeting, 2005. IEDM Technical Digest (5–5 December 2005)* pp 323–6

[15] Choi S-J *et al* 2011 Nonvolatile memory by all-around-gate junctionless transistor composed of silicon nanowire on bulk substrate *IEEE Electron Device Lett.* **32** 602–4

[16] Colinge J-P *et al* 2010 Reduced electric field in junctionless transistors *Appl. Phys. Lett.* **96**

[17] Colinge J P *et al* 2010 Nanowire transistors without junctions *Nat. Nanotechnol.* **5** 225–9

[18] Yan Ny T, Chim W K, Wee Kiong C, Moon Sig J and Byung Jin C 2006 Hafnium aluminum oxide as charge storage and blocking-oxide layers in SONOS-type nonvolatile memory for high-speed operation *IEEE Trans. Electron Devices* **53** 654–62

[19] Lan X *et al* 2013 The effect of thermal treatment induced inter-diffusion at the interfaces on the charge trapping performance of HfO_2/Al_2O_3 nanolaminate-based memory devices *J. Appl. Phys.* **114** 044104

[20] Specht M *et al* 2005 Charge trapping memory structures with Al_2O_3 trapping dielectric for high-temperature applications *Solid-State Electronics* **49** 716–20

[21] Park J *et al* 2010 Improvement of reliability characteristics using the N2 implantation in SOHOS flash memory *2010 IEEE Nanotechnology Materials and Devices Conf. (12–15 October 2010)* pp 364–7

[22] Tan Y N, Chim W K, Cho B J and Choi W K 2004 Over-erase phenomenon in SONOS-type flash memory and its minimization using a hafnium oxide charge storage layer *IEEE Trans. Electron Devices* **51** 1143–7

[23] Huby N, Ferrari S, Guziewicz E, Godlewski M and Osinniy V 2008 Electrical behavior of zinc oxide layers grown by low temperature atomic layer deposition *Appl. Phys. Lett.* **92** 023502

[24] Xu H, Ye Z, Liu N, Wang Y, Zhang N and Liu Y 2017 Low-power transparent RFID circuits using enhancement/depletion logic gates based on deuterium-treated ZnO TFTs *IEEE Electron Device Lett.* **38** 1383–6

[25] Jung Y, Lee K, Kim W-J, Lee W-J, Choi H and Kim K H 2012 Properties of In–Ga–ZnO thin films for thin film transistor channel layer prepared by facing targets sputtering method *Ceram. Int.* **38** S601

[26] Ye Z, Yuan Y, Xu H, Liu Y, Luo J and Wong M 2017 Mechanism and origin of hysteresis in oxide thin-film transistor and its application on 3-D nonvolatile memory *IEEE Trans. Electron Devices* **64** 438–46

[27] Ye Z, Lu L and Wong M 2012 Zinc-oxide thin-film transistor with self-aligned source/drain regions doped with implanted boron for enhanced thermal stability *IEEE Trans. Electron Devices* **59** 393–9

[28] Ye Z and Wong M 2012 Characteristics of plasma-fluorinated zinc oxide thin-film transistors *IEEE Electron Device Lett.* **33** 1147–9

[29] Ye Z and Wong M 2012 Characteristics of thin-film transistors fabricated on fluorinated zinc oxide *IEEE Electron Device Lett.* **33** 549–51

[30] Oruc F B, Cimen F, Rizk A, Ghaffari M, Nayfeh A and Okyay A K 2012 Thin-film ZnO charge-trapping memory cell grown in a single ALD step *IEEE Electron Device Lett.* **33** 1714–6

[31] Yun D-J, Kang H-B and Yoon S-M 2016 Process optimization and device characterization of nonvolatile charge trap memory transistors using In–Ga–ZnO thin films as both charge trap and active channel layers *IEEE Trans. Electron Devices* **63** 3128–34

[32] Kim H-R, Kang C-S, Kim S-K, Byun C-W and Yoon S-M 2019 Characterization on the operation stability of mechanically flexible memory thin-film transistors using engineered ZnO charge-trap layers *J. Phys. D: Appl. Phys.* **52** 325106

[33] Bak J Y *et al* 2015 Effects of thickness and geometric variations in the oxide gate stack on the nonvolatile memory behaviors of charge-trap memory thin-film transistors *Solid-State Electronics* **111** 153–60

[34] Chen S *et al* 2013 Novel Zn-doped Al_2O_3 charge storage medium for light-erasable In–Ga–Zn–O TFT memory *IEEE Electron Device Lett.* **34** 1008–10

[35] Chen S, Zhang W-P, Cui X-M, Ding S-J, Sun Q-Q and Zhang W 2014 Monochromatic light-assisted erasing effects of In-Ga-Zn-O thin film transistor memory with Al_2O_3/Zn-doped Al_2O_3/Al_2O_3 stacks *Appl. Phys. Lett.* **104** 103504

[36] Kim S-J, Lee W-H and Yoon S-M 2015 Nonvolatile memory performances of transparent and/or flexible memory thin-film transistors using IGZO channel and ZnO charge-trap layers *2015 22nd International Workshop on Active-Matrix Flatpanel Displays and Devices (AM-FPD) (1–4 July 2015)* pp 21–3

[37] Kim S-J, Park M-J, Yun D-J, Lee W-H, Kim G-H and Yoon S-M 2016 High performance and stable flexible memory thin-film transistors using In–Ga–Zn–O channel and ZnO charge-trap layers on poly(ethylene naphthalate) substrate *IEEE Trans. Electron Devices* **63** 1557–64

[38] Na S-Y and Yoon S-M 2019 Impacts of HfO_2/ZnO stack-structured charge-trap layers controlled by atomic layer deposition on nonvolatile memory characteristics of In-Ga-Zn-O channel charge-trap memory thin-film transistors *IEEE J. Electron Devices Soc.* **7** 453–61

[39] Zhang W-P, Qian S-B, Liu W-J, Ding S-J and Zhang D W 2015 Novel multi-level cell TFT memory with an In–Ga–Zn-O charge storage layer and channel *IEEE Electron Device Lett.* **36** 1021–3

[40] Her J-L, Chen F-H, Chen C-H and Pan T-M 2015 Electrical characteristics of gallium–indium–zinc oxide thin-film transistor non-volatile memory with Sm_2O_3 and $SmTiO_3$ charge trapping layers *RSC Adv.* **5** 8566–70

[41] Zhang W *et al* 2018 Demonstration of α-InGaZnO TFT nonvolatile memory using TiAlO charge trapping layer *IEEE Trans. Nanotechnol.* **17** 1089–93

[42] Blauwe J D 2002 Nanocrystal nonvolatile memory devices *IEEE Trans. Nanotechnol.* **1** 72–7

[43] Liu Y, Chen T P, Ding L, Zhang S, Fu Y Q and Fung S 2006 Charging mechanism in a SiO_2 matrix embedded with Si nanocrystals *J. Appl. Phys.* **100**

[44] Tiwari S, Rana F, Hanafi H, Hartstein A, Crabbé E F and Chan K 1996 A silicon nanocrystals based memory *Appl. Phys. Lett.* **68** 1377–9

[45] Liu Z, Lee C, Narayanan V, Pei G and Kan E C 2002 Metal nanocrystal memories-part II: electrical characteristics *IEEE Trans. Electron Devices* **49** 1614–22

[46] Chang T-C, Jian F-Y, Chen S-C and Tsai Y-T 2011 Developments in nanocrystal memory *Mater. Today* **14** 608–15

[47] Chiang T-Y, Ma W C-Y, Wu Y-H, Wang K-T and Chao T-S 2010 A novel p-n-diode structure of SONOS-type TFT NVM with embedded silicon nanocrystals *IEEE Electron Device Lett.* **31** 1239-41

[48] Lin Y-H, Chien C-H, Chou T-H, Chao T-S and Lei T-F 2007 Low-temperature polycrystalline silicon thin-film flash memory with hafnium silicate *IEEE Trans. Electron Devices* **54** 531–6

[49] Lo S, Buchanan D A and Taur Y 1999 Modeling and characterization of quantization, polysilicon depletion, and direct tunneling effects in MOSFETs with ultrathin oxides *IBM J. Res. Dev.* **43** 327–37

[50] Lee C, Meteer J, Narayanan V and Kan E 08/01 2005 Self-assembly of metal nanocrystals on ultrathin oxide for nonvolatile memory applications *J. Electron. Mater.* **34** 1–11

[51] Gupta D, Anand M, Ryu S-W, Choi Y-K and Yoo S 2008 Nonvolatile memory based on sol-gel ZnO thin-film transistors with Ag nanoparticles embedded in the ZnO/gate insulator interface *Appl. Phys. Lett.* **93** 224106

[52] Qian S-B, Shao Y, Liu W-J, Zhang D W and Ding S-J 2017 Erasing-modes dependent performance of a-IGZO TFT memory with atomic-layer-deposited Ni nanocrystal charge storage layer *IEEE Trans. Electron Devices* **64** 3023–7

[53] Jang J *et al* 2011 Endurance characteristics of amorphous-InGaZnO transparent flash memory with gold nanocrystal storage layer *IEEE Trans. Electron Devices* **58** 3940–7

[54] Tehrani S, Slaughter J M, Chen E, Durlam M, Shi J and DeHerren M 1999 Progress and outlook for MRAM technology *IEEE Trans. Magn.* **35** 2814–9

[55] Hu G *et al* 2019 Spin-transfer torque MRAM with reliable 2 ns writing for last level cache applications *2019 IEEE Int. Electron Devices Meeting (IEDM) (7–11 December 2019)* pp 2.6.1–4

[56] Nagel N *et al* 2004 New highly scalable 3 dimensional chain FeRAM cell with vertical capacitor *Digest of Technical Papers. 2004 Symp. on VLSI Technology (15–17 June 2004)* 146–7

[57] Mikolajick T *et al* 2019 Next generation ferroelectric memories enabled by hafnium oxide *2019 IEEE Int. Electron Devices Meeting (IEDM) (7–11 December 2019)* pp 15.5.1–4

[58] Kim G, Lee S, Hong S, Baik S J, Hori H and Ahn D 2014 Adjustable voltage dependent switching characteristics of PRAM for low voltage programming of multi-level resistances *2014 14th Annual Non-Volatile Memory Technology Symp. (NVMTS) (27–29 October 2014)* pp 1–3

[59] Wong H S P *et al* 2012 Metal–oxide RRAM *Proc. IEEE* **100** 1951–70
[60] Prall K 2017 Benchmarking and metrics for emerging memory *2017 IEEE Int. Memory Workshop (IMW) (14–17 May 2017)* pp 1–5
[61] Watanabe Y *et al* 2001 Current-driven insulator–conductor transition and nonvolatile memory in chromium-doped $SrTiO_3$ single crystals *Appl. Phys. Lett.* **78** 3738–40
[62] Beck A J, Bednorz J, Gerber C, Rossel C and Widmer D 2000 Reproducible switching effect in thin oxide films for memory applications *Appl. Phys. Lett.* **77** 139–41
[63] Seo S *et al* 2004 Reproducible resistance switching in polycrystalline NiO films *Appl. Phys. Lett.* **85** 5655–7
[64] Rohde C, Choi B J, Jeong D S, Choi S, Zhao J-S and Hwang C 2005 Identification of a determining parameter for resistive switching of TiO_2 thin films *Appl. Phys. Lett.* **86** 262907
[65] Yang J-B *et al* 2014 Dual operation characteristics of resistance random access memory in indium-gallium-zinc-oxide thin film transistors *Appl. Phys. Lett.* **104** 153501
[66] Felfel A *et al* 2020 Quantifying the benefits of monolithic 3D computing systems enabled by TFT and RRAM *2020 Design, Automation & Test in Europe Conf. & Exhibition (DATE) (9–13 March 2020)* pp 43–8

IOP Publishing

Semiconducting Metal Oxide Thin-Film Transistors

Ye Zhou

Chapter 11

Semiconductor metal oxide thin-film transistor for artificial synapse

Jie Jiang, Yanran Li and Xiaoliang Liu

Artificial intelligence has penetrated into all aspects of our lives and provided great convenience for us. Using electronic devices to mimic biological synapses is a new research field. At present, many emerging devices mimicking biological synapses are proposed, such as phase change memories, resistive-random-access memories, electric-double-layer transistor, heterojunction transistor, and ferroelectric transistors. Metal oxide thin-film transistors are attracting wide attention for mimicking artificial synapse due to the good stability, high mobility, and flexible 'see-through' microelectronic characteristics, etc. This chapter introduces the performance of three popular artificial synaptic transistors based on the metal oxide semiconductor films of recent years, providing a great step for learning and realizing more synaptic functions in the future. The important challenges of oxide transistor-based artificial synapses are also discussed.

11.1 Introduction

With transistor technology becoming more and more mature and approaching the limit of Moore's law, it is very urgent to pursue next-generation electronic equipment with higher performance [1, 2]. For this reason, neuromorphic computing has attracted growing interest. Generally, neuromorphic computing is an information processing approach inspired by human brain operation, aiming to imitate the energy-efficiency, high performance, and elasticity of the brain [3, 4]. It is the ultimate goal of artificial intelligence to use neuromorphic computing to mimic the processing of information in the human brain.

In fact, the human brain can be regarded as the most intelligent information-processing system. The close connection of neurons is the pivotal element in the human brain which is composed by 10^{15} synapses, forming a complicated neural system [5, 6]. Each synapse concatenates two adjacent neurons, where the

doi:10.1088/978-0-7503-2556-1ch11

communication is realized through the transmission function of synapse. The transmission and processing of voltage pulse signals by synapses are the basic ways for the brain to realize cognitive and learning functions. The signals generated by external stimuli reach the synapse through the incoming nerve, which is processed by the synapse and then transmitted to the brain. Besides, these synapses can not only store information, but also perform complex operations at the same location, which enables the network to perform intelligent computation in a large-scale parallel framework. As a result, researchers have the incentive to develop neuromorphic computing systems that can match or even exceed the cognitive ability and energy efficiency of the human brain. Therefore, it is of great significance to develop a single electronic device with synaptic function for the construction of a neuromorphic computing system [7–11].

Spiked by this, many synaptic devices have been developed to effectively mimic the operation of the brain, such as memristors, atomic force switches, and transistors [12–18]. Among them, three-terminal neuromorphic transistor has attracted much attention, because it enables the signal to be transmitted with the learning process [19, 20]. The metal oxide thin-film transistor (MOTFT) has been considered to be a significant future direction of evolution for the flat panel display industry. In the last 10 years, MOTFT has been regarded as the most attractive three-terminal transistor device due to their intriguing advantages, such as high mobility (>10 $cm^2 V^{-1} s^{-1}$), low-temperature fabrication process, high transparency, and large-area production [21–23]. At present, MOTFTs are promising for applications in smart sensors, active matrix displays, and active light displays. In addition to the large-area displays, MOTFTs have aroused wide attention due to their growing progress in new fields, such as ionic-liquid-gated correlated-oxide devices, oxide-based electric-double-layer (EDL) transistors, and restickable neuromorphic oxide transistors [24–29]. In this section, we mainly focused on the neuromorphic MOTFT which can be simply divided into three different kinds of artificial synaptic devices (EDL MOTFT, ferroelectric MOTFT, and photoelectric MOTFT).

11.2 Machanism and parameters of MOTFT

MOTFT is a special type of TFT which is based on the large-area thin-film deposition process. TFT is the core device of flat panel display. For the common active matrix liquid crystal display (AMLCD) or active matrix organic light-emitting semiconductors (AMOLED), every pixel depends on TFTs for switching and driving processes [30]. At present, the MOTFT is also deemed to be the promising transistor for the next-generation of flat panel display due to the great advantages of its high mobility, good uniformity in large area, and low-temperature deposition process [31–33].

11.2.1 Structure of MOTFT

TFT controls the channel current between source and drain by applying a gate voltage. It consists of gate layer, dielectric layer, active layer, source and drain electrode [34]. The top gate structure or bottom gate structure of the transistor can be differentiated by the gate position. On the basis of the position of source and drain electrodes, the TFT can be further separated into staggered and coplanar structure according to Weimer's definition (figure 11.1) [35]. Therefore, a TFT is usually divided into the following four structures: bottom gate bottom contact, top gate top contact, top gate bottom contact, and top gate top contact. For the sake of improve the stability of the device, a passivation layer is usually needed [36]. Figure 11.2 shows the representative structure of the MOTFT. Herein, the size of source and drain electrode determines the channel length (L) and width (W) of the TFT.

11.2.2 The principle of MOTFT

For the n-type TFT, the drain current from the drain to the source depends on the presence of accumulation of the electron layer at the channel/insulator interface (figure 11.3).

(1) Cut-off region

If the gate–source voltage (V_{GS}) is less than the turn-on voltage (V_{ON}), there is no drain current because the electrons can not be induced in the channel. As shown in figure 11.3(a). Therefore, drain current of source and drain cannot flow due to the low carrier concentration of the channel.

(2) Linear region

The carrier transport can be exhibited if the V_{GS} is larger than V_{th} (figure 11.3(b)). In practical terms, the electrons of accumulation layer channel are injected from the source into the gate. The injected electrons on transmitting through the low resistance accumulation layer are collected in the drain terminal. The gate overvoltage (V_{GS}–V_{th}) is dependent on the magnitude of the drain current, which determines the density of the charge in the accumulation layer channel. It is worth noting that, compared to

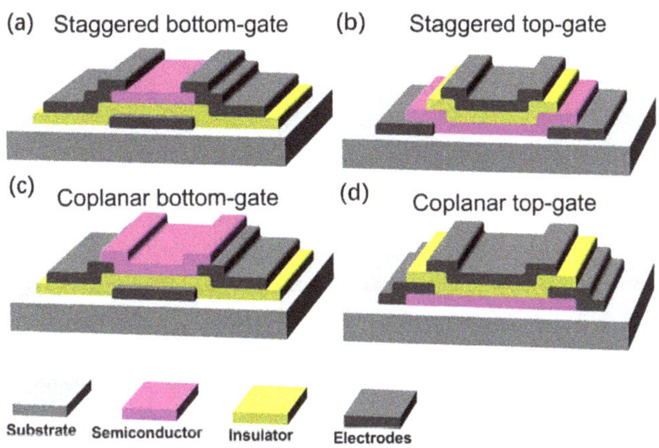

Figure 11.1. (a) Staggered bottom-gate. (b) Staggered top-gate. (c) Coplanar bottom-gate. (d) Coplanar top-gate. [35] John Wiley & Sons. © (2012) WILEY-VCH Verlag GmbH & Co., KGaA, Weinheim.

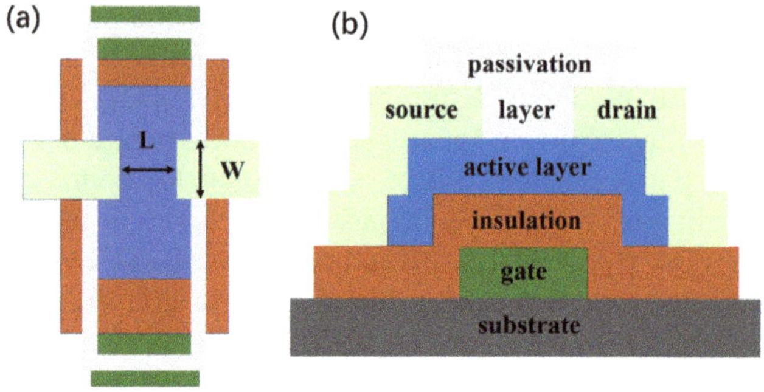

Figure 11.2. (a) The top view of the device structure. (b) The structure diagram of the device.

traditional silicon-based materials, oxide semiconductors do not normally exhibit the inversion layer characters.

(3) Saturation region

If V_{DS} is approaching to the $V_{GS}-V_{th}$, the electronic channel would pinch off, and the drain voltage can be defined to be V_{DSAT}. With the increase of V_{DS}, the pinch off point of the channel stretches over the length of ΔL, as shown in figure 11.3(c). When the V_{DS} continues to increase, the voltage is mainly focused on the pinch off region, while the voltage dropped on the conducting channel remains unchanged. Finally, the I_{DS} tends to be saturated, which is called the saturation region.

11.2.3 Main device parameters of MOTFT

11.2.3.1 Mobility

Field-effect mobility (μ) represents the average drift velocity of carrier in the action of electric field, which can be calculated by the formula (11.1):

$$v_d = \mu|E| \tag{11.1}$$

where E is the electric field intensity, and v_d is the average drift velocity of carrier. In addition, mobility is related to the conductivity of semiconductor materials as follows:

$$\sigma = nq\mu \tag{11.2}$$

where σ is the conductivity, q is the unit charge, and n is the carrier concentration. The on-state current depends on the channel (L and W), the gate insulation (dielectric constant ε_i), and the applied voltage. Using the gradual channel approximation, I_{DS} is able to express by the following equation:

$$I_{DS} = \frac{W}{L}\mu C_i\left(V_{GS} - V_{th} - \frac{1}{2}V_{DS}\right)V_{DS} \tag{11.3}$$

The threshold voltage of V_{th} can be derived by the transfer characteristic curve [37, 38]. Using the above equation, the standard carrier mobility of MOTFT can be finally evaluated.

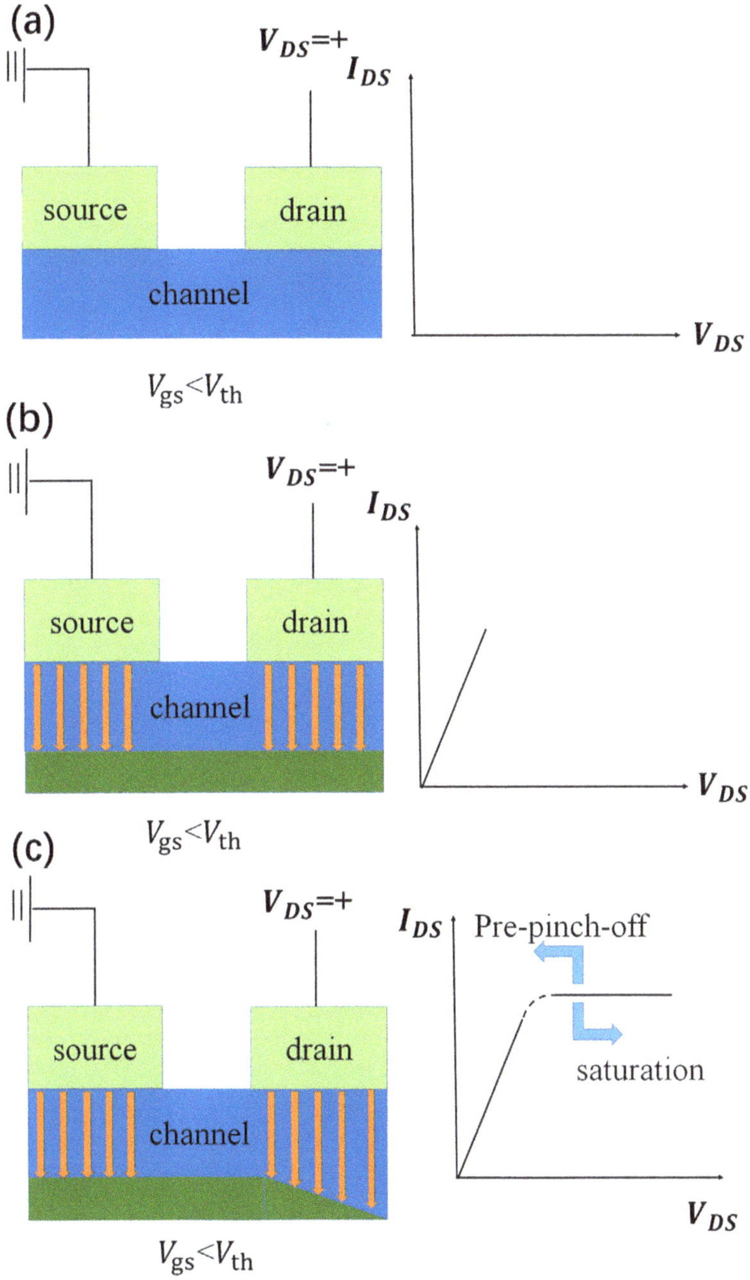

Figure 11.3. (a) Cut-off region. (b) Linear region. (c) Saturation region.

11.2.3.2 Switching ratio

The switching ratio ($I_{on/off}$) of MOTFT indicates the ratio of the on-state current (I_{on}) to the off-state current (I_{off}). The device's ability to turn on and off was expressed by this value. In practical application, I_{off} is an important parameter to determines the minimum power consumption of a MOTFT device.

11.2.3.3 Subthreshold swing

Subthreshold swing (S) is another important parameter of MOTFT, indicating the value of V_{DS} that needs to be increased when log (I_{DS}) is increased by one order of magnitude. The definition is as follows:

$$S = dV_{GS}/\log(I_{DS}). \tag{11.4}$$

11.3 Synapse profile

Synapse refers to the structure in which the impulses of one neuron are transmitted to another neuron or to each other [39]. Synapse is not only the place where neurons have great functional connection, but also the key place of information transmission. According to different transmission signals, synapses can be divided into chemical synapses and electrical synapses. According to whether the signal from presynaptic cells excites postsynaptic cells, the synapses can be divided into two kinds of excitatory and inhibitory synapses, respectively [40, 41]. Figure 11.4 shows the typical neuron diagram, and figure 11.5 shows a synapse diagram.

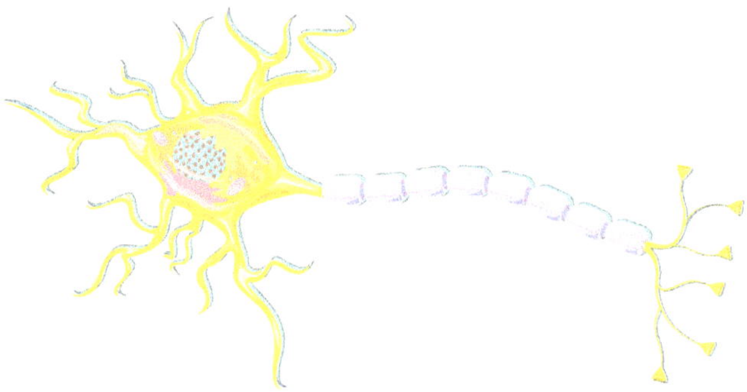

Figure 11.4. Schematic diagram of a neuron.

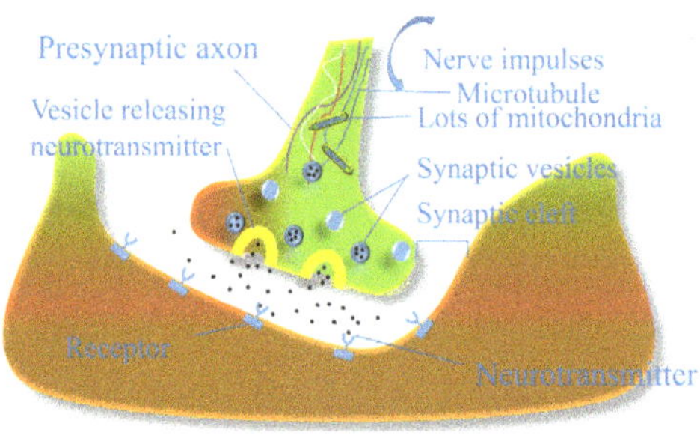

Figure 11.5. Schematic diagram of synaptic structure.

11.3.1 Synaptic structure

The main structural components of neuron are soma, dendrite, and axon. There is only one thick and long axon, which is responsible for receiving information [42, 43]. There are also many dendrites which are thin and short in shape, and responsible for transmitting information. The cell body contains a nucleus which is the core component of the nerve cell. The neuron will receive the signal from the previous neuron, and generate the response in the form of voltage or current pulse [44]. This voltage pulse signal is transmitted from presynaptic neurons through synapses to prominent neurons. A synapse is the gap between the axon terminals of presynaptic neurons and the dendrites of postsynaptic neurons. The information transmitted in the nervous system can pass through synapses in the chemical form of releasing neurotransmitters or in the electrical form of spikes [45]. Chemical synaptic transmission is a complex process. First, electrical signals in the form of spike pulse will open ion channels in presynaptic neurons, release calcium ions, and realize the transformation from electrical signals to chemical signals. Secondly, the migrating calcium ions trigger synaptic vesicles which exist at the end of presynaptic dendrite contacts and release neurotransmitters to the synapses. Finally, the chemical signals in the form of neurotransmitters stimulate the channel receptors on the dendrites of postsynaptic neurons, which make sodium ions migrate to the postsynaptic neurons, and the chemical signals finally change back to electrical signals. When the membrane potential exceeds a certain threshold, the postsynaptic neurons will produce another spike pulse and repeat the previous process to transmit to the next neuron. In neuroscience, the enhanced electrical signals produced by the postsynaptic neurons after receiving neurotransmitters are called the excitatory postsynaptic current or potential according to their different forms [46, 47].

11.3.2 Basical characteristics of synapses

Synaptic plasticity is considered as the construct connections between neurons whose connection strength can be regulated [48]. Synaptic facilitation is the phenomenon that makes certain physiological processes easy to occur through synaptic transmission. The short-term synaptic plasticity mainly includes short-term depression (STD) and potentiation (STP) [49, 50]. And the long-term synaptic plasticity mainly includes (LTP) and long-term depression (LTD) [51].

1. LTP, also known as long-term gain effect, is a kind of continuous enhancement phenomenon occurring in the signal transmission between two neurons because of the synchronous stimulation of two neurons [52, 53].
2. LTD, also called long-term inhibition, refers to the decrease of activity dependence of synaptic efficacy of neurons lasting for hours or longer after long-term stimulation. LTD behavior occurs in many areas of the central nervous system whose mechanism varies with brain area and development process [53, 54].
3. STP is considered to be the enhancement of synaptic transition which lasts only from milliseconds to seconds. In particular, paired-pulse facilitation

(PPF) is one of the most important features of synapse in which excitatory postsynaptic current (EPSC) can be induced by two successive spikes [51, 55].

4. STD is the decrease of synaptic transmission which lasts from milliseconds to seconds. In the human brain, the extent of cortical excitation can be efficiently regulated because the STD of synapse is able to adjust the balance between the cortical excitation and inhibition. Particularly, the realization of various functions in cortical circuits depends on the balance between excitation and depression [56].

11.4 Neural bionic applications of MOTFT

Based on the cognition of the working process, algorithm, and composition of neural system, neuromorphic computing is very promising for the future of an artificial intelligence system [57–68]. In terms of energy efficiency, fault tolerance, self-learning, or compactness, the conventional von Neumann computers have not achieved the robustness of our human brain [69, 70]. Breaking this von Neumann bottleneck is a grand challenge for the next-generation of electronic devices. The nervous system is primarily comprised of neurons which are connected to each other through synapses. The key element of self-learning, dynamic logic, and memory functions of the brain is the transportation of ions (Ca^{2+}, Na^{+}, Mg^{2+}, etc) in the synapses [15, 71]. Since the synapse is widely considered as the key node to achieve neural function, a single neuromorphic device lays the foundation for the grand goal of artificial neural networks. For this purpose, a variety of ion/electron hybrid MOTFTs have been proposed. These devices include two-terminal based memristors (phase change, resistive change, and atomic switch), and three-terminal field-effect transistors (oxide heterojunction, ferroelectric, electric-double-layer (EDL), etc) [72–76].

11.4.1 EDL MOTFT

In recent years, EDL transistors with large capacitance have been widely considered, and some new surface physical phenomena caused by electric field have attracted much attention, such as interface superconductivity and insulator to metal transition [77–81]. The electrochemical and electrostatic modifications of carrier density ascribed from mobile ions are very similar to those in the nervous system, which makes EDL transistors a perfect candidate for synaptic bionics [82].

11.4.1.1 EDL principle

The dielectric polarization mechanism of ionic liquid and polymer electrolyte is generally considered as the EDL mechanism. Figure 11.6 simply depicts the polarization principle of such gate media [83]. When no voltage is applied to the device, the electrolyte contains a large number of free-distributed mobile ions embracing anions and cations. However, at this time, the anions and cations are freely distributed, showing the overall electric neutrality. When the electrode is applied with a positive voltage, the anions/cations in the electrolytes will migrate towards the positively/negatively charged electrode, respectively. These mobile

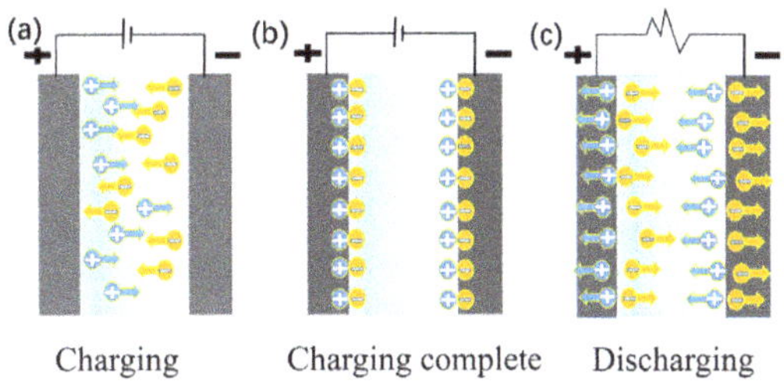

Figure 11.6. Schematic diagram of EDL.

cations would finally reach the interface between the gate and the channel and induce mirror charges with opposite charge polarity on the surface of the channel. There is a strong interface capacitance between the mirror charge and the ions. This capacitance is called the EDL capacitance which is usually more than 1 μF cm^{-2} [84–87].

According to whether the ions in the electrolyte can penetrate into the semiconductor channel layer, the EDL transistor can be divided into electrostatic transistor and electrochemical transistor. Figure 11.6(a) is the sectional diagram of the top-gate EDL transistor. The charging process of ions can be induced under the action of external electric field (figure 11.6(b)). On the other hand, the ions in the electrolyte can be also possibly doped into the semiconductor channel layer, as shown in figure 11.6(c). Under negative bias, the anions in the electrolyte migrate to the gate/electrolyte interface and penetrate from the interface into the semiconductor film, which is called electrochemical doping. It changes the physical properties of semiconductor channel layer, which is even irreversible for a long time [83].

11.4.1.2 Nueromorphic EDL MOTFT

Mimicking the functions of synapses by EDL transistor is of great significance in many synaptic electronic devices and widely used in flexible neuromorphic devices, implantable electronics, and conformable bionic sensors [74, 88, 89]. In fact, a number of important neuromorphic computations have been demonstrated recently, including dynamic filtering, PPF, pattern recognition, etc [76, 90, 91]. In 2018, environment-friendly artificial synapses with learning abilities based on flexible indium tin oxide (ITO) EDL devices were proposed [92]. A flexible ITO-based synaptic transistor with solution-processed chitosan-based polysaccharide electrolyte dielectric was prepared on a polyethylene terephthalate (PET) substrate. Presynaptic spikes were applied on the gate, and the synaptic responses can be measured at the drain terminal. As shown in figure 11.7(a), ITO channel conductance was taken as synaptic weight and the gate voltage pulse as presynaptic stimulus, respectively. A typical EPSC response of ITO-based transistor can be triggered when the ions can be induced by a presynaptic spike on a time scale of a few milliseconds (figurc 11.7(b)). The EPSC gradually decayed back to the basic level because of the concentration gradient of protons. A functional relationship between PPF ratio (A_2/A_1) and pulse time interval (ΔT) was shown in figure 11.7(c). No current facilitation can be observed with a long pulse interval since the prior

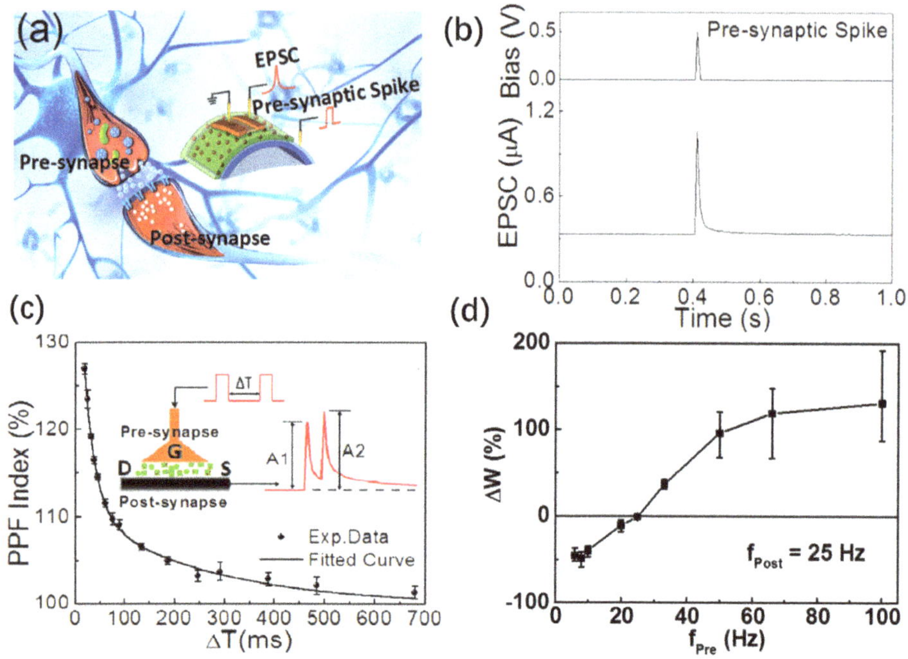

Figure 11.7. (a) Simple schematic diagram of artificial synapse based on an ITO-based EDL transistor. (b) EPSC of the ITO-based synaptic transistor stimulated by the gate spike (0.5 V, 10 ms). (c) PPF ratio (A_2/A_1) as a function of the pulse interval (Δt). Inset: PPF index is measured for 10 times at each ΔT. Reprinted with permission from [92], Copyright (2018) American Chemical Society. (d) Distribution and mean values of synaptic weight change after experiencing SRDP learning rules. Reprinted with permission from [93], Copyright (2016) American Chemical Society.

spike diffused back to the equilibrium position. The proof-of-principle neuromorphic device was reported by Yang *et al* [93] IZO-based, EDL transistor was gated by laterally-coupled grapheme-oxide electrolyte. Dynamic filters and synaptic temporal integration were successfully demonstrated. Moreover, the spike rate-dependent plasticity (SRDP) was also emulated, which plays a key role of long-term plasticity in the neural network [94]. The mean values and distribution of ΔW versus f_{Pre} are plotted in figure 11.7(d). It should be noted that the channel conductance invariably decreases when the input signals are under a critical threshold rate. Nevertheless, when the frequency of excitation went beyond the threshold, the channel conductance increased. The results could lead to a way toward helping to construct a new neuromorphic computing system.

Temporal filtering function was successfully mimicked by the oxide/graphene oxide hybrid neuromorphic device in 2016 [41]. The spatiotemporal dynamic logics based on oxide-based synaptic transistor are illustrated in figure 11.8(a). The different time intervals of two spatiotemporally correlated stimuli play a key role to trigger the EPSCs (figure 11.8(b)). The 'AND' and '$\mathrm{YES}_{\mathrm{V2}}$' logics were realized by two types of spatial summation of presynaptic driving inputs, as shown in figure 11.8(c). In the left local of figure 11.8(c), the 'AND' logic was demonstrated

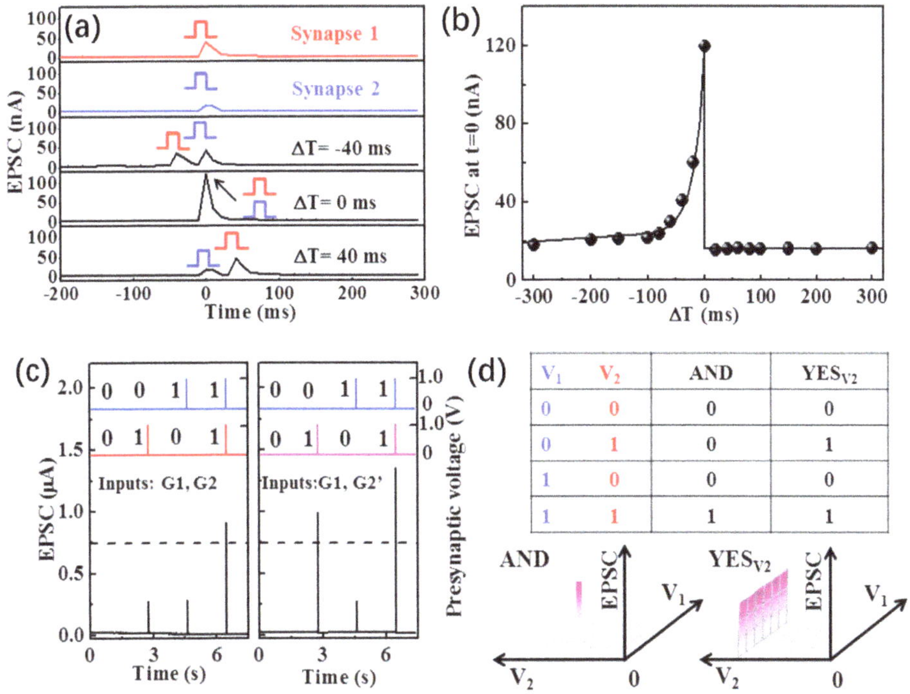

V1	V2	AND	YESV2
0	0	0	0
0	1	0	1
1	0	0	0
1	1	1	1

Figure 11.8. (a) EPSCs triggered by two spatiotemporally correlated inputs (0.2 V, 10 ms) with different time intervals of −40, 0, and 40 ms, respectively. The EPSCs triggered by each individual input are also shown in the top of the figure. (b) The dynamic logic established by the two spatiotemporal correlated spikes. The EPSC amplitudes were. measured at $t = 0$. (c) Input–output characteristics of the 'AND' and 'YES_{V2}' logics from the two combinations of presynaptic input terminals. (d) The truth tables and ideal summation surfaces for the two logics. [41] John Wiley & Sons. © (2016) WILEY-VCH Verlag GmbH & Co., KGaA, Weinheim.

only when the input signals were '11'. On the right region of figure 11.8(c), the 'YES_{V2}' logic was demonstrated if the V_2 input was '1'. The truth tables and the ideal summation surfaces for the two mentioned logics are displayed in figure 11.8(d). Such logic realization is of great significance for capturing the low computing power of neuromorphic computing [95, 96].

In order to obtain an electrolyte gated synaptic transistor with great retention properties, Jin and co-workers designed a $SrCoO_{2.5}$ (BM-SCO) based transistor with O^{2-} ions migration (figure 11.9(a)) [97]. They demonstrated that an antiferromagnetic insulator, i.e. BM-SCO, can be converted into ferromagnetic metal perovskite by electrolyte-gating. The reversible phase transformation of BM-SCO and PV-SCO are shown in figure 11.9(b). The oxygen ions act as neurotransmitters to transfer information from presynaptic neurons to postsynaptic neurons. The LTP and LTD process in a synaptic transistor can be achieved via applying a train of presynaptic spikes of positive voltage (figure 11.9(c)). As shown in figure 11.9(d), the 11 cycling LTP and LTD were used to indicate good repeatability. In the human brain, the weight of synapses between neurons is regulated by STDP which is the basic learning

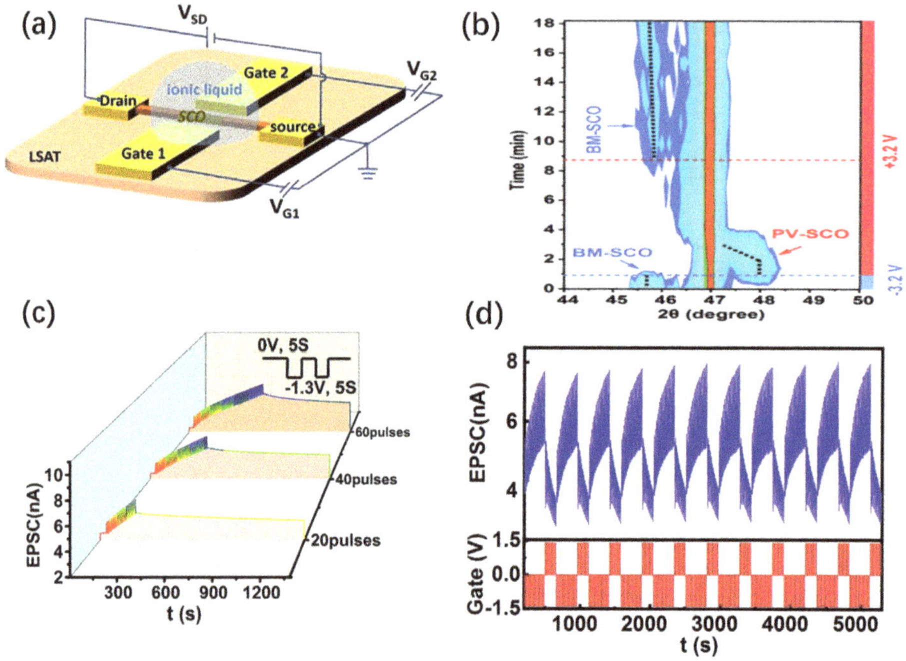

Figure 11.9. (a) Schematic diagram of the three-terminal SCO synaptic device. (b) The transformation from BM-SCO to PV-SCO is realized by applying the gating of +3.2 V; the transformation from PV-SCO to BM-SCO is realized by applying the gating of −3.2 V. (c) The EPSCs triggered by a series of pulses (20, 40, 60 spikes) with the same amplitude of voltage (−1.3 V). (d) Repeatability of synaptic long-term potentiation and depression. [97] John Wiley & Sons. © (2019) WILEY-VCH Verlag GmbH & Co., KGaA, Weinheim.

rule of biological synapses. In canonical STDP, it would result in an LTP behavior if the postsynaptic spikes occurred over tens of milliseconds after presynaptic spikes, whereas the reverse spike activity would induce LTD [98]. Significantly, both STDP and classic conditioning are related to temporally-coupled stimuli [99]. The manufacturing process of the device is demonstrated in figure 11.10(a) [100]. Interestingly, the reported neuromorphic MOTFT was extremely easily dissolved in deionized (DI) water. The Hebbian STDP, anti-Hebbian STDP, symmetrical STDP, and visual STDP are demonstrated (figure 11.10(b)). In Pavlov's dog experiment, feeding food was called unconditioned stimulus (US), while bell ringing and salivation were called neutral stimulus (NS) and unconditioned response (UR), respectively. Originally, bell ringing cannot cause saliva to be secreted, while feeding food (US) can result in salivation (UR). After repeated training, there was a connection between US and NS, i.e. a conditioned response (CR) was triggered. Correspondingly, NS is considered as CS. As shown in figure 11.10(c), paired 'Bell' (NS) and 'Food' (US) signals were applied with Δt_{US-NS} of 20 or −20 ms. The established relation between US and CS experienced extinction when 90 successive CS stimuli were applied after 2 min of the training. The extinction was realized 2 min

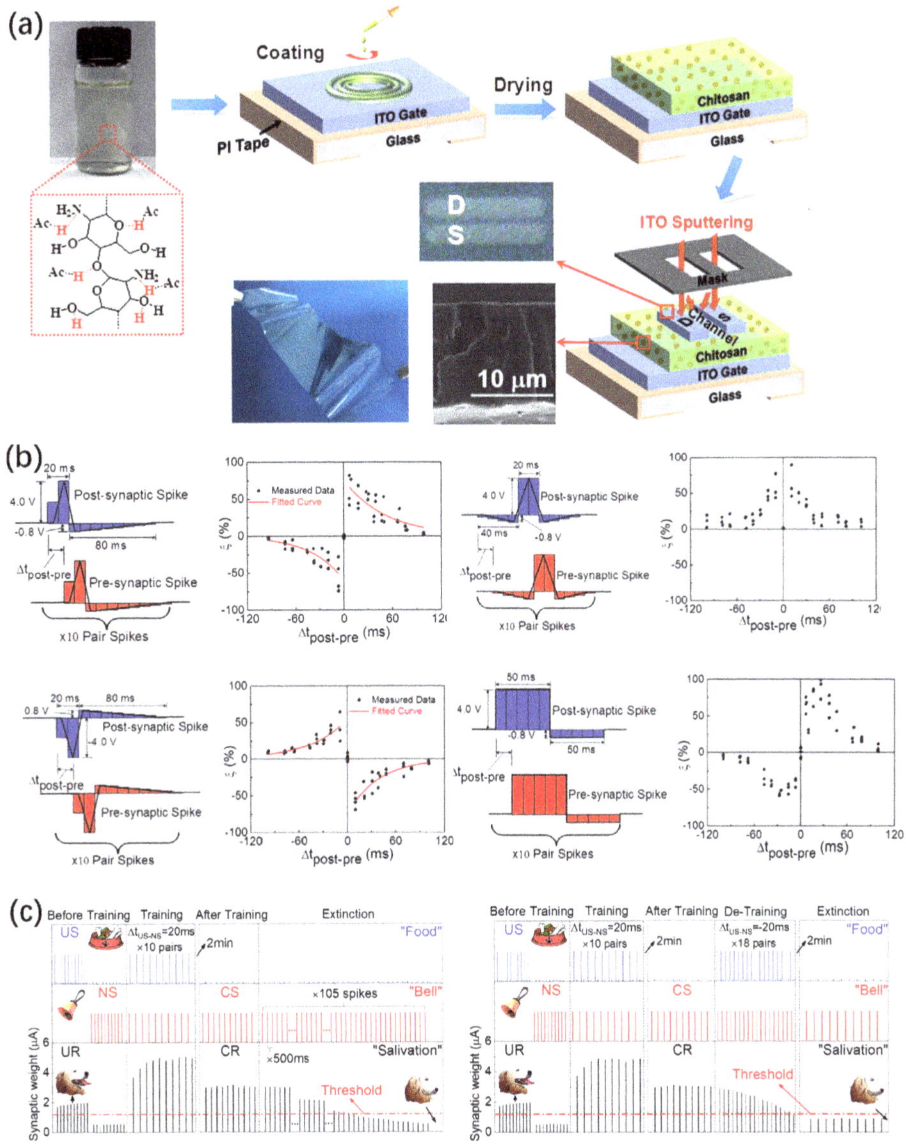

Figure 11.10. (a) Schematic diagram for fabricating chitosan film on polyimide (PI) tape with the ITO channel. (b) Pre- and postsynaptic spikes used in Hebbian STDP measurements (left) and the corresponding STDP forms (right) in artificial synapses. Other forms of STDP such as symmetrical STDP, anti-Hebbian STDP, and visual STDP (left to right). (c) Pavlovian associative learning and the extinction of the proposed neuromorphic transistor. [100] John Wiley & Sons. © (2018) WILEY-VCH Verlag GmbH & Co., KGaA, Weinheim.

later after 18 pairs of CS/US stimuli. Actually, the process of detraining is called conditioned inhibition or reverse regulation in a biological system.

In recent years, the solution-based approach has attracted increasing attention due to its atmospheric processing, lower preparation cost, simplicity, and easy processability and tunable electronics characteristics. In 2019, Shan and co-workers demonstrated a flexible polymer-electrolyte-gated synaptic transistor based on solution-processed In_2O_3 which was fabricated on polyimide (PI) substrate by a nontoxic water-inducement method [101]. Figure 11.11(a) displayed the fabrication process of a flexible synaptic transistor. The In_2O_3 precursor solution was prepared by $In(NO_3)_3{\cdot}H_2O$ which was dissolved in DI water. Then polymer electrolyte solution was prepared by mixing poly (ethylene oxide) (PEO) and lithium perchlorate ($LiClO_4$) in acetonitrile. The weight loss, ascribing to the evaporation of water, occurred while the annealing temperature was raised from 30 to 180 °C (figure 11.11(b)). Figure 11.11(c) shows the acceptable cycle tolerance and stability performances of this kind of MOTFT. The short-term memory and long-term memory were further discussed on the MOTFT by studying the image memorization with 5×5 array (figure 11.11(d)). The symbol of '+' was almost indistinguishable after 5 min. The symbol of 'H' remains highly recognizable after 30 min since the electrochemical doping. It demonstrated that negative pulses can erase the symbol of 'H' and the image can be written/erased once more when the array returns to its initial states. These results indicated potential applications of EDL MOTFT for the neuromorphic cognitive systems.

11.4.2 Ferroelectric MOTFT

At present, a variety of devices have been successfully proved to be synaptic devices to implement neuromorphic computing. However, some of them have suffered from undesirable variations, nonlinear weight-update, small on/off ratio, and poor data retention. Recently, ferroelectric MOTFTs have attracted increasing attention for the great potential in solving these issues [102]. The carrier densities of channel layers depend on the ferroelectric polarization effect of dielectric layer. Generally, the ferroelectric material is a kind of material with spontaneous polarization, which can be reversed even without the external electric field. Some of the spontaneous polarization does not disappear when the external electric field is removed. The remaining polarization is called residual polarization. Because of the existence of residual polarization, when the applied electric field disappears, the channel surface in the transistor can still remain in the state of accumulation or inversion, thus maintaining the switching state of channel [103].

11.4.2.1 The application of ferroelectric MOTFT in neuromorphic electronics

Recently, ferroelectric MOTFTs have been intensively investigated as a new kind of promising platform for mimicking biological synapses owing to their nondestructive readout and low power consumption. For instance, Nishitani *et al* reported a three-terminal ferroelectric synapse device by using ZnO as the semiconductor (figure 11.12(a)) [20]. As shown in figure 11.12(b), non-volatile channel conductance with the multilevel modulation was successfully realized. The schematic cross-sectional picture of the as-fabricated ferroelectric MOTFT can be shown in figure 11.12(c) [103].

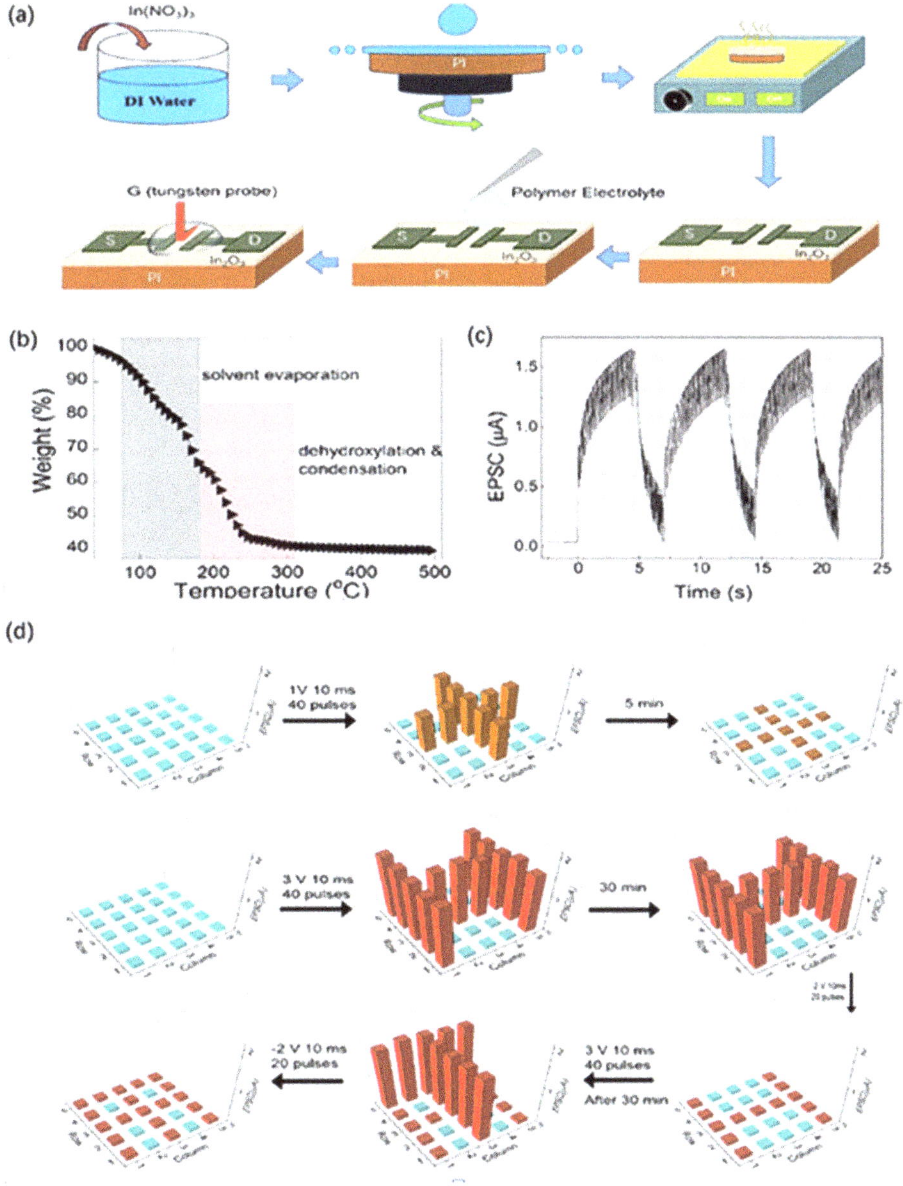

Figure 11.11. (a) Fabrication process for flexible electrolyte gated In_2O_3 synaptic transistor. (b) Thermal behavior of WI In_2O_3 xerogel. (c) Repeatability of long-term potentiation and depression. (d) A 5×5 synapse array for pattern learning. Reprinted with permission from [101], Copyright (2020) American Chemical Society.

The ZnO layer was deposited on the ferroelectric gate Pb $(Zr,Ti)O_3$ which greatly adjusted the channel conductance and got good retention properties. The inset of figure 11.12(d) shows that the changes of conductance with t_D were consistent with the characteristic of symmetric STDP rule.

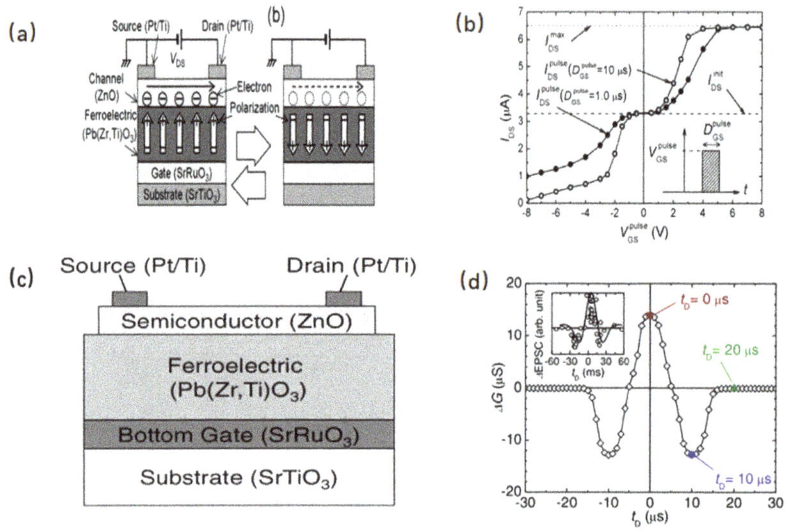

Figure 11.12. (a) Schematic diagram of the ferroelectric-gate synaptic ZnO transistor with upward and downward polarization. (b) The change of the drain current before (I_{DS}^{init}) and after (I_{DS}^{pulse}) applying a square shaped pulse gate voltage with various heights (V_{GS} pulse) and widths (D_{GS} pulse). Reprinted from [20], with the permission of AIP Publishing. (c) Schematic cross-sectional diagram of the FeFET. I_{DS} is measured by applying the drain–source voltage $V_{DS} = 0.1$ V. (d) The change in the conductance ΔG as a function of t_D. The inset displays the measured change in the EPSC of a hippocampal neuron in the rat, which corresponds to changes in the synaptic weight. Reproduced from [103], © 2013 The Japan Society of Applied Physics. All rights reserved.

Recently in 2019, Kim and Lee proposed ferroelectric analog transistors based on nanoscale HfO_x dielectric and IGZO channel [102]. The schematic structure can be shown in figure 11.13(a). Analog memory properties are very important for realizing efficient neural network systems of on-chip storage. As shown in figure 11.13(b), ferroelectric MOTFT displays the great potentiation/depression behaviors and good linearity plasticity. Figure 11.13(c) is a two-layer neural network of perceptron. The good variability of ferroelectric MOTFT is very helpful to improve the recognition accuracy, as shown in figure 11.13(d). These results are meaningful for the development of a neuromorphic hardware system based on ferroelectric MOTFT.

11.4.3 Photoelectric MOTFT

Phototransistor is a photoelectric device composed of three-terminal devices either bipolar transistor or field effect transistor. Light is absorbed in the active region of this kind of device, which produces photocarriers and photocurrent gain through the internal electric amplification mechanism. Phototransistors are mainly divided into bipolar phototransistors and field-effect phototransistors. Herein, photoelectric MOTFs are focused on the latter case.

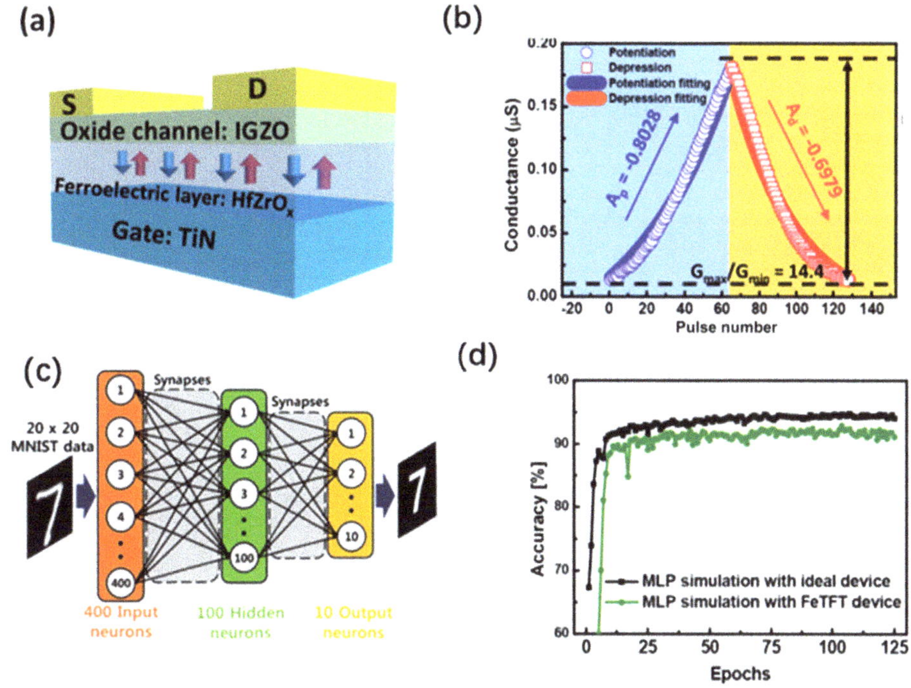

Figure 11.13. (a) Schematic diagram of IGZO FeTFT. (b) Potentiation and depression properties of FeTFT with incremental pulse scheme. (c) Schematic diagram of multilayer perceptron neural network. (d) Pattern recognition accuracy compared to that of an ideal neuromorphic device. Reprinted with permission from [102], Copyright (2019) American Chemical Society.

11.4.3.1 Applications of photoelectric MOTFT in neuromorphic electronics

In daily life, face recognition services are ubiquitous, which is a typical conversion of optical signals into electrical signals. The optical sensors or photoreceptors are an indispensable part of the transformation [104]. Synaptic devices with the optical control are attracting considerable attention, since it exhibits low interconnection energy loss, large bandwidth, and ultrafast signal transmission. A light-stimulated synaptic transistor with the $InGaZnO_x$–Al_2O_3 thin film structure was reported by Lee and co-workers in 2016 [105]. The structure illustration of the IGZO based synaptic device was displayed in figure 11.14(a). The ultraviolet (UV) light pulse was regarded as presynaptic input and the response of postsynaptic current was measured in the drain terminal. A lot of electron–hole pairs was induced after the exposing the UV light in IGZO channel because the photon energy of UV light (3.4 eV) is larger than the bandgap of IGZO (3.36 eV) (the inset of figure 11.14(a)). In the meantime, some of the photogenerated holes were trapped at the IGZO/Al_2O_3 interface and/or in the Al_2O_3 layer. When the UV stimulus is removed, the postsynaptic current will lag behind disappearing as the trapped holes are gradually released. Figure 11.14(b) shows that the postsynaptic current increased continuously with the continuous UV pulses. Optical PPF behavior is also observed in figure 11.14(c). The increases of channel conductance and characteristic relaxation

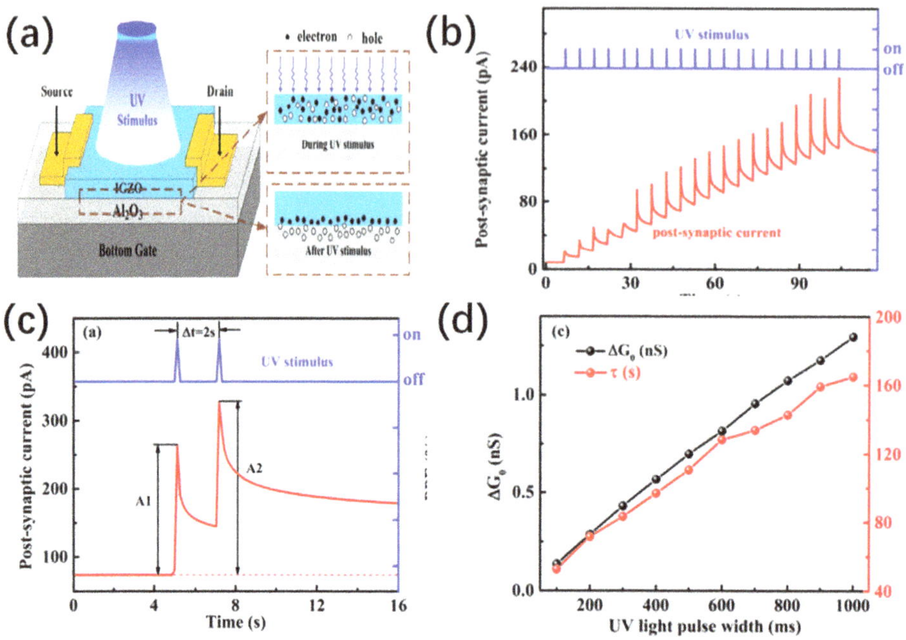

Figure 11.14. (a) The generation of an electron–hole pair in transistors is caused by the stimulation of UV. (b) The interval, intensity and width of the UV pulse train are 5 s, 3 mW cm^{-2}, and 100 ms; and the postsynaptic current was measured at $V_D = 0.5$ V. (c) A pair of presynaptic spikes and the EPSC triggered by two spikes with an inter-spike interval time of 2 s. A_1 and A_2 represent the amplitudes of the first and second EPSCs, respectively. (d) ΔG_0 and τ yielded from the fittings as a function of the UV light pulse width. Reprinted from [105], with the permission of AIP Publishing.

time indicate that raising the pulse width of UV light can result in a typical transition from STM to LTM (figure 11.14(d)).

In 2017, Kim and co-workers reported an optoelectronic synaptic device utilizing photodynamic amorphous IGZO to mimic synaptic functions such as LTM, STM and STDP [106]. The intrinsic persistent photoconductivity (PPC) performance of IGZO plays a key role in emulating synaptic functions successfully. They have demonstrated the light-adjustable optoelectronic neuromorphic transistor array to implement environment-adaptable visual perception function. As shown in figures 11.15(a) and (b), the light-adjustable optoelectronic neuromorphic circuit based on photovoltaic divider (CdSe photosensor + a-IGZO load transistor) and an ionotronic a-IGZO TFT can mimic well an efficient visual system. The demands of the different environmental conditions cannot be satisfied by using the traditional artificial visual systems. Hence, implementing the environment-adaptable function is of great significance for artificial visual systems. Figure 11.15(c) explained the concepts of scotopic and photopic adaptation. Divided voltage can be generated by the photovoltaic divider under different light irradiation and load voltage conditions. In this work, the pattern 'H' with red light illuminance gradation was used to examine the environment-adaptable perception behaviors (figure 11.15(d)). The encoded image patterns show that the gradual patterns can be clearly distinguished

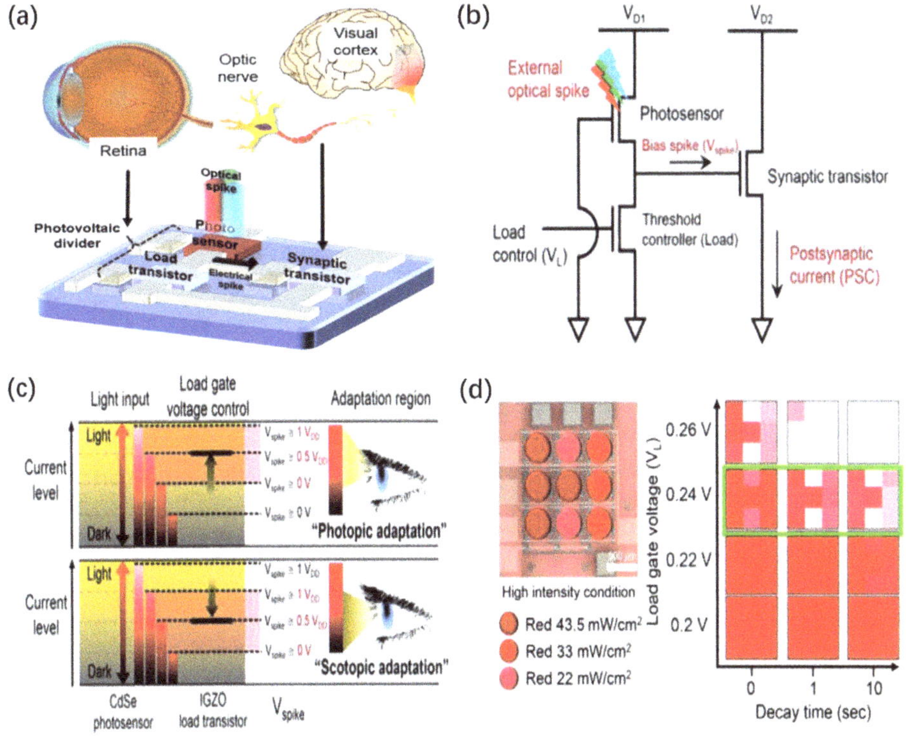

Figure 11.15. (a) Schematic diagram of the light-adjustable optoelectronic neuromorphic circuit and the structure of biological visual perception system. The neuromorphic circuit comprises a photovoltaic divider (CdSe photosensor + a-IGZO load transistor) and an ionotronic synaptic transistor (a-IGZO TFT), functioning as an artificial retina and the optic nerve, respectively. (b) The circuit structure of the light-adjustable neuromorphic circuit. (c) The photopic and scotopic adaptation mechanism of the photovoltaic divider. (d) Under high-intensity (22–43.5 mW cm^{-2}) and low-intensity (10.5–33 mW cm^{-2}), the 3×3 optoelectronic neuromorphic circuit array image and the encoded image in the array are recoded immediately after light irradiation. [106] John Wiley & Sons. © (2019) WILEY-VCH Verlag GmbH & Co., KGaA, Weinheim.

from the visual threshold (V_L), under the condition of higher light intensity, indicating an augmented contrast of the photopic adaptation. However, the disappearance of the encoded patterns was observed under the same adaptation condition at a lower intensity.

Li and co-workers reported a transparent IGZO TFT with chitosan gate dielectrics based on previous artificial synaptic electronics [107]. As shown in figure 11.16(a), EPSC response can be measured under individual and multiple electrical and optical pluses. Figure 11.16(b) shows that the EPSC increased with decreasing of interval time between electrical and optical stimulus when $\Delta t_{\mathrm{preES-preOS}} < 0$. The major reason for this behavior is that the channel conductivity was increased, which is due to the ion migration phenomenon. However, when $\Delta t_{\mathrm{preES-preOS}} > 0$, because light stimulation is not influenced by electrical stimulation, the EPSC was thus in a stable state. The spatiotemporal learning rule (STLR) was further demonstrated by this synergy

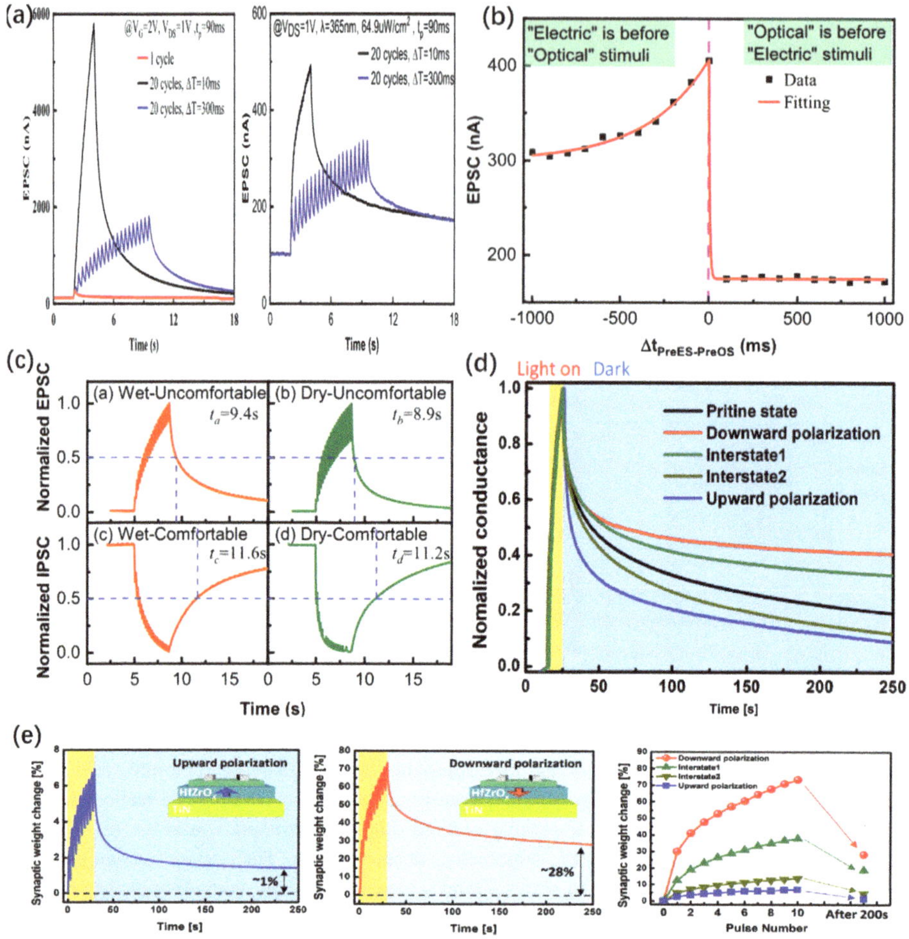

Figure 11.16. (a) The effect of applying the pulse sequence of electrical stimulation with different intervals (ΔT = 10–300 ms, V_G = 2 V, t_p = 90 ms, 20 identical consecutive pulse). (b) Relationship between the channel current and the pulse sequence by optical simulation with different interval times (ΔT = 10–300 ms, t_p = 90 ms, 365 nm, 64.9 μW cm^{-2}, 20 identical consecutive pulses). (c) Neuroprominent behaviors in discomfort, dry stimulation in discomfort, wet stimulation in comfort, and dry stimulation in comfort, respectively. (The spike amplitude is 2/−2 V, ΔT is 100 ms, t_p is 90 ms, and there are 20 identical consecutive pulses.) Reprinted with permission from [107], Copyright (2019) American Chemical Society. (d) The relaxation behaviors of devices with various polarization states in ferroelectric layer. (e) The long-term plasticity characteristics of transistor in upward and downward polarizations. The synaptic weight changes of the device according to number of applied optical pulses in different polarization states. [108] John Wiley & Sons. © (2020) WILEY-VCH Verlag GmbH & Co., KGaA, Weinheim.

approach of electricity and light. The positive and negative postsynaptic signals represented uncomfortable and comfortable states of humanity, respectively (figure 11.16(c)). Wet and dry measurement conditions corresponded to quiet and noise environment stimuli, respectively. The enhanced behaviors of learning and forgetting characteristics were successfully demonstrated. This means that deeper memories can be obtained in the uncomfortable state and quiet surroundings. Conversely, the next two pictures can be considered as the comfortable states of a person. It was found that the memories produced in noisy environments were forgotten significantly. This ability makes our bodies react differently in various emotional states. At the same time, ferroelectric materials also play an important role in photonic synapse devices. Kim and Lee demonstrated a photonic synaptic transistor utilizing oxide semiconductor and ferroelectric material to adjust synaptic functions [108]. Varying decay conduct of a-IGZO TFT was observed due to the different polarization states (figure 11.16(d)). The reason for the difference in conductance amplification seems to be the different polarization states. When the polarization states were changed from upward to downward, the changes of synaptic weight increased accordingly (figure 11.16(e)). The controlled relaxation of PPC characteristics affects the enhancement of LTP performances of photonic synapses. These results indicate that the synaptic function of the artificial photonic synapse in the neuromorphic oxide-based systems can be regulated by the polarization of the ferroelectric layer. These promising studies lay a solid foundation for the development of optoelectronic synaptic devices.

Li and co-workers demonstrated a double-gate a-IGZO neuromorphic TFT with highly reliable electro-photo-sensitive synaptic functions, as shown in figure 11.17(a) [109]. Figure 11.17(b) displayed the offset influence of EPSC and IPSC for this device. Under the electrical pulse, the UV light can enhance the capability of the electron detrapping and in the meantime the PPC phenomenon can be eliminated

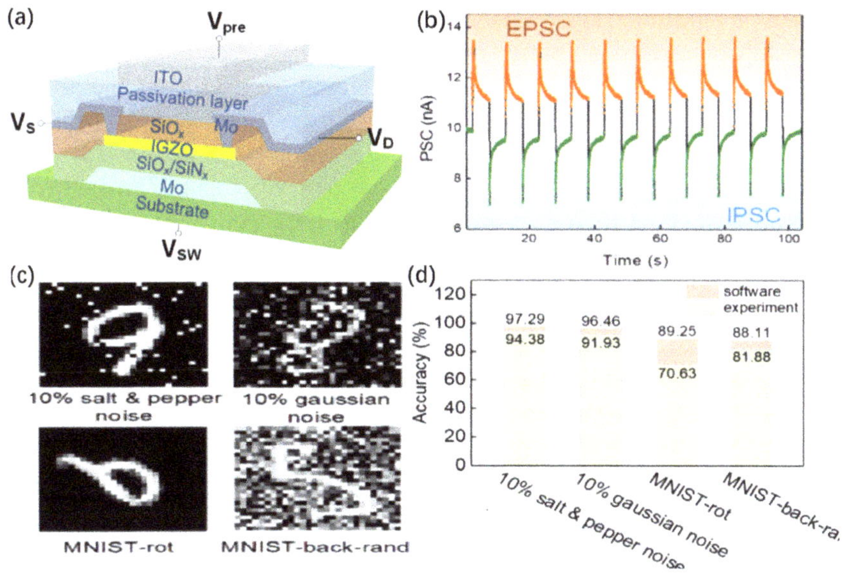

Figure 11.17. (a) Schematic diagram of the device structure of IGZO-based transistor. (b) Interval of alternating application of electric pulse and optical pulse. (c) Schematic illustration of four different noisy testing datasets. (d) The network shows good tolerance to noises and the recognition accuracy remains within acceptable fidelities. Reproduced from [109], with permission of The Royal Society of Chemistry.

via a positive gate voltage under UV light. In addition, they also investigated the influence of inferential datasets with different noises. As shown in figure 11.17(c), 10% salt and pepper noise, 10% Gaussian noise, MNIST-rot and MNIST-back-rand datasets were measured while the trained network suffered from distinct noisy test datasets. It can be seen that great tolerance to noise was observed. And although the recognition accuracy was reduced, it remains to be an acceptable level (figure 11.17(d)). Realization of electro-photo-sensitive artificial synapses with the most advanced a-IGZO TFT technology is of great significance for the integration of the future edge intelligent terminals.

11.5 Summary and outlook

The development of physical synaptic oxide-based devices has made rapid progress in the past few years [110, 111]. The three-terminal/multi-terminal neuromorphic MOTFT has the advantages of multi inputs, which enables parallel data processing and simultaneous learning of the device. Many biological functions of neuromorphic devices, such as voice recognition, sensory perception, and image recognition, have been successfully demonstrated. In this chapter, we have discussed the working principle and the latest development of three kinds of neuromorphic MOTFT, including neuromorphic EDL MOTFT, ferroelectric MOTFT, and optical MOTFT. The neuromorphic EDL MOTFT has attracted the attention of many researchers because of their advantages such as ultra-low operating voltage, low processing cost, and visible light transparency. However, most of these transistors are based on liquid electrolyte dielectric, there are some drawbacks in device stability and system integration [112]. The neuromorphic ferroelectric MOTFT has the characteristics of process stability, fast programming operation, and large switch ratio. However, it usually needs a large working voltage to convert the polarization state of ferroelectric materials. In terms of neuromorphic optical MOTFT, the optical signal can make synaptic transistors achieve large bandwidth and small energy loss of interconnection, thus providing a new emerging neural morphology. Nevertheless, achieving inhibitory synapses with pure optical signals is still a big challenge. Consequently, future efforts need to be made to overcome these weaknesses of neuromorphic MOTFT. At present, there are still several difficulties in the field of neuromorphic MOTFT. For example: (1) the function of synapse is very powerful. However, we only realize a part of synaptic functions, there are still many advanced functions we haven't explored. (2) The device is usually a single device, and there are still few reports on the integration of synaptic devices. Nevertheless, the dream of 'brain-like computing' will eventually become true. In the future, advanced artificial intelligence systems will completely change our lives in medical treatment, education, and transportation, etc.

References

[1] Jiang J, Guo J, Wan X, Yang Y, Xie H and Wan Q 2017 2D MoS_2 neuromorphic devices for brain-like computational systems *Small* **13** 1700933

[2] Wan Q, Sharbati M T, Erickson J R, Du Y and Xiong F 2019 Emerging artificial synaptic devices for neuromorphic computing *Adv. Mater.* **4** 1900037

[3] Wan C, Zhu L, Zhou J, Shi Y and Wan Q 2013 Memory and learning behaviors mimicked in nanogranular SiO_2-based proton conductor gated oxide-based synaptic transistors *Nanoscale* **5** 10194–9

[4] Wan C, Zhu L, Zhou J, Shi Y and Wan Q 2014 Inorganic proton conducting electrolyte coupled oxide-based dendritic transistors for synaptic electronics *Nanoscale* **6** 4491–7

[5] Zhao Y and Jiang J 2018 Recent progress on neuromorphic synapse electronics: from emerging materials, devices, to neural networks *J. Nanosci. Nanotechnol.* **18** 8003–15

[6] Tian H, Mi W, Wang X, Zhao H, Xie Q, Li C, Li Y-X, Yang Y and Ren T-L 2015 Graphene dynamic synapse with modulatable plasticity *Nano Lett.* **15** 8013–9

[7] Li E *et al* 2019 Synaptic transistor capable of accelerated learning induced by temperature-facilitated modulation of synaptic plasticity *ACS Appl. Mater. Interfaces* **11** 46008–16

[8] Shen A M, Chen C-L, Kim K, Cho B, Tudor A and Chen Y 2013 Analog neuromorphic module based on carbon nanotube synapses *ACS Nano* **7** 6117–22

[9] Salonikidou B, Yasunori T, Le Borgne B, England J, Shizuo T and Sporea R A 2019 Toward fully printed memristive elements: a-TiO_2 electronic synapse from functionalized nanoparticle ink *ACS Appl. Electron Mater.* **1** 2692–700

[10] Park Y and Lee J-S 2017 Artificial synapses with short-and long-term memory for spiking neural networks based on renewable materials *ACS Nano* **11** 8962–69

[11] Zhang H, Ju X, Yew K S and Ang D S 2019 Implementation of simple but powerful trilayer oxide-based artificial synapses with a tailored bio-synapse-like structure *ACS Appl. Mater. Interfaces* **12** 1036–45

[12] Jiang S, Nie S, He Y, Liu R, Chen C and Wan Q 2019 Emerging synaptic devices: from two-terminal memristors to multiterminal neuromorphic transistors *Mater. Today Nano* **8** 100059

[13] Ohno T, Hasegawa T, Tsuruoka T, Terabe K, Gimzewski J K and Aono M 2011 Short-term plasticity and long-term potentiation mimicked in single inorganic synapses *Nat. Mater.* **10** 591–5

[14] Chen Y, Yu H, Gong J, Ma M, Han H, Wei H and Xu W 2018 Artificial synapses based on nanomaterials *Nanotechnology* **30** 012001

[15] La Barbera S, Vuillaume D and Alibart F 2015 Filamentary switching: synaptic plasticity through device volatility *ACS Nano* **9** 941–9

[16] Li D, Wu B, Zhu X, Wang J, Ryu B and Liang X 2018 MoS_2 memristors exhibiting variable switching characteristics toward biorealistic synaptic emulation *ACS Nano* **12** 9240–52

[17] Kim S, Kim H, Hwang S, Kim M-H, Chang Y and Park B-G 2017 Analog synaptic behavior of a silicon nitride memristor *ACS Appl. Mater. Interfaces* **9** 40420–7

[18] Lee T-H, Hwang H-G, Woo J-U, Kim D-H, Kim T-W and Nahm S 2018 Synaptic plasticity and metaplasticity of biological synapse realized in a $KNbO_3$ memristor for application to artificial synapse *ACS Appl. Mater. Interfaces* **10** 25673–82

[19] Sun J *et al* 2018 Optoelectronic synapse based on IGZO-alkylated graphene oxide hybrid structure *Adv. Funct. Mater.* **28** 1804397

[20] Nishitani Y, Kaneko Y, Ueda M, Morie T and Fujii E 2012 Three-terminal ferroelectric synapse device with concurrent learning function for artificial neural networks *J. Appl. Phys.* **111** 124108

[21] Xiong W, Zhu L, Ye C, Yu F, Ren Z and Ge Z 2019 Bilayered oxide-based cognitive memristor with brain-inspired learning activities *Adv. Electron. Mater.* **5** 1900439
[22] Barquinha P, Pereira L, Goncalves G, Martins R and Fortunato E 2008 The effect of deposition conditions and annealing on the performance of high-mobility GIZO TFTs *Solid State Lett.* **11** H248–51
[23] Barquinha P, Pereira L, Goncalves G, Martins R and Fortunato E 2009 Toward high-performance amorphous GIZO TFTs *J. Electrochem. Soc.* **156** H161–8
[24] Lin H-C, Shie B-S and Huang T-Y 2014 100-nm IGZO thin-film transistors with film profile engineering *IEEE Trans. Electron Devices* **61** 2224–7
[25] Zhang C *et al* 2018 Ion migration studies in exfoliated 2D molybdenum oxide via ionic liquid gating for neuromorphic device applications *ACS Appl. Mater. Interfaces* **10** 22623–31
[26] Koo J, Yang J, Cho B, Jo H, Lee K H and Kang M S 2018 Nonvolatile electric double-layer transistor memory devices embedded with Au nanoparticles *ACS Appl. Mater. Interfaces* **10** 9563–70
[27] Ji H, Wei J and Natelson D 2012 Modulation of the electrical properties of VO_2 nanobeams using an ionic liquid as a gating medium *Nano Lett.* **12** 2988–92
[28] Xu W, Li H, Xu J-B and Wang L 2018 Recent advances of solution-processed metal oxide thin-film transistors *ACS Appl. Mater. Interfaces* **10** 25878–901
[29] Black J M *et al* 2017 Role of electrical double layer structure in ionic liquid gated devices *ACS Appl. Mater. Interfaces* **9** 40949–58
[30] Yu X, Zhou N, Smith J, Lin H, Stallings K, Yu J, Marks T J and Facchetti A 2013 Synergistic approach to high-performance oxide thin film transistors using a bilayer channel architecture *ACS Appl. Mater. Interfaces* **5** 7983–8
[31] Li Y, Lan L, Sun S, Lin Z, Gao P, Song W, Song E, Zhang P and Peng J 2017 All inkjet-printed metal-oxide thin-film transistor array with good stability and uniformity using surface-energy patterns *ACS Appl. Mater. Interfaces* **9** 8194–200
[32] Hennek J W, Kim M G, Kanatzidis M G, Facchetti A and Marks T J 2012 Exploratory combustion synthesis: amorphous indium yttrium oxide for thin-film transistors *J. Am. Chem. Soc.* **134** 9593–6
[33] Sharma A, Chourasia N K, Pal N, Biring S and Pal B N 2019 Role of electron donation of TiO_2 gate interface for developing solution-processed high-performance one-volt metal-oxide thin-film transistor using ion-conducting gate dielectric *J. Phys. Chem.* C **123** 20278–86
[34] Dou W, Jiang J, Sun J, Zhou B and Wan Q 2011 Low-voltage electric-double-layer TFTs on SiO_2-covered paper substrates *IEEE Electron Device Lett.* **32** 1543–5
[35] Fortunato E, Barquinha P and Martins R 2012 Oxide semiconductor thin-film transistors: a review of recent advances *Adv. Mater.* **24** 2945–86
[36] Sato A, Abe K, Hayashi R, Kumomi H, Nomura K, Kamiya T, Hirano M and Hosono H 2009 Amorphous In–Ga–Zn–O coplanar homojunction thin-film transistor *Appl. Phys. Lett.* **94** 133502
[37] Jiang J, Wan Q, Sun J, Dou W and Zhou B 2011 Junctionless in-plane-gate transparent thin-film transistors *Appl. Phys. Lett.* **99** 193502
[38] Jiang J, Wan Q, Sun J and Lu A 2009 Ultralow-voltage transparent electric-double-layer thin-film transistors processed at room temperature *Appl. Phys. Lett.* **95** 152114
[39] Zucker R S and Regehr W G 2002 Short-term synaptic plasticity *Annu. Rev. Physiol.* **64** 355–405

[40] Zhu L, Wan C, Gao P, Liu Y, Xiao H, Yie J C and Wan Q 2016 Flexible proton-gated oxide synaptic transistors on Si membrane *ACS Appl. Mater. Interfaces* **8** 21770–5

[41] Wan C, Liu Y, Feng P, Wang W, Zhu L, Liu Z P, Shi Y and Wan Q 2016 Flexible metal oxide/graphene oxide hybrid neuromorphic transistors on flexible conducting graphene substrates *Adv. Mater.* **28** 5878–85

[42] Ho V M, Lee J-A and Martin K C 2011 The cell biology of synaptic plasticity *Science* **334** 623–8

[43] Pereda A E 2014 Electrical synapses and their functional interactions with chemical synapses *Nat. Rev. Neurosci.* **15** 250–63

[44] Abbott L and Regehr W G 2004 Synaptic computation *Nature* **431** 796–803

[45] Wen J, Zhu L, Fu Y, Xiao H, Guo L and Wan Q 2017 Activity dependent synaptic plasticity mimicked on indium-tin-oxide electric-double-layer transistor *ACS Appl. Mater. Interfaces* **9** 37064–9

[46] Voglis G and Tavernarakis N 2006 The role of synaptic ion channels in synaptic plasticity *EMBO Rep.* **7** 1104–10

[47] Bi G-Q and Poo M-M 1998 Synaptic modifications in cultured hippocampal neurons: dependence on spike timing, synaptic strength, and postsynaptic cell type *J. Neurosci.* **18** 10464–72

[48] Shen J, Shang D, Chai Y, Wang S, Shen B and Sun Y 2018 Mimicking synaptic plasticity and neural network using memtranstors *Adv. Mater.* **30** 1706717

[49] Wan C, Liu Y, Zhu L, Feng P, Shi Y and Wan Q 2016 Short-term synaptic plasticity regulation in solution-gated indium-gallium-zinc-oxide electric-double-layer transistors *ACS Appl. Mater. Interfaces* **8** 9762–8

[50] Li B *et al* 2018 Mediating short-term plasticity in an artificial memristive synapse by the orientation of silica mesopores *Adv. Mater.* **30** 1706395

[51] Kim M-K and Lee J-S 2018 Short-term plasticity and long-term potentiation in artificial biosynapses with diffusive dynamics *ACS Nano* **12** 1680–7

[52] Pilar-Cuellar F *et al* 2017 Enhanced stress response in 5-HT_{1A}R overexpressing mice: altered HPA function and hippocampal long-term potentiation *ACS Chem. Neurosci.* **8** 2393–401

[53] Markram H, Gerstner W and Sjöström P J 2011 A history of spike-timing-dependent plasticity *Front. Synaptic Neurosci.* **3** 4

[54] Neuhofer D, Lassalle O and Manzoni O J 2018 Muscarinic M1 receptor modulation of synaptic plasticity in nucleus accumbens of wild-type and fragile X mice *ACS Chem. Neurosci.* **9** 2233–40

[55] Li J, Ge C, Lu H, Guo H, Guo E, He M, Wang C, Yang G and Jin K 2019 Energy-efficient artificial synapses based on oxide tunnel junctions *ACS Appl. Mater. Interfaces* **11** 43473–9

[56] Augustin S M and Lovinger D M 2018 Functional relevance of endocannabinoid-dependent synaptic plasticity in the central nervous system *ACS Chem. Neurosci.* **9** 2146–61

[57] Ahmed T, Kuriakose S, Mayes E L H, Ramanathan R, Bansal V, Bhaskaran M, Sriram S and Walia S 2019 Optically stimulated artificial synapse based on layered black phosphorus *Small* **15** 1900966

[58] Chen Y, Qiu W, Wang X, Liu W, Wang J, Dai J, Yuan Y, Gao Y and Sun J 2019 Solar-blind SnO_2 nanowire photo-synapses for associative learning and coincidence detection *Nano Energy* **62** 393–400

[59] He Y, Yang Y, Nie S, Liu R and Wan Q 2018 Electric-double-layer transistors for synaptic devices and neuromorphic systems *J. Mater. Chem. C.* **6** 5336–52
[60] Bao L *et al* 2019 Dual-gated MoS_2 neuristor for neuromorphic computing *ACS Appl. Mater. Interfaces* **11** 41482–9
[61] Jiang J, Hu W, Xie D, Yang J, He J, Gao Y and Wan Q 2019 2D electric-double-layer phototransistor for photoelectronic and spatiotemporal hybrid neuromorphic integration *Nanoscale* **11** 1360–9
[62] Arnold A J, Razavieh A, Nasr J R, Schulman D S, Eichfeld C M and Das S 2017 Mimicking neurotransmitter release in chemical synapses via hysteresis engineering in MoS_2 transistors *ACS Nano* **11** 3110–8
[63] Gou G, Sun J, Qian C, He Y, Kong L, Fu Y, Dai G, Yang J and Gao Y 2016 Artificial synapses based on biopolymer electrolyte-coupled SnO_2 nanowire transistors *J. Mater. Chem.* C **4** 11110–7
[64] Fuller E J *et al* 2017 Li-ion synaptic transistor for low power analog computing *Adv. Mater.* **29** 1604310
[65] Qian C, Kong L, Yang J, Gao Y and Sun J 2017 Multi-gate organic neuron transistors for spatiotemporal information processing *Appl. Phys. Lett.* **110** 083302
[66] Wan C *et al* 2018 An artificial sensory neuron with tactile perceptual learning *Adv. Mater.* **30** 1801291
[67] Sanchez Esqueda I *et al* 2018 Aligned carbon nanotube synaptic transistors for large-scale neuromorphic computing *ACS Nano* **12** 7352–61
[68] Dai S, Wang Y, Zhang J, Zhao Y, Xiao F, Liu D, Wang T and Huang J 2018 Wood-derived nanopaper dielectrics for organic synaptic transistors *ACS Appl. Mater. Interfaces* **10** 39983–91
[69] Kim S, Yoon J, Kim H-D and Choi S-J 2015 Carbon nanotube synaptic transistor network for pattern recognition *ACS Appl. Mater. Interfaces* **7** 25479–86
[70] Yu R, Li E, Wu X, Yan Y, He W, He L, Chen J, Chen H and Guo T 2020 Electret-based organic synaptic transistor for neuromorphic computing *ACS Appl. Mater. Interfaces* **12** 15446–55
[71] Li Z, Wan C, Guo L, Shi Y and Wan Q 2014 Artificial synapse network on inorganic proton conductor for neuromorphic systems *Nat. Commun.* **5** 3158
[72] Hu D, Yang R, Jiang L and Guo X 2018 Memristive synapses with photoelectric plasticity realized in ZnO_{1-x}/AlO_y heterojunction *ACS Appl. Mater. Interfaces* **10** 6463–70
[73] Li G *et al* 2020 Silicon nanomembrane phototransistor flipped with multifunctional sensors toward smart digital dust *Sci. Adv.* **6** eaaz6511
[74] Li J, Jiang D, Yang Y, Zhou Y, Chen Q and Zhang J 2020 Li-ion doping as a strategy to modulate the electrical-double-layer for improved memory and learning behavior of synapse transistor based on fully aqueous solution-processed In_2O_3/AlLiO film *Adv. Electron. Mater.* **6** 1901363.
[75] Halter M, Begon-Lours L, Bragaglia V, Sousa M, Offrein B J, Abel S, Luisier M and Fompeyrine J 2020 Back-end, CMOS-compatible ferroelectric field-effect transistor for synaptic weights *ACS Appl. Mater. Interfaces* **12** 17725–32
[76] Li Y, Fuller E J, Asapu S, Agarwal S, Kurita T, Yang J J and Talin A A 2019 Low-voltage, CMOS-free synaptic memory based on Li_XTiO_2 redox transistors *ACS Appl. Mater. Interfaces* **11** 38982–92

[77] Jiang J, Dai M, Sun J, Zhou B, Lu A and Wan Q 2011 Electrostatic modification of oxide semiconductors by electric double layers of microporous SiO_2-based solid electrolyte *J. Appl. Phys.* **109** 054501
[78] Guo J, Jiang J and Yang B 2018 Low-voltage electric-double-layer MoS_2 transistor gated via water solution *Solid-State Electron.* **150** 8–15
[79] Yuan H, Shimotani H, Ye J, Yoon S, Aliah H, Tzukazaki A, Kawasaki M and Iwasa Y 2010 Electrostatic and electrochemical nature of liquid-gated electric-double-layer transistors based on oxide semiconductors *J. Chem. Soc.* **132** 18402–7
[80] Liu Y, Li B, Zhu L, Feng P, Shi Y and Wan Q 2016 Dopamine detection based on low-voltage oxide homojunction electric-double-layer thin-film transistors *IEEE Electron Device Lett.* **37** 1
[81] Jiang J, Wan Q and Zhang Q 2013 Transparent junctionless electric-double-layer transistors gated by a reinforced chitosan-based biopolymer electrolyte *IEEE Trans. Electron Devices* **60** 1951
[82] Xie D, Jiang J, Hu W, He Y, Yang J, He J, Gao Y and Wan Q 2018 Coplanar multigate MoS_2 electric-double-layer transistors for neuromorphic visual recognition *ACS Appl. Mater. Interfaces* **10** 25943–8
[83] Du H, Lin X, Xu Z and Chu D 2015 Electric double-layer transistors: a review of recent progress *J. Mater. Sci.* **50** 5641–73
[84] Fujimoto T and Awaga K 2013 Electric-double-layer field-effect transistors with ionic liquids *Phys. Chem. Chem. Phys.* **15** 8983–9006
[85] Guo L, Wen J, Zhu L, Fu Y and Xiao H 2017 Humidity-dependent synaptic plasticity for proton gated oxide synaptic transistor *IEEE Electron Device Lett.* **38** 1248–51
[86] Zheng Z, Jiang J, Guo J, Yang J and Gao Y 2016 Bio-inspired coplanar-gate-coupled ITO-free oxide-based transistors employing natural nontoxic bio-polymer electrolyte *Org. Electron.* **37** 474–8
[87] Zheng Z, Jiang J, Guo J, Sun J and Yang J 2016 Chitosan-gated low-voltage transparent indium-free aluminumdoped zinc oxide thin-film transistors *Org. Electron.* **33** 311–5
[88] Zhou J, Liu N, Zhu L, Shi Y and Wan Q 2014 Energy-efficient artificial synapses based on flexible IGZO electric-double-layer transistors *IEEE Electron Device Lett.* **36** 198–200
[89] Hu W, Jiang J, Xie D, Wang S, Bi K and He J 2018 Transient security transistors self-supported on biodegradable natural-polymer membranes for brain-inspired neuromorphic applications *Nanoscale* **10** 14893–901
[90] Zhou Y, Li J, Yang Y, Chen Q and Zhang J 2020 Artificial synapse emulated through fully aqueous solution-processed low-voltage In_2O_3 thin-film transistor with Gd_2O_3 solid electrolyte *ACS Appl. Mater. Interfaces* **12** 980–8
[91] Shao F, Yang Y, Zhu L, Feng P and Wan Q 2016 Oxide-based synaptic transistors gated by sol-gel silica electrolytes *ACS Appl. Mater. Interfaces* **8** 3050–5
[92] Yu F, Zhu L, Gao W, Fu Y, Xiao H, Tao J and Zhou J M 2018 Chitosan-based polysaccharide-gated flexible indium tin oxide synaptic transistor with learning abilities *ACS Appl. Mater. Interfaces* **10** 16881–6
[93] Yang Y, Wen J, Guo L, Wan X, Du P, Feng P, Shi Y and Wang Q 2016 Long-term synaptic plasticity emulated in modified graphene oxide electrolyte gated IZO-based thin-film transistors *ACS Appl. Mater. Interfaces* **8** 30281–6
[94] Rachmuth G, Shouval H Z, Bear M F and Poon C-S 2011 A biophysically-based neuromorphic model of spike rate-and timing-dependent plasticity *Proc. Natl Acad. Sci.* **108** E1266–74

[95] Feng G, Zhao Y and Jiang J 2019 Lightweight flexible indium-free oxide TFTs with AND logic function employing chitosan biopolymer as self-supporting layer *Solid-State Electron.* **153** 16–22
[96] Xie D, Hu W and Jiang J 2018 Bidirectionally-trigged 2D MoS_2 synapse through coplanar-gate electric-double-layer polymer coupling for neuromorphic complementary spatiotemporal learning *Org. Electron.* **63** 120–8
[97] Huang H *et al* 2019 Electrolyte-gated synaptic transistor with oxygen ions *Adv. Funct. Mater.* **29** 1902702
[98] Caporale N and Dan Y 2008 Spike timing–dependent plasticity: a Hebbian learning rule *Annu. Rev. Neurosci.* **31** 25–46
[99] Park H L, Lee Y, Kim N, Seo D G, Go G T and Lee T W 2020 Flexible neuromorphic electronics for computing, soft robotics, and neuroprosthetics *Adv. Mater.* **32** 1903558
[100] Yu F, Zhu L, Xiao H, Gao W and Guo Y 2018 Restickable oxide neuromorphic transistors with spike-timing-dependent plasticity and Pavlovian associative learning activities *Adv. Funct. Mater.* **28** 1804025
[101] Zhu Y, Liu G, Xin Z, Fu C, Wan Q and Shan F 2020 Solution-processed, electrolyte-gated In_2O_3 flexible synaptic transistors for brain-inspired neuromorphic applications *ACS Appl. Mater. Interfaces* **12** 1061–8
[102] Kim M K and Lee J S 2019 Ferroelectric analog synaptic transistors *Nano Lett.* **19** 2044–50
[103] Nishitani Y, Kaneko Y, Ueda M, Fujii E and Tsujimura A 2013 Dynamic observation of brain-like learning in a ferroelectric synapse device *Japan J. Appl. Phys.* **52** 04CE6
[104] Yoo H *et al* 2020 High photosensitive indium-gallium-zinc oxide thin-film phototransistor with a selenium capping layer for visible-light detection *ACS Appl. Mater. Interfaces* **12** 10673–80
[105] Li H, Chen T, Liu P, Hu S, Liu Y, Zhang Q and Lee P S 2016 A light-stimulated synaptic transistor with synaptic plasticity and memory functions based on $InGaZnO_x$–Al_2O_3 thin film structure *J. Appl. Phys.* **119** 244505
[106] Kwon S M, Cho S W, Kim M, Heo J S, Kim Y H and Park S K 2019 Environment-adaptable artificial visual perception behaviors using a light-adjustable optoelectronic neuromorphic device array *Adv. Mater.* **31** 1906433
[107] Peng C, Jiang W, Li Y, Li X and Zhang J 2019 Photoelectric IGZO electric-double-layer transparent artificial synapses for emotional state simulation *ACS Appl. Electron. Mater.* **1** 2406–14
[108] Kim M K and Lee J S 2020 Synergistic improvement of long-term plasticity in photonic synapses using ferroelectric polarization in hafnia-based oxide-semiconductor transistors *Adv. Mater.* **32** 1907826
[109] Duan N *et al* 2019 An electro-photo-sensitive synaptic transistor for edge neuromorphic visual systems *Nanoscale* **11** 17590–9
[110] Huang W *et al* 2020 Zero-power optoelectronic synaptic devices *Nano Energy* **73** 104790
[111] Sangwan V K and Hersam M C 2020 Neuromorphic nanoelectronic materials *Nat. Nanotechnol.*
[112] Dai S, Zhao Y, Wang Y, Zhang J, Fang L, Jin S, Shao Y and Huang J 2019 Recent advances in transistor-based artificial synapses *Adv. Funct. Mater.* **29** 1903700

Chapter 12

Semiconducting metal oxide thin film transistors

Wenfeng Zhang and Xiaokun Wen

Semiconducting metal oxides are historic conduction channel materials for thin-film transistors (TFTs), which have witnessed a rapid expansion of diverse applications. The most notable application of semiconducting metal oxide TFTs is for displays, with the most representative material of amorphous In–Ga–Zn–O (a-IGZO). Other than displays, emerging applications using metal-oxide semiconductor TFTs have also been proposed. In particular, power electronics with the β-Ga_2O_3 channel and low power and cost flexible electronics with ITO short channel TFTs attracts continually increasing attention and has made great advances in recent years. In this chapter, we mainly focus on presenting an overview of the fundamentals, device performance optimization and recent advances of applications for the most representative materials including IGZO, Ga_2O_3 and ITO among all the metal oxides.

12.1 Introduction

Metal-oxide semiconductors, including traditional binary ZnO, In_2O_3, Ga_2O_3, and SnO_2, ternary indium tin oxide (ITO), zinc tin oxide (ZTO), indium zinc oxide (IZO), aluminum zinc oxide (AZO), and representative multicomponent material In–Ga–Zn–O (IGZO) as well as others, are regarded as the suitable TFTs channel materials not only in conventional applications for in-pixel switches and drivers in flat-panel displays, but also in emerging applications for power electronics, large-area flexible, stretchable and/or transparent electronics. Historically, the first metal oxide TFTs can be traced back to 1964, when Klasens and Koelmans demonstrated a TFT comprising an evaporated SnO_2 semiconductor on glass, with aluminium source–drain and gate electrodes and an anodized Al_2O_3 gate dielectric [1]. Then, in 1996, another typical work of the first In_2O_3 non-volatile memory TFTs with ferroelectric gate dielectric was reported [2]. In 2003, Nomura *et al* fabricated a multicomponent IGZO single crystalline layer as TFTs conduction channel, obtaining an impressive effective mobility of 80 $cm^2\ V^{-1}\ s^{-1}$, turn-on voltage of

doi:10.1088/978-0-7503-2556-1ch12

−0.5 V and on/off ratio of 10^6, thus demonstrating the main breakthrough for metal oxide TFTs [3]. Continually, Nomura *et al* further reported that transparent amorphous IGZO TFTs on flexible polyethylene terephthalate (PET) foils exhibited impressive performance of electron mobility ~9 $cm^2 V^{-1} s^{-1}$, turn-on voltage of 1–2 V and on/off ratio of 10^3, which definitely verified the enormous potential of metal oxide TFTs [4]. Following the success of these works, explosive efforts have been devoted and tremendous advances have been achieved on metal oxide semiconductor TFTs in recent years.

Compared with other mainstream conduction channel materials of TFTs including traditional amorphous (a-Si) and low-temperature poly-Si (LTPS), metal oxides show essential advantages [1]. TFTs' electrical performance can be well optimized due to their adjustable multicomponents. Nowadays, high carrier mobility (10–50 $cm^2 V^{-1} s^{-1}$), low off-current (<pA range) and better long-term stability of a-ITZO TFTs has been demonstrated [2]. Metal oxide films generally exhibit large-area uniformity and surface flatness owning to the amorphous structure [3]. Metal oxide films can be fabricated using vacuum or solution processes at quite low fabrication temperature, even at room temperature [4]. Metal oxide films exhibit good transparency (>90%) in the visible region, due to the wide bandgap of oxide semiconductors (>3.5 eV) [5]. Metal oxide films possess large bendability and much lower manufacturing cost. These essential advantages enable their commercialization in 2012 as a driving circuit unit for liquid crystal displays (LCDs) [5]. In addition, these advantages also enable some new emerging applications beyond display applications. For example, metal-oxide TFTs based power electronics is appealing due to their tolerance against high voltage for their large bandgap energy. In particular, β-Ga_2O_3 based power electronics has attracted increasing attention compared to the third generation semiconductors of SiC and GaN in recent years, due to its availability of large and high quality bulk substrates. Another notable example is the large-area flexible, stretchable and/or transparent electronics in which devices are processed over a broad range of large size substrate. Overall, the essential materials properties summarized above enable metal oxide films to be perfect conduction channel candidates of TFTs, and numerous investigations have been performed in this field.

There are already many reviews on semiconducting metal oxide TFTs [6–14]. In this chapter, we mainly focus on presenting an overview of the fundamentals, device performance optimization, and recent advances of applications for the most representative materials including IGZO, Ga_2O_3 and ITO among all the metal oxides.

12.2 Semiconducting InGaZnO (IGZO) thin film transistors

12.2.1 Fundamentals

Amorphous indium–gallium–zinc–oxide (a-IGZO), as the most representative semiconducting metal oxide, has attracted an impressively growing number of investigations as a channel of TFTs since its discovery in 2004 [4]. Initially, Nomura *et al* reported in 2003 that crystalline $InGaO_3(ZnO)_5$ can be fabricated by reactive solid-phase epitaxy (R-SPE) at 1400 °C and used as TFTs channel with amorphous hafnium oxide (HfO_2) as dielectric [3]. The transfer characteristic reveals an electron mobility of ~80 $cm^2\ V^{-1}\ s^{-1}$, threshold voltage ~3 V, and current switching ratio ~10^6, verifying high-performance metal oxide semiconductor TFTs are truly realistic. However, high synthesis temperature would be an obvious obstacle for its possible commercial applications. Then, the milestone work was released by Nomura *et al* in 2004, when amorphous In–Ga–Zn–O (a-IGZO) was synthesized by laser pulse deposition (PLD) on polyethylene terephthalate (PET) substrate at room temperature [4]. As shown in (figure 12.1(a)), a-IGZO is composed of lateral transition metal cations, and the carrier transport path is mainly contributed by the metal s orbital together with less contribution of the oxygen 2p orbital. Because a-IGZO exhibits similar large overlap of adjacent metal s orbitals to crystalline IGZO, the influence on the Hall mobility and carrier concentration of structural changes is limited. Thus, a-IGZO FETs exhibit high mobility of ~10 $cm^2\ V^{-1}\ s^{-1}$, which is an order of magnitude higher than that of hydrogenated amorphous silicon (a-Si:H), together with turn-on voltage of 1–2 V and on/off ratio of 10^3 (figure 12.1(b)). Moreover, performance deterioration of a-IGZO TFTs was not

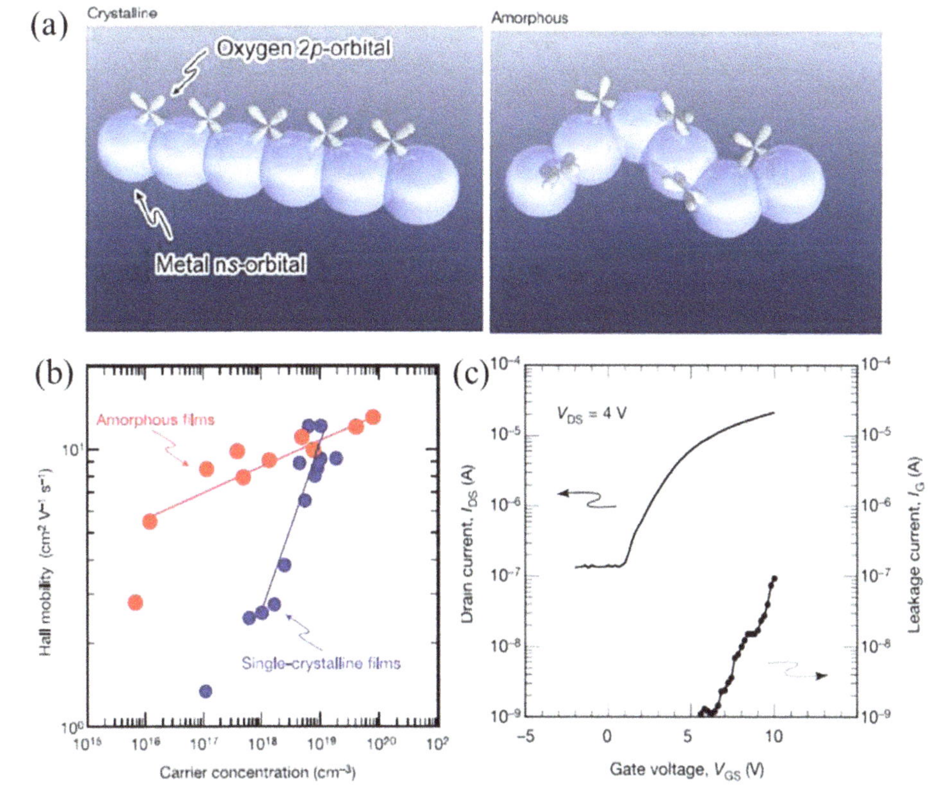

Figure 12.1. (a) Amorphous oxide semiconductors composed of post-transition-metal cations [4]. (b) Relationship between room-temperature Hall mobility and carrier concentration for a-IGZO films [4]. (c) Typical TFT transfer characteristics of a-IGZO devices after bending [4]. Reprinted from [4], with the permission of AIP Publishing.

serious under the condition of a bending radius of 30 mm due to the immunity of disorders for amorphous structure (figure 12.1(c)), implying its promising applications for large area flexible electronics. Nomura's work paves the way for continuously growing investigations on various aspects of a-IGZO, ranging from materials properties, process technologies, to device optimization and applications.

As a multicomponent material, the most distinctive feature of a-IGZO is that the corresponding device electrical performance can be well optimized by changing the element compositions, crystal structure, etc, which has been systematically investigated by Hosono [15], Barquinha *et al* [16], and Iwasaki *et al* [17]. Hosono's investigation summarized the variation of both the amorphous formation region and trends of electron motilities and concentrations with composition for a-IGZO thin film (figure 12.2(a)) [15]. The crystalline tendencies of ZnO and In_2O_3 are disrupted by mixing either with each other or with Ga_2O_3 partly due to the difference in oxygen coordination numbers, thus forming a wide amorphous region. Further

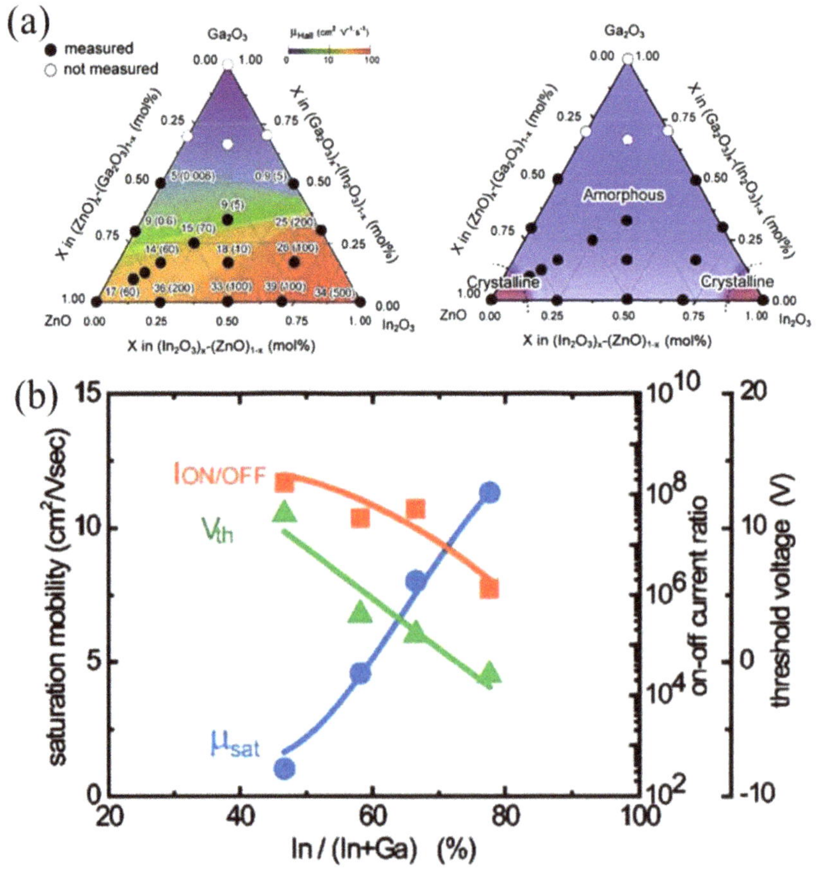

Figure 12.2. (a) The amorphous formation region (right) and the electron motilities and concentrations for the amorphous thin films (left) in the In_2O_3–Ga_2O_3–ZnO system. Reprinted from [15], Copyright (2006), with permission from Elsevier. (b) Saturation motilities, threshold voltages, and on-off current ratios are plotted against channel compositional ratios of In/In+Ga. Reprinted from [17], with the permission of AIP Publishing.

Hall measurements indicated the mobility is primarily determined by the fraction of In_2O_3 content, and the highest value of 40 $cm^2\ V^{-1}\ s^{-1}$ was obtained around the samples containing the maximum In_2O_3 fraction [15]. Meanwhile, Barquinha *et al* also analyzed the role of target composition on the electrical performance of a-IGZO TFTs, where the film was produced by rf sputtering from a range of different composition ceramic targets [16]. Similarly, they found the resistivity increases rapidly by about 6 orders of magnitude when the composition of In:Ga decreases from 4:1 to 2:1, which was attributed to the suppression of carrier generation by compensation of oxygen vacancies with Ga component. The FETs exhibit a clear increase of field-effect mobility, decrease of $I_{on/off}$ ratio, and negative shift of V_{th} with the increase of ratio (In/In+Ga), suggesting a poor channel conductivity modulation. They also revealed that increase of Zn content also favors for enhancing switching behavior, implying proper Zn content is needed. Iwasaki *et al* also fabricated a-IGZO film by using co-sputtering method with three separate oxide targets, and revealed the best performance TFTs at the compositional ratio of In:Ga:Zn=37:13:50 with a saturation mobility of 12 $cm^2\ V^{-1}\ s^{-1}$, threshold voltage of +3 V, on-off current ratio up to 7×10^7, and off current of 1×10^{-12} A (figure 12.2(b)) [17]. These results above verify that the performance of a-IGZO TFTs indeed can be tailored with the chemical composition for specific applications, and the compositional diversity is one of the most distinctive features of the IGZO system. Ultimately, considering the tradeoff between higher mobility, lower I_{off}, V_{on} ~0 V, and moderate carrier concentration, a-IGZO material with composition of In:Ga:Zn=1:1:1 (in atomic ratio) is generally adopted for probing various commercial applications and academic studies.

Next, another attractive advantage of a-IGZO system is its high compatibility with various fabrication methods including vacuum-processed and solution-processed technologies, which enables probing extensive applications. Nowadays, vacuum-processed technologies including pulsed laser deposition (PLD) and sputtering have been well developed for a-IGZO film deposition. For example, Suresh *et al* have reported a-IGZO thin film can be deposited by PLD at room temperature, and the TFTs exhibit high mobility of 11 $cm^2\ V^{-1}\ s^{-1}$ and subthreshold voltage swing of 0.20 V/decade [18]. Further optimization of oxygen partial pressure, channel layer thickness and gate dielectric were verified to further improve device performance with a low off-current of ~10 pA and a drain current on/off ratio of $\sim 5 \times 10^7$. Sputtering including rf sputtering, rf-magnetron sputtering, DC sputtering and pulsed DC sputtering is the predominant vacuum-processed technique to deposit a-IGZO film, because it is the most convenient method among all the vacuum-processed techniques with the advantages of low deposition temperature (typically room temperature), good adhesion and high quality dense film, low cost, and flexibility of optimization via process parameters adjustment such as the power and/or the sputtering pressure. For example, Park *et al* prepared a-IGZO film on a flexible polyethylene naphthalate (PEN) substrate by rf sputtering, and fabricated TFTs devices with a mobility of 15.5 $cm^2\ V^{-1}\ s^{-1}$ and a drain current on/off ratio of 2.2×10^8 (figure 12.3(a)) [19]. Besides vacuum-processed technologies, in recent years, alternative methods of solution-processed methods such as inkjet printing,

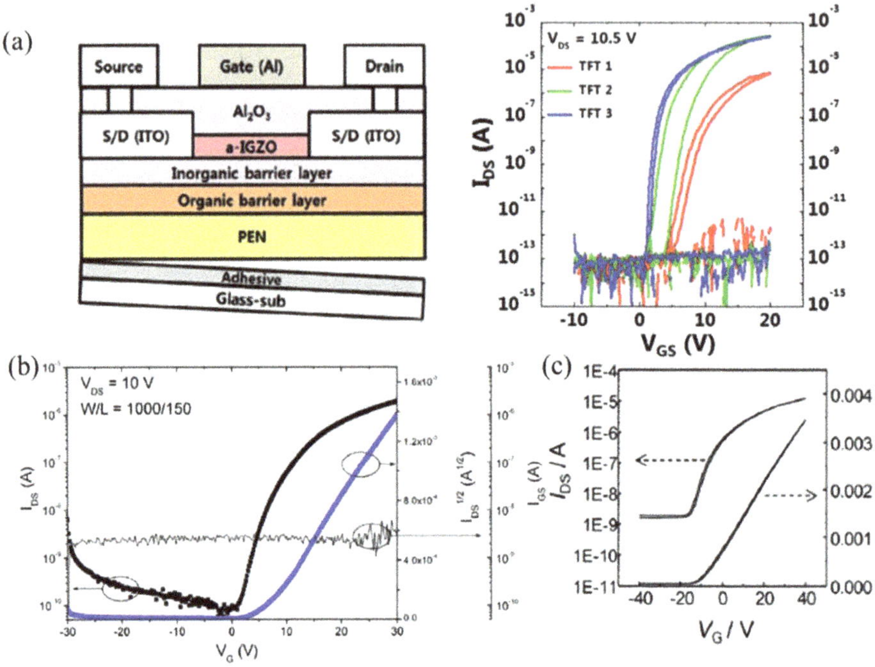

Figure 12.3. (a) Schematic cross-sectional view and sets of transfer curves of the fabricated a-IGZO TFTs on a flexible PEN substrate with no barrier (TFT1), the organic barrier (TFT2), and the organic/inorganic double-layered barrier (TFT3), respectively. The a-IGZO channel was produced by rf sputtering. Reproduced from [19] with permission of The Royal Society of Chemistry. (b) Log transfer characteristics of ~50 nm thick inkjet-printed a-IGZO TFTs. Reprinted from [21], Copyright (2009), with permission from Elsevier. (c) The transfer characteristics of a-IGZO TFTs after annealing at 500 °C, a-IGZO channel was produced by IGZO ink using the IGZO nanoparticles synthesized by co-precipitation. Reproduced from [22] with permission of The Royal Society of Chemistry.

spin coating, or spray pyrolysis with lower cost and higher throughput have attracted more interest [20]. For example, Kim *et al* reported TFTs with a conventional inverted staggered structure using inkjet-printed IGZO as an active channel layer with a field-effect mobility of ~0.03 $cm^2\ V^{-1}\ s^{-1}$ and an on/off ratio of >~10^4 (figure 12.3(b)) [21]. Fukuda *et al* prepared polycrystalline IGZO ink using the IGZO nanoparticles synthesized by co-precipitation method and sintering, then dispersed them in acetic acid, spin-coated onto the substrate and sintered at high temperature to form large area IGZO film. The as-fabricated IGZO TFTs exhibit a mobility of 0.177 $cm^2\ V^{-1}\ s^{-1}$ and an on-off ratio of 6.7×10^3 (figure 12.3(c)) [22]. Liu *et al* used ultrasonic spray pyrolysis deposition (USPD) to deposit SnO_2:F, Al_2O_3, and a-IGZO thin film, respectively, to construct TFTs. The devices show competitive electrical characteristics compared with a sputter-deposited one, with an output current of about 2.28 times higher, as well as higher electron density and field-effect mobility [23]. Compared to vacuum-processed a-IGZO TFTs, although solution-processed technonogies have many advantages, inferior electrical properties of devices are generally revealed due to an inherent relatively low quality of solution-processed thin film. More efforts of both lowering the process temperature and maintaining device electrical performances should be further addressed.

Therefore, since the demonstration of a-IGZO by Nomura *et al*, many research investigations have contributed both to the adjustment of a-IGZO atomic composition and high quality a-IGZO film synthesis via various deposition technologies aiming to achieve satisfied device performance from the viewpoint of materials properties. Meanwhile, further efforts of device structure construction, process optimization and commercial applications have also been conducted, and are summarized below.

12.2.2 Device optimization

Device configuration. Four common planar TFTs configurations were generally adopted for a-IGZO TFTs investigations, typically categorized with respect to the position of the gate and source/drain electrodes relative to the oxide semiconductor and to each other [7]. Among them, bottom-gate (BG) and top-gate (TG) architectures mainly depend on whether the gate electrode is deposited before or after the channel layer. BG and TG devices can be either coplanar or staggered, depending on the gate and the source/drain are the same or opposite side of the channel.

First, BG structure was the most popular configuration used for a-IGZO investigations with commercially available SiO_2/Si substrate, where SiO_2 and heavily doped Si were used as gate dielectric and electrode, respectively. Such configuration is simple, and facilitates for the growth of high quality a-IGZO channel layer on flat and smooth surface of SiO_2. However, both large overlap of the source/drain with the gate causing high parasitic capacitance and direct exposure of the semiconducting back-channel to environmental contaminants would seriously hinder its commercial applications, thus requiring an additional layer (passivation) that protects the back channel from air exposure. For example, Kim *et al* adopted a 100 nm thick SiO_x layer deposited by plasma enhanced chemical vapor deposition as an etch stopper structure [24]. Olziersky *et al* presented a non-vacuum and low temperature process for passivating a-IGZO TFTs, by using the epoxy-based SU-8 resist as an

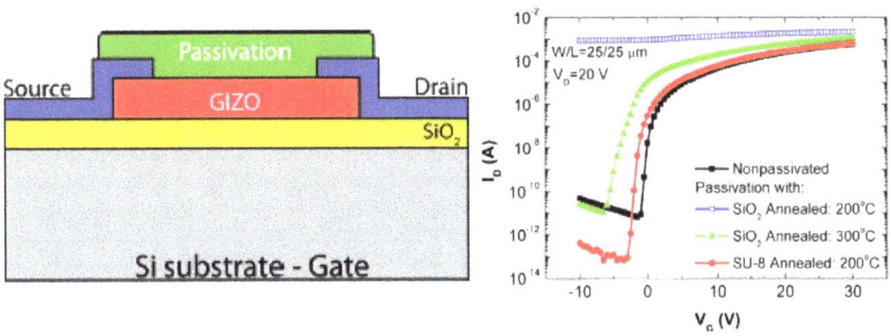

Figure 12.4. Schematic and transfer characteristics of non-passivated and passivated a-IGZO TFTs with SiO_2 and SU-8 after annealing at 200 and 300 °C. Reprinted from [25], with the permission of AIP Publishing.

etch stop layer to prevent device degradation against environmental conditions (figure 12.4) [25]. Park *et al* demonstrated a superior device performance of transparent hafnium indium zinc oxide TFTs adopting single- and double-etch-stopper SiO_x layers, indicating the denser high-temperature SiO_x protects more effectively the underlying semiconductor during the source/drain etch process and also suppresses the generation of defect states [26]. Overall, considering the existing display manufacturing process using a-Si:H TFTs widely adopts BG structure with etch stopper structure, a-IGZO FETs with BG structure is more prone to be integrated into commercial products. In addition, considering a-IGZO TFTs performance degradation has been already verified when exposed to UV light in back-light (AM-LCD) and some AM-OLED displays, opaque gate for BG structure will benefit protecting the channel region from the effects of the back-light, thereby reducing illumination bias stress effects.

Another alternative way to protect the conduction channel is to directly make the gate on top of the channel, also known as top gate configuration, which is mainly used for current commercial LTPS technology. In TG TFTs, the use of a top-gate design allows for a self-aligned gate to minimize parasitic capacitance. For example, Ram *et al* demonstrated that by applying substrate conformal imprint lithography (SCIL) to pattern top-gate self-aligned a-IGZO TFTs down to channel length 450 nm, devices exhibit good scaling behavior with average field-effect mobility of 10 $cm^2\ V^{-1}\ s^{-1}$, SS of 0.3 V/decade, and a near zero turn-on voltage [27]. Zhang *et al* fabricated a self-aligned top-gate a-IGZO TFTs with a 8.62 nm SiO_2 gate dielectric grown by plasma enhanced chemical vapor deposition (PECVD). The TFTs can be used for large-area low-voltage electronics with a low gate leakage current of 10^{-13} A, a high switching ratio of 10^8, and a mobility of 10.13 $cm^2\ V^{-1}\ s^{-1}$ [28]. Another significant advantage of TG configuration is the same as the BG protection channel from the back light, where the top gate can protect the channel from deleterious light such as in OLED displays. For example, Park *et al* demonstrated self-aligned top-gate amorphous a-GIZO TFTs for large size and high resolution displays. The devices with the length of 0.6 μm exhibit a field effect mobility of 5.5 $cm^2\ V^{-1}\ s^{-1}$, a threshold voltage of 1.1 V, and a subthreshold swing of 0.35 V/decade [29]. Finally, concerning staggered and coplanar configuration, we note that BG staggered configuration is the dominant architecture of a-Si:H devices, thus facilitating for the migration to a-GIZO TFTs used in commercial display manufacturing. Meanwhile, coplanar configuration is more suitable for epitaxial grown semiconductor devices.

Defects. Theoretically, carrier mobility, which describes the efficiency of charge carrier transport, is the most crucial parameter of TFTs' performance. Since charge transfer in TFTs is limited to a narrow region close to the gate dielectric/semiconductor interface, several carrier scattering mechanisms including phonon scattering, Coulomb scattering from dielectric charges and interface states, and surface roughness scattering should be considered for mobility improvement. In this regard, the effects of main defects on a-IGZO TFTs device performance in the channel region including oxygen vacancy, metal–metal bonds, hydrogen incorporation should be addressed and are summarized below.

First, oxygen vacancies (V_o) might be the most common defect types for a-IGZO TFTs, which can be formed either in the conduction channel or dielectric during the film deposition process. According to the surrounding environment of V_o in a-IGZO channel, different structures including deep traps (oxygen vacancies coordinated with a small amount of cations), shallow traps (large free space around the oxygen vacancies), and shallow donor states (oxygen vacancies coordinated by a large number of cations with dense edge-sharing networks) can be formed [30]. Among them, both deep and/or shallow traps will reduce carrier concentration, increase subthreshold swing and also contribute to the hysteresis during the device operation, while shallow donor states are the main carrier source of a-IGZO film. As for oxygen vacancies in dielectric, shallow traps are commonly verified and trapped electrons will reduce gate control to the conduction channel with shielding effect. Experimentally, to eliminate oxygen vacancies, both the process optimization of film and post-annealing treatment have been systematically investigated for a-IGZO TFTs. For example, Nomura *et al* revealed that the carrier concentration of a-IGZO film can be accurately controlled between 10^{15} and 10^{20} cm^3 by adjusting the oxygen partial pressure from 0.1 to 7 Pa during the pulsed laser deposition (PLD) process [31]. Yao *et al* also discussed the electrical characteristics of a-IGZO TFTs related to the oxygen vacancies [32]. They found with increase of the oxygen content of the a-IGZO channel by adjusting the oxygen partial pressure during the rf magnetron sputtering deposition, the filling of oxygen vacancies decreases from 14 to 8. The corresponding TFTs' performance degraded with the decrease of the electron density of a-IGZO channel from 7.5 to 3.8×10^{16} cm^{-3}, the saturation mobility from 3.1 to 1.4 cm^2 V^{-1} s^{-1}, the hysteresis-voltage from 3.5 to 2 V and the increase of threshold voltage from 7 to 11 V, the subthreshold slope from 2.4 to 4.4 V/decade for the TFT with a higher interface defect density of 4.9×10^{11} cm^{-2}, and the worst electrical stability of $\Delta V_{th} \sim 10$ V, and a hysteresis-voltage decrease from 3.5 to 2 V. Instead of process optimization, Kikuchi *et al* examined low-temperature annealing process to improve the performances of a-IGZO TFTs fabricated with pulsed laser deposition (PLD) and rf magnetron sputtering at room temperature [33]. For PLD TFTs, electrical characteristics were improved significantly by wet annealing at the annealing temperature of 200 °C (figure 12.5(a)). For sputtered TFTs, electrical characteristics were improved when annealing at $\geqslant$150 °C for dry annealing process, while very large negative threshold voltage (V_{th}) shift was observed for wet annealing at $\geqslant$150 °C, which was speculated to originate from the donor states compensated by deep trap states (figure 12.5(a)). Meanwhile, oxygen vacancies are usually accompanied by the formation of metal–metal bonds to eliminate dangling bonds, thus metal–metal bonds are also called neutral oxygen vacancies [34]. These metal–metal bonds can induce additional deep sub-gap states below the conduction band minimum, which would significantly increase the leakage current in the off state. Experimentally, as for eliminating oxygen vacancies, similar routes including both the film process optimization and post-annealing treatment have also been investigated to reduce the metal–metal bonds, verifying much more difficult elimination compared with oxygen vacancies due to their higher bond energy.

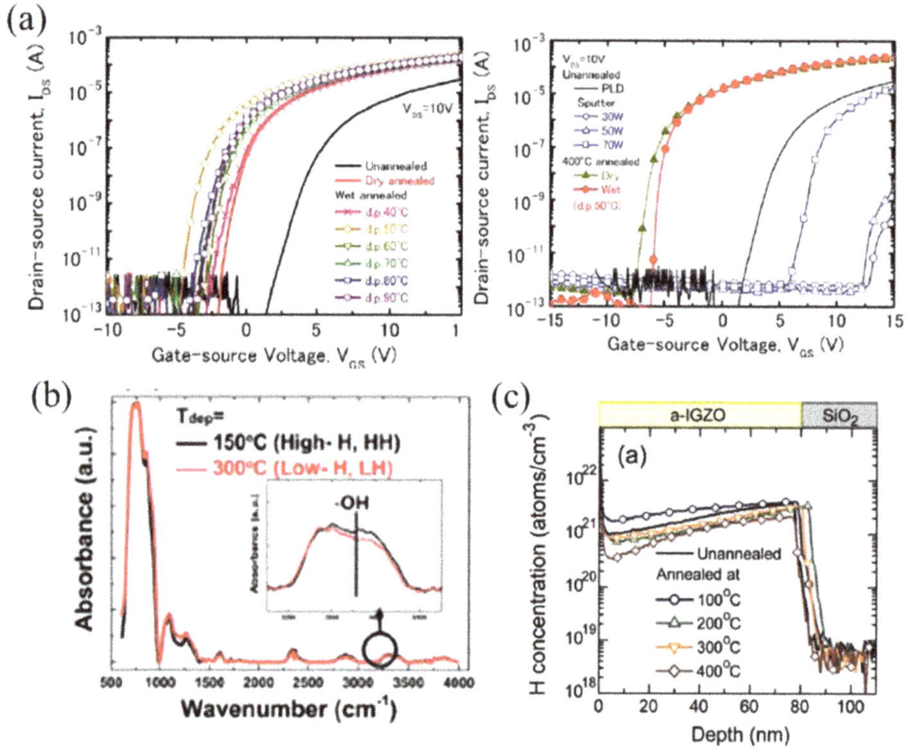

Figure 12.5. (a) Transfer curves of PLD-deposited TFTs (left) and sputtered TFTs (right) unannealed, dry annealed, and wet annealed. Reprinted from [33], Copyright (2010), with permission from Elsevier. (b) ATR-FTIR spectra of the ALD-Al_2O_3 layer verifies hydroxyl-OH in a-IGZO TFTs. Reproduced from [37] with permission of The Royal Society of Chemistry. (c) Variation of hydrogen SIMS depth profiles for a-IGZO films as-deposited and annealed in 5%-H_2/95%-N_2 at 100 °C–400 °C for 1 h [35].

Finally, besides oxygen vacancies and associated metal–metal bonds, hydrogen incorporation generally occurs during a-IGZO TFTs processes, such as deposition, etching, and annealing [35]. Systematic investigations have revealed that hydrogen either bonds to weakly bound oxygen to create –OH or replaces oxygen at vacancies sites to passivate these defects under the condition of low concentration, thus facilitating improving device performance [36]. For example, Nam *et al* revealed the effect of hydrogen diffusion on a-IGZO TFTs with Al_2O_3 gate insulator. ATR-FTIR spectra (figure 12.5(b)) verified hydrogen exists in the form of hydroxyl-OH in a-IGZO TFTs, and has a beneficial effect on the TFTs' performance in terms of defect passivation [37]. In contrast, when the hydrogen concentration is high (above $\sim 2 \times 10^{21}$ cm^{-3}), excess hydrogen would form a conductive channel near the dielectric interface, causing a negative shift of the turn-on voltage [37]. Compared with other defects, incorporated hydrogen is more difficult to eliminate with the post-annealing treatment [35]. Hydrogen SIMS (H-SIMS) depth profiles for a-IGZO film as-deposited and annealed in a 5%H_2/95%N_2 atmosphere at 100 °C–400 °C are shown in figure 12.5(c), revealing the hydrogen concentration was nearly unchanged

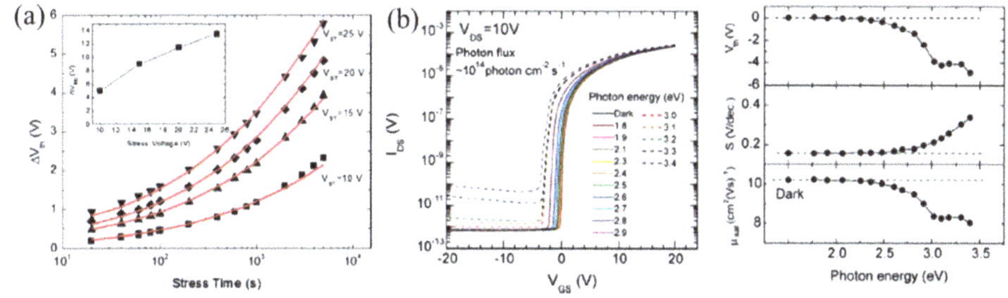

Figure 12.6. (a) Time dependence of V_{th} of a-IGZO TFTs under different positive gate bias stresses. The measured data are well fitted with a stretched-exponential equation. Reprinted from [38], with the permission of AIP Publishing. (b) Transfer curves for wet-annealed of a-IGZO TFTs in dark and under monochromated light illumination with various photon energies from 1.8 to 3.4 eV (left), and variation of TFT parameters (V_{th}, S value, and μ) in dark and under light illumination (right). [41] John Wiley & Sons. © 2012 WILEY-VCH Verlag GmbH & Co., KGaA, Weinheim.

upon annealing treatment. Clearly, more thorough understanding and control methodologies for hydrogen incorporation in a-IGZO TFTs deserve further investigation.

Stability. The stability of device performance is another important issue of a-IGZO TFTs for their mass production and versatile applications. Generally, the TFTs' stability is described by the static I–V parameters (i.e. SS, V_{th}, I_{on}, I_{off}, μ) variation under or after exerting voltage stress during the device operation. Thus, electrical stress measurements including positive bias stress (PBS), negative bias stress (NBS), and negative bias illumination stress (NBIS) are crucial to understand instability mechanism of a-IGZO TFTs, as well as to estimate whether the devices satisfy the requirement of stable operation within the periods of time and cycles. First, for PBS testing of a-IGZO TFTs, a previous report has revealed a stretched-exponential time dependence of threshold voltage (V_{th}) shift, which is attributed to the electron injection from the channel into interface/dielectric traps (figure 12.6(a)) [38]. Nomura *et al* proposed multiple origins for V_{th} shifts, including shallow traps causing large V_{th} shifts ~10 V, which can be removed remarkably by thermal annealing in both dry and wet O_2 annealing, and deep charged traps causing the V_{th} shifts as small as ~1 V, which are hardly removed by thermal annealing [39]. Meanwhile, an exponential dependence of V_{th} on the duration was also explained by charge tunneling mechanism, which results in trapped negative charge screening the applied gate voltage [40]. They further compared the V_{th} shifts of a-IGZO TFTs subjected to PBS and NBS test, revealing that the negative gate bias stress does not change the threshold voltage. Such observation implies that electrons at the channel/dielectric interface were depleted and no charge trapping and tunneling process occurs when applying negative bias. However, under illumination conditions, NBS has been verified to have a significant impact on device stability (NBIS). In particular, considering illumination is indispensable for the most important commercial display application of a-IGZO TFTs, NBIS issue should be well addressed for their practical usage. Nomura *et al* revealed that the photo-response of a-IGZO TFTs begins at 2.3 eV of photon excitation, which is lower than the bandgap

(3.1 eV), thus corresponding to sub-gap states [41]. Meanwhile, illumination also increases I_{off} and induces negative parallel shifts of V_{th} (figure 12.6(b)). They proposed a model and interpreted that the negative V_{th} shifts are caused by the holes photo-excited in the inversion region. These holes diffusively drift and are trapped by the defects at the channel-gate–insulator interface, where migration of the holes is enhanced by the electric field formed by the negative gate bias. In practice, a-IGZO TFTs' reliability under NBIS can be improved by optimizing electrode capping structure. Tu *et al* deposited a layer of SiO_x/SiN_x (250/250 nm) film to optimize contact capping area by plasma-enhanced chemical vapor deposition (PECVD) [42]. The a-IGZO TFTs with capped source electrode exhibited excellent stability after NBIS, due to the lack of holes injecting into the gate oxide, verifying the validity of these strategies.

12.2.3 Display applications

Currently, a-IGZO TFTs have envisioned a wide range of possible applications such as displays, logic circuits, sensors, data storage, and RFID/NFC communication, which was summarized by many reviews [6, 7, 10, 20, 30, 43–45]. Among these versatile applications, display technology is the most important commercial application of a-IGZO TFTs, which has always been the main driving force for the development of a-IGZO TFTs. Since the first reported transparent and flexible a-IGZO TFTs in 2004 [4], LG Electronics (LGE) group soon reported the first active matrix (AM)-OLED display in 2006 [46]. Then, Samsung SDI and Samsung Advanced Institute of Technology (SAIT) also reported AM-OLED displays in 2007 [47]. From 2008, more companies started activities in this area and demonstrated progressively larger displays, as well as LCD back panels. Commercial mass-produced IGZO LCDs were released by Sharp in 2012, the first time in the world [5]. In 2018, Hara *et al* released IGZO TFTs technology for large-screen 8K display in their G8 factory [48]. Figure 12.7(a) summarizes the technology evolution of IGZO TFTs, where the latest $IGZO_5$ process achieved high mobility ~15 $cm^2\ V^{-1}\ s^{-1}$ (more than twice as high as $IGZO_2$-TFTs). The devices also satisfy TFT static characteristics and long-term reliability, demonstrating the mass production capability of developing LCD with high definition, narrow frame, high frequency, and in-cell touch panel.

Besides flat-panel displays, large area flexible and transparent displays are more appealing for a-IGZO TFT applications, considering the special features of transparency, stretchability, conformability, dissolubility, and mechanical activity. For example, Munzenrieder *et al* evaluated electrical behavior of a-IGZO TFTs on flexible substrates with tensile and compressive strains down to bending radius <10 mm under dark and light conditions [49]. Without illumination, the mechanical strain caused little change in the TFTs' performance, but has a considerable influence on device performance with illumination. In 2015, they further presented a-IGZO TFTs based stretchable electronics in a wavy layout on a locally reinforced elastic composite substrate which can be stretched by more than 200%. The flexible a-IGZO devices exhibit excellent performance and cycle stability, with only reduced

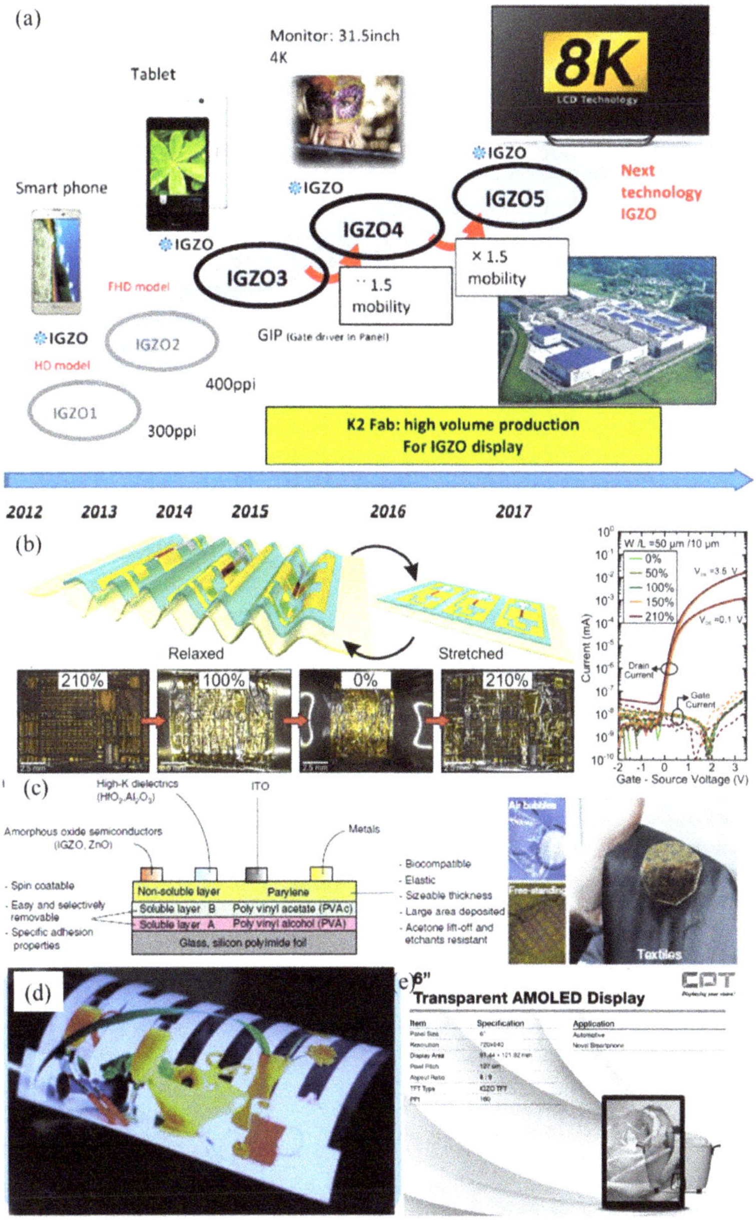

Figure 12.7. (a) a-IGZO TFTs evolution in Sharp Corporation. [48] John Wiley & Sons. © 2018 WILEY-VCH Verlag GmbH & Co., KGaA, Weinheim. (b) Realization of stretchable electronics using a wavy surface for a-IGZO TFTs, and the formation of wrinkles at different levels of strain and transfer characteristics of a TFT (initial strain: 210%) measured at different strain values down to 0%. [50] John Wiley & Sons. © 2015 WILEY-VCH Verlag GmbH & Co., KGaA, Weinheim. (c) The proposed a-IGZO TFTs based ultra-flexible electronics and photonics (left). The electronic membrane can be transferred onto flexible and elastic foils, textiles, biological tissues and implantable devices (right). [51] (2014) Copyright. With permission of Springer. (d) Photograph of a 2.78 in. 1058 ppi ultra-high-resolution flexible OLED display using CAAC-IGZO FETs in the bending state (curvature radius R 17.5 mm). [55] John Wiley & Sons. © 2016 WILEY-VCH Verlag GmbH & Co., KGaA, Weinheim. (e) The specification of high-transparent 6 in. a-IGZO TFTs AM-OLED display. [56] John Wiley & Sons. © 2015 WILEY-VCH Verlag GmbH & Co., KGaA, Weinheim.

V_{th} by ~90 mV and mobility <2% under the condition of substrate changes >210% (figure 12.7(b)) [50]. Salvatore *et al* also proposed a wafer-scale methodology to fabricate electronics on an engineered substrate, which can be transferred to any object, surface and biological tissue (figure 12.7(c)). Analogue amplifiers operating at 12 V and above 1 MHz were successfully realized with the a-IGZO and high-k dielectric [51]. Beyond simply being flexible, a-IGZO outperforms other competitive materials such as organic semiconductors, a-Si:H, and LPTS because of its transparency. For example, Wu *et al* fabricated high-performance, highly transparent a-IGZO TFTs with IGZO source/drain electrodes on glass substrate. Such TFTs not only show high transparency, but also exhibit a mobility of 8.3 $cm^2\ V^{-1}\ s^{-1}$, a switching ratio of 3.1×10^6, and a subthreshold swing of 0.44/decade [52]. For commercial products, it took just 1 year for the first demonstration of flexible black and white e-paper by Toppan Printing Co. Ltd in 2005 [53] since the discovery of a-IGZO TFTs. LGE report the first flexible AM-OLED on stainless steel foil in 2007 [54]. In 2016, Nagata *et al* successfully fabricated a 2.78 in. 1058 ppi flexible OLED display ($R_{bending}$ = 1 mm) using *c*-axis-aligned crystalline In–Ga–Zn–O (CAAC-IGZO) TFTs (figure 12.7(d)). The display exhibits the highest pixel density in the world among flexible displays using TFTs [55]. For transparent displays, Samsung SDI reported a 4.1 in. transparent dual-emission AM-OLED using a-IGZO TFTs with 20% transparency in 2007 [47]. Lee *et al* demonstrated a 6 in. IGZO TFTs AM-OLED display with 23% light transmittance in 2015 (figure 12.7(e)) [56].

Overall, semiconducting a-IGZO material, as the most appealing metal oxide, has made enormous advances since its discovery in 2004. Comprehensive understanding of the fundamental material properties and device process optimization of improving device performance, especially focusing on optimizing device configuration, understanding and eliminating various types defects, and improving device stability of a-IGZO TFTs have been widely explored and developed. The device improvements further promote a-IGZO TFTs for versatile practical applications, especially for displays including flat-panel displays and flexible and transparent displays.

12.3 Semiconducting Ga_2O_3 thin film transistors

12.3.1 Fundamentals

Gallium oxide (Ga_2O_3), as a basic IGZO composition, is generally classified into ultra-wide bandgap semiconductor and has emerged as a potential candidate material for next generation high-power electronics in recent years. Ga_2O_3 has commonly identified five different polymorphs, designated as the α-Ga_2O_3, β-Ga_2O_3, γ-Ga_2O_3, δ-Ga_2O_3, and ε-Ga_2O_3, which can be either insulators or conductors relying on the detailed growth control. Among them, the β-Ga_2O_3 is the most stable under normal conditions, and has been the most widely studied and utilized. Currently, there are already some published reviews on Ga_2O_3 [13, 57–65], hereby we will only present a specific summary of the main achievements concerning the fundamentals, MOSFETs channel formation, device performance and applications of the β-Ga_2O_3.

β-Ga_2O_3 belongs to monoclinic system, C2/m space group, and its lattice constants are $a = 1.22$ nm, $b = 0.30$ nm, and $c = 0.58$ nm. The unit cell of β-Ga_2O_3 is composed of two types of gallium ions and three types of oxygen ions, which leads to an anisotropy of various properties [65]. In addition, β-Ga_2O_3 also exhibits different orientations for common use, such as ($\bar{2}01$), (010), and (001) planes, for example, cutting the bulk along (010) orientation will produce a flat ($\bar{2}01$) surface with very low roughness for wafer fabrication and device application. Meanwhile, different exposure surfaces differ significantly in terms of their dangling bond densities of oxygen, which would also impact the device process and performance. The electrical conductivity of β-Ga_2O_3 is limited by its ultrawide bandgap (4.7–4.9 eV), and 4.8 eV is commonly quoted in the reports. The band structure of β-Ga_2O_3 calculated from density functional theory verifies that the top of the valance band is mainly made up of the anionic O 2p states, which exhibit a small dispersion and fairly flat characteristic, indicating high density of states of valance band and a large hole effective mass [66]. Thus, holes tend to trap at a lattice distortion to form localized small polarons and hardly move freely, which leads to the solid demonstration of p-type conductivity and limit the possible devices with unipolar conductivity. Meanwhile, the bottom of the conduction band is mainly derived from the cationic states (Ga 4s states), and the electron effective mass has been calculated as 0.34 m_0, where m_0 is the free electron mass. In fact, similar to ZnO, many studies have revealed that the intrinsic conductivity of β-Ga_2O_3 mainly originates from the free electrons generated from defect energy levels existing in the bandgap. Thus, oxygen vacancies are crucial for the electric conductivity of β-Ga_2O_3, and have demonstrated the potential application for gas sensors to detect H_2, CH_4, CO, O_2, etc.

Recently, β-Ga_2O_3 has been the research focus in power electronics. Compared to SiC and GaN, which are common semiconductors for power electronics, the ultrawide bandgap of β-Ga_2O_3 gives it a much higher critical breakdown electrical field ($E_{br} \approx 8$ MV cm^{-1}), which is about three times larger than that of SiC or GaN. The much higher E_{br} allows high voltage operation of β-Ga_2O_3, benefiting for the reduction of device size and enhancement of the integration level of power modules. Moreover, various figures-of-merit (FOM) including Baliga's figure-of-merit (BFOM) estimating DC conduction losses can be calculated to evaluate the suitability of different semiconductors for power device applications. Ga_2O_3 shows a much larger BFOM than SiC and GaN (four times larger than GaN) [67]. Meanwhile, at the same V_{br}, the conduction loss of Ga_2O_3 devices is significantly one order of magnitude lower than those of SiC and GaN devices, indicating its great potential in unipolar devices. Therefore, the excellent material properties of β-Ga_2O_3 enable it to be a promising candidate for next-generation power electronics.

At present, field effect transistors (FETs) are the most prevailing unipolar device configuration for high-power operation. FETs are mainly used in converters and inverters, by modulating the resistance between source and drain electrodes with gate electrical field. The first single-crystal β-Ga_2O_3 metal-semiconductor FET (MESFET) was demonstrated by Higashiwaki *et al* in 2012 [67]. Circular MESFETs with a gate length of 4 μm and a source–drain spacing of 20 μm were fabricated with a Sn-doped Ga_2O_3 molecular-beam (MBE) epitaxial layer on a semi-insulating β-Ga_2O_3 (010) substrate. The device showed an ideal transistor action, with a high breakdown voltage (>250 V) and high on/off drain current ratio (~10^4), although the off state leakage current was quite serious (~3 μA). Afterwards, Ga_2O_3 based transistor devices made solid advances, and we will introduce some typical works.

12.3.2 Channel fabrication

A distinctive interest of β-Ga_2O_3 among wide-bandgap semiconductors relies in the large-diameter single-crystal wafers, which are required for mass production of vertical devices and nanoflakes exfoliation, and can be synthesized cost-effectively by using melt growth techniques such as floating-zone (FZ) [68], Czochralski (CZ) [69], Bridgman (horizontal or vertical, HB and VB) [70], and edge-defined film-fed growth (EFG) [71] methods. Meanwhile, high-quality β-Ga_2O_3 homoepitaxial film grown on native substrates can be also realized by various vacuum methods such as molecular beam epitaxy (MBE), halide vapor phase epitaxy (HVPE), metal-organic chemical vapor deposition (MOCVD), mist chemical vapor deposition (mist-CVD), and other CVD techniques. Thus, driven by the high quality conduction channel fabrication technologies, β-Ga_2O_3 has emerged as a promising competitor to SiC and GaN toward next generation power devices, and attracts more and more research focus.

First, typical examples of bulk β-Ga_2O_3 single crystals growth were shown in figure 12.8. In the early stages, FZ method was often used to synthesize β-Ga_2O_3 due to its relatively easy process. Large-size single crystals of β-Ga_2O_3, 1 in. in diameter, have been grown by the FZ technique (figure 12.8(a)), although typical crystals often contain high densities of twin boundaries [68]. CZ method is another attractive approach for commercial production of large diameter and high quality β-Ga_2O_3 single crystals. β-Ga_2O_3 cylindrical crystals with 2 in. diameter and weight up to 1 kg have been demonstrated when using oxygen concentration between 8 and 35 vol.% (figure 12.8(b)) [69]. In addition, Hoshikawa *et al* have demonstrated β-Ga_2O_3 single crystals 25 mm in diameter can be grown in Pt–Rh (70%–30%) alloy crucible without a crystal seed by using VB method in ambient air (figure 12.8(c)) [70]. Currently, EFG method is the leading established melt growth method and has

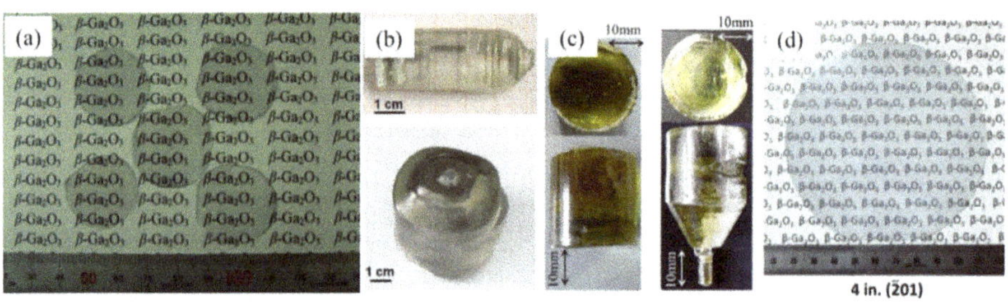

Figure 12.8. (a) β-Ga_2O_3 single crystalline wafers of 1 in. grown by FZ method. Reprinted from [68], Copyright (2004), with permission from Elsevier. (b) Semiconducting β-Ga_2O_3 crystals 5 cm in diameter grown at high oxygen concentration by CZ method [69]. (c) β-Ga_2O_3 crystals grown by the VB method in the full-diameter crucible or in the conical crucible. Reprinted from [70], Copyright (2016), with permission from Elsevier. (d) Photograph of 4 in. diameter single-crystal Ga_2O_3 wafer grown by EFG method [71].

already been used by the commercial suppliers of bulk Ga_2O_3 crystals. By using the EFG process, β-Ga_2O_3 substrate containing no twin boundaries with sizes up to 4 in. diameter were fabricated (figure 12.8(d)). Si was the main residual impurity and the dislocation density was on the order of 10^3 cm^{-3} in the EFG grown crystals [71].

High quality β-Ga_2O_3 bulk growth is crucial both for obtaining high quality nanoflakes mechanically exfoliated from bulk wafer, and also for subsequent epitaxial thin film growth. Typical examples of homoepitaxial β-Ga_2O_3 growth on native substrates were summarized in figure 12.9. First, MOCVD is the most popular thin-film growth technique for mass production of various compound semiconductors. Alema *et al* reported β-Ga_2O_3 was grown on c-plane sapphire with

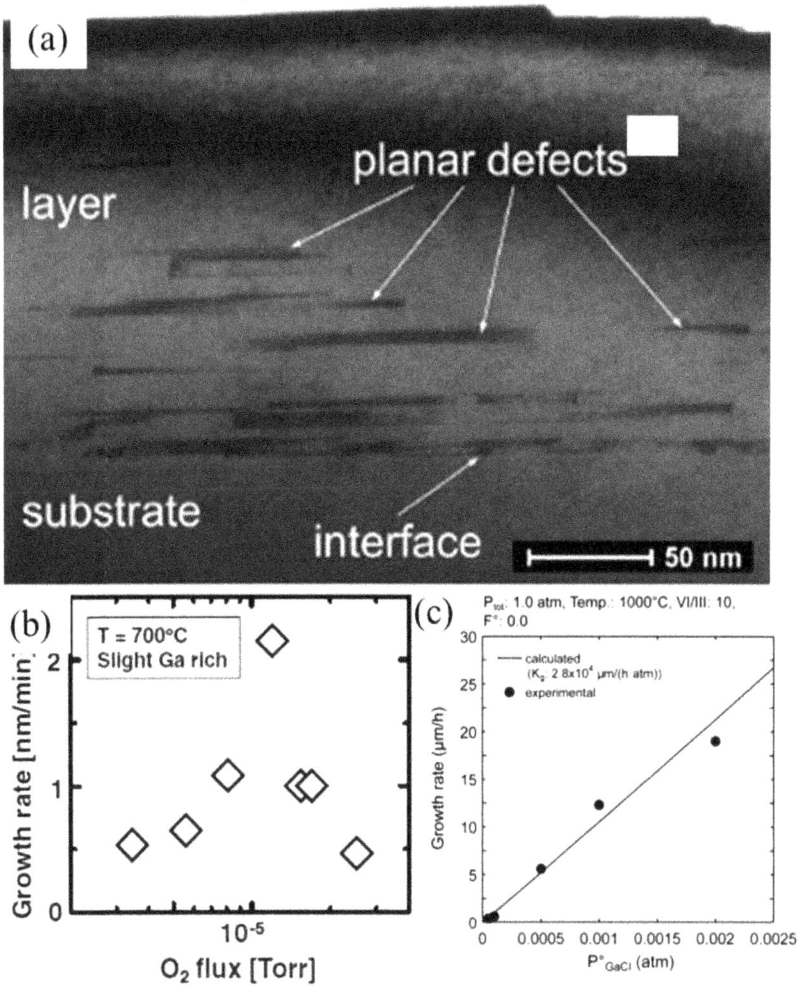

Figure 12.9. (a) Bright field TEM image of epitaxial β-Ga_2O_3 layer grown by MOCVD. [73] (2015) Copyright. With permission of Springer. (b) Relation of the growth rate of β-Ga_2O_3 to the oxygen flux at 700 °C under slightly Ga-rich condition grown by MBE [74]. (c) Calculated and experimental growth rates for HVPE. Reprinted from [76], Copyright (2016), with permission from Elsevier.

high rate growth up to ~10 μm h^{-1} by using trimethylgallium (TMGa) precursors at a substrate temperature of 900 °C in a close coupled showerhead (CCS) MOCVD reactor, thus suggesting a pathway to overcoming the long-standing challenge of realizing fast growth rates for Ga_2O_3 using the MOCVD method [72]. The most serious issue of MOCVD-grown Ga_2O_3 thin films is the poor control of electrical conductivity. N-type Sn-doped β-Ga_2O_3 epitaxial layer was demonstrated to be successfully grown on [98] β-Ga_2O_3 substrates with MOCVD method by Baldini *et al* aiming to solve this problem (figure 12.9(a)). Sn was revealed to be homogeneously incorporated throughout the whole layer with a flat profile at concentration levels ranging from 2×10^{17} to 3×10^{19} cm, and the maximum mobility of 41 cm^2 V^{-1} s^{-1} was achieved with free carrier concentration of 1×10^{18} cm^{-3} [73].

MBE is another popular approach among all the Ga_2O_3 epitaxial growth techniques. The main advantage of MBE is capability of growing very high quality Ga_2O_3 film with the precise control over growth parameters and impurity incorporation into epitaxial film, while the drawbacks such as low growth rate and high production cost are also obvious. In 2014, Okumura *et al* performed β-Ga_2O_3 homoepitaxial growth under Ga-rich condition by plasma-assisted molecular beam epitaxy. A high growth rate of 2.2 nm min^{-1} was achieved by optimizing the O flux between 650 and 750 °C (figure 12.9(b)) [74]. Similar to MOCVD, Sn-doped β-Ga_2O_3 thin film homoepitaxial grown on Ga_2O_3(010) substrates was also demonstrated by Sasaki *et al* [75]. Ga_2O_3 film with an atomically smooth surface were obtained within a quite narrow growth temperature window of 550 °C–570 °C, implying it is difficult to precisely control temperature during the growth of thick n-doped Ga_2O_3 layer for vertical power devices.

HVPE was also considered as a promising technique for future mass production of Ga_2O_3 epitaxial film due to the high-speed growth of high-quality and thick Ga_2O_3 film with well controlled electrical conductivity. Nomura *et al* performed thermodynamic analysis and selected GaCl and O_2 as precursors for Ga_2O_3 growth by HVPE [76]. $SiCl_4$ dopant gas was simultaneously supplied for growing n-type Ga_2O_3 with a growth temperature ~1000 °C and the highest growth rate ~20 μm h^{-1} (figure 12.9(c)). Murakami *et al* also demonstrated thick high-purity β-Ga_2O_3 epitaxial layer with smooth surface, high crystalline quality, and low effective donor concentration (N_d–N_a < 10^{13} cm^{-3}) [77]. These results suggest HVPE is certainly capable of growing thick β-Ga_2O_3 epitaxial layer with acceptable morphology and high purity for power electronics.

Other vacuum-processed thin film deposition techniques were also investigated to grow β-Ga_2O_3 epitaxial film. For example, pulsed laser deposition (PLD) was often used to create n-type doped β-Ga_2O_3 layer due to its relatively low deposition temperature. Leedy *et al* investigated the homoepitaxial Si-doped β-Ga_2O_3 film on (010) β-Ga_2O_3 substrate by PLD [78]. An average conductivity of 732 S cm^{-1} with a mobility of 26.5 cm^2 V^{-1} s^{-1} and a carrier concentration of 1.74×10^{20} cm^{-3} as determined by Hall effect measurements was achieved for film deposited at 550 °C. The high conductivity implies this approach might be useful in making low resistance Ohmic contact layer in β-Ga_2O_3 devices. In addition, the mist CVD is a low-cost approach that employs mist particles containing source precursors.

Epitaxial stabilization of the metastable γ-Ga_2O_3 film was achieved by mist CVD firstly, and Ga_2O_3 film was gradually transformed to the stable β-Ga_2O_3 with increasing growth temperatures [79]. Atomic layer epitaxy (ALE)/atomic layer deposition (ALD) was also used for growth studies of Ga_2O_3 film. Altuntas *et al* reported 7.5 nm β-Ga_2O_3 thin film can be obtained by depositing on p-type (111) silicon wafer using plasma enhanced atomic layer deposition (PEALD) and annealing at 600, 700, and 800 °C [80]. Overall, various vacuum-processed film deposition techniques have been investigated to grow high quality Ga_2O_3 epitaxial film, among which MOCVD and MBE show the most promising potential for future mass production.

12.3.3 Device performance and applications

The FET is a significant component of power electronics. For β-Ga_2O_3 transistors, the semiconducting channel can be varied from Si, Sn or Ge-doped n-type epitaxial β-Ga_2O_3 film grown on its native substrate to the micromechanically exfoliated nanoflakes summarized above. Meanwhile, for device fabrication, different high-k dielectrics of SiO_2, HfO_2, and Al_2O_3 have been demonstrated, and the structure has evolved from the first MESFETs to planar MOSFETs, then to vertical configuration. To date, most of the β-Ga_2O_3 transistors work in depletion mode, and a few enhancement (e-mode) operation devices were realized recently.

Although p-type doping is difficult to achieve for β-Ga_2O_3 due to the low activation energy of acceptors, n-type doping has been well demonstrated to control its electrical conductivity. In particular, common n-type dopants, such as Si, Ge and Sn with different doping concentration to modulate the carrier concentration have been well reported. First, Krishnamoorthy *et al* reported that by using silicon delta-doped β-Ga_2O_3 as conduction channel grown by plasma-assisted molecular beam epitaxy, the transistors exhibit high drain current (I_{Dmax} = 236 mA mm^{-1}) and transconductance (g_m = 26 mS mm^{-1}) [81]. Afterwards, with further efforts to improve source–drain ohmic contacts formed by patterned regrowth of n-type Ga_2O_3, the fabricated MESFETs have a peak drain current of 140 mA mm^{-1}, transconductance (g_m) of 34 mS mm^{-1}, and three-terminal off-state breakdown voltage of 170 V [82]. Wong *et al* also demonstrated a vertical β-Ga_2O_3 MOSFET exhibiting an off-state breakdown voltage of 755 V, a high drain current on/off ratio of over 10^9, and stable high temperature operation against 300 °C thermal stress with a selective-area Si ion implanted Ga_2O_3 conduction channel [83]. Except for Si, Sn is another common n-type dopant for β-Ga_2O_3. For example, in 2012, Higashiwaki *et al* reported the first β-Ga_2O_3 based MESFET device, showing an ideal transistor action with a high breakdown voltage (>250 V) by adopting Sn-doped n-Ga_2O_3 film grown by molecular-beam epitaxy as channel layer [67]. Afterwards, the device performance was further improved with 20 nm Al_2O_3 formed by atomic layer deposition as the gate dielectrics in 2013, and the device showed an extremely low off-state drain leakage of less than a few pA mm^{-1}, leading to a high drain current on/off ratio of over 10^{10}. The breakdown voltage was also increased to 370 V [84]. In 2016, Green *et al* achieved a tremendously high breakdown electrical field of 3.8 MV cm^{-1} for Ga_2O_3 MESFETs, which was the highest reported value for Ga_2O_3 transistors and surpassed SiC and GaN theoretical limits, by adopting a high quality Sn-doped (100) β-Ga_2O_3 epitaxial layer grown via metal-organic vapor phase epitaxy onto a single crystal Mg-doped semi-insulating (100) β-Ga_2O_3 substrate [85]. In 2017, aiming to improve the source/drain (S/D) contact, tin (Sn) doped spin-on glass (SOG) S/D doping technology was reported by Zeng *et al*, and the fabricated MOSFETs devices achieved an increased peak output drain current density of 40 mA mm^{-1}, the maximum transconductance (g_m) of 1.23 mS mm^{-1}, large on/off ratio of 10^8 and breakdown voltage of 382 V with L_g = 2 μm [86]. Another typical example of using Sn-doped (100) β-Ga_2O_3 epitaxial layer as conduction channel is enhancement mode wrap-gate fin-array field-effect transistors (finFETs) fabricated by top-down BCl_3 plasma etching on a native semi-insulating Mg-doped (100) β-Ga_2O_3 substrate by Chabak *et al* [87]. The device exhibits a three-terminal breakdown voltage exceeding 600 V with 21 μm gate–drain spacing. Later, they also reported recessed-gate enhancement-mode β-Ga_2O_3 MOSFETs exhibiting improved performance with an on current of 40 mA mm^{-1} with on/off ratio of ~10^9 and V_{br} of 505 V [88]. Finally, similar to Sn and Si, Ge-doped β-Ga_2O_3 can also be commonly used as conduction channel. The corresponding FET devices were reported to exhibit high drain current on/off ratios >10^8, and a three-terminal breakdown voltage of 479 V with a gate–drain spacing of 5.5 μm [89].

Besides epitaxial β-Ga_2O_3 film, two-dimensional (2D) β-Ga_2O_3 nanoflakes or nanomembranes can also be mechanically exfoliated from bulk wafers due to their lattice constant value in the (100) direction being much larger than that in (001) and (010). These exfoliated nanoflakes or nanomembranes can be transferred to various substrates for further device fabrication. A unique advantage of 2D β-Ga_2O_3 lies in providing a low-cost device thermal solution with the capacity of large-area and controllable transfer of exfoliated Ga_2O_3 nanoflakes onto high thermal conductivity substrates, which is crucial for β-Ga_2O_3 based FETs considering thermal management is one of the most serious issues of β-Ga_2O_3 due to its low thermal conductivity. Experimentally, early in 2014, back-gated high-voltage FETs with intrinsic β-Ga_2O_3 nanomembrane as conduction channel have been fabricated and characterized [90]. The devices show high on/off ratio, low subthreshold slopes and high breakdown field. In particular, the nanomembrane topology offers opportunities for efficient heat removal. Later, in 2016, Ahn *et al* reported that 2D β-Ga_2O_3 based FETs with SiO_2 and Al_2O_3 as dielectric for the back and front sides, respectively, exhibited better channel modulation (on/off ratios of ~10^5) with both front and back gates compared to either front or back-gating alone [91]. The maximum transconductance was ~4.4 mS mm^{-1} with front and back-gating, and a maximum drain source current density of 60 mA mm^{-1} was achieved at a drain–source voltage of 10 V. Also in 2016, Kim *et al* demonstrated the exfoliation of a 2D β-Ga_2O_3 nanobelt and subsequent processing into a TFTs structure [92]. This β-Ga_2O_3 nanobelt based transistor displayed an on/off ratio of 10^7 over the operating temperature range of 20 °C to 250 °C. No electrical breakdown was observed up to V_{DS} = +40 V and V_{GS} = −60 V between 25 °C and 250 °C (figure 12.10(a)). Zhou *et al* demonstrated the depletion/enhancement-mode back-gate β-Ga_2O_3 nanomembrane

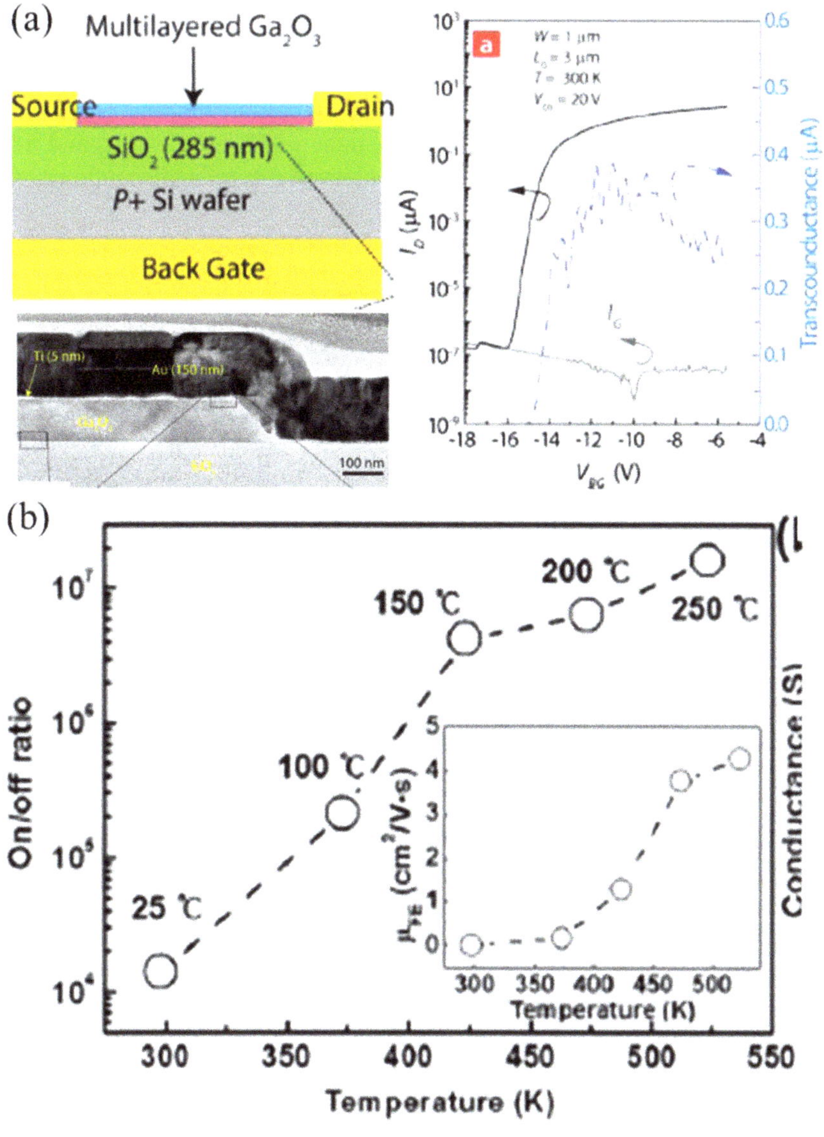

Figure 12.10. (a) On/off ratio (inset: field-effect mobility) at the temperatures ranging from 25 °C to 250 °C of β-Ga_2O_3 nanomembrane FETs. Reproduced from [92] with permission of The Royal Society of Chemistry. (b) Transfer characteristics of top gate GOOI FETs on SiO_2/Si ad sapphire substrates with different thermal conductivities. Reprinted with permission from [95], Copyright (2017) American Chemical Society.

transistors on insulators with high drain currents (ID) of 600/450 mA mm^{-1} [93]. The threshold voltage (V_T) can be modulated by varying the thickness of the β-Ga_2O_3 films, and the E-mode FETs can be simply achieved by shrinking the β-Ga_2O_3 film thickness. An average breakdown electric field of 2 MV cm^{-1} was realized in E-mode devices with a breakdown voltage of 185 V, implying its great promise for power electronics. They further demonstrated a record high drain current density of 1.5/1.0 A mm^{-1} can be achieved for depletion/enhancement-mode β-Ga_2O_3 FETs, together with high on/off ratio of 10^{10}, and low subthreshold slope of 150 mV/decade even with 300 nm thick SiO_2, by utilizing a highly doped β-Ga_2O_3 nanomembrane as the channel In 2017 [94]. In addition, they also investigated top-gate β-Ga_2O_3 nanomembrane FETs on a higher thermal conductivity sapphire substrate rather than a SiO_2/Si substrate, and maximum drain current density of 535 mA mm^{-1} was achieved due to significantly reduced self-heating (figure 12.10(b)) [95].

In summary, both Si, Sn or Ge-doped n-type epitaxial β-Ga_2O_3 film and nanoflakes micromechanically exfoliated from bulk wafer have been investigated to be used as conduction channel for TFTs fabrication. The device structure has evolved from the first MESFETs to planar MOSFETs, then to vertical configuration. The device performance suggests promising potential of β-Ga_2O_3 based TFTs for power electronics. Further device fabrication process and performance optimization deserves more investigation.

12.4 Semiconducting ITO thin film transistors

Except for the widely addressed a-IGZO and Ga_2O_3 thin film transistors, various other metal oxides such as ZnO, In_2O_3, SnO_2, indium tin oxide (ITO) have also been probed for TFT applications. Among them, although ITO is traditionally used as the transparent conductive film electrode due to its high optical transmittance and electrical conductivity, semiconducting ITO channel transistors have attracted continually increasing attention in recent years.

In the early stage, ITO is rarely used as an active channel material due to its unsatisfactory switching performance due to its inherent high carrier density and polycrystalline nature. Early in 2010 [96], Lu *et al* developed a simple self-assembling method for ITO-based low-voltage coplanar homo-junction TFT fabrication at room temperature with one metal shadow mask. In this process, an ITO channel layer can be self-assembled between ITO source/drain electrodes during magnetron sputtering deposition. The devices showed a small subthreshold swing of 0.12 V/decade and a large on/off ratio of 10^6. Although this method is promising for low-cost ITO TFTs fabrication, however, it is not compatible with the mainstream flat-panel display technologies. Later, they further fabricated junctionless flexible TFTs with channel and source/drain electrodes of ITO film on paper substrates at room temperature. TFTs show good device performance with a small S of 0.21 V/decade and a large $I_{on/off}$ of 2×10^6 [97]. Then, in 2014, Shao *et al* proposed an 'All by ITO' technology in which the source/drain electrodes, channel layers, the gate electrodes, together with the pixel electrodes of display devices are all made of ITO material [98]. They demonstrated homo-junction TFTs with source/drain and channel layers all made of the same ITO layer with the anodic oxidation treatment process to modulate the electrical, optical and structural characteristics of the ITO films. The TFTs devices showed a high on/off current ratio (1×10^8), a saturation mobility of 29.0 cm^2 V^{-1} s^{-1} and a subthreshold swing of 0.20 V/decade. Such an approach is appealing to low temperature transparent and flexible large area electronics.

A remarkable breakthrough of ITO TFTs was published in 2019. Li *et al* demonstrated short-channel active transistors with an ultrathin (~4 nm) ITO channel and a high-quality lanthanum-doped hafnium oxide dielectric of equivalent oxide thickness (EOT) of 0.8 nm [99]. Short-channel immunity, with a nearly ideal subthreshold slope (66 mV/decade), the lowest off-current (<100 fA μm^{-1}) and the highest on/off ratio (5.5×10^9) were achieved for the transistors with 40 nm channel length (figure 12.11(a)). A logic inverter with high gain of 178 at low-supply power of 0.5 V (figure 12.11(b)) and high-performance rf transistors with f_T and f_{max} both >10 GHz (figure 12.11(c)) were also demonstrated. Such results provide prospects of ITO transistors for the next generation of low-cost and low-power electronics. Then, they also found the 15 nm long ITO transistors exhibit high performance with record-high on-state current of 970 $\mu A\ \mu m^{-1}$, transconductance of 400 $\mu S\ \mu m^{-1}$ and on/off ratio near 10^{11}. NAND, NOR, and SRAM based on enhancement/depletion mode (E/D) inverters have achieved full rail-to-rail output characteristics and stable memory function. Finally, they demonstrated a five-stage ring oscillator with record-fast stage delay of 0.49 ns s^{-1} ta^{-1} g^{-1} e^{-1}, which is the best among all results based on metal-oxides transistors. Later, they further fabricated flexible high-frequency ITO rf transistors with an inverted gate structure [100]. The 160 nm channel length devices exhibit a record-high extrinsic cut-off frequency (f_T) of 2.1 GHz and an extrinsic maximum oscillation frequency (f_{max}) of 3.7 GHz. A high conversion gain of −20.9 dB was achieved for the gigahertz ITO mixer. The devices also show excellent stability under extremely harsh bending conditions up to 50 000 bending cycles or down to 1 mm bending radius, indicating the remarkable potential of flexible ITO rf transistors for high-frequency applications. They also further improved the devices performance of high-speed ultrathin-body (3.5 nm) ITO transistors [101]. These systematic investigations fully demonstrate ultrathin ITO channel TFTs behave with remarkable operating characteristics, and have enormous potential applications in high performance electronics [102].

In addition to the vacuum-fabricated ITO channel, in 2020, Wang *et al* demonstrated TFTs using solution-processed ultra-thin ITO film channel exhibiting a high mobility (15 $cm^2\ V^{-1}\ s^{-1}$) and a high on-off ratio of 10^8. They further introduced moderate Ga doping to form Ga-doped ITO TFTs to improve device stability, and the devices exhibited a high apparent mobility (>40 $cm^2\ V^{-1}\ s^{-1}$) and

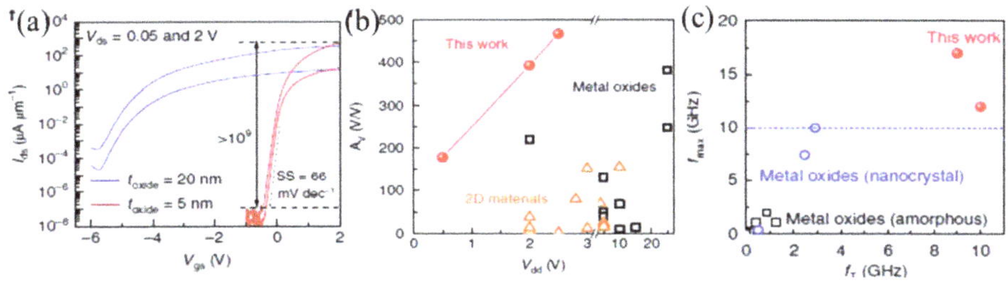

Figure 12.11. (a) Transfer characteristics of ITO transistors with channel thickness 4 nm and dielectric 5 or 20 nm HfLaO. (b) Benchmark of ITO inverter gain as a function of V_{dd}. (c) Benchmark of f_T and f_{max} of ITO transistors. [99] (2019) Copyright. With permission of Springer.

maintained an on-off ratio of more than 10^4 during operation at small V_D values, implying good potential for low cost and low power electronics [103].

Besides functioning as sole conduction channel of TFTs, research interests for semiconducting ITO thin film also focus on its use as stabilization layer of other channel materials such as ZnO [104–107] and titanium doped zinc oxide (TiZO) [108], by forming a dual-layer channel. In 2015, Lu and Huang examined the performance of ZnO/ITO dual-channel TFTs, and revealed that the performance of ZnO TFTs were improved considerably after ITO channel deposition with an on/off ratio of 10^6, a subthreshold swing of 0.1 V/decade, and a mobility of 21 $cm^2\ V^{-1}\ s^{-1}$ (figure 12.12) [104]. Zhong *et al* fabricated bottom-gate ITO-stabilized ZnO TFTs by co-sputtering of ZnO and ITO target, and also found ITO favors for improving device performance with the saturation field-effect mobility up to 21 $cm^2\ V^{-1}\ s^{-1}$, the subthreshold swing as low as 0.29 mV/decade and the threshold voltage only −0.12 V [105]. Deng *et al* also reported ITO-stabilized ZnO thin film with a hybrid-phase microstructure can effectively suppress the grain boundary and native defects in the pristine polycrystalline ZnO. The corresponding staggered top-gate TFTs exhibited fairly high electrical characteristics, especially with a field effect mobility of nearly 20 $cm^2\ V^{-1}\ s^{-1}$ and a subthreshold swing as low as 0.115 V/decade [106]. Except for the common device performance, Liu *et al* investigated the low-frequency noise (LFN) in hybrid-phase microstructure ITO-stabilized ZnO TFTs. The measured LFN is nearly $1/f^{0.9}$ at various bias conditions, and is dominated by the channel resistance under low current intensities and dominated by contact resistance under high current intensities [107]. Except for ZnO channel, the device performance of single layer TiZO TFTs and the dual layer ITO/TiZO TFTs was also compared, indicating that the I_{on}/I_{off} was increased by approximately three orders of magnitude and the subthreshold swing was improved from 0.66 to 0.15 V/decade for ITO/TiZO TFTs [108]. These investigations convincingly suggest a crucial role of ITO as stabilization layer in dual-channel metal oxide FETs.

Overall, ITO is traditionally a highly conductive semiconductor with a wide bandgap of ~3.5 eV and mobility of ~30 $cm^2\ V^{-1}\ s^{-1}$, and is an indispensable part of

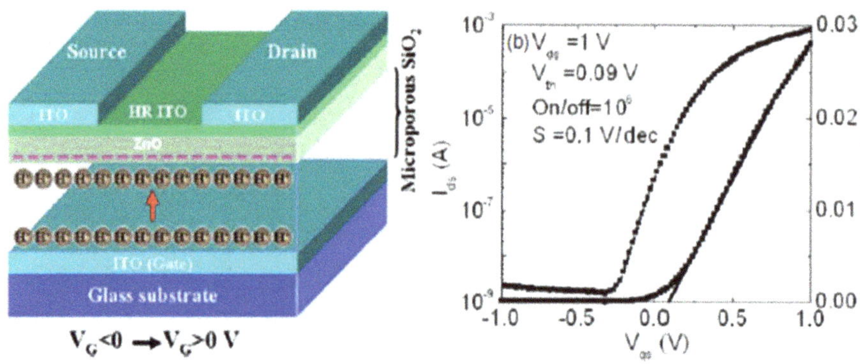

Figure 12.12. (a) Schematic of electric-double-layer and transfer characteristics of the ZnO/ITO dual-channel TFTs [104].

modern display devices as electrodes. In recent years, both sole ITO channel TFTs and dual-channel TFTs functioning as stabilization layer have attracted continually increasing attention. In particular, reducing the thickness of an amorphous ITO layer down to a few nanometers enabled the realization of TFTs with benchmarking operating characteristics in 2019. High performance of solution-processed ITO TFTs further suggests their promising potential for low cost and low power flexible electronics.

12.5 Summary

In recent years, metal oxide semiconductors have seen substantial progress as the suitable TFTs channel materials not only in conventional applications for in-pixel switches and drivers in flat-panel displays, but opening doors to emerging applications for power electronics and low power and cost flexible electronics. Compared with other mainstream channel materials of TFTs, metal oxides show essential advantages including adjustable device electrical performance, large area uniformity and surface flatness, compatibility with various process technologies, and good transparency and mechanical bendability. At present, many companies such as Samsung Electronics, Samsung Mobile Display, Toppan, AU Optronics, Sharp, and LG Display have already demonstrated mass production of flat-panel displays using metal oxides TFTs, with the display market as the most foremost application area. Moreover, power electronics, low power and cost flexible electronics based on metal oxide TFTs are becoming hot topics for potential mass fabrication and applications.

Among all the metal oxides, a-IGZO is the most representative material since its discovery in 2004, and has realized commercialization in 2012 as a driving circuit unit for liquid crystal displays (LCDs). Currently, vacuum-processed a-IGZO still shows the best TFTs performance. Meanwhile, solution-processed a-IGZO film synthesis and the corresponding device fabrication process are essential for emerging low cost, large area flexible, stretchable and/or transparent electronics, which still requires more fundamental investigation. Except for a-IGZO, Ga_2O_3 is gradually recognized to be an attractive option for power electronics compared to SiC and GaN, due to its excellent materials properties and the availability of large and high quality bulk substrates. Currently, remarkable progress in both Ga_2O_3 material and device technologies has been achieved to realize high-performance Ga_2O_3 TFTs. However, there are still many issues to be solved in moving Ga_2O_3 device technology forward into commercialization. These issues include but are not limited to high quality mass production of bulk crystals and large area epitaxial films, successful realization of p-type conductivity, novel thermal management, fundamental understanding of device physics and interfacial engineering to suppress various charges of MOS devices, novel device architecture and others. Ga_2O_3 FETs indeed hold great potential of advancing the power semiconductor industry. Ultimately, ITO is the indispensable part of modern display devices among all the metal oxides, and an 'All by ITO' technology has been proposed, which would enable simplified and cost effective fabrication process. A remarkable breakthrough has been demonstrated for short-channel active transistors based on an ultra-thin (down to 4 nm) ITO channel and a high-quality, lanthanum-doped hafnium oxide dielectric of equivalent oxide thickness of 0.8 nm, with performance comparative to that of existing metal oxides and emerging two-dimensional materials. The appealing result gives ITO TFTs great potential for further commercial low power and cost flexible electronics applications.

Overall, semiconducting metal oxide TFTs hold great potential of replacing the conventional silicon technology not only in display applications but also new areas of applications. Although many challenges still remain for metal oxide TFTs, with the continuous efforts from multidisciplinary cooperation of researchers, a wide range of future applicability of metal oxide TFTs technology is inspiring.

References

[1] Klasens H A and Koelmans H 1964 A tin oxide field-effect transistor *Solid-State Electron.* **7** 701–2

[2] Seager C H, McIntyre D C, Warren W L and Tuttle B A 1996 Charge trapping and device behavior in ferroelectric memories *Appl. Phys. Lett.* **68** 2660–2

[3] Nomura K, Ohta H, Ueda K, Kamiya T, Hirano M and Hosono H 2003 Thin-film transistor fabricated in single-crystalline transparent oxide semiconductor *Science* **300** 1269–72

[4] Nomura K, Ohta H, Takagi A, Kamiya T, Hirano M and Hosono H 2004 Room-temperature fabrication of transparent flexible thin-film transistors using amorphous oxide semiconductors *Nature* **432** 488–92

[5] Sharp Begins Production of World's First LCD Panels Incorporating IGZO Oxide Semiconductors 2012 (http://wwwsharp-worldcom/corporate/news/120413html)

[6] Petti L, Münzenrieder N, Vogt C, Faber H, Büthe L, Cantarella G, Bottacchi F, Anthopoulos T D and Tröster G 2016 Metal oxide semiconductor thin-film transistors for flexible electronics *Appl. Phys. Rev.* **3** 021303

[7] Troughton J and Atkinson D 2019 Amorphous InGaZnO and metal oxide semiconductor devices: an overview and current status *J. Mater. Chem.* C **7** 12388–414

[8] Park J W, Kang B H and Kim H J 2020 A review of low-temperature solution-processed metal oxide thin-film transistors for flexible electronics *Adv. Funct. Mater.* **30** 1904632

[9] Park J S, Maeng W J, Kim H S and Park J S 2012 Review of recent developments in amorphous oxide semiconductor thin-film transistor devices *Thin Solid Films* **520** 1679–93

[10] Fortunato E, Barquinha P and Martins R 2012 Oxide semiconductor thin-film transistors: a review of recent advances *Adv. Mater.* **24** 2945–86

[11] Kimura M 2019 Emerging applications using metal-oxide semiconductor thin-film devices *Jpn. J. Appl. Phys.* **58** 090503

[12] Xu W Y, Li H, Xu J B and Wang L 2018 Recent advances of solution-processed metal oxide thin-film transistors *ACS Appl. Mater. Interfaces* **10** 25878–901

[13] Pearton S J, Yang J C, Cary P H, Ren F, Kim J, Tadjer M J and Mastro M A 2018 A review of Ga_2O_3 materials, processing, and devices *Appl. Phys. Rev.* **5** 011301

[14] Myny K 2018 The development of flexible integrated circuits based on thin-film transistors *Nat. Electron.* **1** 30–9

[15] Hosono H 2006 Ionic amorphous oxide semiconductors: material design, carrier transport, and device application *J. Non-Cryst. Solids* **352** 851–8

[16] Barquinha P, Pereira L, Gonçalves G, Martins R and Fortunato E 2009 Toward high-performance amorphous GIZO TFTs *J. Electrochem. Soc.* **156** H161
[17] Iwasaki T, Itagaki N, Den T, Kumomi H, Nomura K, Kamiya T and Hosono H 2007 Combinatorial approach to thin-film transistors using multicomponent semiconductor channels: an application to amorphous oxide semiconductors in In-Ga-Zn-O system *Appl. Phys. Lett.* **90** 242114
[18] Suresh A, Wellenius P, Dhawan A and Muth J 2007 Room temperature pulsed laser deposited indium gallium zinc oxide channel based transparent thin film transistors *Appl. Phys. Lett.* **90** 123512
[19] Park M J *et al* 2015 Improvements in the bending performance and bias stability of flexible InGaZnO thin film transistors and optimum barrier structures for plastic poly(ethylene naphthalate) substrates *J. Mater. Chem.* C **3** 4779–86
[20] Choi C H, Lin L Y, Cheng C C and Chang C H 2015 Printed oxide thin film transistors: a mini review *ECS J. Solid State Sci. Technol.* **4** P3044–51
[21] Kim G H, Kim H S, Shin H S, Du Ahn B, Kim K H and Kim H J 2009 Inkjet-printed InGaZnO thin film transistor *Thin Solid Films* **517** 4007–10
[22] Fukuda N, Watanabe Y, Uemura S, Yoshida Y, Nakamura T and Ushijima H 2014 In–Ga–Zn oxide nanoparticles acting as an oxide semiconductor material synthesized via a coprecipitation-based method *J. Mater. Chem.* C **2** 2448
[23] Liu H Y, Hung C C and Hsu W C 2018 Deposition of oxide thin films by ultrasonic spray pyrolysis deposition for InGaZnO thin-film transistor applications *IEEE Electron Device Lett.* **39** 1520–3
[24] Yabuta H, Sano M, Abe K, Aiba T, Den T, Kumomi H, Nomura K, Kamiya T and Hosono H 2006 High-mobility thin-film transistor with amorphous $InGaZnO_4$ channel fabricated by room temperature rf-magnetron sputtering *Appl. Phys. Lett.* **89** 112123
[25] Olziersky A, Barquinha P, Vilà A, Pereira L, Gonçalves G, Fortunato E, Martins R and Morante J R 2010 Insight on the SU-8 resist as passivation layer for transparent Ga_2O_3–In_2O_3–ZnO thin-film transistors *J. Appl. Phys.* **108** 064505
[26] Park J S *et al* 2010 High-performance and stable transparent Hf-In-Zn-O thin-film transistors with a double-etch-stopper layer *IEEE Electron Device Lett.* **31** 1248–50
[27] Ram M S, Kort L, Riet J, Verbeek R, Bel T, Gelinck G and Kronemeijer A J 2019 Submicrometer top-gate self-aligned a-IGZO TFTs by substrate conformal imprint lithography *IEEE Trans. Electron Devices* **66** 1778–82
[28] Zhang Y, Yang H, Peng H, Cao Y, Qin L and Zhang S 2019 Self-aligned top-gate amorphous InGaZnO TFTs with plasma enhanced chemical vapor deposited sub-10 nm SiO_2 gate dielectric for low-voltage applications *IEEE Electron Device Lett.* **40** 1459–62
[29] Park J C *et al* 2009 High performance amorphous oxide thin film transistors with self-aligned top-gate structure *2009 IEEE Int. Electron Devices Meeting (IEDM) (7–9 December 2009)* (https://doi.org/10.1109/IEDM.2009.5424391)
[30] Kamiya T, Nomura K and Hosono H 2010 Present status of amorphous In-Ga-Zn-O thin-film transistors *Sci. Technol. Adv. Mater.* **11** 044305
[31] Nomura K, Takagi A, Kamiya T, Ohta H, Hirano M and Hosono H 2006 Amorphous oxide semiconductors for high-performance flexible thin-film transistors *Japan J. Appl. Phys.* **45** 4303–8
[32] Yao J *et al* 2011 Electrical and photosensitive characteristics of a-IGZO TFTs related to oxygen vacancy *IEEE Trans. Electron Devices* **58** 1121–6

[33] Kikuchi Y, Nomura K, Yanagi H, Kamiya T, Hirano M and Hosono H 2010 Device characteristics improvement of a-In-Ga-Zn-O TFTs by low-temperature annealing *Thin Solid Films* **518** 3017–21

[34] Noh H K, Chang K J, Ryu B and Lee W J 2011 Electronic structure of oxygen-vacancy defects in amorphous In-Ga-Zn-O semiconductors *Phys. Rev.* B **84** 115205

[35] Nomura K, Kamiya T and Hosono H 2013 Effects of diffusion of hydrogen and oxygen on electrical properties of amorphous oxide semiconductor, In-Ga-Zn-O *ECS J. Solid State Sci. Technol.* **2** P5–8

[36] Han K L, Ok K C, Cho H S, Oh S and Park J S 2017 Effect of hydrogen on the device performance and stability characteristics of amorphous InGaZnO thin-film transistors with a $SiO_2/SiN_x/SiO_2$ buffer *Appl. Phys. Lett.* **111** 063502

[37] Nam Y, Kim H O, Cho S H and Ko Park S H 2018 Effect of hydrogen diffusion in an In-Ga-Zn-O thin film transistor with an aluminum oxide gate insulator on its electrical properties *RSC Adv.* **8** 5622–8

[38] Lee J M, Cho I T, Lee J H and Kwon H I 2008 Bias-stress-induced stretched-exponential time dependence of threshold voltage shift in InGaZnO thin film transistors *Appl. Phys. Lett.* **93** 093504

[39] Nomura K, Kamiya T, Hirano M and Hosono H 2009 Origins of threshold voltage shifts in room-temperature deposited and annealed a-In-Ga-Zn-O thin-film transistors *Appl. Phys. Lett.* **95** 013502

[40] Suresh A and Muth J F 2008 Bias stress stability of indium gallium zinc oxide channel based transparent thin film transistors *Appl. Phys. Lett.* **92** 033502

[41] Nomura K, Kamiya T and Hosono H 2010 Interface and bulk effects for bias-light-illumination instability in amorphous-In-Ga-Zn-O thin-film transistors *J. Soc. Inf. Disp.* **18** 789

[42] Tu Y *et al* 2020 Improving a-InGaZnO TFTs reliability by optimizing electrode capping structure under negative bias illumination stress *IEEE Electron Device Lett.* **41** 1221–4

[43] Kim S J, Yoon S and Kim H J 2014 Review of solution-processed oxide thin-film transistors *Jpn. J. Appl. Phys.* **53** 02BA

[44] Thomas S R, Pattanasattayavong P and Anthopoulos T D 2013 Solution-processable metal oxide semiconductors for thin-film transistor applications *Chem. Soc. Rev.* **42** 6910–23

[45] Kwon J Y, Lee D J and Kim K B 2011 Transparent amorphous oxide semiconductor thin film transistor *Electron. Mater. Lett.* **7** 1–11

[46] Lee H N, Kyung J W, Kang S K, Kim D Y, Sung M C, Kim S J, Kim C N, Kim H G and Kim S T 2006 *Proc. Int. Display Workshop 2006 (Otsu, Japan, 6–8 December 2006)* p 663

[47] Jeong J K *et al* 2007 *47th Int. Meeting on Information Display (Daegue, Korea, 27–31 August 2007)* pp 9–4

[48] Hara Y, Kikuchi T, Kitagawa H, Morinaga J, Ohgami H, Imai H, Daitoh T and Matsuo T 2018 IGZO-TFT technology for large-screen 8K display *J. Soc. Inf. Disp.* **26** 169–77

[49] Munzenrieder N, Cherenack K H and Troster G 2011 The effects of mechanical bending and illumination on the performance of flexible IGZO TFTs *IEEE Trans. Electron Devices* **58** 2041–8

[50] Münzenrieder N *et al* 2015 Stretchable and conformable oxide thin-film electronics *Adv. Electron. Mater.* **1** 1400038

[51] Salvatore G A, Münzenrieder N, Kinkeldei T, Petti L, Zysset C, Strebel I, Büthe L and Tröster G 2014 Wafer-scale design of lightweight and transparent electronics that wraps around hairs *Nat. Commun.* **5** 2982

[52] Wu H C and Chien C H 2014 Highly transparent, high-performance IGZO-TFTs using the selective formation of IGZO source and drain electrodes *IEEE Electron Device Lett.* **35** 645–7
[53] Ito M, Kon M, Okubo T, Ishizaki M and Sekine N 2005 *Proc. Int. Display Workshop/Asia Display (Takamatsu, Japan, 6–9 December 2005)* 845
[54] Sung M C, Lee H N, Kim C N, Kang S K, Kim D Y, Kim S J, Kim S K, Kim S K, Kim H G and Kim S T 2007 *7th Int. Meeting Information Display (Daegue, Korea, 27–31 August 2007)* 9–1
[55] Nagata T *et al* 2016 A 2.78-in 1058-ppi ultra-high-resolution flexible OLED display using CAAC-IGZO FETs *Dig. Tech. Pap. SID Int. Symp.* **47** 1052–5
[56] Lee C T *et al* 2015 A novel highly transparent 6-in. AMOLED display consisting of IGZO TFTs *Dig. Tech. Pap. SID Int. Symp.* **46** 872–5
[57] Guo D, Guo Q, Chen Z, Wu Z, Li P and Tang W 2019 Review of Ga_2O_3-based optoelectronic devices *Mater. Today Phys.* **11** 100157
[58] Higashiwaki M, Kuramata A, Murakami H and Kumagai Y 2017 State-of-the-art technologies of gallium oxide power devices *J. Phys. D: Appl. Phys.* **50** 333002
[59] Huan Y W, Sun S M, Gu C J, Liu W J, Ding S J, Yu H Y, Xia C T and Zhang D W 2018 Recent advances in β-Ga_2O_3-metal contacts *Nanoscale Res. Lett.* **13** 246
[60] Xue H, He Q, Jian G, Long S, Pang T and Liu M 2018 An overview of the ultrawide bandgap Ga_2O_3 semiconductor-based Schottky barrier diode for power electronics application *Nanoscale Res. Lett.* **13** 290
[61] Fu B, Jia Z, Mu W, Yin Y, Zhang J and Tao X 2019 A review of β-Ga_2O_3 single crystal defects, their effects on device performance and their formation mechanism *J. Semicond.* **40** 011804
[62] Galazka Z 2018 β-Ga_2O_3 for wide-bandgap electronics and optoelectronics *Semicond. Sci. Technol.* **33** 113001
[63] Higashiwaki M, Murakami H, Kumagai Y and Kuramata A 2016 Current status of Ga_2O_3 power devices *Japan J. Appl. Phys.* **55** 1202A1
[64] Dong H, Xue H, He Q, Qin Y, Jian G, Long S and Liu M 2019 Progress of power field effect transistor based on ultra-wide bandgap Ga_2O_3 semiconductor material *J. Semicond.* **40** 011802
[65] Higashiwaki M, Sasaki K, Murakami H, Kumagai Y, Koukitu A, Kuramata A, Masui T and Yamakoshi S 2016 Recent progress in Ga_2O_3 power devices *Semicond. Sci. Technol.* **31** 034001
[66] He H, Blanco M A and Pandey R 2006 Electronic and thermodynamic properties of β-Ga_2O_3 *Appl. Phys. Lett.* **88** 261904
[67] Higashiwaki M, Sasaki K, Kuramata A, Masui T and Yamakoshi S 2012 Gallium oxide (Ga_2O_3) metal-semiconductor field-effect transistors on single-crystal β-Ga_2O_3 (010) substrates *Appl. Phys. Lett.* **100** 013504
[68] Víllora E G, Shimamura K, Yoshikawa Y, Aoki K and Ichinose N 2004 Large-size β-Ga_2O_3 single crystals and wafers *J. Cryst. Growth* **270** 420–6
[69] Galazka Z *et al* 2016 Scaling-up of bulk β-Ga_2O_3 single crystals by the Czochralski method *ECS J. Solid State Sci. Technol.* **6** Q3007–11
[70] Hoshikawa K, Ohba E, Kobayashi T, Yanagisawa J, Miyagawa C and Nakamura Y 2016 Growth of β-Ga_2O_3 single crystals using vertical Bridgman method in ambient air *J. Cryst. Growth* **447** 36–41

[71] Kuramata A, Koshi K, Watanabe S, Yamaoka Y, Masui T and Yamakoshi S 2016 High-quality β-Ga_2O_3 single crystals grown by edge-defined film-fed growth *Japan J. Appl. Phys.* **55** 1202A2

[72] Alema F, Hertog B, Osinsky A, Mukhopadhyay P, Toporkov M and Schoenfeld W V 2017 Fast growth rate of epitaxial β-Ga_2O_3 by close coupled showerhead MOCVD *J. Cryst. Growth* **475** 77–82

[73] Baldini M, Albrecht M, Fiedler A, Irmscher K, Klimm D, Schewski R and Wagner G 2016 Semiconducting Sn-doped β-Ga_2O_3 homoepitaxial layers grown by metal organic vapour-phase epitaxy *J. Mater. Sci.* **51** 3650–6

[74] Okumura H, Kita M, Sasaki K, Kuramata A, Higashiwaki M and Speck J S 2014 Systematic investigation of the growth rate of β-Ga_2O_3 (010) by plasma-assisted molecular beam epitaxy *Appl. Phys. Express* **7** 095501

[75] Sasaki K, Higashiwaki M, Kuramata A, Masui T and Yamakoshi S 2014 Growth temperature dependences of structural and electrical properties of Ga_2O_3 epitaxial films grown on β-Ga_2O_3 (010) substrates by molecular beam epitaxy *J. Cryst. Growth* **392** 30–3

[76] Nomura K, Goto K, Togashi R, Murakami H, Kumagai Y, Kuramata A, Yamakoshi S and Koukitu A 2014 Thermodynamic study of β-Ga_2O_3 growth by halide vapor phase epitaxy *J. Cryst. Growth* **405** 19–22

[77] Murakami H *et al* 2014 Homoepitaxial growth of β-Ga_2O_3 layers by halide vapor phase epitaxy *Appl. Phys. Express* **8** 015503

[78] Leedy K D *et al* 2017 Highly conductive homoepitaxial Si-doped Ga_2O_3 films on (010) β-Ga_2O_3 by pulsed laser deposition *Appl. Phys. Lett.* **111** 012103

[79] Oshima T, Nakazono T, Mukai A and Ohtomo A 2012 Epitaxial growth of γ-Ga_2O_3 films by mist chemical vapor deposition *J. Cryst. Growth* **359** 60–3

[80] Altuntas H, Donmez I, Ozgit-Akgun C and Biyikli N 2014 Electrical characteristics of β-Ga_2O_3 thin films grown by PEALD *J. Alloys Compd.* **593** 190–5

[81] Krishnamoorthy S, Xia Z, Bajaj S, Brenner M and Rajan S 2017 Delta-doped β-gallium oxide field-effect transistor *Appl. Phys. Express* **10** 051102

[82] Xia Z *et al* 2018 Delta doped β-Ga_2O_3 field effect transistors with regrown ohmic contacts *IEEE Electron Device Lett.* **39** 568–71

[83] Wong M H, Nakata Y, Kuramata A, Yamakoshi S and Higashiwaki M 2017 Enhancement-mode Ga_2O_3 MOSFETs with Si-ion-implanted source and drain *Appl. Phys. Express* **10** 041101

[84] Higashiwaki M, Sasaki K, Kamimura T, Hoi Wong M, Krishnamurthy D, Kuramata A, Masui T and Yamakoshi S 2013 Depletion-mode Ga_2O_3 metal-oxide-semiconductor field-effect transistors on β-Ga_2O_3 (010) substrates and temperature dependence of their device characteristics *Appl. Phys. Lett.* **103** 123511

[85] Green A J *et al* 2016 3.8-MV/cm breakdown strength of MOVPE-grown Sn-doped β-Ga_2O_3 MOSFETs *IEEE Electron Device Lett.* **37** 902–5

[86] Zeng K *et al* 2017 Ga_2O_3 MOSFETs using spin-on-glass source/drain doping technology *IEEE Electron Device Lett.* **38** 513–6

[87] Chabak K D *et al* 2016 Enhancement-mode Ga_2O_3 wrap-gate fin field-effect transistors on native (100) β-Ga_2O_3 substrate with high breakdown voltage *Appl. Phys. Lett.* **109** 213501

[88] Chabak K D *et al* 2018 Recessed-gate enhancement-mode β-Ga_2O_3 MOSFETs *IEEE Electron Device Lett.* **39** 67–70

[89] Moser N *et al* 2017 Ge-doped β-Ga_2O_3 MOSFETs *IEEE Electron Device Lett.* **38** 775–8

[90] Hwang W S *et al* 2014 High-voltage field effect transistors with wide-bandgap β-Ga_2O_3 nanomembranes *Appl. Phys. Lett.* **104** 203111

[91] Ahn S, Ren F, Kim J, Oh S, Kim J, Mastro M A and Pearton S J 2016 Effect of front and back gates on β-Ga_2O_3 nano-belt field-effect transistors *Appl. Phys. Lett.* **109** 062102

[92] Kim J, Oh S, Mastro M A and Kim J 2016 Exfoliated β-Ga_2O_3 nano-belt field-effect transistors for air-stable high power and high temperature electronics *Phys. Chem. Chem. Phys.* **18** 15760–4

[93] Zhou H, Si M, Alghamdi S, Qiu G, Yang L and Ye P D 2017 High-performance depletion/enhancement-ode β-Ga_2O_3 on insulator (GOOI) field-effect transistors with record drain currents of 600/450 mA/mm *IEEE Electron Device Lett.* **38** 103–6

[94] Zhou H, Maize K, Qiu G, Shakouri A and Ye P D 2017 β-Ga_2O_3 on insulator field-effect transistors with drain currents exceeding 1.5 A/mm and their self-heating effect *Appl. Phys. Lett.* **111** 092102

[95] Zhou H, Maize K, Noh J, Shakouri A and Ye P D 2017 Thermodynamic studies of β-Ga_2O_3 nanomembrane field-effect transistors on a sapphire substrate *ACS Omega* **2** 7723–9

[96] Lu A, Sun J, Jiang J and Wan Q 2010 One-shadow-mask self-assembled ultralow-voltage coplanar homojunction thin-film transistors *IEEE Electron Device Lett.* **31** 1137–9

[97] Jiang J, Sun J, Dou W and Wan Q 2012 Junctionless flexible oxide-based thin-film transistors on paper substrates *IEEE Electron Device Lett.* **33** 65–7

[98] Shao Y, Xiao X, Wang L Y, Liu Y and Zhang S D 2014 Anodized ITO thin-film transistors *Adv. Funct. Mater.* **24** 4170–5

[99] Li S, Tian M, Gao Q, Wang M, Li T, Hu Q, Li X and Wu Y 2019 Nanometre-thin indium tin oxide for advanced high-performance electronics *Nat. Mater.* **18** 1091–7

[100] Wang M *et al* 2019 High performance gigahertz flexible radio frequency transistors with extreme bending conditions *2019 IEEE Int. Electron Devices Meeting (IEDM) (7–11 December 2019)*

[101] Li S *et al* 2019 BEOL compatible 15-nm channel length ultrathin indium-tin-oxide transistors with ion = 970 μA/μm and on/off ratio near 10^{11} at V_{ds} = 0.5 V *2019 IEEE Int. Electron Devices Meeting (IEDM) (7–11 December 2019)*

[102] Anthopoulos T D 2019 Ultrathin channels make transistors go faster *Nat. Mater.* **18** 1033–4

[103] Wang Y F, Wang Z G, Huang K R, Liang X C, Liu C N, Chen C D and Liu C 2020 Solution-processed ITO thin-film transistors with doping of gallium oxide show high on-off ratios and work at 1 mV drain voltage *Appl. Phys. Lett.* **116** 219901

[104] Lu A X and Huang H M 2015 Low-voltage transparent thin-film transistors with ZnO/ITO double-channel layers *Jpn. J. Appl. Phys.* **54** 106502

[105] Zhong W, Li G Y, Chen R S, Lan L F, Zhang X, Pei W H, Chen H and Deng S 2017 A study on the bottom-gate ITO-stabilized ZnO thin-film transistors *Int. Conf. Electron Devices and Solid-State Circuits (EDSSC) (Hsinchu, Taiwan, 18–20 October 2017)*

[106] Deng S B, Chen R S, Li G J, Xia Z H, Zhang M, Zhou W, Wong M and Kwok H-S 2016 High-performance staggered top-gate thin-film transistors with hybrid-phase microstructural ITO-stabilized ZnO channels *Appl. Phys. Lett.* **109** 182105

[107] Liu Y, Deng S B, Chen R S, Li B, En Y F and Chen Y Q 2018 Low-frequency noise in hybrid-phase-microstructure ITO-stabilized ZnO thin-film transistors *IEEE Electron Device Lett.* **39** 200–3

[108] Yu W *et al* 2019 Performance enhancement of TiZO thin film transistors by introducing a thin ITO interlayer *IEEE J. Electron Devices Soc.* **7** 1302–5

www.ingramcontent.com/pod-product-compliance
Ingram Content Group UK Ltd.
Pitfield, Milton Keynes, MK11 3LW, UK
UKHW051905100926
8556IPUK00005B/768

9 780750 325547